C0-ALY-727

System Design with
The MC68020, MC68030, and MC68040
32-BIT MICROPROCESSORS

System Design with The MC68020, MC68030, and MC68040 32-BIT MICROPROCESSORS

Asghar Noor

VNR VAN NOSTRAND REINHOLD
New York

Library of Congress Catalog Card Number 92-12923
ISBN 0-442-31886-3

ITP Van Nostrand Reinhold is an International Thomson Publishing company.
ITP logo is a trademark under license.

Printed in the United States of America

Van Nostrand Reinhold
115 Fifth Avenue
New York, NY 10003

ITP Germany
Königswinterer Str. 418
53227 Bonn
Germany

International Thomson Publishing
Berkshire House,168-173
High Holborn, London WC1V 7AA
England

International Thomson Publishing Asia
38 Kim Tian Rd., #0105
Kim Tian Plaza
Singapore 0316

Thomas Nelson Australia
102 Dodds Street
South Melbourne 3205
Victoria, Australia

International Thomson Publishing Japan
Kyowa Building, 3F
2-2-1 Hirakawacho
Chiyada-Ku, Tokyo 102
Japan

Nelson Canada
1120 Birchmount Road
Scarborough, Ontario
M1K 5G4, Canada

16 15 14 13 12 11 10 9 8 7 6 5 4 3 2 1

Library of Congress Cataloging in Publication Data
Noor, Asghar.
System design with the MC68020, MC68030, and MC68040 32-bit microprocessors / Asghar Noor.
p. cm.
Includes bibliographical references and index.
ISBN 0-442-31886-3
1. Motorola 68000 series microprocessors. 2. System design.
I. Title.
TK7895.M5N66 1992
621.39'16—dc20

92-12923
CIP

Contents

Preface

Throughout history, key technological innovations have triggered dramatic changes in the way we work and live. The invention of the microprocessor is such an event. When Intel Corp. introduced its microprocessor in 1970, it declared that the device had launched a new era in computer technology. Intel was right, and today we see that the microprocessor is an integral part of computer technology.

Motorola introduced its microprocessor at a later date, learning from Intel's mistakes, and presented a better product. MC68000 is a second generation processor from Motorola. Its architects foresaw the need for a flexible architecture that would enable them to create new products by adding features and functions. This basic architecture has become the foundation for many high performance micros like MC68020, MC68030, and MC68040. Today, these processors are most widely used in the design of high performance workstations, computers, systems, and multiprocessors; they are also found in a wide variety of other electronic equipment.

As microprocessor technology spreads, certain design principles and practices should be understood by anyone who would use this technology or be affected by it, including technology managers, system, hardware and software designers, and application developers.

This book describes the principles behind the CISC and RISC microprocessors, and how these principles are implemented in MC68020 and its successors. The book then explores how to design software and the interfacing techniques used in the design of MC680x0 based systems. Anyone who seeks an understanding of the basic design principles used in 32-bit microprocessor technology should read this book. The implementation within modern day microprocessors of various operating system principles (memory management, resource sharing, interrupt management), communication principles (interfacing, busy schemes, exception processing), coprocessor interfacing principles, and multiprocessing principles, are also delineated.

The book explains all key features, including unique architectures, modes of operations, pipelines to facilitate software execution, coprocessors and memory management strategies. Advanced programming techniques demonstrate how basic instructions are utilized to implement programs and how they can be used to support structured programming techniques normally found in modern programming languages.

Hardware interfacing principles are described in detail with various examples. The examples range from interfacing simple devices to designing complex multiprocessor systems using VME backplane.

This book also explores various memory management techniques and IEEE floating point standards and how they are implemented in modern peripherals such as the MC68851 (Memory Management Unit) and MC68881 (Floating Point Unit), concentrating on how these functions are added to MC68030 and MC68040 to make them very versatile and powerful processors. Various practical examples are given to help designers and users better understand how to harness the power of these powerful micros to solve application problems. The book also addresses software porting issues from previous to later generation processor-based systems.

In conclusion, I want to acknowledge the contribution made by various people and organizations. I am specially thankful to Motorola Inc. and their educational service for permitting me to use diagrams and materials. Without these the book would have been very difficult to complete. I am also indebted to my publisher and editors for suggestions. My special thanks goes to the editor who encouraged me to complete the book during the last three years, which were marked by family tragedies and misfortunes. I also thank the unknown reviewers for valuable suggestions and criticisms.

I need to thank several other people who were catalytic agents in the production of this book. My daughter, Sharmeen, who was three at the time I began the book, has given me inspiration and has listened patiently as I have read materials from the manuscript aloud, thinking, perhaps, that I was reading her a story which she did not understand. I also want to thank Dr. M. Islam and Mrs. F. Hussain, who assisted me in revising the manuscript, and I am deeply grateful to many other people who helped me in numerous ways.

The acknowledgment I make to my wife is not the usual kind. She has maintained only an objective interest to this book, but she organized and maintained a strict regime for me, keeping me on an even keel despite the many torments and inconveniences of life. Only through the intense efforts of the people mentioned here and many unnamed could a project of this complexity reach fruition.

Asghar I. Noor
Natick, Massachusetts

Chapter 1

Introduction to Microprocessors and Microcomputers

The study of the microprocessor is the study of its organization, interconnection of components, and the programming of it to solve problems. System designers construct systems using basic building blocks, namely the microprocessor, memories, interfacing devices, and buses. Using these building blocks, system designers then build various application-specific systems, ranging from a simple embedded core processor to the latest high-performance work stations, even using multiprocessors. The major differences lie in the way that the modules are connected together, the characteristics of the components, and the application software that controls the system. In summary, microprocessor-based system design is a discipline devoted to the design of highly specific and individual computers from a collection of common building blocks.

The plan for this book is to take the reader through important topics of microprocessor-based system design. Starting with the general concept of the microprocessor, we will introduce the MC68020 itself, to lay the foundation for further reading. This will then lead to more specific topics involving such areas as hardware design using the MC68020, coprocessors and their interfacing, programming of the system, design of the memory system, etc. The book then explores the next-generation processors of this family, the MC68030 and MC68040.

In Chapter 1, we will present the most elementary structure of the microprocessor. Here, we will cover the various attributes used to classify the microprocessor and the evolving trends of their architectures. We will then introduce the MC68000 processor family and highlight the various differences among the members, the MC68000, MC68008, MC68010, MC68020, MC68030, and its newest member, MC68040.

Chapter 2 introduces the design goals for the MC68020. The unique architecture of the processor is described here, including its mode of operation, internal architecture, and flexibility.

Chapter 3 treats the software capabilities of the processor. The pipeline technique is extensively used in the MC68020 to enhance its capabilities. Data organization supported by this processor is then followed, using the virtual machine concept supported by this machine. Instruction sets are then introduced, along with their classification. The multiprocessor capability is then introduced to end the chapter.

In Chapter 4, we will introduce the hardware aspects of the MC68020, namely the pin configurations, their functions, and how they are to be used to interface memory and other devices. Although the hardware and system designers are mostly concerned with the electrical and mechanical aspects of the processor, the software designers also need to become familiar with these topics to help them port their software and debug it. The interrupts, bus arbitration mechanism, and system control features are also described in this section.

Chapter 5 builds on the knowledge acquired in Chapter 3. It shows how the instructions are utilized to implement programs, and how they can be used to support the structured programming techniques that are normally found in high-level languages. Examples are provided on how to write position-dependent and independent software. The multiprocessor support provided by the MC68020 and its cousins is described here. Examples are provided on how to exploit this feature using software.

In Chapter 6, we will use the concepts developed in Chapters 3, 4, and 5 and describe the system design paradigm. The paradigm includes three views of design; the system designer's concerns, the hardware concerns, and the software designer's concerns, including the synergies that exist between various areas. Designers are continuously called upon to make various tradeoffs during the design of a memory system, the interface for peripheral devices, and the programming of them. The MC68020 was designed to be used in a multiprocessor-based system. As a result, it incorporates various features, like an indivisible read-modify-write cycle, a locking mechanism, and instructions. These features are illustrated through examples.

Computer buses are becoming important, because of the role they play during hardware and system design. By adopting commercially available buses as a system backbone, the system designers will be able to use off-the-shelf boards, components, or peripheral devices such as hard disks, tape backups, or printers. The most popular buses, like Multi-Bus, VME-Bus, and Future-Bus, are described here, including their protocols. Various features are then compared, which will help designers to make informed decisions.

Chapter 7 deals with the exception processing of the MC68020. Exceptions are entered when the processor deviates from its normal operation. Two levels of exceptions are supported in the MC68020, which are designed to facilitate the implementation of a multitasking operating system like UNIX. This elegant exception processing, when coupled with multiprocessing support, creates a powerful environment for a multitasking, fault-tolerant multiprocessing system. These possibilities are explored through examples.

The MC68020 supports a virtual memory management system. The virtual memory allows the software designers to develop their program without the burden of worrying whether the program will fit within the limited physical memory. The foundation for this virtual memory system is developed in Chapter 8.

In order to bridge the performance gap that exists between the processor and main memory system, designers often utilize a logical and physical cache. The MC68020 includes a physical cache within itself. Chapter 8 also deals with the organization of the cache memory.

Chapter 9 deals with the concept of the coprocessor, including the protocol supported by MC68020. The instructions provided in support of this protocol are described here. Examples are given on how to use these instructions to configure the coprocessor so that it can work in tandem with the MC68020.

Chapters 11 and 12 deal with the memory management coprocessor MC68851 and the floating-point arithmetic unit MC68881/MC68882. These coprocessors allow the system designer to devise a high-performance computer system using the MC68020.

The designers of the MC68000 family of processors are regularly enhancing the basic architecture of the MC68020. In recent years, they have added two more processors to the family. The MC68030 is discussed in Chapter 13, and the MC68040 is discussed in Chapter 14. Without them, any discussion of the MC68020 will remain incomplete. The MC68030 includes the MC68020 core, the floating-point coprocessor MC68881, extended cache memory, and additional software capability.

The MC68040 took the MC68030 core and added the MC68851 (memory management unit), extended cache, and additional software functions. These make the processor extremely powerful.

In the remaining part of this introductory chapter, we will discuss the concept of the digital system, the evolution of the microprocessor, and the MC68000 family of processors. The architectural differences between CISC and RISC processors are discussed, along with their inherent bottlenecks. The chapter then concludes by highlighting the various features of the MC68020 and other members of the MC68000 family.

1.1 THE STRUCTURE OF DIGITAL COMPUTERS

It is generally said that a microprocessor *is* a computer. Hence, an understanding of a classic structure of a computer will help us to know about microprocessor architecture. The classic structure illustrated in Figure 1.1 is often identified as von Neumann architecture for reasons that are not clearly shown by history.

This architecture is made up of four building blocks:

- Central Processing Unit (CPU)
- Memory
- Input
- Output

Each block has a very well-defined purpose. The CPU unit implements general functions, namely:

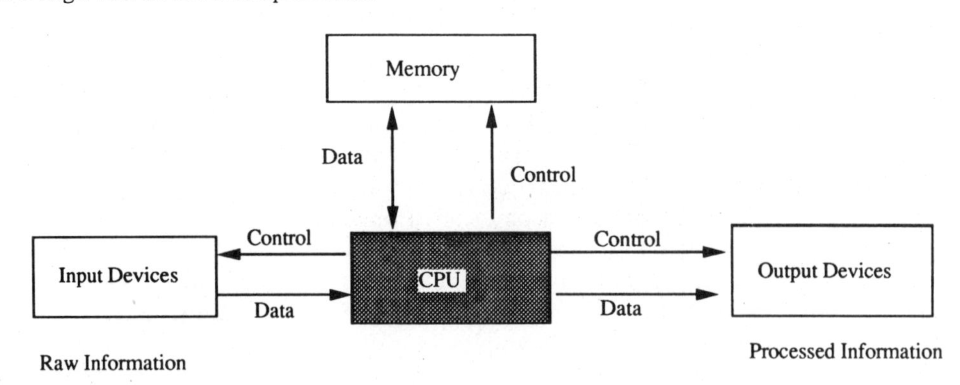

FIGURE 1.1. Architecture of the digital computer.

1. It controls the operation of the whole system by issuing timing and control signals to all parts of the system.
2. Signals are generated according to the instruction received by the CPU. The CPU decodes the instruction and acts accordingly. Instructions are part of the program stored in memory.
3. CPU performs all computation, such as arithmetic, logic, and control operations on numerical or nonnumerical data.
4. It also holds the temporary results produced during the computation process.

The memory acts as a repository for the programs that are accessed by the CPU. It also bridges the gaps that exist between the CPU and the slower input and output devices. Various levels of memory are used today. They range from high-speed static to dynamic devices to slower disk-based devices, from hard disks to optical disks. For back-up purposes, tape device memory systems are also utilized.

The last portion of the architecture is the Input/Output (I/O) unit. The von Neumann architecture was probably less influential in this area than in others. The main purpose of the I/O system is to be the interface between the computer and the external world, and to transform data into a form understandable to the CPU.

1.2 THE GENERAL ARCHITECTURE OF MICROCOMPUTERS

A microprocessor includes a control unit, arithmetic logic unit, and the associated circuitry, all scaled down to fit on a silicon chip. Thus the microprocessor is often called a CPU on a chip. The term microprocessor was first used in 1972, although the era began in 1971.

The architecture of the microprocessor has evolved from the 4-bit chip to a complex 32-bit chip within a single decade. In the case of mainframe computers,

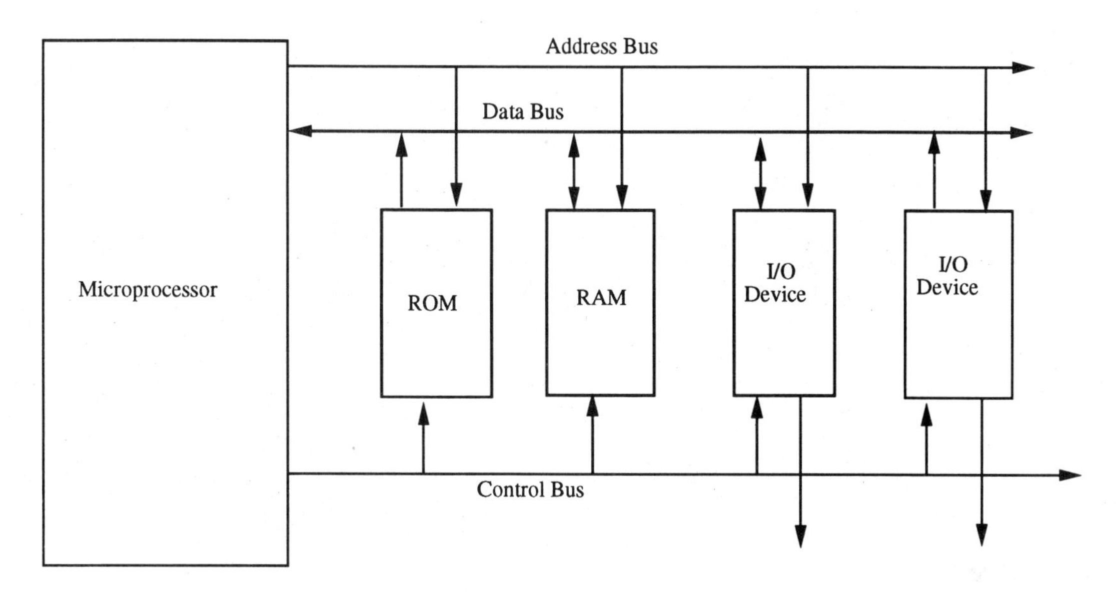

Address Lines : Unidirectional (direct addressing capability)

Data Lines : Uni/Bidirectional (maximum size of data bits)

Control Lines : Uni-directional (synchronous and asynchronous signals during data transfer)

FIGURE 1.2. The elements of a microprocessor-based system.

the equivalent development spanned over two decades. The pace of micro development is even more impressive, incorporating innovative concepts like object-oriented architecture, memory management, and floating-point arithmetic and cache memory within the chip, including advanced interfacing techniques. It is important to acknowledge that many of the above ideas were taken from the domain of the mainframe, and then integrated with the microprocessor by taking advantage of the latest VLSI technology.

However, the basic building blocks have not radically changed. They still include the CPU, memory, and I/O elements, connected via a system bus (Figure 1.2), just like the model proposed by von Neumann in 1940. The changes that occurred involved a reduction in size, migration of various operating system functionality into the hardware, and the integration of pipelining and parallelism in the architecture.

1.3 CLASSIFICATION OF MICROPROCESSORS

Microprocessors can be characterized by three major attributes:

- Chip technology
- Word size

- Number of ICs used to design a CPU

Due to the complex nature of the microprocessor, one should add additional attributes, including:

- Clock frequency
- Microprogramming
- Number of instructions
- Addressing capabilities
- Number of internal registers
- Support for various data types
- Memory management features
- Multiprocessing capabilities
- Communication supports

1.3.1 Chip Technology

Most manufacturers are using MOS technology over the older bipolar technology. At present, the most popular technology is nMOS, because of its high packaging density and switching speed. CMOS consumes less power than circuits implemented with traditional pMOS and nMOS. The main disadvantages, however, are a lower packing density and speed. Today, manufacturers are using HCMOS, which includes the features of nMOS and CMOS. Using new fabrication technology, chip manufacturers are enhancing the performance at high gate counts. New breakthroughs are occurring in fabrication technology, and today 1 to 4 million transistors on a single chip is becoming commonplace.

1.3.2 Word Size

Word size reflects the basic unit of work for the microprocessor. A larger word size implies more processing power and addressing capabilities. In the first and second generation micros, the size of the registers, the size of the external instruction, and the data path were identical. This is no longer true. As an example, the MC68000 has 32-bit internal paths and registers, but a 16-bit external data path. In today's 32-bit microprocessor, they are all 32 bits in size. It is the external data bus size that is generally used to identify the nature of the microprocessor as 4, 8, 16, 32, or 64 bits.

1.3.3 Types of Microprocessors

Most of today's microprocessors execute all bits of one word in parallel. Others work with 'slices' of data and/or instruction words, and are called "bit-slice architecture." Here, several identical chips are used to execute different slices in parallel, and are termed "bit-slice microprocessors."

1.3.4 Microprogramming

Early microprocessors used hardware logic to implement control circuitry. In recent years, such design strategy is going through a constant shift. To save valuable chip space, designers started to use software techniques called microprogramming to implement the control logic. The control logic performs various functions, such as: fetching, examining, executing conventional machine instructions (including directing other parts of the machine), etc.

Under microprogramming these activities are implemented as software interpreters called microoperations. The control variables are coded by strings of 0s and 1s, known as control words, which are stored as Read Only Memory (ROM) inside the processor. Each control word contains within it a microinstruction. The microinstruction specifies one or more microoperations for the processor. A sequence of microinstructions constitutes the microprogram.

A microprogrammed control unit has many advantages. The ROM control store is dense and much less complex than its random logic counterpart. The regularity decreases the interconnection complexity, and therefore the size of the control circuit. Again, to minimize the control store area, the designers often utilize a multilevel storage structure. However, the multilevel structure is slow. As a result, the compromise is often a two-level storage structure.

In a two-level structure, each processor instruction is implemented via a sequence of microinstructions. The microinstructions are narrow, consisting primarily of pointers to nanoinstructions and branching information. The nanoinstructions are wide, providing fairly direct decoded control of the execution unit. The nano-control store contains an arbitrarily ordered set of unique machine-state control words. A given control state is implemented only once in the nano-ROM, and its use is "shared" by several microwords.

As the sharing ratio (i.e., the number of microwords/number of nanowords) increases, the total microcode and nanocode ROM area decreases. In order for a two-level structure to be cost effective, the total number of nanocontrol states implemented must be a small fraction of the number of possible control states. Often, designers utilize a nanoword width of 128 bits, implying a possible 2^{128} control states. However, for many cases, the actual implementation of the full control sequence requires less than 512 (i.e., 2^9) unique control states. Thus, a high sharing ratio, combined with few required nanocontrol states, results in a saving of significant chip area.

1.3.5 Addressing Capability

Because of small word size, early micros could reference only limited memory space. Large word size enables direct addressing of larger memory space. Today's 32-bit micros generally have a direct addressing capability of 16 Gbyte. In addition, virtual addressing modes have become popular. This capability has been extended through the inclusion of virtual memory within or outside of the chip.

1.3.6 Registers

Registers are vital to the performance of micros. They are used to hold the results of arithmetic and logic operations, as pointers, stacks, and temporary data, so that the CPU does not have to go frequently to the memory. Most modern micros have 16 or more registers. RISC micros can have as many as 132 registers.

1.3.7 Data Types

All micros support data in the form of bytes, words, and even long and quad-words. However, some also support data in the form of bits, binary coded decimals, floating-point numbers, and character strings.

1.3.8 Multiprocessing Capabilities

Today, it has become necessary in many pursuits to integrate several micros to form a single system. The notion of connecting several microprocessors together to create a powerful computing system is a problem in system design. This appearance is deceptive, however, because the implementation of such an idea requires a thorough consideration of architecture, hardware, operating system, and (ultimately) applications. Two primary architectures exist for the design of such a system, namely:

- Tightly coupled multiprocessors, where data can be communicated from one micro to any other micro at rates on the order of the bandwidth (i.e., the rate at which information can be transferred) of the memory.
- A loosely coupled multiprocessor system, where the communication delays between the microprocessors depend on whether the CPUs are locally connected to each other or are connected via one or more layers of interconnecting paths.

The selection of the architecture depends on various application environments and performance issues. However, the overall throughput and efficiency of any architecture is directly related to the hardware and software interconnection mechanisms supported by the basic micro. Some microprocessors support multiprocessor capability within the hardware, and others do the same via instructions. We will address some of these issues in Chapter 6, when we discuss the multiprocessing capabilities of the MC68020 microprocessor.

1.4 GENERAL TRENDS IN MICROPROCESSOR ARCHITECTURE

As technology has advanced, hardware costs have declined rapidly. Meanwhile, the overall cost of software has increased dramatically. This raises the question "If hardware cost is not a factor, which software functions should be carried on by the hardware," where they can be performed faster and more reliably. This gives birth to the notion of Complex Instruction Set Architecture (CISC).

1.4.1 CISC Architecture

The general trend towards CISC architecture was fueled by considerations such as:

- New models are often required to be downward compatible with existing models in the same microprocessor family. This results in a proliferation of features.
- Designers are always striving to reduce the semantic gap between high-level languages and instruction sets. To reduce software costs, they are creating easily programmable machines that use the instruction sets most often used by high-level languages.
- Microprocessor designers are constantly moving functions from software to microcode and from microcode to hardware to develop faster machines.
- Rapid advancement in fabrication technology has accelerated the inclusion of various complex addressing modes and innovative instructions.
- Designers are also incorporating pipelining and multiprocessing techniques to overcome the instruction execution bottleneck.

1.4.1.1 The Implementation Paradigm in CISC

Today, the microprocessor has become a proving ground for testing many of the above concepts. Many traditional nonprocessor features like interrupt processing, bus arbitration, memory management, cache memory, etc. are becoming a standard part of CISC architecture.

Thus, to evaluate the micros, one has to consider implementation-level attributes, like: architectural features, number of registers, size of the registers, size of the data path, ALU size, pipeline and parallelism, multiprocessing features, restartability and reconfigurability, etc.

The instruction-level features, such as: data structure, addressing mode, high-level language support, special I/O instruction, special instruction for privileges, and instruction for interfacing should also be criteria.

Routine-level supports, like: interrupts, protection, and task switching, including other operating system functions, are also important.

Task-level supports, namely: multilevel queue management, protection among different tasks, task switching, etc. should also be considered during evaluation.

1.4.2 Reduced Instruction Set Computer (RISC)

Instruction traces from CISC machines consistently show that a limited number of the available instruction sets and addressing modes are used by the programs. This led IBM's John Cocke, in the early 1970s, to comtemplate a new architecture. The research group developed a machine called the 801 (named after the building they were working in), based on three design principles [Radin 82];

- The instruction sets are to be such that they cannot be moved at the compile time.
- The compiler should be intelligent enough to understand the intent of the pro-

gram, to create optimal object code, and to map them to the underlying architecture.

- The instruction set should be implemented in random logic.

These concepts became the foundation for a number of development activities at the University of California at Berkeley, Stanford University, and at various start-up companies. UC-Berkeley called their machine the RISC machine [Patterson 82, Lioupis 83]. These architectures share a number of common attributes:

1. Single-cycle operation: The instruction sets are designed in such a manner that they can be executed during each machine cycle. This facilitates the rapid execution of instructions.
2. Load/Store operation: The desire for single-cycle operation for all tasks led to the creation of the Load/Store operation.
3. Hardwired control: This implements the control in hardware, because microprograms generally lead to slower control paths and to interpretation overhead.
4. Reduced instructions and addressing modes: These are designed to facilitate a fast and simple implementation of instructions by control logic.
5. Fixed instruction format: This helps with the decoding of instructions, and speeds up the control path.
6. Intelligent compile: This offers an opportunity to explicitly move run-time complexity into the compiler.

The advantages of RISC architectures come from a close interaction between the architecture and implementation, where simplicity of architecture leads to simple implementation. The advantages gained from this include:

1. Simplified instruction formats, allowing fast instruction decoding. This reduces the length of the pipeline.
2. Most instructions can be made to execute within a single cycle. The large register sets, when included in the architecture, facilitate this.
3. This leads to the streamlining of the architecture. The overhead for each instruction can be reduced, allowing a shorter clock cycle.
4. The fewer low-level instruction sets, provide the best targets for an optimized compiler. Nearly every transformation done by the optimizer on the intermediate form will result in an improved execution time, because the transformation will eliminate instructions.
5. The simplified instruction set provides an opportunity to eliminate a level of translation at run time, in favor of translation at compile time. Thus, the microcode of a compiler instruction set will replace the compiler's code generation function.

Again, RISC architecture suffers from certain drawbacks, these being memory bandwidth and additional software requirements.

A simplified instruction set will utilize more instructions to perform the same task. Thus, the same program in a RISC machine will require more memory than its CISC cousin. Also, RISC will access memory more often, resulting in the need for a higher memory bandwidth.

A smart compiler will be required to create optimized instructions for a high-level language. The software debugging will be more difficult, due to the depth of the pipelines.

However, the simplified RISC architecture allows the organization to be streamlined. This, in turn, reduces the overhead on each instruction, thus shortening the execution time. In addition, six synergistic elements seem essential to the RISC architectures:

- Single-cycle operation
- Load/Store design
- Hardware control
- Relatively few instructions and addressing modes
- Fixed instruction format
- More compilation time effort than a CISC machine.

1.4.3 The RISC Versus CISC Paradigms

Most of today's microprocessors follow the CISC architecture, where designers attempt to squeeze the most performance out of any given architecture. Since it is possible to make microcode execute much faster than external instructions, the overriding concern of a CISC architect is to build into a processor high semantic content instructions, which reduce the number of external memory accesses. Again, microcode allows the inclusion of varied addressing modes, and several control points, in the microprocessor's internal hardware.

RISC architecture has two accepted meanings, one being that of a reduced instruction-set computer, and the other is that of a reusable instruction-set computer. Both imply that RISC has something to do with optimizing the instruction set. The main tenet is the investigation of the assignment of system functionality within an architecture. As a result, the instruction and addressing modes remaining on the RISC CPU are those that are frequently used by code generators embedded in compilers, are most advantageous to a language, and are more efficient if implemented in hardware. This allows the designers to utilize the silicon for a large number of registers (typically greater than 100), special hardware (barrel shifters, floating hardware, cache memories, or multipliers), or special functionality (interprocess communication mechanisms).

The large number of registers allows the program variables required by a given procedure to reside within the chip. The overlapping windows make a procedure call/return a simple matter of updating a window pointer and changing the program counter. This means that no data transfers to the external world actually take place.

It is very difficult to predict the future architectural trends of microprocessors. Since designers must maintain the instruction compatibility of a microprocessor family, they may have very few choices. On the other hand, the current CISC architectures are hitting a performance bottleneck. Thus, we may see more and more emphasis on RISC architecture or on multiprocessor capability in CISC architecture. But both architectures face the inherent restrictions of the von Neumann bottleneck.

1.4.4 Organization and Implementation Issues

The interdependence between microprocessor architecture and its implementation technology (MOS, BiCMOS, ECL, GAAS, etc.), has always had a profound influence on the cost/performance ratios attainable for the architecture. Many of the techniques used to obtain high performance in conventional computer design have found their way into today's microprocessor design. Designers often use techniques such as pipelining and parallelism to enhance the execution speed.

1.4.5 Pipelining

A classical technique for enhancing the performance of a processor is pipelining. Here, a task is subdivided into various subtasks in such a way that they can be executed by specific hardware operating in sequence. Once the pipeline is filled, the task is completed in a shorter time. The performance enhancement is determined by the depth of the pipeline. That is, if the maximum rate at which operations can be executed is 'r', and the pipeline has a depth of 'd', then the maximum execution rate is given by 'r*d'. However, this performance rate depends on the division of a task into a proper sequence of subtasks and the rate at which the pipeline can be kept busy.

The execution of a CPU instruction involves two activities [Stone 1980]: the Instruction Processing (IP), and the Instruction Execution (IE). The task IP can be broken down into a number of subtasks, namely:

The *instruction fetch* obtains a copy of the instruction from the memory.
The *instruction decode* examines the instruction and initializes the control signals required to execute the instruction.
The *address generation* computes the effective address of the operands as specified by the instruction.
The *operand fetch* acquires the operand from the memory.

The task IE can also be implemented as a sequence of subtasks:

The *execution* executes the instruction in the microprocessor.
The *operand store* returns the result to the memory for WRITE operations.
The *update program counter* updates the program counter and generates the address of the next instruction.

It is quite feasible to design a CPU by creating a pipeline for each subtask outlined above. Such a design is illustrated in Figure 1.3. If we assume that each pipeline consumes a single clock cycle, and a typical instruction execution takes seven cycles, then the throughput would be seven times faster. However, such

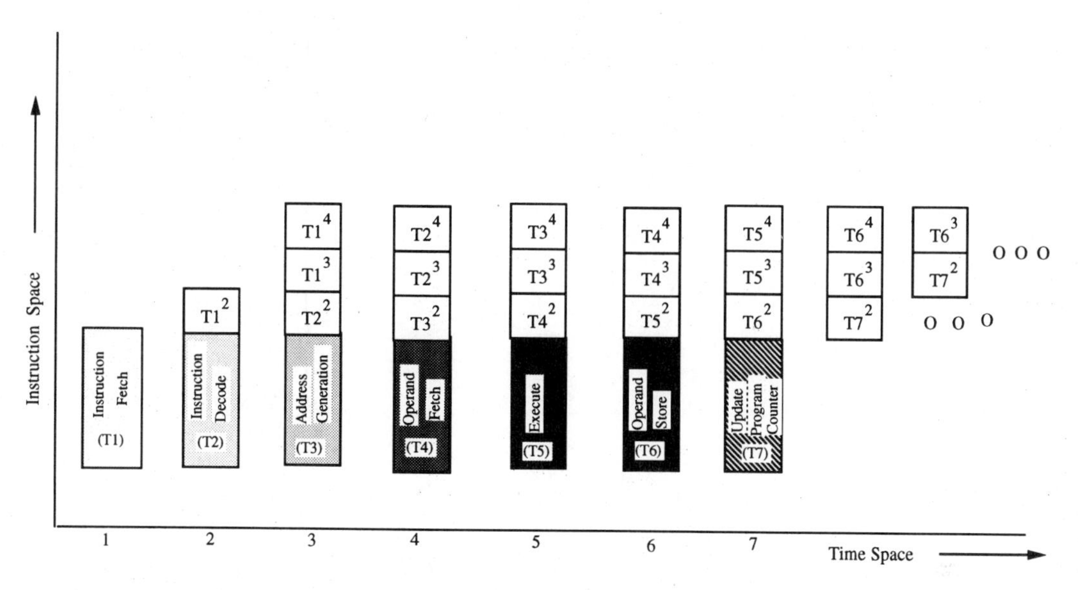

FIGURE 1.3. The linear pipeline processor, with overlapped processing.

performance cannot be achieved in practice. Several factors play against achieving such a high rate of performance, and they are:

First, delays are introduced as soon as data required for the execution of an instruction remains in the pipeline.

Second, branch instructions disrupt the pipeline operation. A branch requires that the CPU calculate the effective address of the destination of the branch and fetch instructions from that location. In the case of a conditional branch, this cannot be done without delaying the subtask in the pipeline for at least one cycle. Otherwise, it is required to fetch the successor of the branch or to predict the outcome of the operation before the execution.

Third, the complexity of managing the pipeline and handling breaks adds additional overhead to the control logic of the CPU.

In the pursuit of performance enhancement, microprocessor designers often increase the number of pipeline stages per instruction. For example, MC68020 uses a four-stage pipeline. The pipeline introduces other problems.

Not all instructions require the same number of pipe stages. Many instructions, in particular the simplest ones, fit best in pipelines of length of two, three, or at most four stages. On the average, longer pipelines will waste cycles that are equal to the difference between the number of stages in the pipeline and the average number of stages required by the instruction. This could lead to the creation of more complex instructions, which would use more pipeline units. Again, complex instructions introduce additional problems, such as branch frequency and operand hazards.

The frequency of branches in compiled code limits the length of the pipeline,

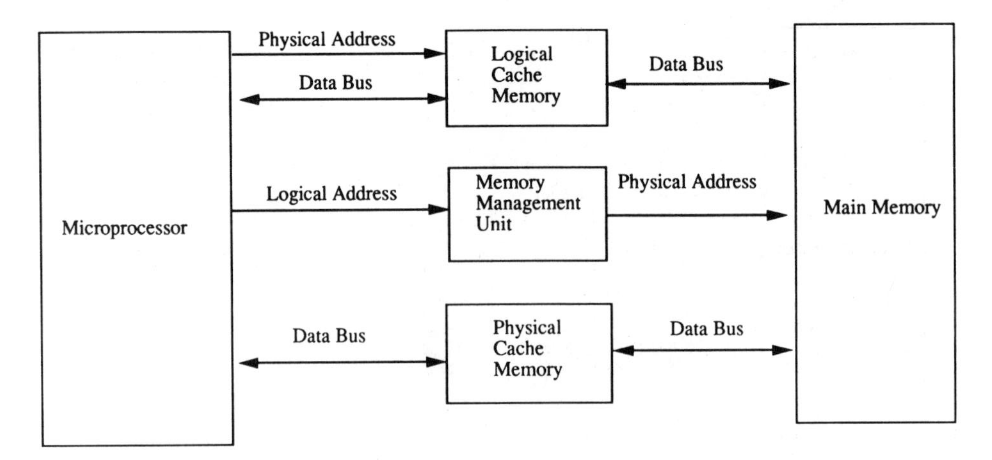

FIGURE 1.4. The memory management system in von Neumann architecture.

because they determine the average number of instructions that are executed before the pipeline can be flushed. Operand hazards cause more difficulty for architectures that use powerful instructions and shorter run lengths, where scheduling may be ineffective.

To solve these problems, the CPU keeps a buffer of sequential instruction up to and including the branch instruction. The CPU examines all of the instructions from the buffer in parallel to decide if they are ready for execution. A part of an instruction is executed as soon as an operand becomes available. The designers also buffer the intermediate results of the instruction, until all the previous parts are completed. This solution becomes complex, especially if an instruction generates results longer than a word [Stone 80].

1.4.6 von Neumann Bottleneck

A problem common to both RISC and CISC architectures is the von Neumann bottleneck, where the CPU processes information faster than the memory can supply it. The most traditional approach to solving this problem is to implement some form of cache memory that acts as a buffer between the main memory and the microprocessor. Figure 1.4 shows this model.

Another approach is to implement the Harvard architecture, which utilizes separate paths for instructions and data to the memory system. Here, the CPU has to wait less often for the memory subsystem, and the effect of other intelligent devices on the main processor is less dominant. Figure 1.5 illustrates the Harvard architecture.

1.5 INTRODUCTION TO THE MC68000 MICROPROCESSOR FAMILY

In 1979, Motorola introduced the first member of its 16-bit microprocessor family, the MC68000. It was based on a 16-bit external data bus and a 24-bit external

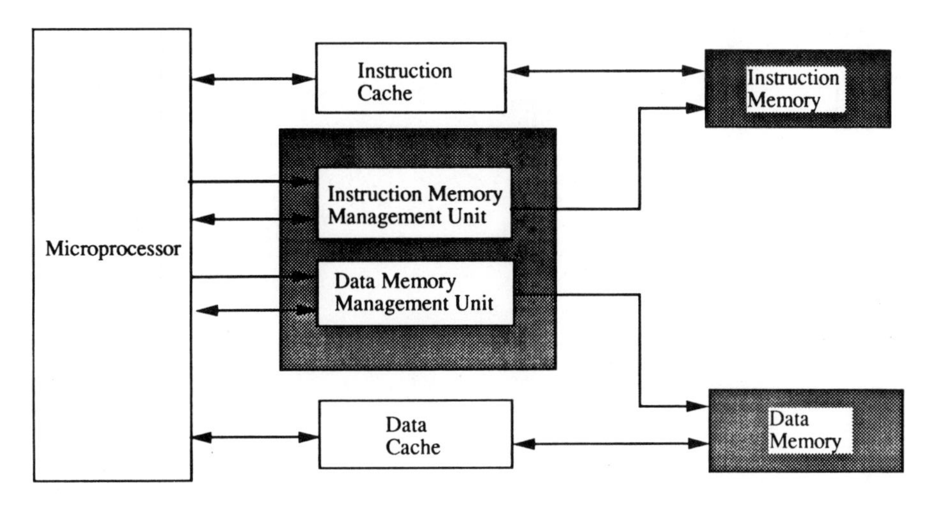

FIGURE 1.5. The CPU and memory linkage in Harvard architecture.

address bus. The product history of the family is illustrated in Figure 1.6. The initial version, operating at 4 MHz, was introduced in 1979. By 1980, the 6MHz and 8MHz versions became available for general use. The 12.5MHz version was available in 1982. The reduced version of the architecture, known as the MC68008, with an 8-bit data bus and 20-bit address bus, was introduced in early 1983. Motorola extended the core architecture by incorporating the virtual memory feature, and introduced two new microprocessors, the MC68010 and MC68020. The MC68020 was introduced in 1984, and the MC68030 in 1988. Motorola introduced the latest member of the family, the MC68040, in 1990. The 33MHz version of the MC68040 was expected to be released in 1991. Motorola has committed to adding new members to the family in the future. The new member, the MC68060, will be introduced in 1993, and will be downward compatible with the previous generations of processors.

From the programmer's perspective, the MC68000 and MC68008 are identical, with the exception that the MC68000 can directly access 16 Mbytes, using its 24 address pins, and the MC68008 can directly access 1 Mbyte, through its 20 bits of address. The MC68010/MC68012 have much in common with the first two devices (the MC68010 has 24 address pins and the MC68012 has 30 address pins, giving an addressing range of 2 Gbytes), but also possess various additional instructions and registers, as well as full virtual machine/memory capability. Virtual memory is a term used to highlight a feature of the machine that gives the illusion to the programmer that he/she has the total memory (i.e., 16 Mbytes in the case of MC68010) available, although in reality the physical memory capacity is smaller than that. The MC68020/MC68030, with their 32 address pins, provide addressing capabilities of 4 gigabytes.

The Motorola processors are fabricated using HMOS and HCMOS technologies. The HMOS is known as a high-density, short-channel MOS, originally de-

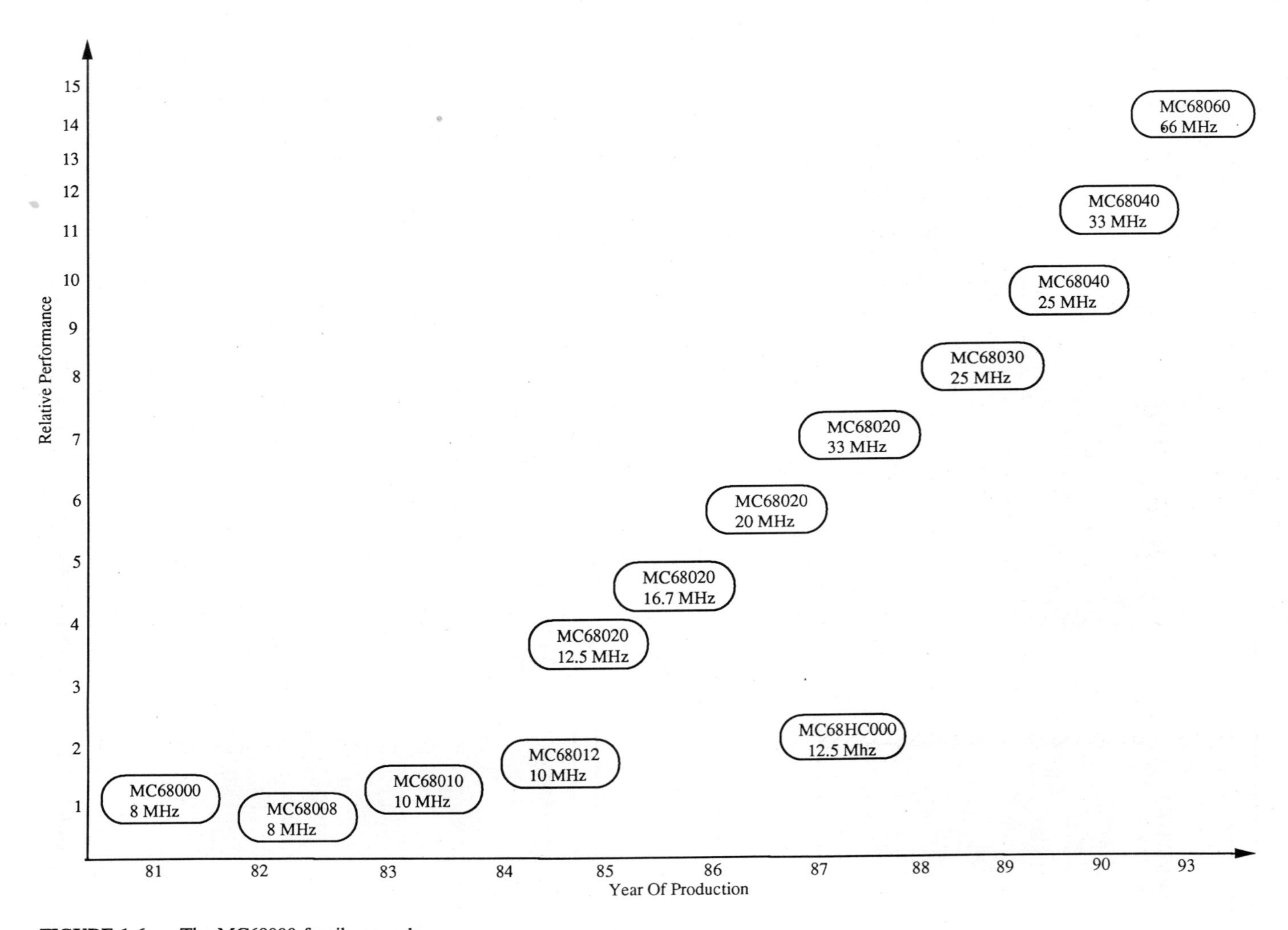

FIGURE 1.6. The MC68000 family genealogy.

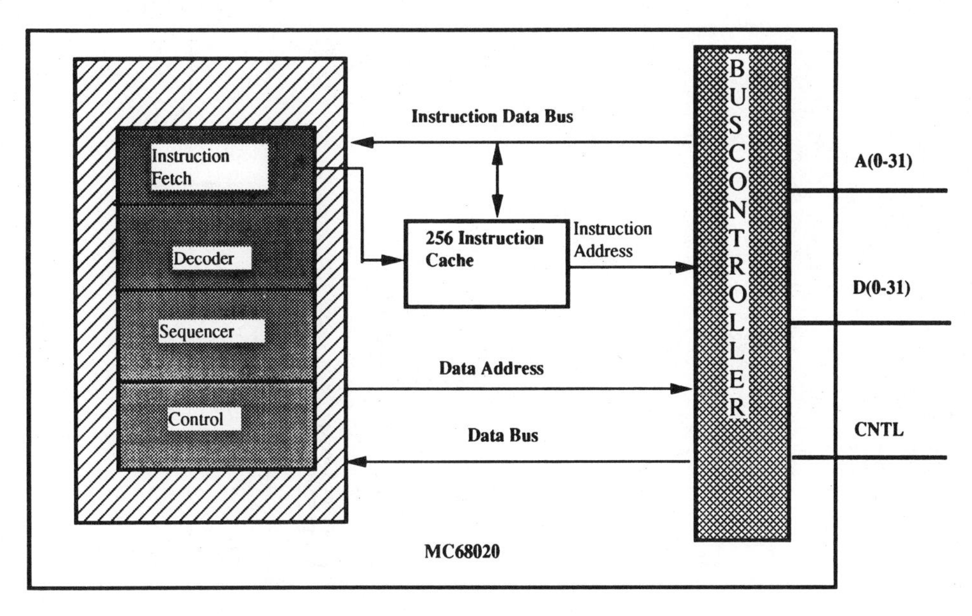

FIGURE 1.7a. The architecture of an MC68020 processor.

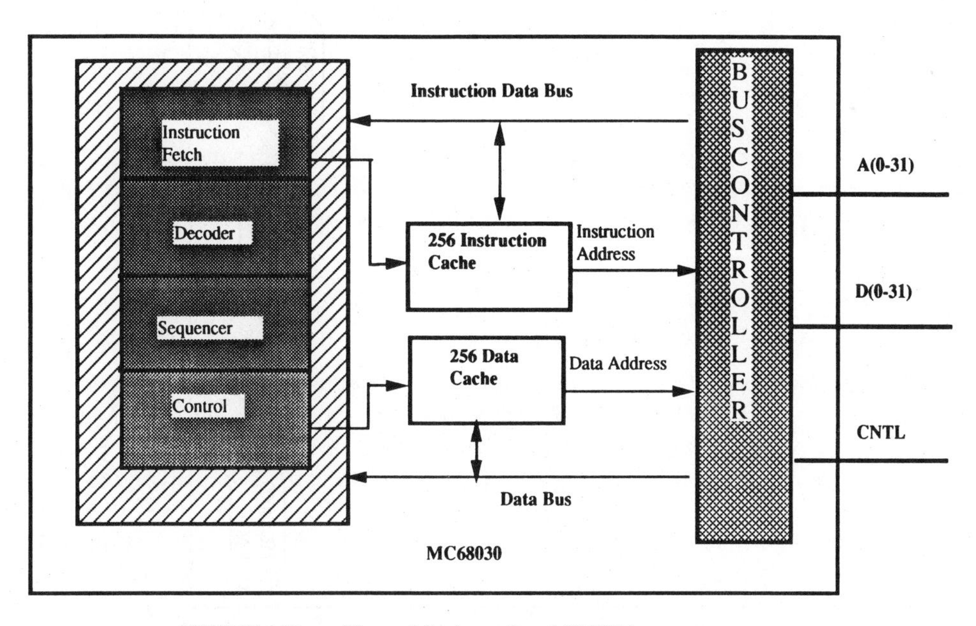

FIGURE 1.7b. The architecture of an MC68030 processor.

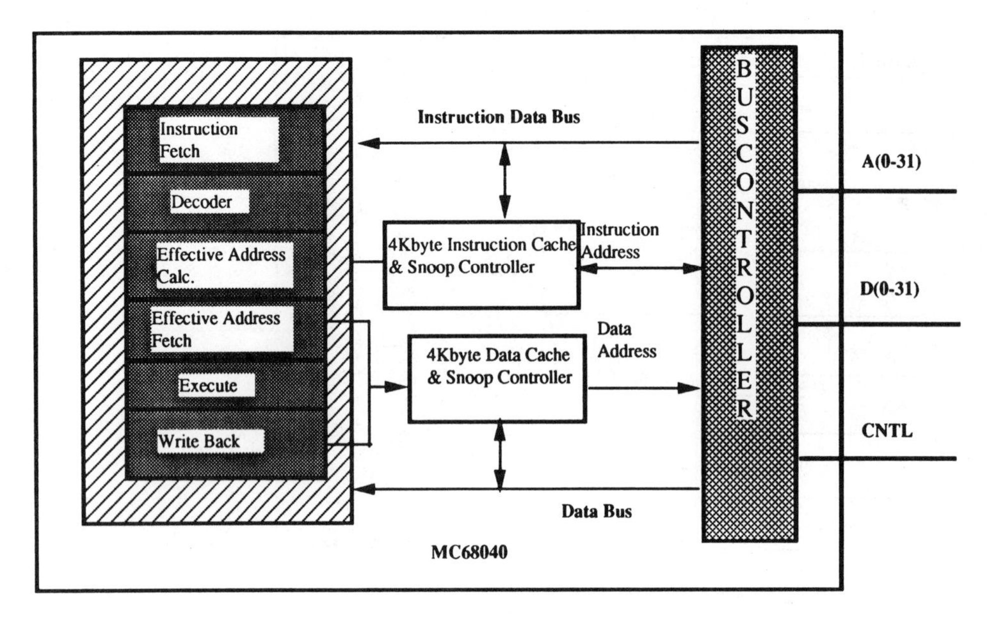

FIGURE 1.7c. The architecture of an MC68040 processor.

veloped by Intel Corporation and used in the MC68000. The technology provides circuit densities twice those of standard NMOS, and a speed-power product four times better than that of standard NMOS. The actual MC68000 chip is about 246 by 281 mils in size, which is about 68,000 square mils in area, 1 mil being one-thousandth of an inch.

The physical size of the MC68020 is 350 by 375 mils, and it contains about 200,000 transistor equivalents, based on HCMOS technology. It is housed in a 122-pin pin gate array package and consumes 1.75 watts of power.

1.5.1 Functional Differences Between the Members of the MC68000 Family

The functional differences between the various members of the MC68000 family are highlighted in Table 1.1 and Figure 1.7a–c. The designers have always maintained compatibility with the previous members of the family, enhancing the direct addressing capability of the processor, and including various operating system functionalities into the architecture.

The MC68020/30/40 all use a three-stage instruction pipeline. The size of each pipeline unit is a trade-off between increasing the performance of the on-chip execution unit for interpreting microcode and the frequency of branches in high-level languages.

The MC68020 included 256 bytes of instruction cache, and the MC68030 added an additional 256 bytes of data cache. The caches are arranged as 16 lines of 4 long-words each, with each long-word separately accessible. The MC68040

TABLE 1.1. Characteristics of the MC68000 Family.

Attribute	MC68000	MC68008	MC68010	MC68020	MC68030	MC68040
Data Bus Size	16	8	16	8, 16, 32	8, 16, 32	32
Address Bus Size	24	20	24	32	32	32
Memory Addressing Capacity	16 Mbyte	1 Mbyte	16 Mbyte	16 GByte	16 GByte	16 GByte
Instruction Cache (Byte)	---	---	3	256	256	4096
Data Cache (Byte)	---	---	---	---	256	4096
Clock Speed - MHz	8, 10, 12.5, 16.67	8, 10	8, 10, 12.5	12.5, 16.67, 20, 25, 33.33	16.67, 20, 25	25
Min. Speed - MHz	4	4	4	4	8	8
Pin Count	64, 68	48, 52	64, 68	122	132	179
Package Type	DIP, QUAD, PGA	DIP, QUAD	DIP, QUAD, PGA	PGA	PGA	PGA
Number of Power Pins	2, 2	1, 1	2, 2, 2	10	10	27
Number of Ground Pins	2, 4, 2	2, 2	2, 4, 2	13	14	40

DIP - Dual In-Line Package
QUAD - Lead Quad Pack
PGA - Pin Grid Array

Attribute	MC68010	MC68020	MC68030	MC68040
Virtual Memory/Machine	yes	yes	yes	no
Virtual Memory	no	yes	yes	yes
Bus Error & Fault Recovery	yes	yes	yes	yes
On-Chip MMU	Interfaced Externally	Interfaced Externally	Internal to Chip	Internal to Chip
On-Chip Floating-Point Unit	Interfaced Externally	Interfaced Externally	Interfaced Externally	Internal to Chip
Multiprocessing Support (Indivisiable Bus Cycle)	Use $\overline{AS}$ Signal	Use $\overline{RMC}$ Signal	Use $\overline{RMC}$ Signal	Use $\overline{LOCK}$ & $\overline{LOCKE}$
Context Switching Support (Stack Frames)	Format $0 & $8	Support $0, $1, $2, $9, $A, $B	Support $0, $1, $2, $9, $A, $B	Support $0, $1, $2, $3, $7
Control Register	SFC, DFC, VBR	SFC, DFC, VBR, CACR, CAAR	SFC, DFC, VBR, CACR, CAAR, CRP, SRP, TC, TT0, TT1, MMUSR	SFC, DFC, VBR, CACR, URP, SRP, TC, DTT0, DTT1, ITT0, ITT1, MMUSR
Status Register Bits	T, S, I0/I1/I2, X/N/Z/V/C	T0, T1, S, I0/I1/I2, X/N/Z/V/C	T0, T1, S, I0/I1/I2, X/N/Z/V/C	T0, T1, S, I0/I1/I2, X/N/Z/V/C

includes 4 Kbytes of physical instruction cache and 4 Kbytes of physical data cache, accessible simultaneously.

Memory management techniques are supported by the MC68010 and later members of the family. The Memory Management Unit (MMU) chip resides outside the CPU in the case of the MC68020, and for the MC68030/40 it is included inside the CPU. The Floating Point Arithmetic Unit (FPU) has been added into the MC68040 chip, and resides outside for the other processors.

Chapter 2

Introduction to the MC68020 Microprocessor

The microprocessor with a 32-bit internal data path and register size, and 16-bit external data bus has been in existence since 1980. However, the era of true 32-bit micros began in the early 1980s, with the introduction of the iAPX 432. Due to technological constraints, it was implemented as a three-chip set. A number of micros have been introduced into the marketplace by various vendors, which extended that architecture to include various features described in Chapter 1.

The MC68020 is the first true 32-bit microprocessor from Motorola. It is implemented with a 32-bit register, along with 32-bit internal and external data paths, 32-bit addresses, and a rich instruction set supported by versatile addressing modes.

This chapter deals with the internal architecture of the MC68020, its mode of operation, and its implementation. System designers must understand these features in order to be able to use the processor in their design, and/or to program its operation and debug the software/hardware interface.

2.1 DESIGN GOALS AND ENHANCEMENTS

The MC68020 is object-code compatible with the other members of the MC68000 family. The MC68020, a product of 60 man-years of effort, is based on High-density Complementary Metal Oxide Semiconductor (HCMOS) technology, which provides an important combination of high-speed performance and low power consumption. This technology allows CMOS and HMOS gates to be combined on the same device. CMOS technology is used where speed and low power is required, and HMOS structures are used where minimum silicon area is desired. The first implementation operated at 12.5 MHz. Today, one can get a CPU that runs at 16.67, 20, 25, or 33.33 MHz. It consists of 200,000 actual transistors in a die size of 0.375 by 0.350 inches, and is packaged in a 114 PGA (Pin Grid Array) that dissipates less than 1.75 watts.

The design goals of the MC68020 were:

- To maintain compatibility with the MC68000 via:
  - Adherence to the register-memory interaction philosophy
  - Maintaining eight general-purpose data registers
  - Eight general-purpose address registers
  - Extending the large linear address space of the MC68000
- To provide enhancements to the MC68000 through:
  - Expanding the instruction set
  - Improving register set orthogonality
  - Enhancing the addressing modes
  - Improving the coprocessor interface
- Design trade-offs through:
  - Increased clock frequency
  - On-chip instruction cache
  - Enhanced pipeline structure
  - Increasing bus architecture to a full 32 bits
  - Increased internal parallel activity
  - A fast barrel shifter

The MC68020 supports 32-bit nonmultiplexed address and data buses. The processor supports dynamic bus sizing, a mechanism that allows the transfer of operands to or from external devices, while automatically determining their port size on a cycle-by-cycle basis. This eliminates the bus alignment restrictions of the previous members of the family.

In addition to meeting the above design goals, the design of this powerful micro provides additional desirable features:

- Virtual memory/machine support
- Sixteen 32-bit wide, general-purpose data/address registers
- A total of three stack pointers
- A full 32-bit program counter
- Five control registers
- A large direct-addressing capability, up to 4 Gbyte
- 18 addressing modes
- Over 100 instructions
- Hardware support for coprocessor interfacing
- On-chip cache memory
- Hardware support for multiprocessing/multitasking
- Complete memory management and floating-point support.

2.2 THE MC68020 AS AN INTEGRATED CIRCUIT

This section provides a brief description of the MC68020 signal lines, logic requirements, packaging constraints, the operating frequency, and the signaling constraints. The signal level at all pins are TTL compatible, i.e., logic state 1 (high voltage) is represented by a level from 2.4V through 5V, and logic level 0 (low voltage) is by a voltage level between 0.8V and 0V. Any voltage level beyond that is undefined. However, it is important to understand the meanings of two terms 'assertion' and 'negation' in the context of MC68020 operation. These terms are used to specify the forcing of a signal to a particular state, independent of the voltage level that it represents.

The term 'assertion,' or 'assert,' is used to specify that a signal is active or true, irrespective of whether that level is represented by a high or low voltage. Whereas, the term 'negation,' or 'negate,' is used to specify that a signal is inactive or false.

In the previous chapter, we have seen that the microprocessor signals can be functionally organized into three groups, namely:

- Address lines
- Data lines
- Control lines

The control lines of the MC68020 can be grouped into eight classes, to identify their unique functionalities. The resultant ten functional signal groups of MC68020 are illustrated in Figure 2.1, and they are the:

- Address bus
- Data bus
- Processor status (function codes)
- Transfer size
- Asynchronous bus control
- Bus arbitration control
- Interrupt control
- System control
- Cache control
- Excitation control

2.2.1 Address Lines (A0-A31)

These 32-bit, unidirectional, three-state signal lines form the *address bus*. The address bus enables the MC68020 to directly address 4 Gbytes of different locations. The independent address pins eliminate the need for external demultiplexing circuitry during the design of memory and I/O subsystems. The pin A31 represents the Most Significant Bit (MSB) and A0, the Least Significant Bit (LSB) of the addresses. The address pins get asserted during all currently defined cycles except the CPU space references. During the CPU-space references, the address

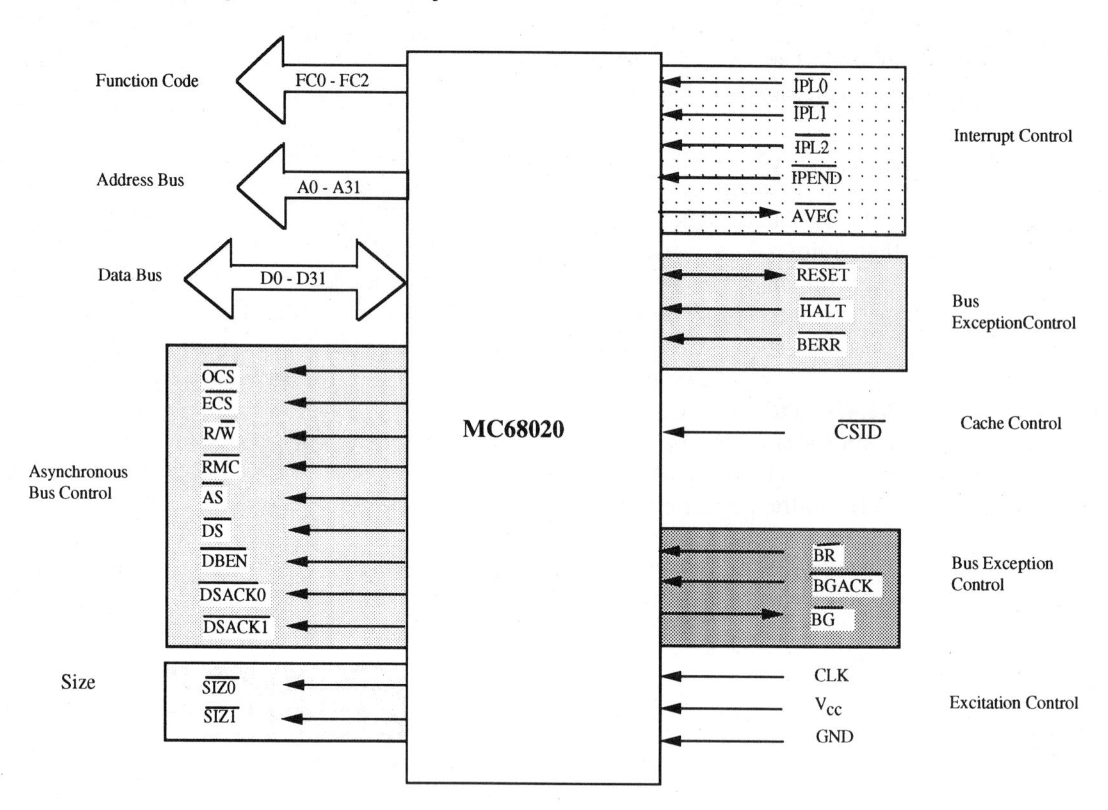

FIGURE 2.1. Pin diagram of the MC68020.

bus carries specific CPU-related information. During the active state, the bus is capable of driving 3.2 ma (milliampere) current, which is equivalent to 2 TTL loads.

2.2.2 Data Lines (D0-D31)

The 32 bidirectional, 3-state signal lines are the general-purpose data path. It is used to transfer and accept data in either long-word, word, or byte size. The dynamic bus sizing capability of the MC68020 allows it to negotiate the size of the data to be received from, or transmitted to, a device on a cycle-by-cycle basis. The bus is capable of driving 3.2 ma current, which is equivalent to 2 TTL loads during the active state. Thus, it is important to boost the drive capabilities when the microprocessor is interfaced with other devices in the system.

2.2.3 Processor Status (FC0-FC2)

The MC68000 family of CPUs utilizes these three common signals to declare its addressing mode to the outside world. These three-state output (FC0, FC1, and

FC2) lines are called function codes. By decoding these signals, a memory system can be designed to utilize the full 4 Gbyte address range for each address space.

2.2.4 Asynchronous Bus Control Signals

The MC68020 performs a transaction in an asynchronous mode. This means that, once a bus cycle is initiated, it is not completed until a signal is returned from the external world or watchdog circuitry. The signal lines can supply 5.3 ma current and are capable of driving 4 TTL loads. The asynchronism control lines are listed in the following subsections (Figure 2.2).

2.2.4.1 $\overline{AS}$ (Address Strobe)

The processor asserts the address strobe ($\overline{AS}$) over this pin as it places a valid address on the address bus. This three-state output signal validates the function code, address, size, and R/$\overline{W}$ state information on the bus.

2.2.4.2 $\overline{DS}$ (Data Strobe)

During a read cycle, this three-state output indicates that an external device should place valid data on data lines and drive the data bus. During a write cycle,

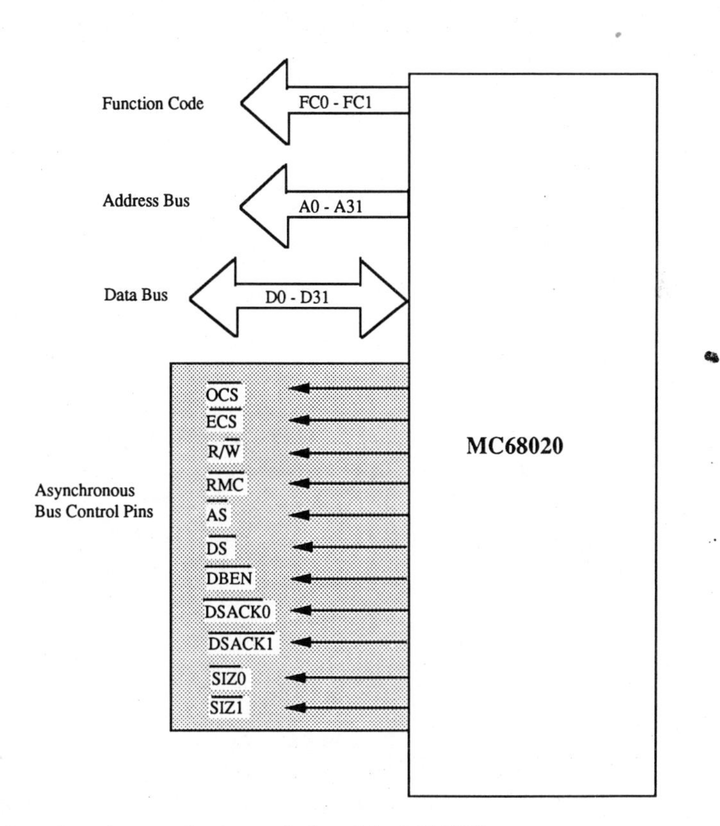

FIGURE 2.2. Asynchronous bus control pins of the MC68020.

the data strobe indicates that the MC68020 has placed a valid data for the receiving device.

2.2.4.3 R/$\overline{W}$ (Read/Write)

This signal defines the nature of the data transfer (i.e., a read or write operation). During a read cycle, the processor reads data from the data bus and asserts a logic 1 on this line. A logic 0 is asserted when data is supplied to memory or I/O devices.

2.2.4.4 $\overline{ECS}$ (External Cycle Start)

This output signal initiates the beginning of a bus cycle. It is generated by the MC68020 during the first one-half clock of every bus cycle, to give the earliest indication of an impending bus cycle. This signal must be validated later with $\overline{AS}$, since the CPU may start an instruction fetch cycle, and then abort it, because the information is found in the cache. The MC68020 continues to drive the address, size, and function code lines as it aborts a bus cycle due to cache hit.

2.2.4.5 $\overline{OCS}$ (Operand Cycle Start)

$\overline{OCS}$ has the same timing as $\overline{ECS}$, and indicates the beginning of an external bus cycle for an instruction prefetch or operand transfer. It is not asserted for subsequent cycles, which are performed due to dynamic bus sizing or operand misalignment.

2.2.4.6 $\overline{DBEN}$ (Data Buffer Enable)

This three-state output signal is designed to enable external data buffers, and may not be required in all systems. The signal allows the R/$\overline{W}$ line to change without the probability of external buffer contention. However, the timing of this signal may preclude its use in a system that supports two-clock synchronous bus cycles.

2.2.4.7 $\overline{DSACK0}$ and $\overline{DSACK1}$ (Data Transfer and Size Acknowledgment)

Under the asynchronous bus cycle, external devices or circuitry must signal the processor that the bus cycle can be completed. When the processor recognizes $\overline{DSACKx}$ during a read cycle, it latches the data and terminates the bus cycle. On the other hand, the CPU terminates the bus cycle on recognizing the $\overline{DSACKx}$.

2.2.4.8 $\overline{RMC}$ (Read-Modify-Write Cycle)

This three-state output indicates that the current bus cycle needs to be treated as an indivisible read-modify-write operation. It remains asserted during all three of the activities. $\overline{RMC}$ should be used to ensure the integrity of an operation that requires the exclusive use of a resource. This signal is often used to implement a hardware semaphore.

2.2.4.9 SIZ0 and SIZ1

These three-state output signal lines identify the remaining number of bytes to be transferred for the current bus cycle. These signals, along with A0, A1, DSACK0, and DSACK1, define the number of bytes transferred over the data bus.

2.2.5 Bus Arbitration Control Signals

The bus arbitration control signals support the protocol developed for the MC68020 by which the control of system bus is transferred between devices. A device currently having the control of the bus is known as the *bus master.* The bus master always controls the address, data, and control buses. If other devices (the DMA controller or attached processors) want to become the bus master, they use these control lines and an arbitration protocol (discussed in Chapter 7) to request the use of the system bus. The bus arbitration control signal lines are:

BR: Bus Request
BG: Bus Grant
BGACK: Bus Grant Acknowledgment

2.2.5.1 BR (Bus Request)

Any would-be bus master requests the bus by asserting a signal on this line. This informs the current bus master that some other device desires to become the bus master. All potential bus masters "wire-ORed" (but do not need to be constructed from open collector devices) their requests to this line.

2.2.5.2 BG (Bus Grant)

After receiving the BR, the processor responds by asserting an active low on this output line. This signals the potential bus masters that the MC68020 will release the control of the system bus at the end of current bus cycle.

2.2.5.3 BGACK (Bus Grant Acknowledgment)

The requesting device signals its willingness to accept the control of the system bus, using this line. This signal should not be asserted until the following four conditions are met:

a. A bus grant has been received through the bus arbitration mechanism.
b. The address strobe is negated, which indicates that the CPU is not using the bus.
c. The data transfer acknowledgment is inactive, which indicates that neither the memory nor the peripherals are using the bus.
d. The bus grant acknowledgment is negated, which indicates that no other device is still claiming bus ownership.

Upon receiving the BGACK, the MC68020 responds by removing the bus grant (BG) signal. This completes the bus arbitration handshake mechanism. The requesting device can now become the bus master. It releases control of the bus by negating $\overline{\text{BGACK}}$ for rearbitration, or returns the bus to the processor at the completion of the task.

2.2.6 System Control

The system control lines are used either to control the operation of the processor or to indicate its internal state. The signals are:

$\overline{\text{BERR}}$: Bus error
$\overline{\text{RESET}}$: Reset
$\overline{\text{HALT}}$: Halt

2.2.6.1 $\overline{\textbf{BERR}}$ (Bus Error)

This input line is used to tell the processor that an error condition has occurred during the current bus cycle. An error may occur due to:

a. Nonresponding devices.
b. Interrupt vector number acquisition failure.
c. Illegal access request, as determined by a memory management unit (e.g., an access fault due to a protected memory scheme or a page fault in the case of a virtual memory system).
d. Noncorrectable (hard) parity errors.
e. Other application dependent errors.

$\overline{\text{BERR}}$ is asserted after the CPU completes the following activities:

- Abort the current bus cycle
- Ignore the content of the data bus
- Tri-states the address and data bus
- Negate the bus control lines

At the end, the CPU examines the $\overline{\text{HALT}}$ signal to determine whether the current bus cycle should be reexecuted, or an exception processing should be initiated. Exception processing is discussed in detail in Chapter 7.

2.2.6.2 $\overline{\textbf{RESET}}$ (Reset)

This is a bidirectional open-drain signal, which can be driven internally or externally. It is driven externally to invoke a system initialization sequence. It is driven internally when the MC68020 executes a $\overline{\text{RESET}}$ instruction. The $\overline{\text{RESET}}$ instruction will 'reset' external devices (whose reset lines are connected to processor's $\overline{\text{RESET}}$ line), but will not change its internal states.

An externally generated $\overline{\text{RESET}}$ signal must meet the following timing constraints:

During a power-on reset: $\overline{\text{RESET}}$ and $\overline{\text{HALT}}$ signals must be asserted low and be held low for greater than or equal to 100 milliseconds.
During a single-pulse reset: $\overline{\text{RESET}}$ and $\overline{\text{HALT}}$ signals must be asserted for

greater than or equal to a 10-clock period, in order to allow the CPU to respond to this event.

2.2.6.3 $\overline{\mathrm{HALT}}$ (Halt)

This is a bidirectional open-drain signal. Upon detection of a valid status on this line, the MC68020 stops all bus activities after the completion of the current bus cycle. The CPU then negates all control signals. However, address, R/$\overline{\mathrm{W}}$, size, and function code signals retain their previous bus cycle values, with the exception of the data bus. $\overline{\mathrm{HALT}}$ does not stop processor execution, only the external bus activity. However, the CPU will continue its execution if a cache hit occurs and the external bus is not required.

If the 'halt' state is caused by a double-bus fault, then the CPU will drive the $\overline{\mathrm{HALT}}$ line to indicate to the external devices that it has 'stopped' its operations.

2.2.7 Interrupt Control Signals

In order to facilitate the implementation of real-time systems, designers of the MC68020 provided a number of signal lines and a protocol. The protocol is implemented via the signal lines:

$\overline{\mathrm{IPL0\text{-}2}}$: Interrupt priority levels
$\overline{\mathrm{IPEND}}$: Interrupt pending
$\overline{\mathrm{AVEC}}$: Autovector

2.2.7.1 $\overline{\mathrm{IPL0\text{-}2}}$ (Interrupt Priority Levels)

The three input signals together present an encoded value of the interrupt condition to the CPU. The pins can indicate seven levels of priority, where level seven has the highest priority and cannot be masked. Level zero indicates the absence of an interrupt request. The most significant bit is $\overline{\mathrm{IPL2}}$, and the least significant bit is $\overline{\mathrm{IPL0}}$.

2.2.7.2 $\overline{\mathrm{IPEND}}$ (Interrupt Pending)

This output signal, when asserted, indicates that the MC68020 has internally registered an interrupt request whose priority exceeds the current interrupt priority mask of the Status Register (SR). Interrupt pending signal can be utilized by the external devices (e.g., another bus master or a coprocessor) to predict CPU operation during instruction boundaries.

2.2.7.3 $\overline{\mathrm{AVEC}}$ (Autovector)

This input signal is utilized by nonintelligent external devices. When asserted, the MC68020 automatically generates a vector number during an interrupt acknowledgment cycle. The vector number corresponds to the encoded interrupt level found on the interrupt signal lines.

TABLE 2.1. Power and Ground Lines.

CPU Type	Number of Power Lines	Number of Grounds	Minimum Frequency (MHZ)	Maximum Frequency (MHZ)
MC68000	1	2	4	16.67
MC68008	1	2	4	12.5
MC68010	2	2	4	12.5
MC68020	10	15	8	33.33
MC68030	12	17	12.5/20	20/50
MC68040	25	35	—	33.33

2.2.8 Emulator Support Signal

This signal can be utilized to support emulation. It provides a mechanism for an "emulator" to disable the on-chip cache memory, and thus identify the internal status information of the MC68020 to an emulator.

2.2.8.1 $\overline{CDIS}$ (Cache Disable)

The internal cache of the MC68020 is disabled after this line is asserted externally. This forces the CPU to fetch instructions from the primary memory. The cache will become operational once $\overline{CDIS}$ is negated and gets synchronized internally. The cache entries remain unaltered while the cache is disabled. They again become available once $\overline{CDIS}$ is negated.

2.2.9 Excitation controls

The excitation control pins are the life lines of the CPU. They supply power, provide grounding, and clock the signal. Table 2.1 lists the number of power and ground lines, including the minimum and maximum operating frequencies, of various members of the MC68000 family.

2.2.9.1 Clock Controls

Clock generator circuitry is not provided to the MC68020, and must be generated externally. This TTL-compatible signal is buffered internally to generate the internal clocks needed by the CPU. The clock should not be gated off at any time, and must conform to requirements for minimum and maximum periods, including the pulse width. The clock requires a 50% duty cycle.

2.3 THE PROGRAMMING MODEL

The MC68020 is the first true 32-bit microprocessor from Motorola. One of the primary design goals for this processor was to maintain strict user object code compatibility with the rest of the MC68000 family. Object code compatibility is becoming a major factor, because software development costs always exceed the hardware costs. The designers achieved that by utilizing two-tier programming

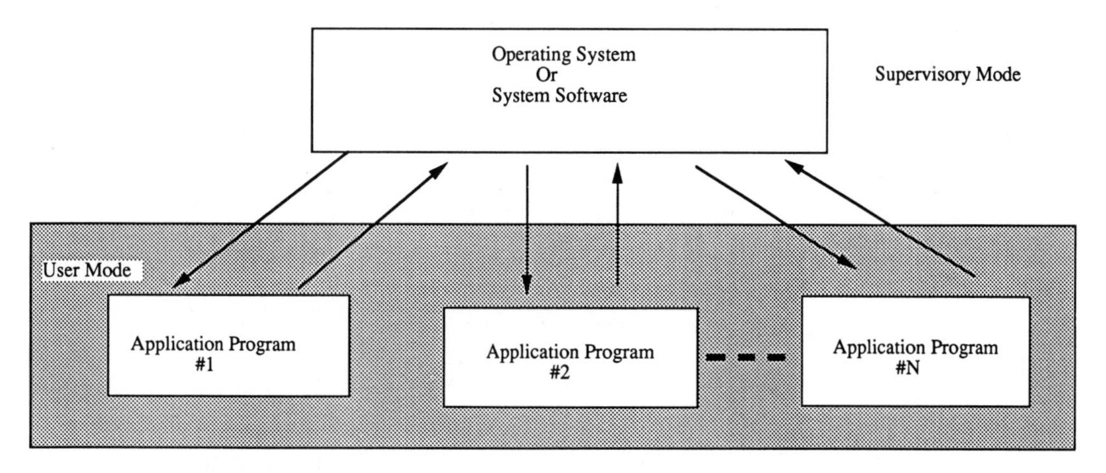

FIGURE 2.3. Software execution models.

models for the MC68000 processor family (Figure 2.3). One is called the *user programming model* and the other is the *supervisory programming model.* The user programming model (also called the User Model), remains largely the same for all members of the MC68000 family. This model uses the CPU resources via the following registers (Figure 2.4):

- General-purpose 32-bit registers (D0-D7, A0-A7)
- A 32-bit Program Counter (PC)
- An 8-bit Condition Code Register (CCR)

The supervisory programming model allows the executing program to utilize all CPU resources. It includes the User Model, plus additional resources that can only be accessed via eight control registers (Figure 2.5);

- Two 32-bit supervisory Stack Pointers (ISP and MSP)
- A 16-bit Status Register (SR)
- A 32-bit Vector Base Register (VBR)
- Two 32-bit alternate (Source and Destination) Function Code registers (SFC and DFC)
- One 32-bit Cache Control Register (CACR)
- One 32-bit Cache Address Register (CAAR)

The supervisory models are different for each member of the family. They also provide the personality of the processor. The processor designers utilized the supervisory model to enhance the power of the processor.

2.4 MODES OF OPERATION

The MC68020 executes instructions in one of two modes (Figure 2.3), either the supervisor mode (state), or the user mode (state).

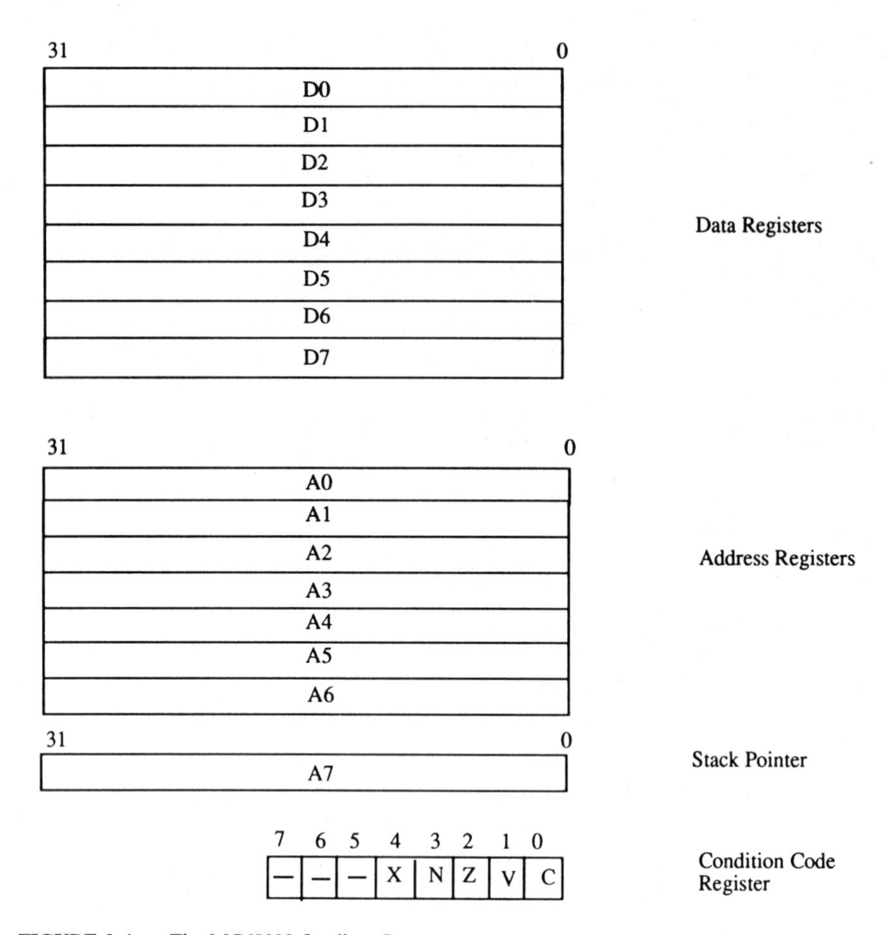

FIGURE 2.4. The MC68000 family—Common resources.

This built-in architectural feature provides security and determines which operations are legal and which are not. The supervisory mode is intended for the operating system and imposes restrictions on the usage of system resources. The user mode allows the execution of the majority of instructions required for application programs. In the supervisory mode, the processor is allowed to execute all of the instructions available in user mode, including privilege instructions. The processor leaves an internal footprint on the status register to indicate its operating mode. It also asserts function control pins to indicate its mode of operation to the outside world.

Based on the M bit of the status register, the MC68020 performs two types of supervisory activities. The purpose of the M bit is to separate the I/O-related supervisory tasks from rest of the supervisory tasks. This feature facilitates the implementation of the multitasking operating system, where it is more efficient to have a supervisory stack space associated with each user task and separate stack space for interrupt-related tasks.

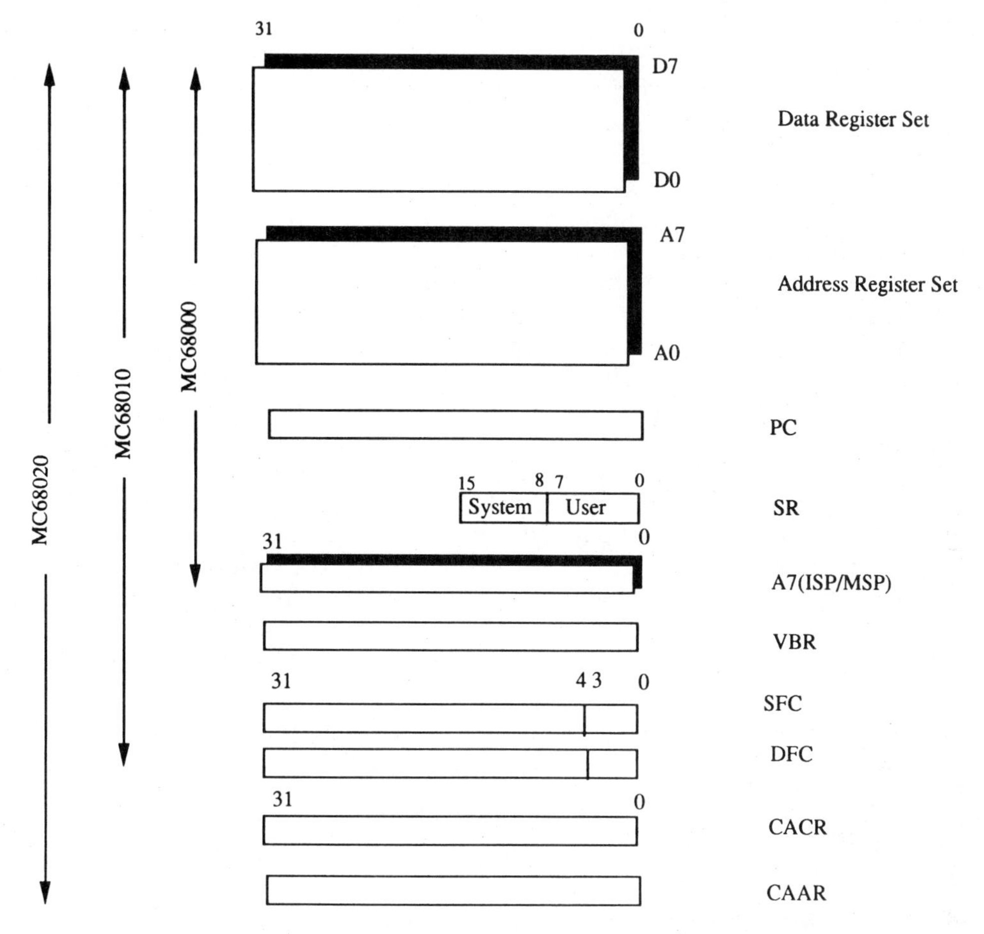

FIGURE 2.5. The MC68000/MC68010/MC68020 supervisory programming models.

2.5 INTERNAL ARCHITECTURE OF THE MICROPROCESSOR

The MC68020 offers sixteen 32-bit wide, general-purpose registers. The registers are subdivided into groups. Eight of them form the data register group, and are identified as D0 to D7. The data registers are generally used to perform data operations. The second set of seven registers, A0 through A6, are used during addressing operations. The address register A7 is used as a software stack pointer. When the processor operates under supervisory mode, A7 is utilized to access the supervisory stack area, and is known as the Supervisory Stack Pointer (SSP). During interrupt operation, A7 is labeled as the Interrupt Stack Pointer (ISP), or A7''. While the processor executes a user program/application program, the stack points to user stack area, and is known as the User Stack Pointer (USP), or A7'. Any of the address or data registers may be used as index registers.

2.5.1 Data Registers (Dx)

The data registers D0–D7 are 32 bits wide. The Most Significant Bit (MSB) is bit 31, and the Least Significant Bit (LSB) is bit 0. The organization of the data in the data register is illustrated in Figure 2.6. The long-word operand occupies the entire 32 bits, a word operand occupies the least significant 16 bits, and a byte operand utilizes the least significant 8 bits. During bit-field operations, the MSB is addressed as bit zero, and the LSB is designated as the width of the field minus one. However, if the width of the field plus the offset becomes greater than 32, then the bit field wraps around within the register. When a data register is used as either a source or destination operand during program execution, then only the appropriate low-order byte or word (i.e., in byte or word operations, respectively) is used or changed. The remaining high-order portion is neither used or changed.

Quad-word data consists of two long-words, and are used during the 32-bit multiply and divide operations. Quad-words can be placed on any two registers without the restrictions of any order or pairs. Instructions are not provided for the management of this data type, although the MOVEM instruction can be used to move a quad-word to and from registers into memory.

Binary Coded Decimal (BCD) data represents decimal numbers in binary

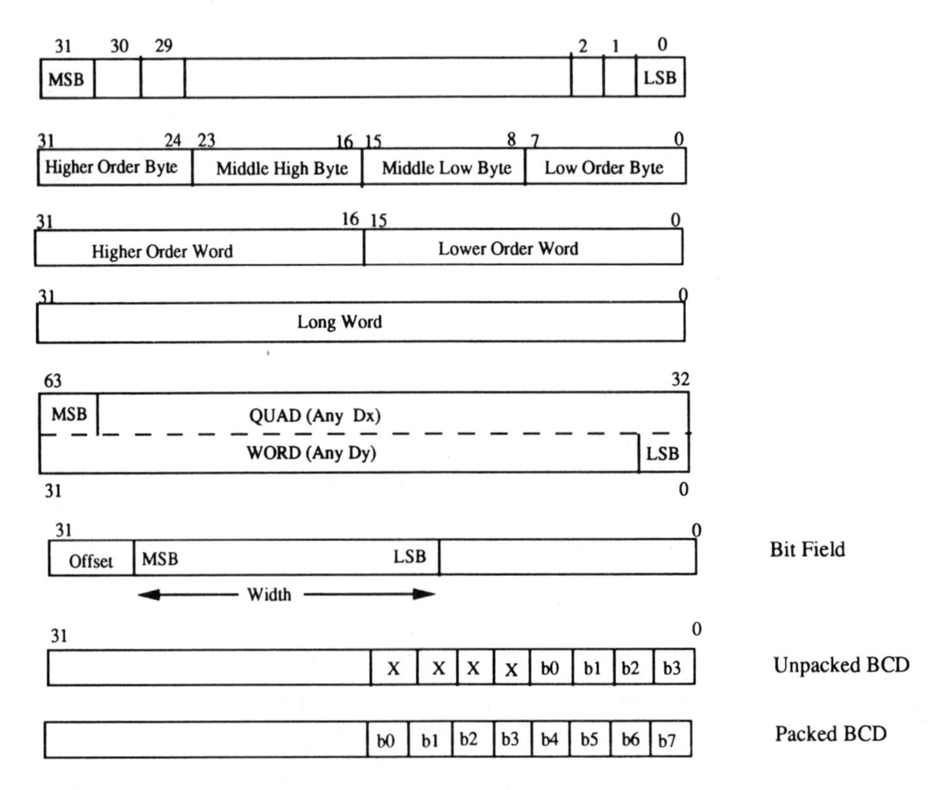

FIGURE 2.6. Data organization in data registers.

form. The BCD format utilizes four bits to describe decimal numbers. The MC68000 family supports two BCD formats: *unpacked BCD* and *packed BCD.* In unpacked BCD, a byte contains only one digit. Here, the four least significant bits contain the binary value of the digit and the most significant bits are undefined. In a packed BCD, a byte contains two digits, where the least significant four bits contain the least significant digit, and the most significant digit occupies the most significant four bits.

2.5.2 Address Registers (Ax)

The address registers A0–A7 are 32 bits wide and can hold a 32-bit address. They generally hold source operands. The data organizations within the addresses registers are shown in Figure 2.7. Address registers are primarily used to hold addresses and support address computation. As a result, they cannot be used for byte-sized operands. The size of the operand determines whether the least significant word or the long-word of the address register will be utilized. However, the entire long-word is affected when used as destination operands. If the operation size is a word, then the processor-first sign extends the operand to 32 bits before its execution. Instructions are available that can add to, subtract from, compare against, and move the contents of the address registers.

2.5.3 Control Registers

The control registers retain control information relevant to supervisory functions. The exception is the 16-bit wide Status Register (SR). It is logically divided into two parts: the lower byte is known as the Condition Code Register (CCR), and the higher byte is called the *system byte.* All control registers are privileged resources, and are thus accessed only by the instructions at the supervisor privilege level (In the case of the MC68000 and the MC68008, the CCR can be accessed from the user mode).

The status register contains all of the processor-relevant information at all times. Only 12 bits of the SR are defined. The remaining 4 undefined bits are reserved by Motorola for future use. These 4 bits are read as zeros and should not be written into, for future compatibility. All operations to the SR and CCR are word-sized operations. However, in case of the CCR, the upper byte is read

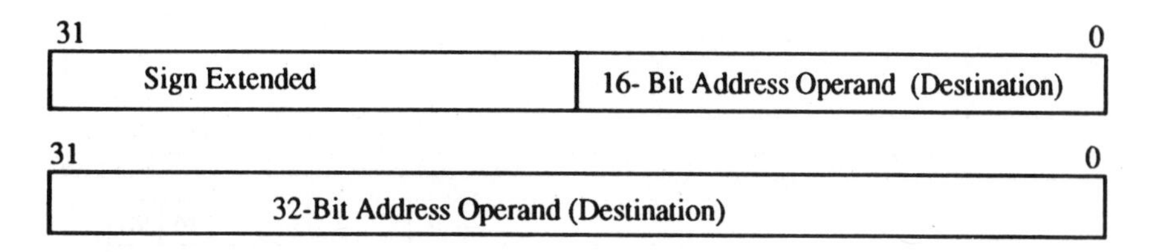

FIGURE 2.7. Address organization in address registers.

as zeros and are ignored during update (e.g., written), regardless of the privilege level.

The CCR flag bits are (Figure 2.8):

The Carry Flag (C): This bit indicates any carry generated by an addition instruction, or any borrow produced as a result of a subtract operation. This flag will take the value of the MSB or LSB bit after a shift operation.

The Overflow Flag (V): This bit is used for operations involving signed numbers. It is set if an add or subtract operation produces a result that exceeds two's complement range of numbers, otherwise the flag is set to 0.

The Zero Flag (Z): This bit is set to 1 if the execution of an instruction produces a zero result.

The Negative Flag (N): This bit is set to 1 if the most significant bit of an operand is 1 (negative numbers). Otherwise, it is set to 0, to indicate a positive number. Note that the operand can be 8, 16, or 32 bits long.

The Extend Flag (X): This bit is set during a multiprecision operation to reflect the carry. Otherwise, X takes the value of the carry flag.

The flags within the CCR are utilized by conditional branch instructions. Depending on the status of the flags, the Program Counter (PC) gets loaded with the address of the next instruction or the address of the branch.

The system byte of the SR is composed of three logical subfields (Figure 2.8) namely:

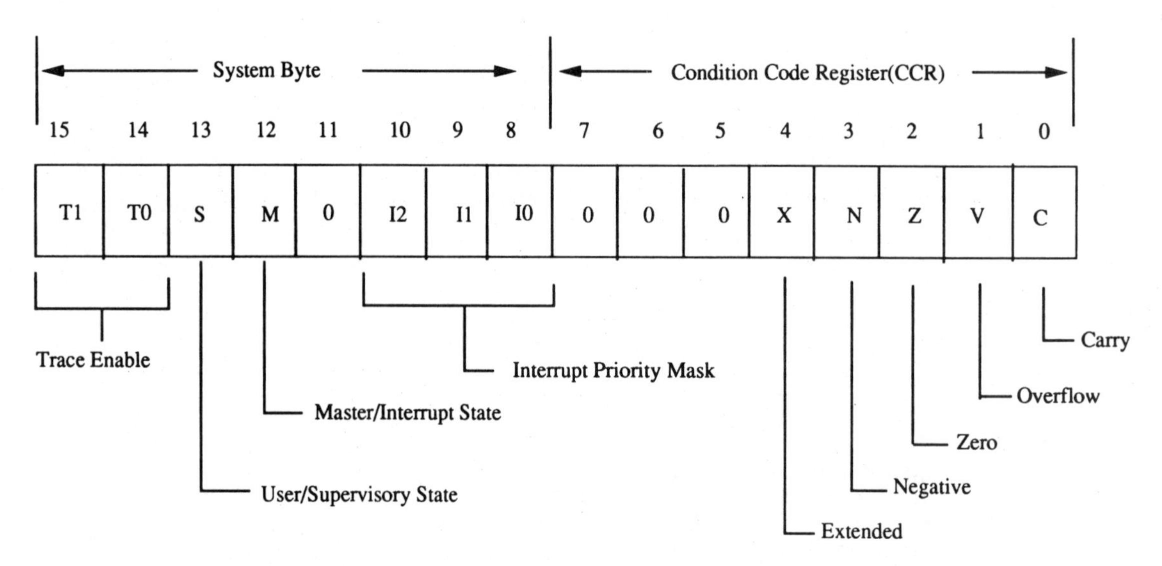

FIGURE 2.8. Bit description in the status register.

- Interrupt masks (bits 8, 9, and 10)
- Status masks (bits 12 and 13)
- Trace masks (bits 14 and 15)

The *interrupt masks* (I0, I1, I2) provide a numeric value for the priority of the interrupt request. All interrupts presented to the CPU get compared against the numeric value of this subfield. If the compare produces a positive value, then the interrupt gets serviced. Otherwise, it is ignored by the CPU.

The *supervisory flag* (S) is set whenever processors enter into the supervisory mode. In this mode, the processor may access all system resources, including the complete instruction set. When the S bit is cleared, the processor enters into the user mode.

The *master/interrupt flag* (M) bit becomes important when the MC68020 enters into a supervisory state. It indicates which stack the MC68020 will currently utilize. The MC68020 enters into the interrupt state when M is set to zero, and will operate like the MC68000, MC68008, MC68010, and MC68012's supervisory state. The MC68020 uses the Interrupt Stack Pointer (ISP) to reference the system stack area. When the M bit is set to 1, the CPU enters into the master state and uses the MSP to reference the system stack area.

The interrupt state and the master state are designed to facilitate the implementation of multitasking operating system. The operating system utilizes this feature to separate task-related activities from I/O activities. In a multitasking operating system, it is convenient to allocate a supervisory stack space for each user task, and an independent stack space for interrupt-associated tasks. The master stack may be used to manage task-control information for the currently executing user tasks, while the interrupt stack is used for interrupt-related task-control information and the assignment of temporary storage. When the CPU switches from one user task to another, the master stack pointer gets loaded with a new value that points to the new task context. This allows the operating system to maintain a valid independent task space for interrupts.

The trace mode flags (T0-T1), when set to 1 (both are not, however, set at the same time), enable the internal debugging facilities of the processor. In the case of $T1=1$ and $T0=0$, the processor will single-step through each instruction. After the execution of each instruction, the processor will enter into supervisory mode, and jump to a user-written trace routine. The routine could be used to find the contents of specific registers or memory, or any other desired debug operations. If $T1=0$ and $T0=1$, the processor will execute the trace routine, in the event of any changes in the instruction execution flows. When $T1=0$ and $T0=0$, the tracing operations are disabled. System designers should not set $T0=1$ and $T1=1$ in their software, to maintain future software compatibility.

2.5.4 Program Counter (PC)

The 32-bit program counter keeps track of the next instruction in the program. It is generally modified via JUMP and branch instructions, subroutine linkages and interrupts, etc.

2.5.5 Vector Base Register (VBR)

The 32-bit vector base register is used to determine the location of the exception vector table in the memory. The operating system may support multiple vector tables by loading different values in this register. This feature facilitates the implementation of the multitasking environment. It is important to note than an externally generated $\overline{\text{RESET}}$ will clear the contents of the VBR.

2.5.6 Source Function Code Register (SFC) and Destination Function Code Register (DFC)

The 32-bit function code registers allow the supervisor to have access to any address space, and are used only with the MOVES instruction. These registers are not a part of the previous members of the MC68000 family. As a result, all supervisor accesses were made to either the supervisor program or to the data space. During a read operation, the 29 MSB of the two registers are read as zero, and ignored when written. When the MOVES instruction is executed, the contents of the alternate function code registers are put on the function code lines (FC0-FC2), for either the source or destination operand. This allows the supervisor to extend the linear address from 4 Gbytes to as many as eight 4-Gbyte linear address spaces. The function code lines can then be utilized to implement the hardware protection mechanism for the eight different address spaces.

The alternate function code registers allow the operating system program or supervisory program to access user data space or to emulate CPU space cycles.

2.5.7 Cache Control Register (CACR) and Cache Address Register (CAAR)

The 32-bit cache registers are provided to allow the software to manage the on-chip instruction cache. The CACR contains bits to clear, freeze, enable the cache, or clear a specific entry in the cache. The address of the entry is contained in the cache address register during the clearing of a cache entry. All operations involving the CACR and CAAR are long-word operations, regardless of whether or not these registers are used as the source or destination operand.

2.6 IMPLEMENTATION OF THE ARCHITECTURE

Figure 2.9 illustrates the internal architecture of the MC68020. It consists of three main sections:

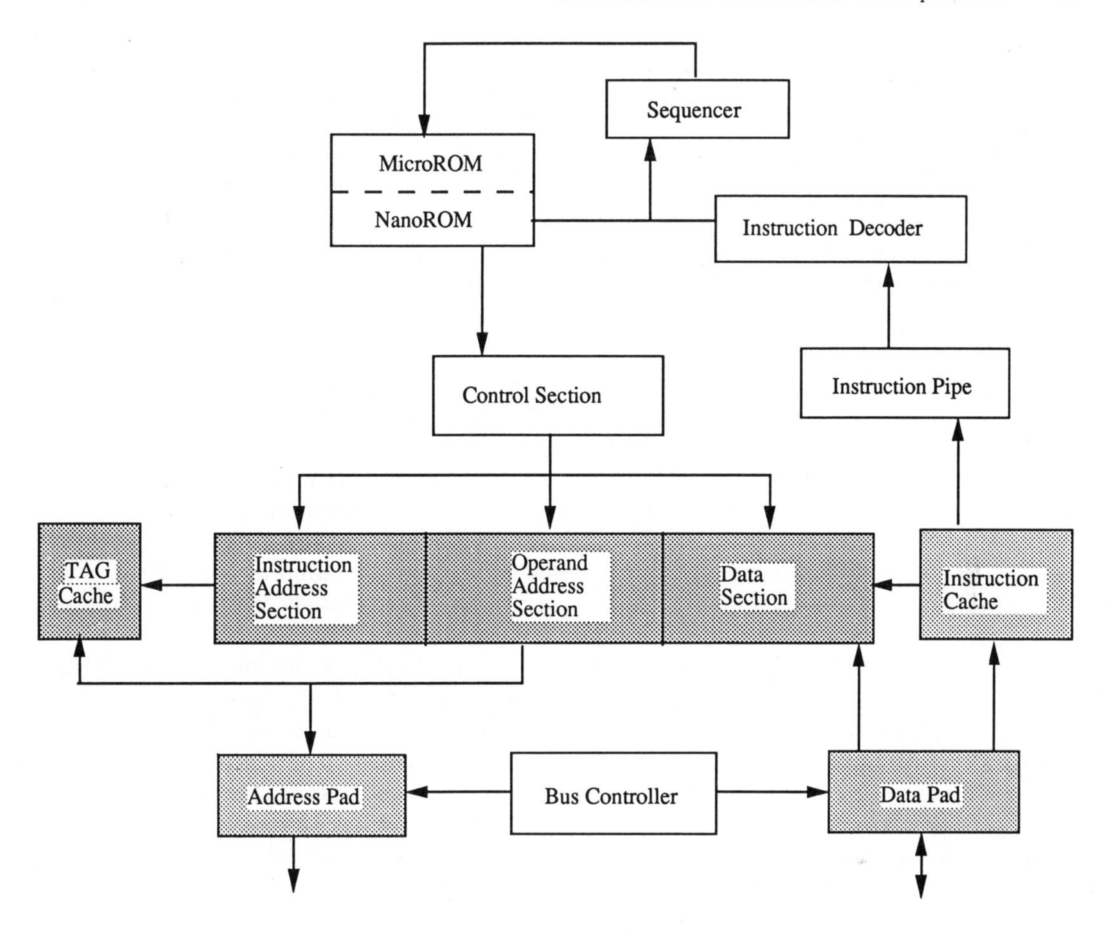

FIGURE 2.9. The internal data flow within the MC68020.

- The bus controller
- The micromachine
- The miscellaneous area

This division reflects the sectional autonomy and their concurrent modes of operation. The bus controller comprises the address, data pads, and multiplexers required to support dynamic bus sizing. The macro bus controller schedules the bus cycle on the basis of priority. Two micro bus controllers are used in the implementation. One of these is used to control the bus cycles for operand access, and the other to control the bus cycles for instruction accesses, including the instruction cache with its associated control.

The micromachine consists of an execution unit, ROM control stores, decode PLAs (Programmable Logic Arrays), and instruction pipelines and their control sections.

The execution unit consists of three sections: the instruction address, the operand address, and the data section. The microcode is implemented by a modified two-level hierarchy. The micro-ROM sits on the top level, and nano-ROM at the bottom level.

The instruction execution unit utilizes three pipeline units and provides for the decoding of instructions (Figure 2.10) [Motorola 85]. Instructions that have been prefetched from either the internal cache or from external memory are loaded into Stage B. The instructions are then sequenced from Stage B to Stage D. Stage D contains a decoded instruction ready for execution. If the instruction had contained an extension word, it would be available in Stage C at the time that the operation code would be in Stage D.

The instruction pipeline units enhance the performance of the CPU. This is because it is always possible to fetch three words in two accesses, because of the possibility of branching to an odd-word address, in which case two accesses are required to branch. This is illustrated in the example shown in Figure 2.11, which depicts two memory configurations for the instruction stream A, B, C, D, and E. In the first case, if a branch is made (even address aligned), the CPU will have to make one access to fetch two words. Once it is determined that two accesses are required to fill the pipe, it is then easy to see that the CPU will need two accesses to fetch the first three words.

To improve performance, the MC68020 utilizes parallelism in the execution unit. The execution unit is composed of three 32-bit adders (Figure 2.9). The instruction addresses are calculated and the pointers are stored in the instruction address section. The operand addresses are calculated in the operand address

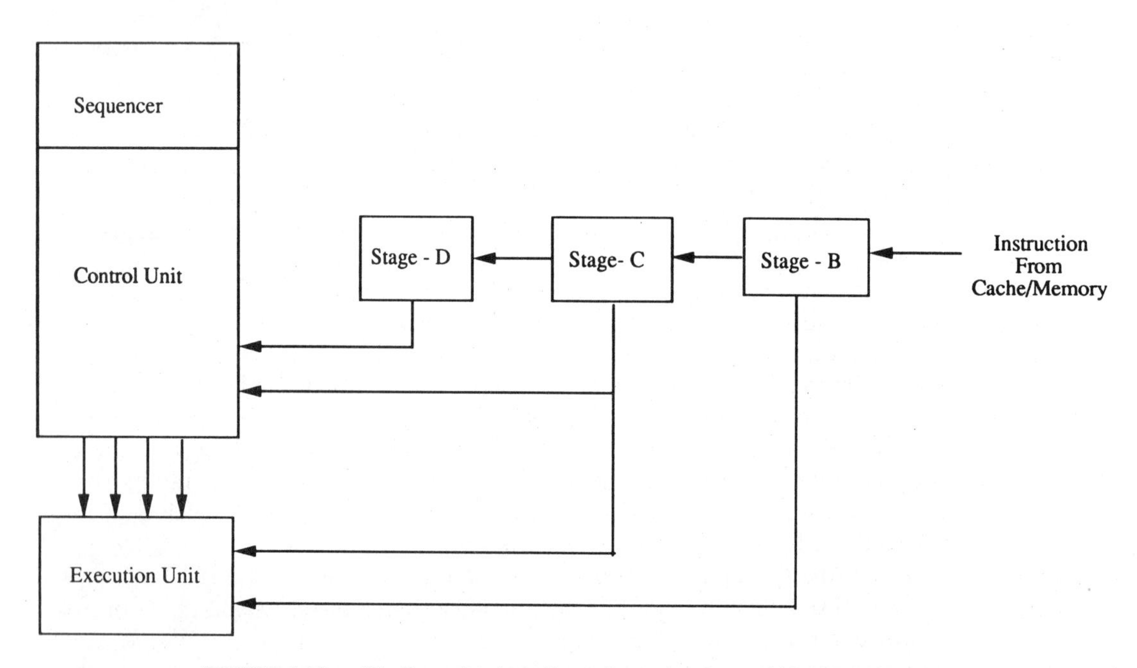

FIGURE 2.10. Pipeline units, including information flow, within the MC68020.

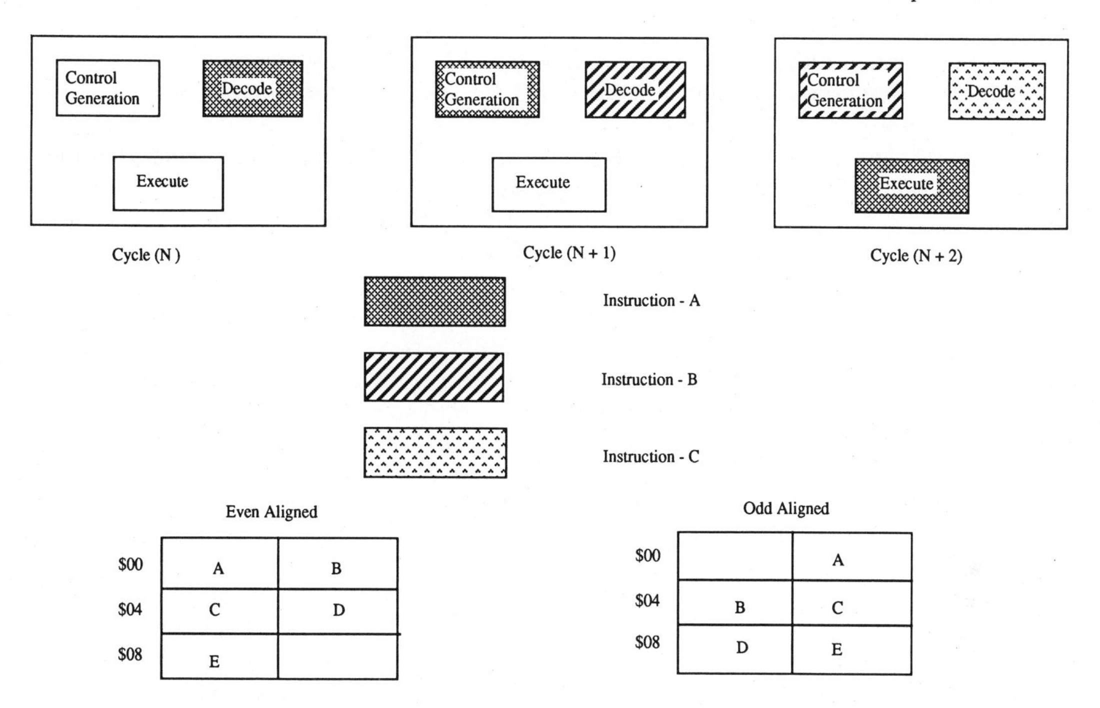

FIGURE 2.11. The stages of execution and memory configuration for the instruction stream A, B, C, D, and E.

section and are utilized to fetch the operand from memory. The data operand calculations take place in the data operand section.

Although sections are optimized for their primary functions, they can also operate as a general computation unit when required. This is possible because one unit can be connected to its neighbor via a high-speed bus. This feature is used during the signed and unsigned multiplication operation. For example, a 32-by-32 bit multiplication will use the computation units available within the operand and data address sections to produce a 64-bit result.

Two semiautonomous machines exist within the MC68020, which are the bus controller and the micromachine. The bus controller schedules and controls all bus activities. The micromachine controls the execution of instructions by requesting bus activity and controlling the execution unit and the instruction pipe.

An 80-bit barrel shifter is added to this unit to enhance performance. The barrel shifter is used for instructions that can't be performed efficiently by the ALU. The bit field and similar instructions are good examples. The barrel shifter also performs multiplications of two bits per microinstruction. Another significant addition is the scaling unit. During the evaluation of an effective address, this allows the scaling of index values without any additional delays.

A significant amount of parallelism exists between these two machines (i.e., the bus controller and micromachine). For example, when an instruction has completed execution within the execution unit, an updating of the results to the memory may be pending. In such cases, the micromachine will proceed with the execution of subsequent instructions, provided that they do not require the bus resources to complete the write operation. This is highlighted in the example shown in Figure 2.11. It also illustrates what happens when the instructions are aligned in odd or even long-word boundaries.

At clock cycle-N, instruction A enters into the decode module. During the next cycle, A gets passed to the control unit, and appropriate control signals are generated as instruction B enters into the decoder. During the third cycle, Instruction C enters into the decoder, B moves into the control unit, and A enters into the execution unit. Operation pipelining continues until a branch causes an instruction to be fetched from an odd-word address. If a branch is made to the even-aligned address (first example), then only one access is required to fetch two words. However, for an odd-aligned address (second example), two accesses will be required to fetch the first two words. Once it is determined that two accesses are required to fill the pipeline, it can be observed that it is always possible to fetch the first three words, utilizing two accesses.

Chapter 3

Software Capabilities of the MC68020

The cost of designing software is becoming the main concern of system designers. These concerns have been addressed by the designers of the MC68020, who equipped this processor with powerful instructions, versatile addressing capability, an interrupt protocol, instructions for multitasking, etc. to facilitate the software development process. The processor's elegant data set can help the compiler and device driver developers to implement efficient solutions. Thus, it is often said that the MC68000 family was designed by programmers for programmers. This chapter addresses the software features of the processor.

3.1 THE PROGRAMMING MODEL AND THE PIPELINED ARCHITECTURE

We have already seen that the MC68020 executes programs in either the user mode or the supervisory mode. Programs that run under the user mode have the lowest level of privilege, and can access the following resources (Figure 3.1):

- Eight general-purpose 32-bit address registers (A0–A7)
- Eight general-purpose 32-bit data registers (D0–D7)
- One 32-bit program counter (PC)
- An 8-bit Condition Code Register (CCR)

The supervisory programming models of the MC68000 family differ from one another. The differences reflect the migration of functionalities from software to

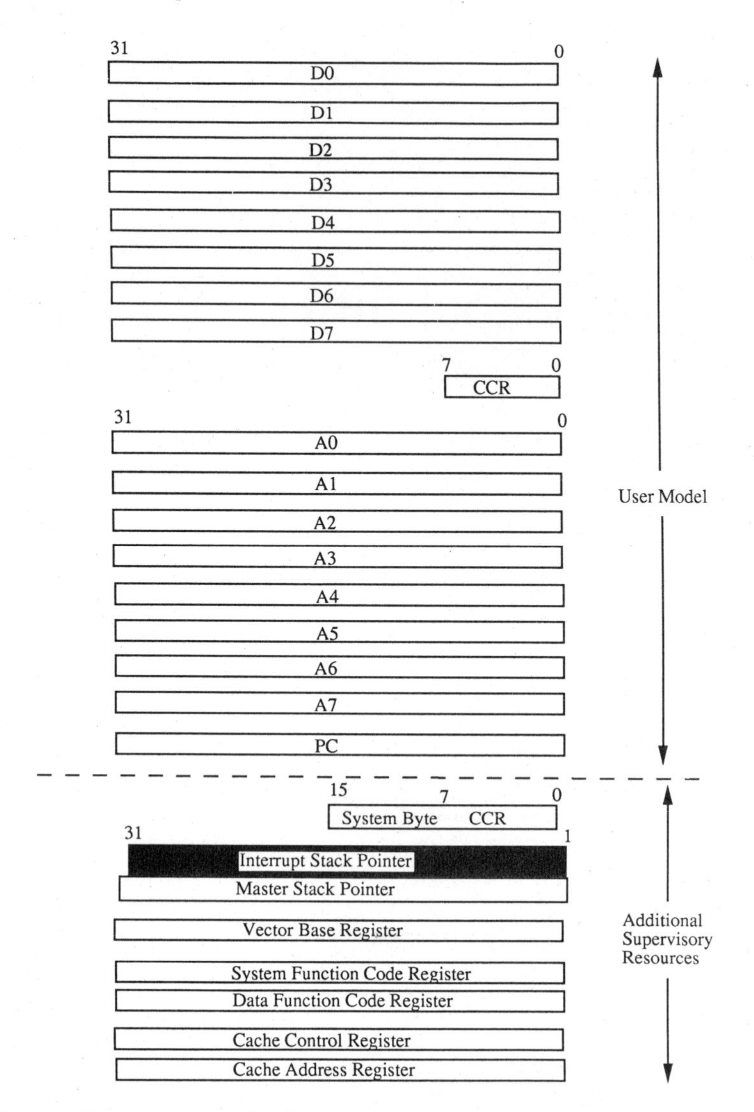

FIGURE 3.1. Programming model of the MC68020.

hardware and feature enhancements. In the MC68020, programs running under the supervisory mode can use resources available under the user mode, plus the features associated with seven control registers. The supervisory resources are utilized by system programmers to implement operating systems functions, I/O control, memory management, coprocessor interfacing, and various restrictive software tools. All controls to access and enable special features of the MC68020 are included in the supervisory mode. This segregation allows all application software (that ran at a lower-privileged user mode) developed for the previous generation of the MC68000 family of processors to automatically migrate to MC68020-based hardware.

3.2 DATA ORGANIZATION

The MC68020 supports seven basic data types, namely:

- Bits
- Bit fields
- BCD digits (4 bits each, generally packed in pairs)
- Byte integers (8 bits)
- Word integers (16 bits)
- Long-word integers (32 bits)
- Quad-word integers (64 bits)

The data registers, D0 through D7, support the complete range of basic data types. The address registers, A0 through A7, including the stack pointer, only support word and long-word data types. However, data-type support by the six control registers SR, VBR, SFC, DFC, CACR, and CAAR depend on the register involved during the operation. The processor also supports the variable size operands required by the coprocessor. The user can also create his/her own data type, not supported by the MC68020.

In the MC68020-based system, the memory is organized as byte addressable units (Figure 3.2). Bytes are individually addressable with higher order byte having an even address and lower order byte at odd address. Thus, byte-0, word-0 and long-word-0 are the most significant part of the respective data structure. Bytes can have either even addresses (i.e., byte 0, 2, 4, etc.) or odd addresses (i.e., byte 1, 3, 5, etc). But words, long-words, and quad-words generally have even addresses. Therefore, if a word is located at address N (where N is even), the next word is located at address N+2. Similarly, if a long-word is located at address N, the next long-word will be located at address N+4, and for quad-words, that will be N+8.

3.2.1 Memory Access Rules

The Motorola MC68000 processor family obeys a well-defined set of rules during memory access, which are:

1. Bytes can be accessed from either an even or odd address.
2. Instruction code must reside on an even-address boundary.
3. Words, long-words, and quad-words should be accessed from the even address for data transfer efficiency.

However, the MC68020 or later members of the family can access a word, long-word, or quad-word starting at any even or odd byte address.

3.2.2 Bit Operand

A bit operand is specified by a base address and a bit identifier. The base address selects one byte in memory, and the bit identifier selects a specific bit within this

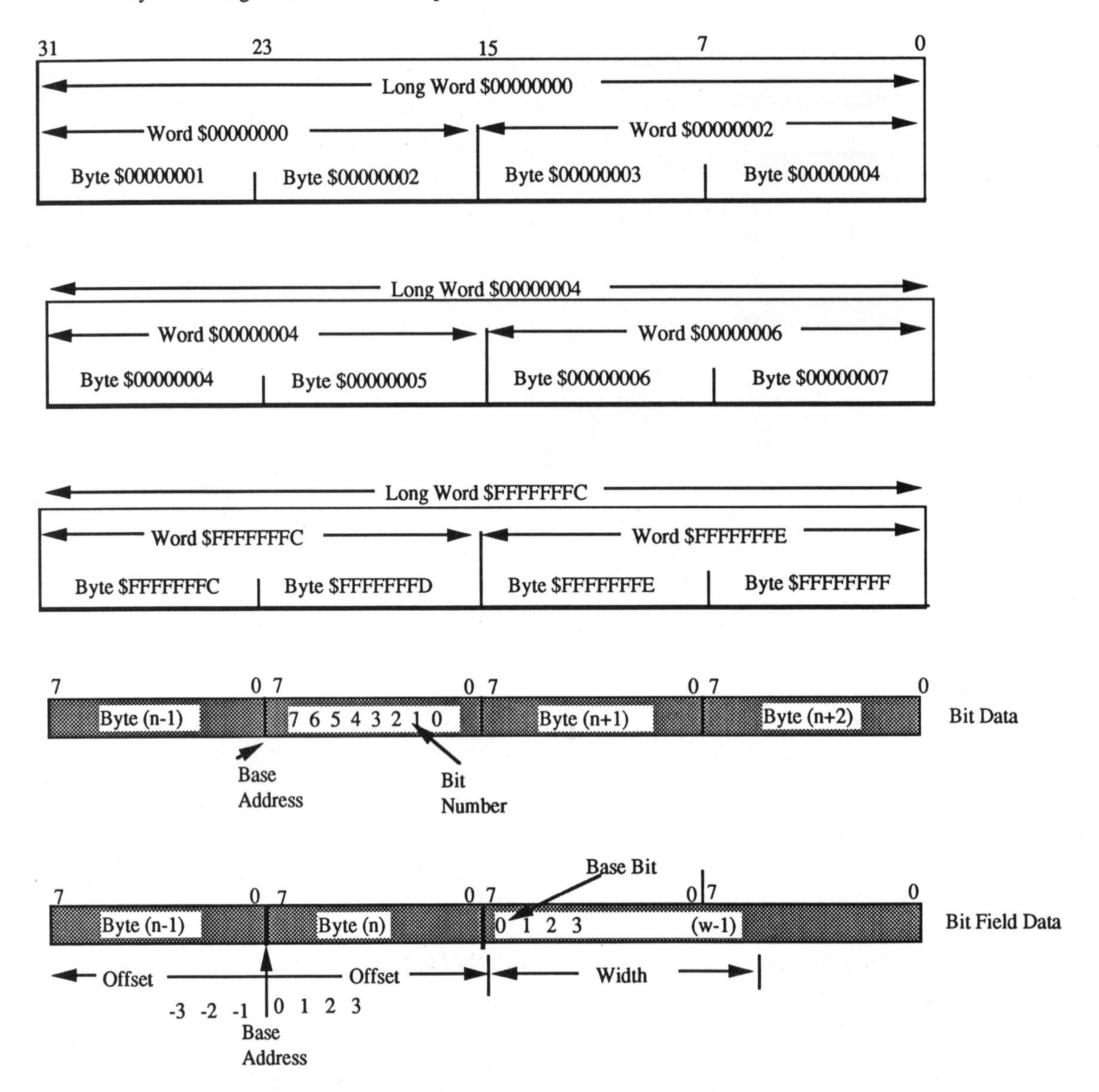

FIGURE 3.2. Data organization in memory.

byte. The most significant bit of the byte is bit 7. A bit operand is determined by:

1. A base address that points to one byte in memory.
2. A bit-field offset that specifies the leftmost (base) bit of the field in relation to the most significant bit of the base byte.

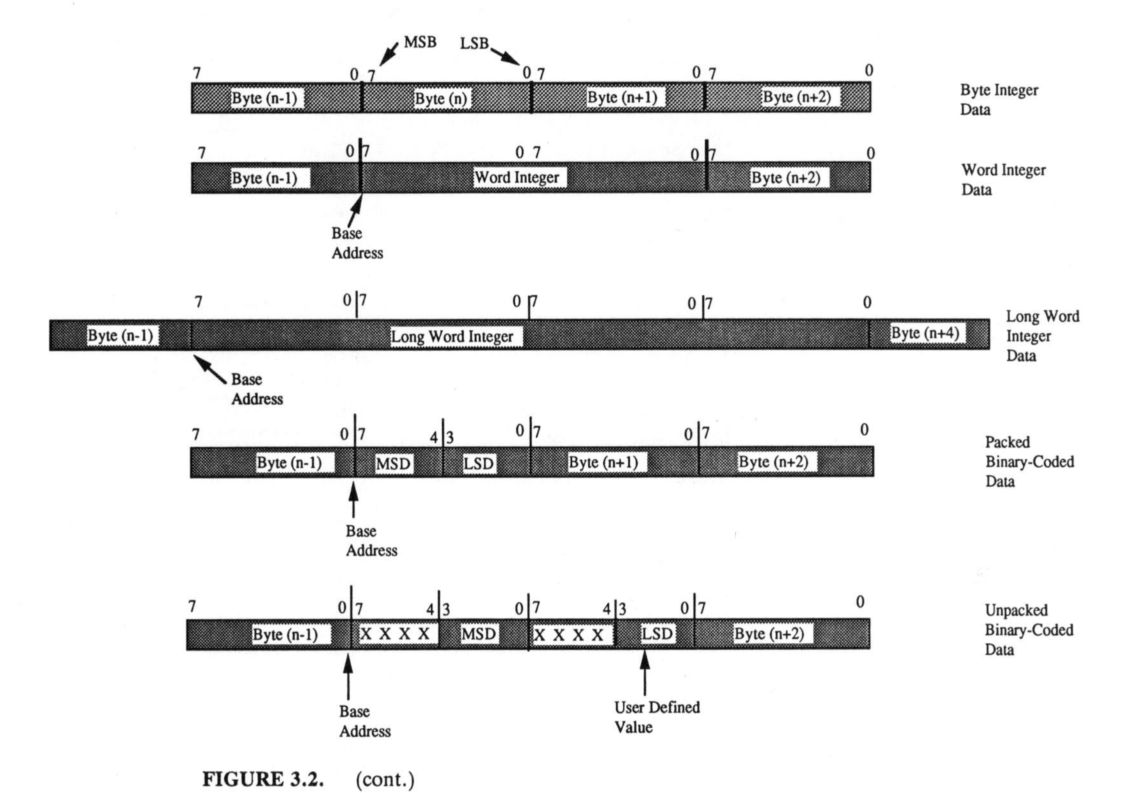

FIGURE 3.2. (cont.)

3. A bit-field width determines how many bits to the right of the base bit are to be considered in the bit field.

The most significant bit of the base byte is bit offset 0, and the least significant bit of the base byte is bit offset 7. That is, the least significant bit of the previous byte in memory is at bit offset -1. The range of bit offsets fall between -2^{31} to 2^{31} -1, whereas the bit-field widths range between 1 and 32.

3.2.3 Binary Coded Decimal (BCD) Data

BCD data represents decimal number into binary format. Two BCD formats are supported by the MC68020, which are the unpacked and packed BCD. A byte contains one digit in the unpacked BCD format, where the four least significant bits contain the binary value, and the four most significant bits remain undefined. The packed BCD format contains two digits. The least significant four bits contains the least significant digit, and the most significant four bits contains the most significant digit.

3.2.4 Separation of Program and Data Memory

The MC68020, and the later members of the family, separate memory into two groups, namely, *program references* and *data references.* During program execution, the CPU accesses the program area, and the system designer can use EPROM for an embedded system or can deploy a hardware protection mechanism, so that a user program may not overwrite into this area by mistake. Data references refer to that section of the memory that contains the program's data, and the CPU generally reads the operands from, and writes the data into, this area. The exception to this rule is due to specific instructions, such as MOVES. We will elaborate these concepts over various chapters of this book.

3.2.5 Data Organization in Address Registers

All of the eight address registers and stack pointers are 32 bits wide. The entire 32 bits are brought out to the address pins A0-A31. As a provider of the source operands, the lower 16 bits, or the entire 32 bits, of the address registers get utilized. When an address is used as a destination operand, the entire 32 bits are affected, regardless of the size of the operand. For example:

```
MOVE.L  #$80, A0
```

This states that the CPU should move the Hex number 80 into the address register A0 and, at the end of the execution, A0 should contain:

00 00 00 80	→ A0

Since the address register does not support byte operation as a destination register, the CPU copies the most significant bit of the source operand (which is 1, in this case), into the rest of the bits (bits 8 through 31 in this case) and into the A0 register. The resultant value in A0 will be:

FF FF FF 80	→ A0

In the following example,

```
MOVE.L  #$7BCD, A1
```

the CPU should move the value "7BCD" into address register A1:

7B CD	→ A1

However, the suffix at the end of MOVE instruction states that the entire A1 register should be changed. As a result, the CPU will take the value of bit 15 (which is 0, in this case), and copy that value into bits 16 through 31. The result of the operation is shown below:

00 00 7B CD	→ A1

3.2.6 Data Organization in Data Registers

Data registers are used for bit, bit-field, BCD, byte, word, long- word, and quad-word operations. The operations on data registers generally affect the condition code register, but do not cause the results in the register to be sign-extended to the full 32 bits. When a data register is used for either source or destination operands, only the appropriate low-order byte or word is used or changed, and the remaining portion remains unaffected.

For bit fields, the most significant bit is addressed as bit zero, and the least significant bit is addressed as the width of the field minus one. The quad-word data type is utilized during 32-bit multiply and divide operations and occupies two long-words. Any two data registers can be used during this operation. No explicit instruction exists for the management of quad-word data types, and MOVEM is the only one available to move a quad-word into and out of the registers from/to memory.

3.3 VIRTUAL MACHINE CONCEPTS

A virtual machine is a model of a real machine. It is created by a virtual-machine operating system, which makes a real machine appear to be several machines to the users. By combining hardware and software, many virtual machines can be simulated on a single real machine (Figure 3.3). The software to accomplish this is known as a *governing operating system.* The protection advantages of several address spaces are extended one level further, so that each user appears to have its own machine, on which even an unprotected operating system can be run without any possibility of interfering with other users. The virtual machine approach is useful in testing software on many different virtual hardware configurations, without the need to acquire the corresponding physical hardware.

In essence, a governing operating system works by interpreting each user's program. Mapping each virtual-machine memory address to a real-machine address, mapping each virtual machine to a real machine copy, and mapping each virtual-machine I/O device to a real-machine I/O device. In addition, this simulates each instruction, so that their effects are reflected in the copies of the memory and hardware state for the appropriate virtual machine. Thus, any attempt by the new software to use virtual resources that are not physically present are trapped and forwarded to the governing operating system, and performed by this software.

In MC68020 implementation (Figure 3.4), the governing operating system executes at the supervisory privilege level. Any attempt by the new operating system to access supervisor resources or to execute privileged instructions causes a trap

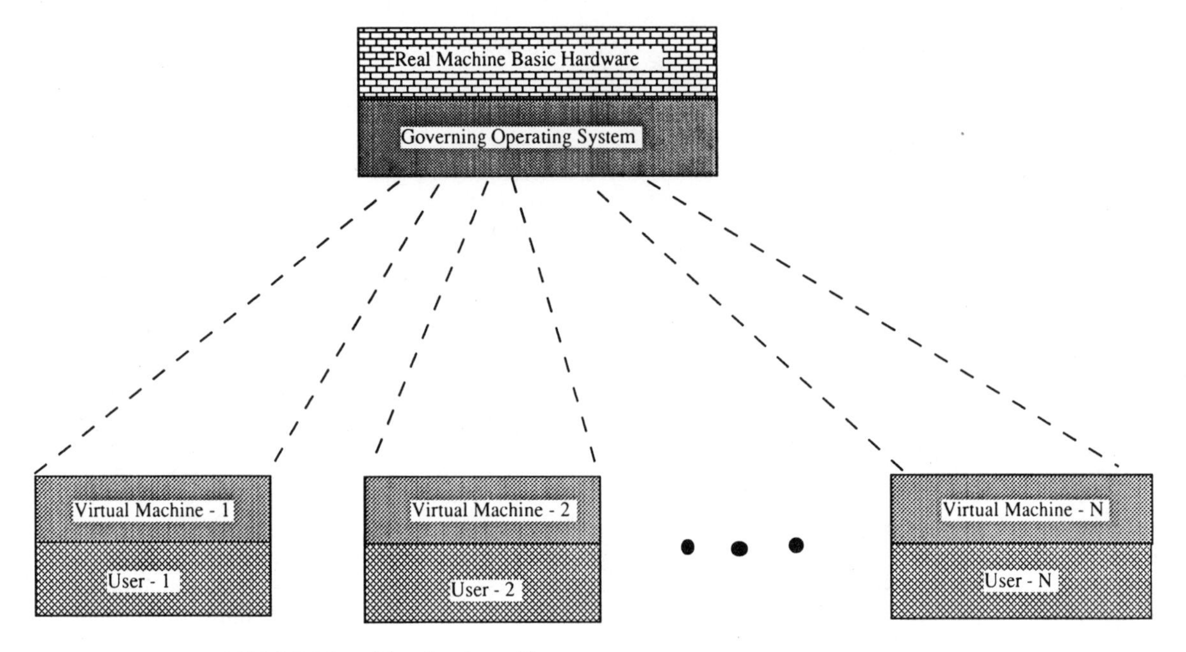

FIGURE 3.3. The virtual machine concept.

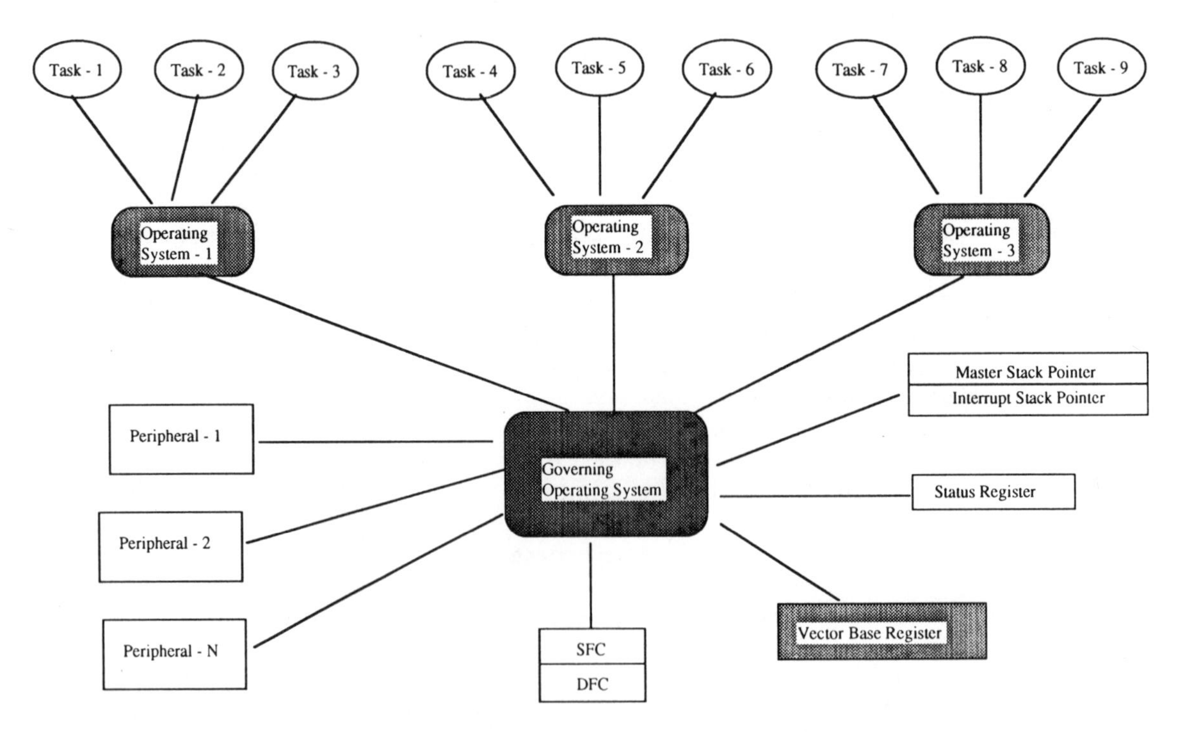

FIGURE 3.4. The implementation of a virtual machine utilizing features available in the MC68020. (Courtesy Motorola Inc.).

to the governing operating system. The governing OS then executes the virtual resources, which eventually get queued for physical resources. The general OS runs at the user mode. The execution of privileged instructions will cause an error, which will be trapped to the governing OS. Thus, the virtual application must run at the user privilege level.

The instruction continuation is supported by the MC68020, and this feature is used to support virtual I/O devices in memory-mapped I/O systems. The control and data registers for the virtual devices are simulated in the memory map. By loading a value in the vector-based register, one can create various copies of user operating system, including their own virtual devices.

3.4 ADDRESSING MODES AND INSTRUCTION FORMATS

Instructions for the MC68020 can occupy from one to eleven words in the memory. The general format for instruction is shown in Figure 3.5. The first word of the instruction, called the *Operation word* (Op-word), specifies the operation to be performed and the length of the instruction. The remaining words, known as *extension words,* contain the remainder of the instruction, including the associated operands. Operands can be found as a part of the Op-word, or they are to

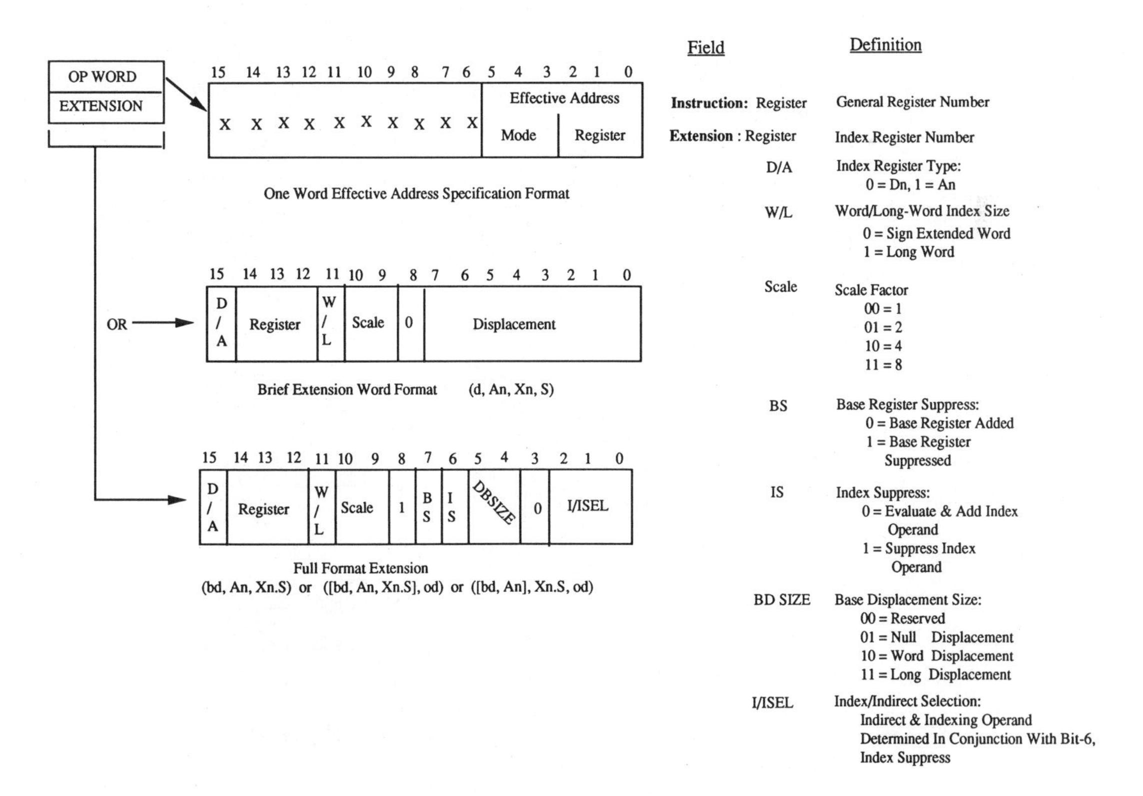

FIGURE 3.5. The effective address specification format.

be found using their Effective Address (EA). The EA may require additional information to specify the operand address. This extra information, called the effective address extension, is contained in additional words, and is considered as part of the instruction.

The operand size of each instruction is either explicitly encoded in the instruction, or implicitly defined by the operation of the instruction. Extension words are required for operand addressing modes that use constants (immediate values), absolute addresses, relative addresses or displacements and/or offsets, etc. For example, a two-word immediate operand and a two-word absolute address in an instruction will be assembled into high-word/low-word order in memory. Thus, if the high-order word of the operand is stored at N, then the low-order word will be stored at N+2.

All explicit instructions support the basic data operands. The implicit instructions support a limited subset of the data structures.

3.4.1 Addressing Modes

Instructions are composed of two kinds of information, the functions to be performed and the location or locations of the one or more operands on which the function is to be performed.

The addressing mode of an instruction can identify the operand's location in three possible ways:

Register specification: The register is uniquely identified in the register field of the instruction.
Effective address: Selects one from the various effective addressing modes.
Implicit reference: The definition of certain instructions implies the use of specific registers.

An assembler syntax is available for each addressing mode. For example, the general instruction structure for MOVE is given by:

MOVE.X (source effective address), (destination effective address)

where X is called the attribute of the instruction and can take various values, namely:

X = B (byte) ; operation will be performed on an 8-bit data
X = W (word) ; operation will be performed on a 16-bit data
X = L (long-word) ; 32-bit operation on a 32-bit data

Note that the MOVE instruction always has two operands—one operand addresses the memory location or register that contains the source data to be moved, the other addresses the memory location or register that will receive the destination data.

The general format for MOVE is shown as an example:

15	14	13 12	11 10 9	8 7 6	5 4 3	2 1 0
X	X	size	Destinaton		Source	
			Register	Mode	Mode	Register

The source/destination effective field specifies the addressing modes for an operand that can use one of the various defined modes. The EA (Effective Address) is composed of two fields, the mode field and the register field, each three bits wide. The value of the mode field selects the addressing modes from a valid set. The register field designates a register for the mode, or for a submode that does not use registers.

3.4.2 Effective Address

The MC68020 supports 18 addressing modes. They are designed to provide ways for a programmer to generate operands. In general, operands referenced by an effective address reside either in one of the internal registers or in external memory, and use one of the three formats shown in Figure 3.5. The single effective addressing format is generally encoded within the instruction word, where the register field contains the general register number and the mode field identifies the addressing mode. The brief format is composed of the instruction word followed by an extension word, which is used for the indexed and extended mode. The third type consists of the instruction word, including the full format of the extension word. Here, the Index Suppress (IS) bit and the Index/Indirect Selection (I/ISEL) field determine the type of indexing and indirect addressing to be performed.

The effective addressing modes are grouped according to their usage, and are classified as:

Data: A data-address effective addressing mode is one that refers to data operands.

Memory: A memory-address effective addressing mode finds the operands from memory.

Alterable: An alterable effective addressing mode is one that refers to alterable operands (i.e., the operands whose value is changed in the memory).

Control: A control-effective addressing mode is one that refers to memory operands without an associated size.

The various addressing modes of the MC68020 are shown in Figure 3.6. The modes are also used to form a new class that is alterable in memory and data.

3.4.3 Register Direct Addressing

Register direct addressing modes (Figure 3.6a) are used when the data or address register holds the desired operand for the instruction. When the data register contains the desired operand, then it is known as data register direct addressing, and the notation used is Dn, where D stands for the data register and n identifies the specific register. For example, if we want to clear the D0 register, then we use the instruction:

```
CLR.L   D0
```

This clears the contents of the D0 register. If D0 contains a result, and we want to keep it for future reference, then we may use the register direct addressing mode to retain a copy in D3 register before clearing D0, using the previous instruction.

```
MOVE.L   D0, D3
```

Here, MOVE will copy the value of D0 into D3.

The address register direct mode (Figure 3.6b) is similar to the data register direct mode, except that an address register is used. For example, if A1 contains the starting address of a table in memory, and we want to retain that pointer for future reference then, we may use the following instruction:

```
MOVEA.L   A1, A3
```

At the end of the execution of the instruction, A3 will retain the value of A1, (i.e., if A1 = $5000, then A3 = $5000).

3.4.4 Register Indirect Addressing

In the register indirect addressing modes, the contents of an address register 'points to' the operand (Figure 3.6c). That is, the specified address register holds a base address which CPU utilizes to calculate the effective address of the oper-

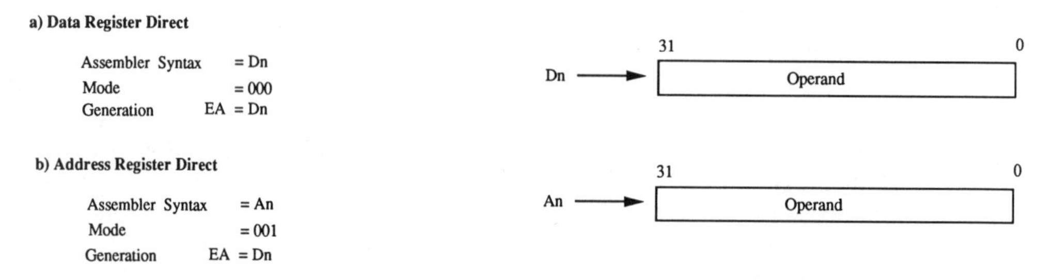

FIGURE 3.6. Addressing modes of the MC68020.

and. The relationship between the base address and the effective address depends on which of the following five addressing modes are being employed:

Register indirect	EA = (An)
Postincrement register indirect	EA = (An), An <- An + M
Predecrement register indirect	An <- An-M, EA = (An)
Register indirect with offset	EA = (An) + d16
Index register indirect with offset	EA = (An) + (Rx) + d16

The register indirect address is the simplest of all indirect addressing modes where, the address register holds the effective address. As an example, let us assume that register A5 contains $5000 and the contents of the memory locations are:

```
5000--->   |11|
5001--->   |22|
5002--->   |33|
5004--->   |44|
5005--->   |55|
5006--->   |66|
```

The MOVE instructions with a register indirect address will then produce the following results:

```
MOVE.B    (A5), D3   ; moves 11 to the low 8 bits of D3
MOVE.W    (A5), D3   ; moves 1122 to the lower 16 bits of D3
MOVE.L    (A5), D3   ; moves 11223344 to the D3 register.
```

Here, A5 points to the address location in memory. The actual operand size, byte, word, or long-word, is identified by the size of the attributes. Note that the comments are preceded by ";", a convention used by many compilers of MC68000 processors.

Today, users frequently write software that involves operations on a block of contiguous data in memory, such as a table or a string. In the case of many microprocessors, this involves accessing an operand, incrementing (or decrementing) the pointer, and keeping a count of how much data has been moved. The designers of the MC68020 have created the postincrement and predecrement modes as a part of the address register indirect addressing to facilitate those tasks.

In a register indirect address with postincrement (Figure 3.6d), the operand is in memory and the address of the operand is in the address register specified by the register field. The processor first finds the operand and then increments the address register by 1, 2, or 4, based on the size of the operand. However, if the address register happens to be the stack pointer, then the CPU will add 2 to keep the stack pointer aligned to a word boundary. Again, in the case of a coprocessor-

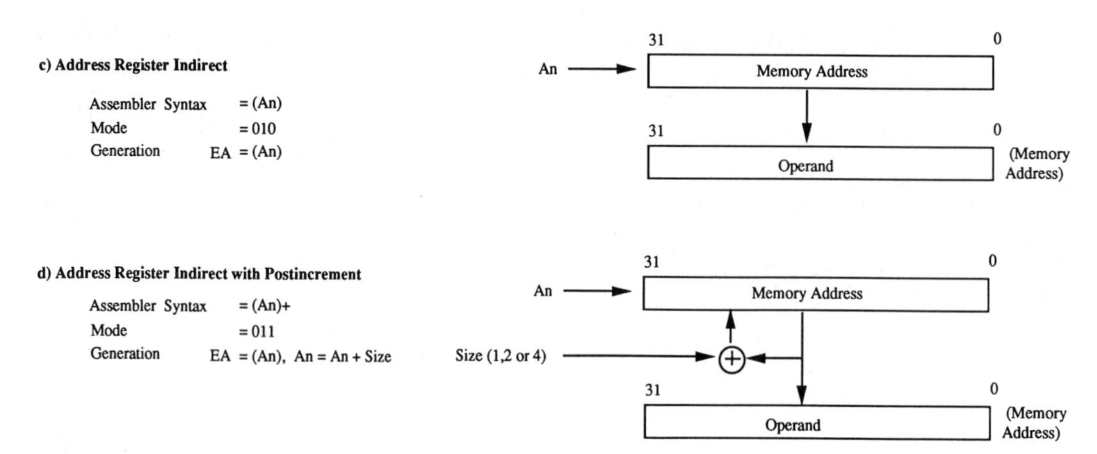

FIGURE 3.6. (cont.)

related task, the address register may get incremented by a value of from 1 through 255.

The operation of this mode is illustrated in the following example:

Example: Suppose that memory location $5000 contains the long-word 11112222, and $5004 has the value 33334444. Then the following program will add the numbers, and store the result at location $5008.

```
MOVE.W   #$5000, A3    ; put hex 5000 into A3
MOVE.L   (A3)+, D3     ; 33334444 to 11112222 and increment A3 to
ADD.L    (A3)+, D3     ; increment A3 to $5004. add
                       ; 33334444 to 11112222 and increment A3 to
                       ; 5008
MOVE.L   D3, (A3)      ; place result from D3 into address
                       ; 5008
```

Here, the processor increments the address register automatically, by either 1, 2, or 4 bytes, depending on the data size used in the Op-code.
Note: The Comments always start with ";" and continue until the end of the line. Other assemblers may use another delimitor.

Example: String manipulation. There is a string of ASCII bytes at a location whose address is given by the contents of A5. The end of the string is indicated by the ASCII null (i.e., 00). We want to copy the string into another part of memory, whose starting address resides in A6.

```
LOOP:   TST.B     (A5)          ; contents of address in string
                                ; pointed to by A5
```

```
          BEQ       DONE            ; string copies yes, branch to DONE
          MOVE.B    (A5)+,(A6)+     ; no copy byte from address
                                    ; pointed to by the contents of A5
                                    ; into location pointed to by (A6)
                                    ; and then increment each pointer by
                                    ; 1 to point to the next character
                                    ; in the string to be copied from
          BRA       LOOP            ; branch to LOOP and continue
DONE:                 ....
                      ....
```

The BEQ is a conditional branch based on something equal to 0. BEQ tests the Z flag in SR, which would be set if the result were 0 in a test, arithmetic, compare, etc.

Indirect addressing with predecrement (Figure 3.6e) accesses the operand byte, word, or long-word at the address pointed to by An. However, not before first decrementing the contents by 1, 2, or 4, based on the operand size specified with the instruction. This is illustrated by the following example, where A5 points to the memory location $5000.

```
CLR.B   -(A5)   ; Clears the byte at location 4FFF.
CLR.W   -(A5)   ; Clears the word at 4FFE, i.e., high
                ; byte 4FFE and low byte 4FFF.
CLR.L   -(A5)   ; Clears the long-word at 4FFC.
```

Using -(An), we can perform a string manipulation or table scan from the end to the beginning. However, we must set the pointer An to one position (byte, word or long-word, as the case may be) after the end of the string or table, so that the CPU can 'zero-in' on the last entry in the string or table. This concept is illustrated through the following example:

Example: Stack Implementation.

```
MOVE.W  D1, -(SP)   ; decrement SP by 2 and then push word
                    ; from D1 to the address SP-2
MOVE.W  (SP)+, D1   ; remove the word from address
                    ; SP (now SP-2) into lower 16-bit of D1
                    ; and then increment the SP by 2 to
                    ; restore the SP value to point to the
                    ; original location
```

The *register indirect with displacement* (Figure 3.6f) is a powerful addressing mode and is designed to process data structures, like tables. It provides the flexibility to access or alter entries of the table.

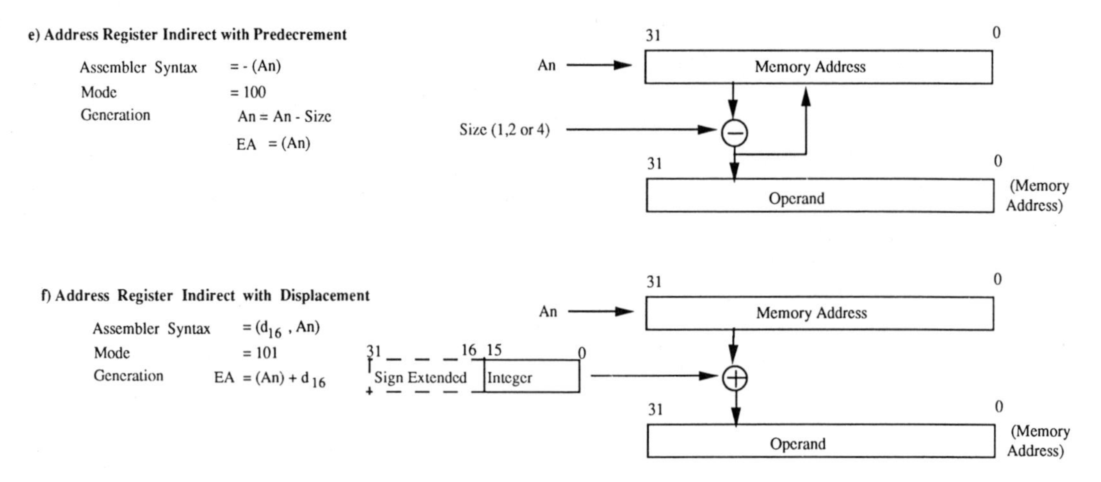

FIGURE 3.6. (cont.)

In *address register indirect* mode, the address of the operand is calculated by adding the contents of the address register to the 16-bit displacement found in the extension word. Displacements are always sign-extended to 32 bits prior to their use in effective address calculation.

For example, if A0 points to the starting address of the table, then the instruction

```
MOVE.L (8,A0), D0
```

will fetch the entry at A0+8 (the third long-word of the table) and move it into the D0 register. The table has its first long-word at (0,A0) = (A0), the second long-word at (4,A0) = (A0+4), the third long-word at (8,A0) = (A0+8), etc. The prefixes are 0, 4, 8, 12 . . . , and are called the offsets (into the table, i.e., offset from the base address (A0). It is worth noting that A0 does not change, unlike -(An) or (An)+. The usage of this mode is shown in the following example:

Example: Interchange the third and fifth entries of the table shown below:

```
5000---> |00|  <---first entry: (0,A5) ; value = 0011
         |11|
         |22|  <---second entry: (2,A5) ; value = 2233
         |33|
         |44|  <---third entry: (4,A5) ; value = 4455
         |55|
         |66|  <---fourth entry: (6,A5) ; value = 6677
         |77|
```

```
|88|   <---fifth entry: (8,A5) ; value = 8899
|99|
|AA|   <---sixth entry: (10,A5) ; value = AABB
|BB|
|CC|
```

The following program will exchange the value of the fifth entry (8899) with the third entry and the value of the third entry (4455) with the fifth entry.

```
MOVE.W   (4,A5), D0       ; place entry-3 into D0 register
MOVE.W   (8,A5), (4,A5)   ; replace value into entry-3
MOVE.W   D0,   (8,A5)     ; update the entry-5 value
```

At the end of the execution, (A5) remains unchanged and still points to the base of the table, i.e., $5000.

Register indirect with index (8-bit displacement) mode (Figure 3.6g): This addressing mode uses an extension word that identifies the index register and an 8-bit displacement. The index register indicator includes the 'size and scale' information, whose format is:

"Xn.SIZE*SCALE"

Where

Xn —selects any data/address register as the index register.
SIZE —specifies the index size (may be W for word and L for long-word).
SCALE —specifies the index register value to be multiplied by a value of 1, 2, 4, or 8.

The displacement is sign-extended to 32 bits prior to being used in effective address calculation. The user has to specify the address register, index register, and the displacement value in this mode.

This addressing mode is extremely useful where multidimensional arrays or tables of data have to be accessed from memory. For example, if An points to the base of a table array, and the offset d_{16} is assigned to represent the column number, then (d_{16}, An, Xi.W*SCALE) will access the data cell that lies at the intersection of row d_{16} and column Xi in the array. The displacement d_{16} can select columns from -65536 through 65535 relative to An, and Xi.L can identify the rows from -2 Gbytes to 2 Gbytes on either side of An, without the scaling factor. The use of this mode is shown in the following example:

Example: Interchange the third and fifth entries in the previous table, using this addressing mode.

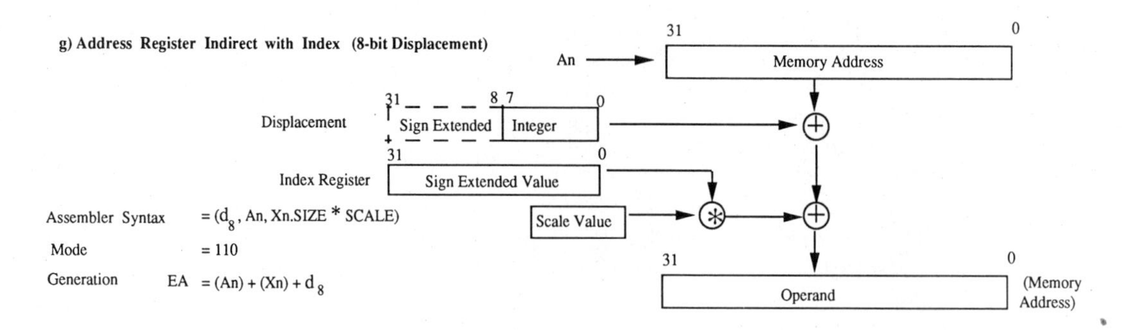

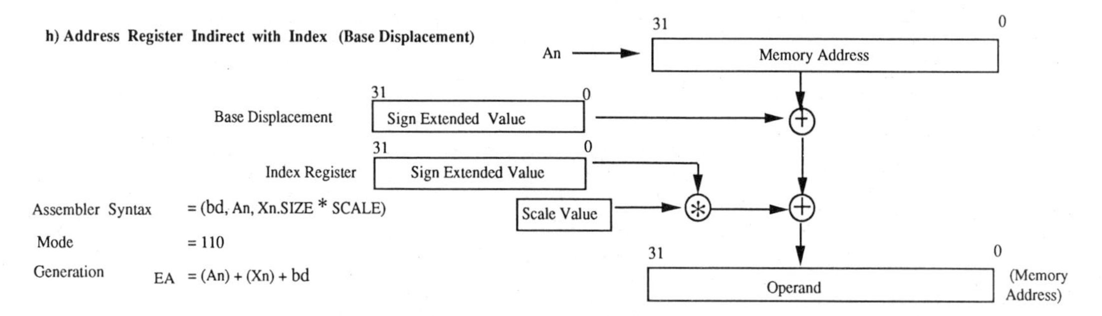

FIGURE 3.6. (cont.)

```
MOVEQ.L   #3, D0                       ; set index register D0 = 3
MOVE.W    (1,A5,D0.W), D1              ; place the third entry value in D1
MOVEA.W   #1, A6                       ; set index register A6 = 1
MOVE.W    (7,A5,A6.W), (1,A5,D1.W)     ; place the fifth entry at
                                       ; A5+7+1 = A5+8 into the third entry at
                                       ; A5+1+3 = A5+4.
MOVE.W    D0, (7,A5,A6.W)              ; place the value into the fifth entry.
```

Register indirect with index (base displacement) mode (Figure 3.6h): This mode requires an index register indicator and an optional 16- or 32-bit sign-extended base displacement, including a size and scaling information. The effective address of the operand is obtained by adding the content of the address register to the scaled contents of the sign-extended index register and the base displacement.

The address register, the index register, and the displacement are all optional, and in their absence the effective address becomes zero. However, when no address register is specified and the data register becomes an index register, then this addressing mode becomes equivalent to data register indirect addressing. The following example shows how this instruction mode can be used to clear a long-word:

```
CLR  ($6000, A0,D0)
```

3.4.5 Memory Indirect Addressing

This addressing mode is introduced in the MC68020 microprocessor for the first time, and can also be found in MC68030 and MC68040 processors. They are introduced to handle link list, multidimensional arrays commonly found in high-level languages. Here, the operand and the address reside in memory, and require one to five words of extension. The square brackets ([]) in the assembly syntax separates it from the address register indirect mode.

The location of the square brackets determines the user-specified values to be used during the calculation of an immediate memory address. An address operand is fetched from the specified memory address, and is then used to calculate the effective address. As a result, the memory locations can be considered as 32-bit "address registers." The index operand may be updated after the intermediate memory access (postindexed) or before the intermediate memory access (preindexed). The instruction mode is shown in Figure 3.7, where the instruction

CLR ([A0])

fetches the effective address from a memory location pointed to by A0, and then fetches the operand from location $0AB1.

The memory-indirect addressing mode is useful in accessing the device drivers from the device table. This is shown in Figure 3.8. The instruction, JSR ([10*4, A0]) instructs the CPU to add to the content to A0 decimal 40 and fetch the effective address and jump to the appropriate subroutine. Here, the system programmer has the freedom to change the device routine without changing the rest of the program. Such a scheme facilitates the implementation of operating system.

Two types of memory-indirect addressing modes are supported by the

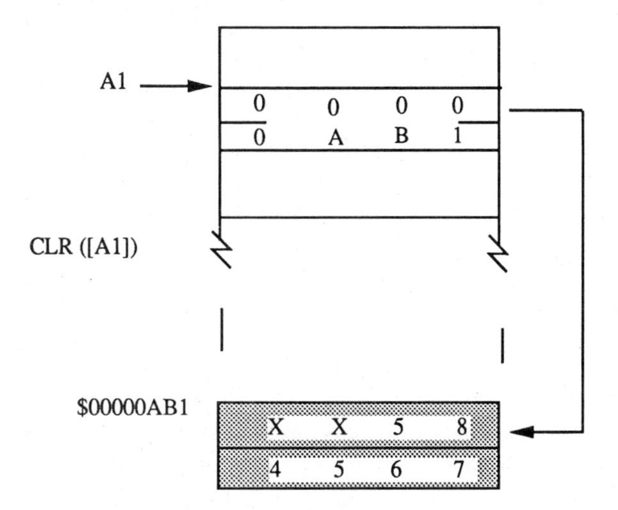

FIGURE 3.7. The indirect memory addressing concept.

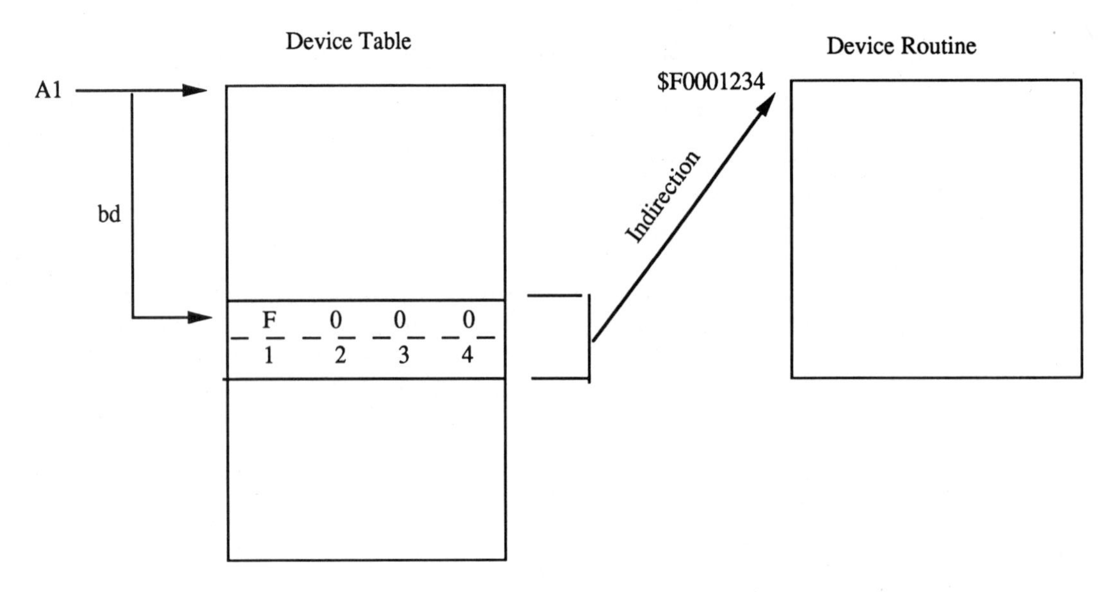

FIGURE 3.8. Device routine access from the device table, utilizing indirect memory addressing.

MC68020, and they are the *memory indirect postindexed* and *memory indirect preindexed* modes.

In the *memory-indirect postindexed* addressing mode, an intermediate memory address is calculated using the base register (An) and the base displacement (bd). This address is then utilized to access a long-word from memory. The CPU adds the index operand (Xn.SIZE*SCALE) and optional outer displacement (od) to the long-word accesses from memory to calculate the final effective address of an operand. This mode is illustrated in Figure 3.9. A1 points to the address of the table, and 'bd' identifies the memory location that contains the indirect address. D1 is used for postindexing, and added to 'od' to fetch the operand from memory. This instruction mode is convenient for implementing the virtual machine and compiler. The program below shows the usage of this mode.

```
        TLENGTH   EQU   10                  ; get the length of the table
        MOVE.W    #TLENGTH, D0              ; initialize the counter
LOOP:   CMP.B     ([0,A0], D0*4,1), D1      ; check the table entry
        DBEQ      D0, LOOP                  ; if equal then stop
        END
```

In the *memory-indirect preindexed* mode, the operand and its address are also fetched from memory. The MC68020 calculates an intermediate indirect memory address using the base register (An) and base displacement (bd), including the index operand (Xn.SIZE*SCALE). The CPU then fetches a long-word from this address and adds the outer displacement, to calculate the effective address. This

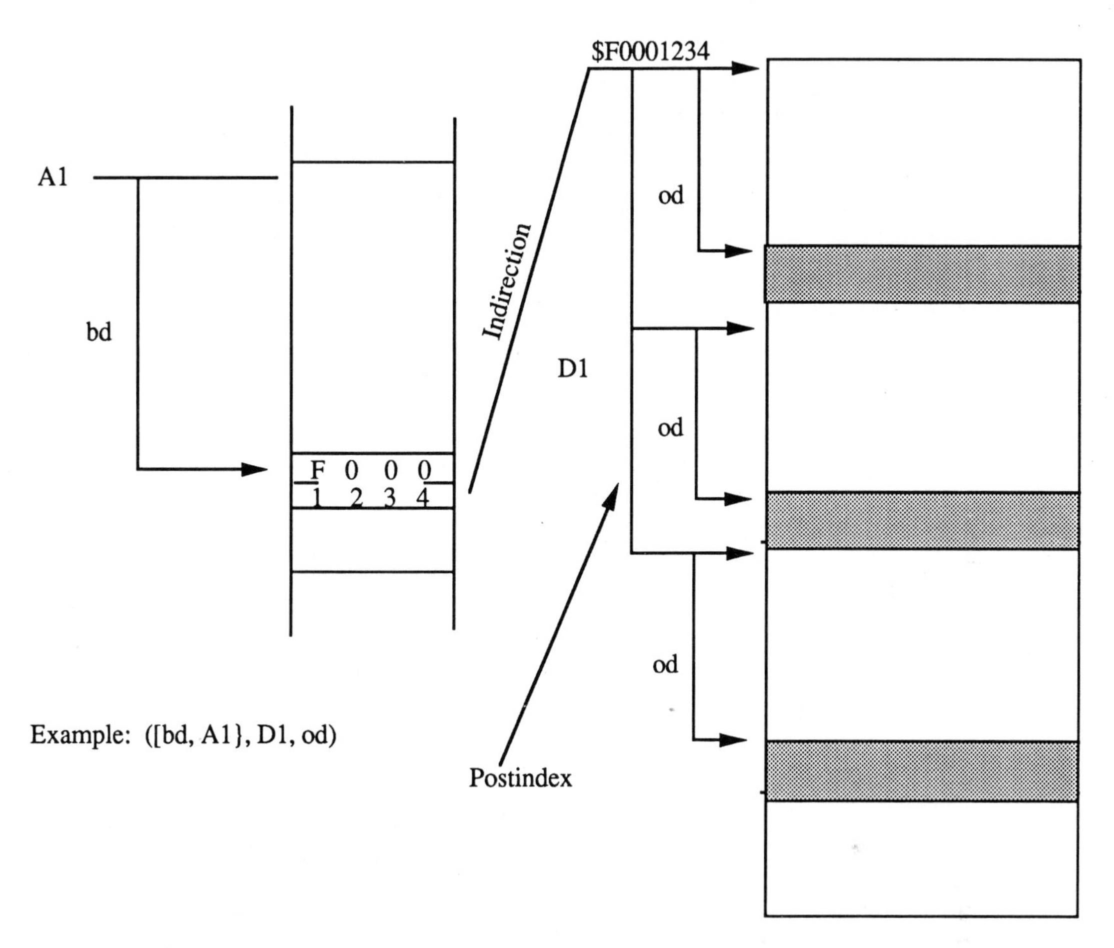

FIGURE 3.9. Memory indirect addressing mode with postindexing.

effective address is then utilized to find the operand from memory (Figure 3.10a). It is important to note that both displacements and the index register contents are sign extended to 32 bits.

The assembler syntax is given by ([bd, An, Xn.SIZE*SCALE], od). The elements within the inner brackets are used to calculate the intermediate memory address, and all four user-specified elements: bd, An, Xn, and SIZE*SCALE are optional. Both the base and outer displacements may be null, word or long-word. When a displacement is omitted or an element is suppressed, its value is taken as zero during the effective address calculation.

The use of the memory-indirect preindexing mode is shown in Figure 3.10b. Here, the device address table contains the addresses of the devices. The processor can access the devices if they are not busy, otherwise it will have to try again. The program to achieve this is shown below:

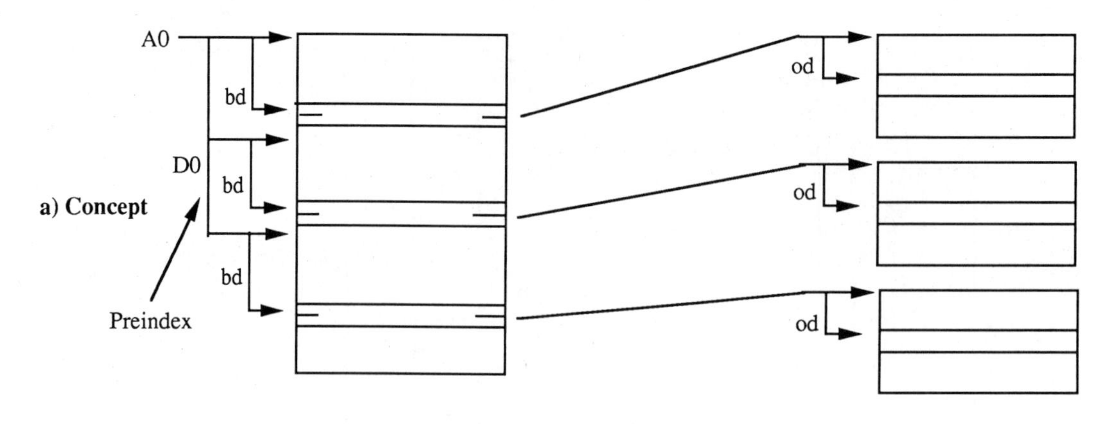

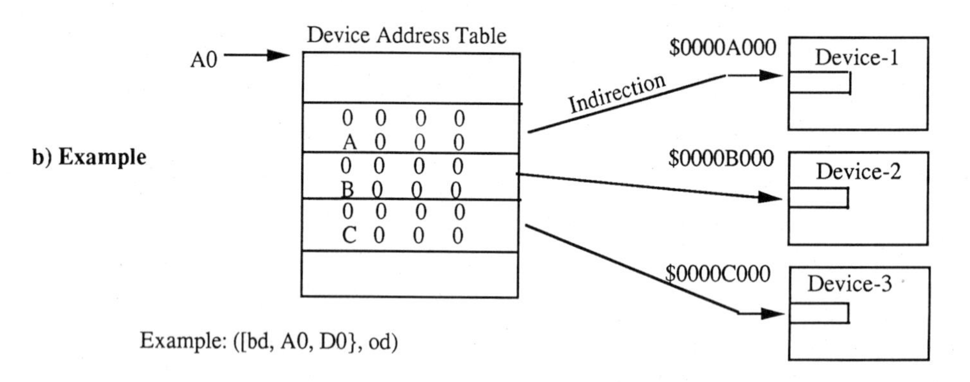

FIGURE 3.10. Example of memory indirect addressing with preindexed instruction.

```
        MOVEQ   #3, D0                  ; load the number of devices to be polled
LOOP:   BTST    #7,([0,A0,D0*4],2)      ; check the Z-bit in SR
        DBNE    D0, LOOP                ; finished checking every device if 1
                                        ; else look at the next device
        END                             ; finish the task
```

3.4.6 PC Relatives

It is advisable to write programs so that they can run in any area of the MC68020's address space. Programs having such a feature are called position independent. Once written and assembled, they can be executed in any system based on the same processor. Programs sold in ROM generally have this type of characteristic [Stark 88]. In a relocatable program, references must be made relative to addresses within the program itself, which are not known in advance. When a program is loaded in memory and is running, the instructions must have enough information to enable the processor to calculate the effective address of each operand, so that the processor can locate, fetch, and update them. The key to all of this is the Program Counter (PC), which holds the absolute address of the current instruction. The PC relative addressing mode can calculate that the operand resides at 'so many locations' relative to the current position of the PC. The MC68020 supports five classes of PC relative addressing modes, and they are:

- Program Counter Indirect (PCI) with displacement
- Program Counter Indirect (PCI) with index (8-bit displacement)
- Program Counter Indirect (PCI) with index (base displacement)
- Program counter Memory indirect postindexed
- Program counter memory indirect preindexed

PCI with displacement (Figure 3.11a) is the simplest of all PC indirect addressing modes. Here, the effective address is calculated by adding to the content of the PC a sign-extended 16-bit displacement found in the extension word of the instruction, i.e., EA = (PC) + d16. The programmer may rely on the assembler to calculate the displacement. For example:

```
MOVE   LOOP, D0
```

will instruct the assembler to calculate the displacement between the extension word of the move instruction and the location LOOP, and insert that value in the extension word. During the execution, the MC68020 will locate the contents of the location LOOP into the lower 16-bit location of the D0 register.

It is important to note that this MOVE instruction provides no clue as to whether the assembler will use program counter relative addressing mode or absolute addressing. However, labels such as LOOP, when preceded by an assembler directive "RORG," will cause the assembler to utilize indirect addressing modes relative to a program counter during the generation of operand locations in memory, whereas labels preceded by an "ORG" directive will cause the assembler to utilize absolute addressing during the generation of operand locations in memory.

PCI with index (8-bit displacement) (Figure 3.11b). Here, the effective address is the sum of three terms—the address in the program counter, the sign-extended

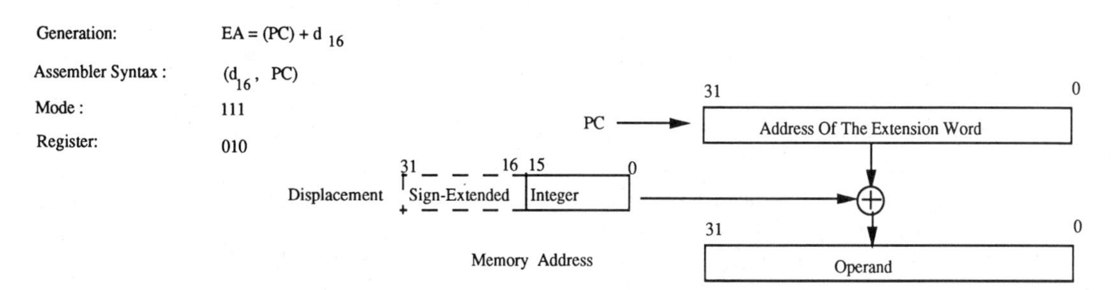

FIGURE 3.11a. Program counter indirect addressing with displacement. (Courtesy Motorola Inc.).

Generation:	EA = (PC) + (Xn) + d_8
Assembler Syntax:	(d_8, PC, Xn.SIZE * SCALE)
Mode:	111
Register:	011

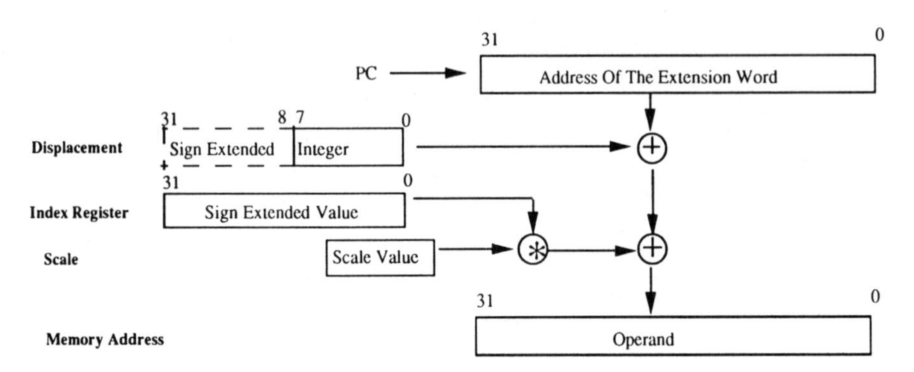

FIGURE 3.11b. Program counter indirect addressing with index (8-bit displacement). (Courtesy Motorola Inc.).

displacement integer in the lower eight bits of the extended word, and the size, including the scaled signed-extended index operand. The EA is:

$$EA = (PC) + (Xn) + d_8$$

This mode is particularly useful for accessing values from a list or data table. The user must include the displacement, the PC, and the index register to use this addressing mode. The following example will calculate the average of the first two entries in a table, using this addressing mode.

```
CLR.L   D1                  ; clear D1, which will hold the answer
CLR.W   D0                  ; clear the index register
MOVE    TABLE(PC,D0), D1    ; get the first data element
ADDQ    #2, D0              ; increment the pointer to access the next entry.
ADD     TABLE(PC,D0), D1    ; add the data word from the second
                            ; entry of the table to the previous contents of D1
ASR     #1, D1              ; divide the sum by 2 and store the result in D1.
```

PCI with index (base displacement) (Figure 3.11c). This powerful PC indirect addressing mode requires additional extension words to indicate the presence of the index register and the optional 16/32-bit base displacement (which is always sign-extended to 32 bits). The effective address is found by adding the contents

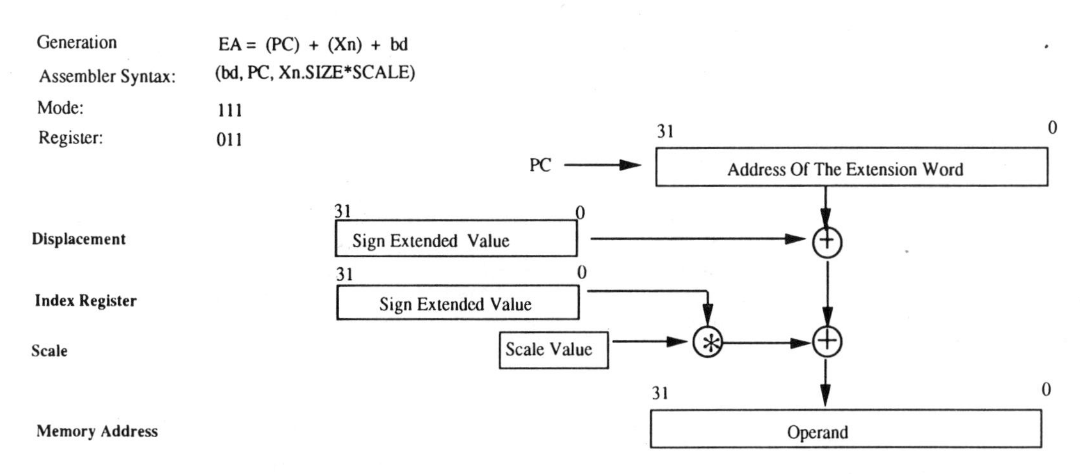

FIGURE 3.11c. Program counter indirect addressing with index and base displacement. (Courtesy Motorola Inc.).

of the PC to the scaled value of the sign-extended index register and the base displacement. The EA can be calculated using the formula:

$$EA = (PC) + (Xn) + bd$$

Program counter memory indirect post-indexed (Figure 3.11d). In this mode, the PC is used as a base register, and its value can be adjusted by adding an optional base displacement. An index register specifies an index operand and, finally, an optional outer displacement can be added to the address operand to determine the effective address.

Like the memory indirect mode, the square brackets are also used to instruct the assembler to calculate the intermediate memory address. The content of this memory address is then used in the final calculation. The 32-bit memory locations can be considered as address pointers.

Here, the MC68020 calculates an intermediate indirect memory address using the base register (An) and the base displacement (bd). The CPU then fetches a long-word from this address, and adds the index operand (Xn.SIZE*SCALE) and the outer displacement to determine the effective address. Like previous cases, the displacement and index register values are sign-extended to 32 bits before their usage. This can be seen as:

$$EA = (bd + PC) + Xn.SIZE*SCALE + od$$

The user can omit any or all four user-specified values. Again, the base and outer displacements may be null, words, or long-words. This is shown in the following example:

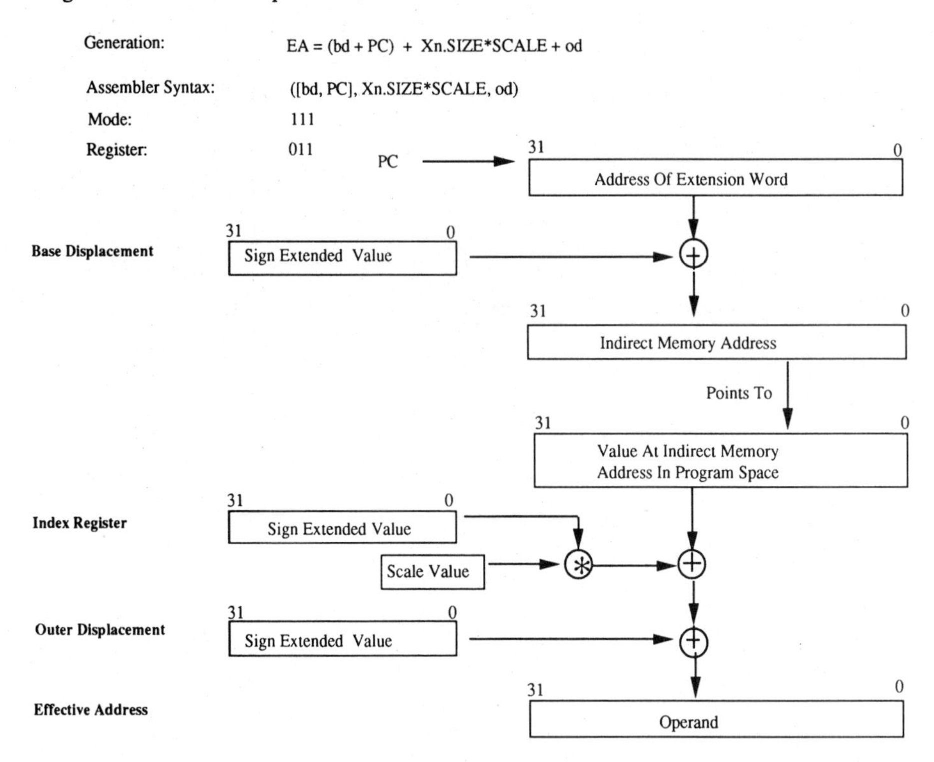

FIGURE 3.11d. Program counter memory indirect addressing with postindexing. (Courtesy Motorola Inc.).

```
00020000               ORG    $20000         ; select the origin
4EBB01710000           JSR    ([CALC,PC])    ; jump to the PC
00020200               ORG    $20200         ; select new origin
                                             ; for data area
00020400   CALC:       DC.L   $20400         ; reserve space
```

Program counter memory indirect preindexed (Figure 3.11e). In this addressing mode, the operand and its address are also in memory. The MC68020 calculates an intermediate memory address utilizing the base register (An), a base displacement (bd), and the index operand (Xn.SIZE*SCALE). The CPU then fetches a long-word from this address, and adds an outer displacement to yield the effective address. Again, the index register content is sign-extended to 32 bits. Thus, the effective address is calculated as:

$$EA = (bd + PC + Xn.SIZE{*}SCALE) + od$$

The PC indirect addressing mode is summarized in Figure 3.12. The MC68000 supported two types, PCI with displacement, and PCI with index and displace-

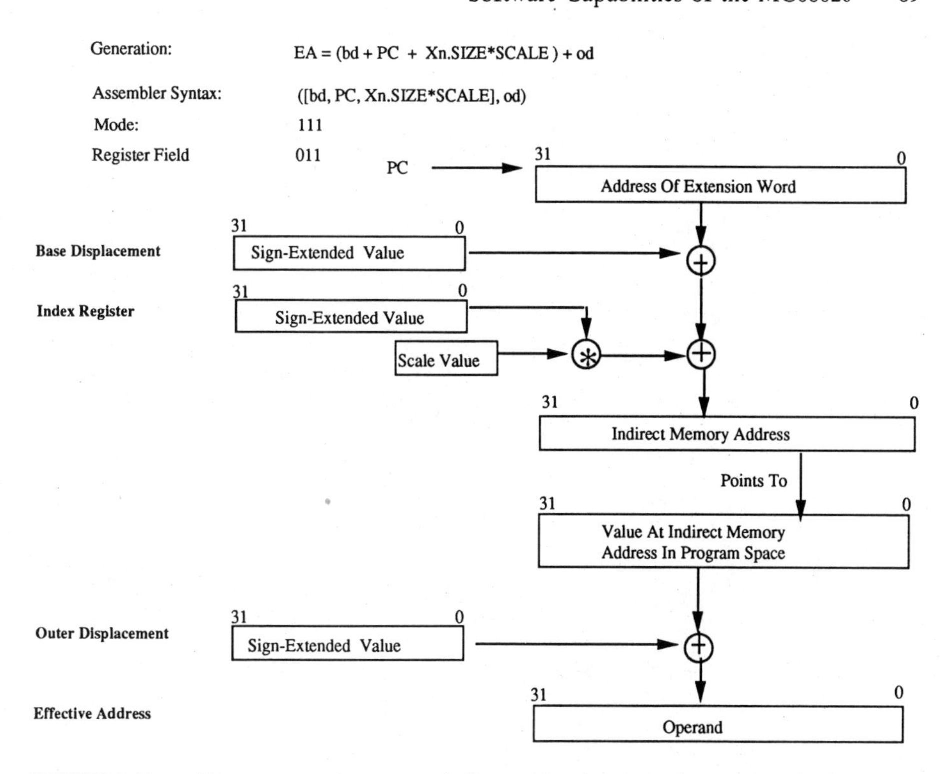

FIGURE 3.11e. Program counter memory indirect addressing, including postindexing and outer displacement. (Courtesy Motorola Inc.).

ment. Three additional modes are added to MC68020, namely *PCI with index and base displacement, PC memory indirect post index,* and the last one is *PC memory indirect pre index.* These five PC relative addressing modes give a very powerful tool to users to develop 'relocatable' programs.

3.4.7 Immediate Data Modes

Immediate data addressing is useful to specify a constant value as a source operand. The value is included within the instruction, rather than in a register or a memory location. There are two immediate addressing modes, *absolute short address* and *absolute long address.*

In absolute short address mode (Figure 3.13a), the address of the operand is encoded in the extension word. The 16-bit data is sign-extended to 32 bits if the destination is an address register. If the destination is a data register, then the data will not be sign-extended. For example,

MOVE #$ABCD, D0

The CPU will load $ABCD into the least significant word of D0. The most significant word portion will remain unchanged. However, in case of

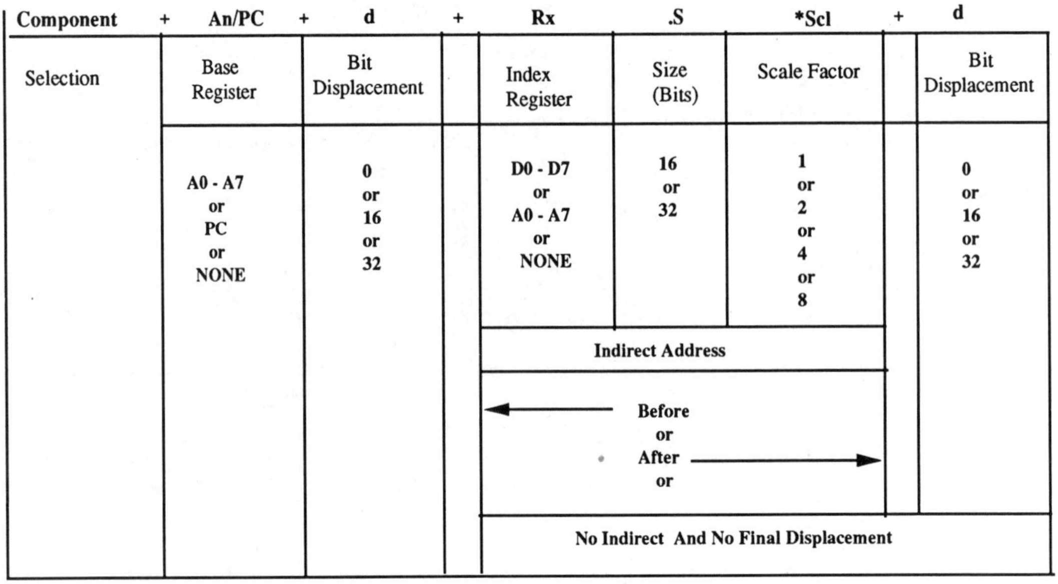

Component	+ An/PC	+ d	+ Rx	.S	*Scl	+ d
Selection	Base Register	Bit Displacement	Index Register	Size (Bits)	Scale Factor	Bit Displacement
	A0 - A7 or PC or NONE	0 or 16 or 32	D0 - D7 or A0 - A7 or NONE	16 or 32	1 or 2 or 4 or 8	0 or 16 or 32
			Indirect Address			
			← Before or After → or			
			No Indirect And No Final Displacement			

<ea> = ([d, An, Rx.S*Scl], d)

FIGURE 3.12. Summary of program counter indirect addressing modes. (Courtesy Motorola Inc.).

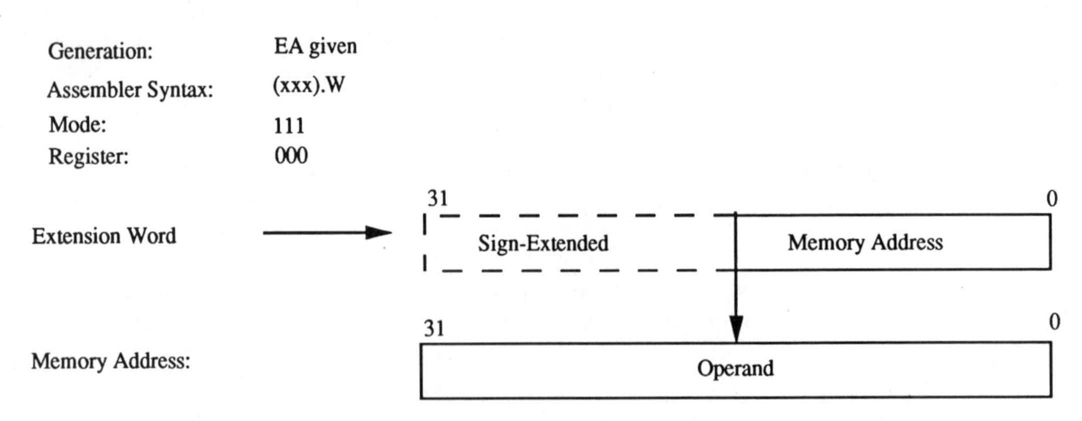

FIGURE 3.13a. Immediate addressing data modes: Absolute short address. (Courtesy Motorola Inc.).

MOVE #$ABCD, A0

the CPU will load the value $FFFFABCD into A0, affecting the entire 32-bits.

The absolute long addressing mode requires two words of extension (Figure 3.13.b). The address of the operand is calculated by concatenating the two 16-bit numbers to 32 bits. The high-order part of the address is the first extension word, and the low-order part of the address is the second extension word. The example of this addressing mode is

MOVE.L #$ABCDEF, A0

To enhance performance, the programmer may decide not to utilize the entire address space, but rather to use the short 16-bit mode. In the absolute short mode, the processor sign extends the single extension word to form a 32-bit address—this is much faster than the absolute long addressing mode. The short mode accesses the addresses between the range of $00000000 through $00007FF (the lower 32K of memory), or from $FF80000 through $FFFFFFFF (the upper 32K of memory).

Assemblers differ considerably in the way that they handle the short and long modes. Some assemblers will automatically create the optimum mode, while oth-

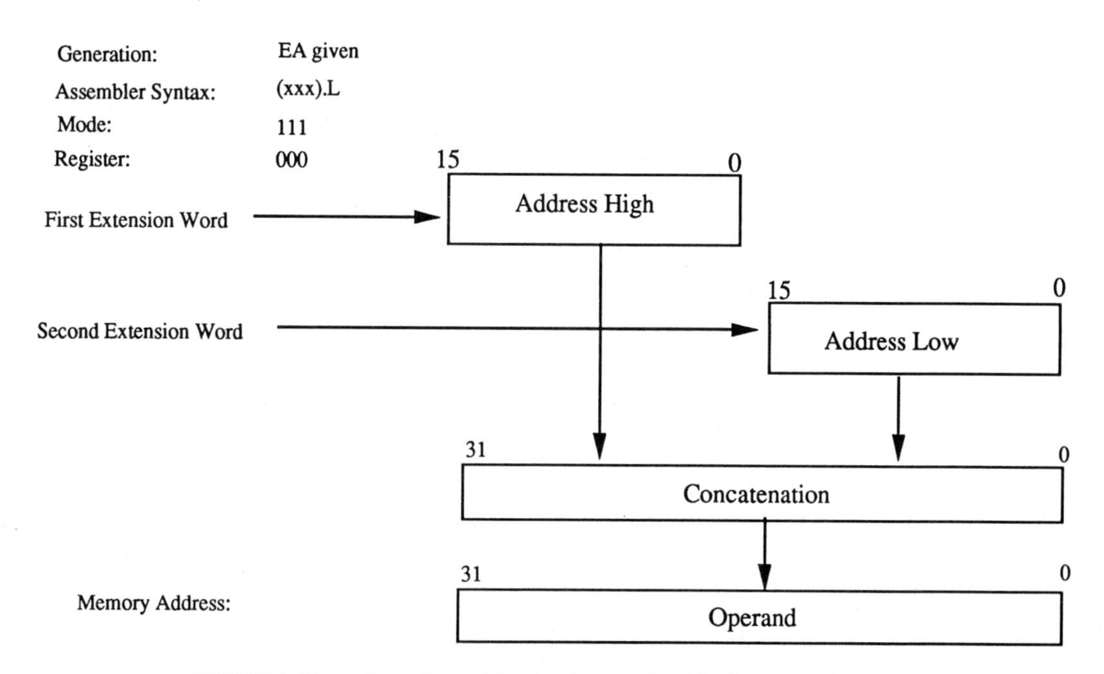

FIGURE 3.13b. Immediate addressing data modes: Absolute long address. (Courtesy Motorola Inc.).

ers require that programmers specify the "L" (long) or "W" (short) attributes as part of the instruction.

3.5 INSTRUCTION SETS AND THEIR CLASSIFICATIONS

The instructions allow programmers to tell the processor how to utilize its resources to complete a task. The instructions can be classified as:

- Data movement
- Integer arithmetic
- Logical operations
- Shift and rotate operations
- Bit manipulation operations
- Bit field operations
- Binary coded decimal arithmetic
- Program control
- System control
- Multiprocessor support

3.5.1 Data Movement Operations

The data movement operations facilitate the transfer of byte, word, and long-word operands from: memory to memory, register to memory, memory to register, and register to register. MOVE is the most versatile data movement instruction available to the MC68000 processor family. To ensure the proper calculation of the effective address, MOVEA or MOVE involving an address register can only transfer word and long-word operands. The MC68020 supports several additional data movement instructions, namely: MOVEM—to move operands to and from the memory and registers; MOVEP—to transfer peripheral data; MOVEQ—to quickly move data into a register; EXG— to interchange the contents of registers; LEA—to load an effective address into a register; PEA—a push-effective address; LINK—to link a stack; UNLK—to unlink a stack. Figure 3.14 summarizes the operations of these instructions. We will explore the usage of many of these instructions throughout this book.

3.5.2 Integer Arithmetic Operations

The MC68020 provides several instructions for performing binary arithmetic operations. They support both 'signed' and 'unsigned' byte, word, long-word, or sometimes quad-word data structures. The integer arithmetic instructions are shown in Figure 3.15. The ADD, CMP, and SUB instructions are applicable to both address and data operations. All data types are valid for data operands. Address operations, however, are limited to 16- and 32-bit operands. The CLR and various negate instructions may be used on all sizes of data operands.

Instruction	Operand Syntax	Operand Size	Operation
EXG	Rn, Rn	32	Rn ◄► Rn
LEA	<ea>, An	32	<ea> → An
LINK	An, #<d>	16, 32	SP - 4 → SP, An → (SP); SP → An; SP + d → SP
MOVE MOVEA	<ea>, <ea> <ea>, An	8, 16, 32 16, 32 → 32	Source → Destination
MOVEM	List, <ea> <ea>, List	16, 32 16, 32 → 32	Listed registers → Destination Source → Listed registers
MOVEP	Dn, (d_{16}, An)	16, 32	Dn[31:24] → (An + d); Dn[23:16] → (An + d + 2); Dn[15:8] → (An + d + 4); Dn[7:0] → (An + d + 6); (An + d); → Dn[31:24]; (An + d + 2) → Dn[23:16] (An + d + 4) → Dn[15:8]; (An + d + 6) → Dn[7:0];
MOVEQ	#<data>, Dn	8 → 32	Immediate data → Destination
PEA	<ea>	8 → 32	SP - 4 → SP; <ea> → (SP)
UNLK	An	32	An → SP; (SP) → An; SP + 4 → SP

FIGURE 3.14. Data movement instructions. (Courtesy Motorola Inc.).

However, the extended instructions ADDX, SUBX, EXT, and NEGX are designed to perform multiprecision and mixed-size arithmetic.

The arithmetic operations set the bits in the Condition Code Register (CCR) to reflect the result of the operations. The bits are set according to the following criteria:

The N bit is set if the result is negative, and cleared otherwise.
The Z bit is set if the result of the operation produces 0, else it is cleared.
V is set in the case of an overflow, else it is cleared, i.e., V=0.
X and C are set when a carry is generated or a borrow takes place, and are cleared otherwise.

All CCR bits are affected by the ADD, SUB, and NEG instructions. In the case of the MUL, DIV, and EXT instructions, the flags V and C are always cleared, N and Z are set or cleared based on the result, and X remains unchanged.

Multiplication: The MC68020 provides two basic multiplication instructions that utilize the same format. The signed multiplication format is:

Instruction	Operand Syntax	Operand Size	Operation
ADD ADDA	Dn, <ea> <ea>, Dn <ea>, An	8, 16, 32 8, 16, 32 16, 32	Source + Destination → Destination
ADDI ADDQ	#<data>, <ea> #<data>, <ea>	8, 16, 32 8, 16, 32	Immediate data + Destination → Destination
ADDX	Dn, Dn -(An), -(An)	8, 16, 32 8, 16, 32	Source + Destination + X → Destination
CLR	<ea>	8, 16, 32	0 → Destination
CMP CMPA	<ea>, Dn <ea>, An	8, 16, 32 16, 32	Destination - Source
CMPI	#<data>, <ea>	8, 16, 32	Destination - Immediate Data
CMPM	(An)+, (An)+	8, 16, 32	Destination - Source
CMP2	<ea>, Rn	8, 16, 32	Lower Bound < = Rn < = Upper Bound
DIVS/DIVU DIVSL/DIVUL	<ea>, Dn <ea>, Dr:Dq <ea>, Dq <ea>, Dr:Dq	32/16 → 16 : 16 64/32 → 32 : 32 32/32 → 32 32/32 → 32 : 32	Destination / Source → Destination (Signed or Unsigned)
EXT EXTB	Dn Dn Dn	8 → 16 16 → 32 8 → 32	Sign Extended Destination → Destination
MULS/MULU	<ea>, Dn <ea>, Dl <ea>, Dh : Dl	16 * 16 → 32 32 * 32 → 32 32 * 32 → 64	Source * Destination → Destination (Signed or Unsigned)
NEG	<ea>	8, 16, 32	0 - Destination → Destination
NEGX	<ea>	8, 16, 32	0 - Destination - X → Destination
SUB SUBA	<ea>, Dn Dn, <ea> <ea>, An	8, 16, 32 8, 16, 32 16, 32	Destination - Source → Destination
SUBI SUBQ	#<data>, <ea> #<data>, <ea>	8, 16, 32 8, 16, 32	Destination - Immediate Data → Destination
SUBX	Dn, Dn - (An), - (An)	8, 16, 32 8, 16, 32	Destination - Source - X → Destination

FIGURE 3.15. Integer arithmetic instructions. (Courtesy Motorola Inc.).

MULS <source>, Dn

It performs the multiplication of two 16-bit signed numbers and places the 32-bit result in the destination register Dn. The other instruction format is:

MULU <source>, Dn

This instruction operates the same way as before, but performs multiplication on two unsigned 16-bit numbers. The <source> can be accessed utilizing any addressing mode, with the exception that it cannot be in an address register.

The CCR register flags set by MULU and MULS are listed below:

Flag	X	N	Z	V	C
MULU/MULS	—	*	*	0	0

The V and C flag bits are always cleared, and the X bit retains its previous value. The Z flag is set if the result becomes "0", and N gets the value of bit 31 of the destination register. Since MULU performs multiplication without any regard to the sign, N will indicate whether there is a 0 or 1 in bit position 31 of Dn after the completion of the operation. This is illustrated through two examples:

```
MOVE.L   #$FFFFFFFE, D0
MULU     #2, D0
```

At the end of the execution of the operation, D0 will contain $0001FFC. In the next case:

```
MOVE.L   #$FFFFFFFE, D0
MULS     #2, D0
```

Here, D0 will contain $FFFFFFFC at the end of the execution. The use of the multiplication instruction is illustrated further in the example program.

Example: Let D0 contain gross pay for a year, and D1 contain the tax rate. We wish to calculate the tax amount due to the tax collector. We also want to retain the tax rate and the tax amount due for future usage.

```
MOVE.W   D1, -(SP)     ; save D1 on the stack
MULS     D0, D1        ; D1(32-bits) = D0(16-bits)*D1(16-bits)
                       ; = Gross*Tax Rate
BMI      REFUND        ; if negative result (N flag = 1) go to
                       ; branch to REFUND (do you believe that?)
BEQ      NADA          ; if not minus, is result 0?
```

```
                                     ; YES then go to NADA. Otherwise we are
                                     ; even (we made it—bingo!!)
MOVE.L     D1, D2                    ; not negative, not equal to zero. I owe
                                     ; to the tax collector (rats!!)
                                     ; save the amount owed to D2 (do it $$)
MOVE.W     (SP)+, D1                 ; restore D1 (Tax Rate) from the stack
           :
           :                         ; let us do what we have to do
BRA        DONE                      ; complete (what a relief)
REFUND:    MOVE.W (SP)+, D1          ; restore D1 from the stack back to D1
           :
           :
           :                         ; do other things (let us have a party)
BRA        DONE                      ; branch to DONE
NADA       MOVE.W (SP)+, D1          ; restore D1 from the stack
           :
           :                         ; do other things
DONE:      :                         ; wind up the program
```

Division: Like multiply instructions, there are two divide instructions:

```
DIVS   <Source>, Dn   ; Signed divide
DIVU   <source>, Dn   ; Unsigned divide
```

Here, the 32-bit value in the destination register Dn is divided by the 16-bit value obtained from the <source>. The <source> cannot be an address register. However, one can divide the contents of a data register by another data register, including the contents of a memory location. The result will be placed in Dn using the following convention. The 16-bit quotient will be stored in the least significant word of Dn, and the 16-bit remainder will be stored in the most significant word of Dn.

Both DIVS and DIVU alters the CCR. It clears the C flag without changing the X flag. Other flags are set to reflect the result. During the divide operation, if the divisor is zero, an exception occurs. If the processor detects an overflow, i.e., V---> 1, then the operands remain unchanged.

An overflow can occur when the range of the dividend is 0 to (2^{32}) -1 and the quotient 16 bits long. It is obvious that dividing an integer greater than $(2^{16}-1)$ by 1 will result in an overflow, since the dividend exceeds the divisor by a magnitude of 2^{16} or greater. An example of the divide instruction is given in the following.

Example: The first entry of the TABLE contains the total number of PC-boards produced. Find the average number of PC-boards produced in the month of June. We are also required to store the average into the second entry, the quotient in the fourth entry, and the remainder in the sixth entry. It is given that A0 is the pointer to the table and the entries in the table are 16 bits wide.

```
        LEA       TABLE, A0    ; load the address into A0
        MOVE.L    D0, -(SP)    ; save the value of D0 in the stack
        MOVE      (A0), D0     ; get the production total
        DIVS      #30, D0      ; calculate the average for June
        BVS       OVFL         ; report an error if the total is zero
        MOVE      D0, 6(A0)    ; save the quotient in the fourth entry
        SWAP      D0           ; interchange the quotient and
                               ; remainder in D0; now the remainder
                               ; occupies the least significant 16 bits.
        MOVE      D0, 10(A0)   ; save the remainder in the sixth entry
        MOVE      (SP)+, D0    ; retrieve the original value of D0
OVFL:             ..           ; process the error condition
```

Note: LEA first calculates the effective address of the TABLE and then loads the sign-extended 32-bit value into A0. The Instruction SWAP works only on a data register, and interchanges the most significant word into the least significant word and vise versa, i.e., Dn(31-16)---> Dn(15-0) and Dn(15-0)---> Dn(31-16).

3.5.3 Logical Operations

Logical instructions support functions like AND, OR, Exclusive OR, and NOT. Various forms of these instructions, including operand sizes and operations, are shown in Figure 3.16. During the execution of the logical instruction, the CPU clears the V and C flags, sets/resets the N and Z flags to reflect the result, and retains the previous contents of X. TST performs an arithmetic comparison of an operand against 'zero' and sets the CCR flags to reflect the result.

The user should be careful while performing logical operations on the SR and CCR registers as destinations. Only the operating system (i.e., programs running in supervisory modes) should perform logical operations on the SR. A user program attempting to access the 'system byte' portion of the SR register will trigger an exception processing (this concept will be described in Chapter 7). However, a user can set CCR flags utilizing the ANDI, ORI, or EORI instructions.

The following example shows the use of the logical instruction.

Example: Clear bit 7 in each ASCII character in a string of such characters stored in memory. Assume that A6 contains the starting address of the string in memory. The last byte of the string is Null(00).

```
LOOP:   ANDI.B  #$7F, (A6)+    ; clear bit 7.
        TST.B   (A6)           ; test for null on next character
        BNE     LOOP           ; not 00, repeat the loop.
        :                      ; perform rest of the program
        :
```

Instruction	Operand Syntax	Operand Size	Operation
AND	<ea>, Dn Dn, <ea>	8, 16, 32 8, 16, 32	Source ∧ Destination → Destination
ANDI	#<data>, <ea>	8, 16, 32	Immediate data ∧ Destination → Destination
EOR	Dn, <ea>	8, 16, 32	Source ⊕ Destination → Destination
EORI	#<data>, <ea>	8, 16, 32	Immediate data ⊕ Destination → Destination
NOT	<ea>	8, 16, 32	-- Destination → Destination
OR	<ea>, Dn Dn, <ea>	8, 16, 32 8, 16, 32	Source V Destination → Destination
ORI	#<data>, <ea>	8, 16, 32	Immediate data V Destination → Destination
TST	<ea>	8, 16, 32	Source — 0 to Set Condition Codes

FIGURE 3.16. Logical instructions. (Courtesy Motorola Inc.).

3.5.4 Shift and Rotate Operations

The shift and rotate instructions change the bit positions of the data bits in an operand. The instructions are efficient, to multiply or divide a given number by a power of 2, check the status of specific bits in an operand, or simply shift the position of data bits in a register or memory location (useful for a memory check or to determine the status of drivers). The shift and rotate operations are shown in Figure 3.17.

Shift and rotate operations on registers are supported for all operand sizes. A shift count from 1 to 8 is encoded in the instruction word. A shift count greater than 8 is specified as a modulo 64 number and requires a register. Memory shifts and rotates are valid for word operands, and allow single-bit shifts or rotates. Shifts can be used to multiply or divide by 2, 4, 8, etc., but when dealing with ‘signed’ numbers, care must be taken not to lose the sign bit, which is the most significant bit of the data. The arithmetic shifts are useful for signed numbers, and logical shifts for unsigned numbers. The performance of shift/rotate instructions are enhanced in the MC68020 over the MC68000, so that fast-byte swapping is possible using the ROR (Rotate Right) and ROL (Rotate Left) instructions, with a shift count of 8. The effect on the flag bits in the CCR for arithmetic and logical shifts are shown in the following:

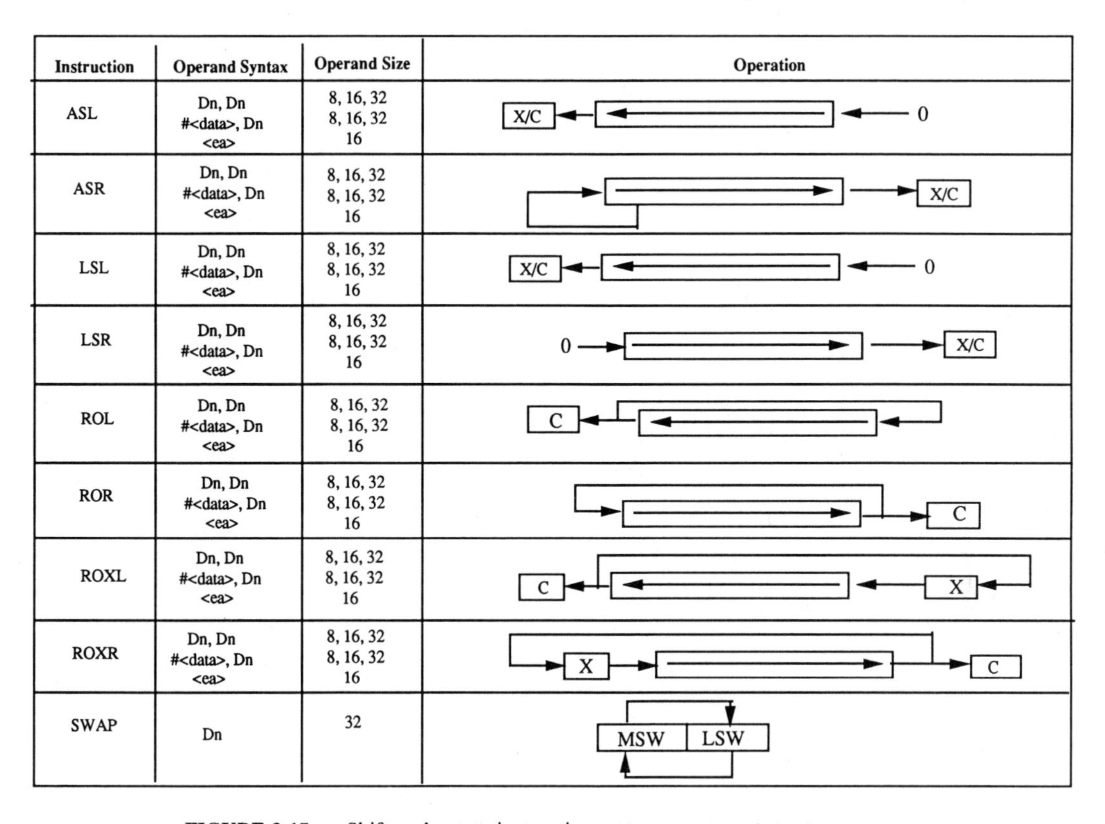

Instruction	Operand Syntax	Operand Size	Operation
ASL	Dn, Dn #<data>, Dn <ea>	8, 16, 32 8, 16, 32 16	X/C 0
ASR	Dn, Dn #<data>, Dn <ea>	8, 16, 32 8, 16, 32 16	X/C
LSL	Dn, Dn #<data>, Dn <ea>	8, 16, 32 8, 16, 32 16	X/C 0
LSR	Dn, Dn #<data>, Dn <ea>	8, 16, 32 8, 16, 32 16	0 X/C
ROL	Dn, Dn #<data>, Dn <ea>	8, 16, 32 8, 16, 32 16	C
ROR	Dn, Dn #<data>, Dn <ea>	8, 16, 32 8, 16, 32 16	C
ROXL	Dn, Dn #<data>, Dn <ea>	8, 16, 32 8, 16, 32 16	C X
ROXR	Dn, Dn #<data>, Dn <ea>	8, 16, 32 8, 16, 32 16	X C
SWAP	Dn	32	MSW LSW

FIGURE 3.17. Shift and rotate instructions. (Courtesy Motorola Inc.).

ASL, ASR affects X, N*, Z, V, C flags, and are set to reflect the result of the operation.

LSL, LSR affects X, N*, Z, C flags and always sets V to 0.

Note: * with ASL/ASR: N is set if the Most Significant Bit (MSB) of the result is set, and cleared otherwise.
with LSL/LSR, N is set if the result is negative, and clears otherwise.

The operations of arithmetic and logical shifts are illustrated via a number of examples.

Example: Check the number of 1s in an 8-bit ASCII character, for the purpose of parity check. Set D0 = 0 if there are an even number of 1s, otherwise set D0 = 1.

```
LOOP:   CLR.L   D0      ; clear the register
        TST.B   D1      ; test if D1 = 00
        BEQ     DONE    ; if D1=00, all shift out of D1
        LSL.B   #1,D1   ; not a NULL, then shift left once
```

```
                                ; and MSB is moved to carry.
        BCC      LOOP           ; if the bit shifted out was 0, then
                                ; continue to loop.
        EORI.B   #1,D0          ; if the bit shifted out was 1, set D0
                                ; to 1. If the next shifted bit is 1,
                                ; this instruction will clear D0.
                                ; Thus D0=1 for an odd number of 1s,
                                ; D0-0 for an even number of 1s.
        BRA      LOOP           ; Repeat the shift operation and test
                                ; for Bcc (branch) condition until D1=00,
                                  i.e.,
                                ; all 0s shifted in from right.
DONE:   TST.B    D0             ; Is D0=0?
        BNE      ERROR          ; If D0 is not equal to 0, then branch
                                ; to the ERROR routine. The parity is odd
                                ; and an error condition has occurred.
        :
        :
ERROR:  :                       ; ask for retransmission or take another
                                ; action.
```

Example: Calculate the average of two signed numbers. The numbers can be found in registers D0 and D1, and the result should be placed in the D2 register.

In such a situation, we are to consider three cases:

Case I. Both are positive numbers. Let
D0 = $1234 D1=$5677
then, D0+D1 = $68AB is still a positive number
D2=$68AB/2 = $3455 (with the roundup) is still positive.

Case II. Both are negative numbers. Let
D0=-3 = $FFFD & D1=-9 = $FFF7
D0+D1 = -3 + -9 = -12 = $FFF4 (no overflow)
D2 = $FFF4/2 = $FFFA

Case III. Both are positive numbers. Let
D0=$7000 D1=$7002
D0+D1 = $E002, which can be interpreted as a negative number thus resulting in an error.

Thus, any program must be able to resolve such conditions. The solution is given in the following:

```
        MOVE     D0,D2          ; place D0 in D2
        ADD      D1,D2          ; the sum is now in D2
        BVS      ERROR          ; if the V flag is set, then there is an
                                ; overflow and branch to the error
```

```
                                ; condition.
          ASR      #1,D2        ; divide D2 by 2 and ignore the
                                ; remainder.
          :
          :
ERROR:    Resolve the error condition
```

The MC68020 has four rotate instructions. Rotating the bits in a register is very much like a logical shift, except that the displaced bits are fed back into the register at the other end. The syntax of the rotate instructions are:

```
ROL/R   Dx, Dy       ; low 6 bits in Dx(0-63) determines
                     ; the number of rotations to be
                     ; performed in Dy.
ROL/R   #data, Dy    ; # data = 1 to 8 = the number of
                     ; rotations to the left or right in Dy.
ROL/R   <ea>         ; a word in memory at the effective
                     ; address (EA) is rotated 1 bit to the left
                     ; or right
```

The X-bit (extended bit) is not included in the ROL/R. The instructions ROXL/R are to be used to include the X-bit. The rotate instructions change the flags as follows:

```
ROL, ROR   ; affects the N, Z, C flags, V is always set to 0, and X
           ; remains unchanged.
```

The X, N, Z, C flags reflect the results of the ROXL, ROXR instructions, but V is always set to 0.
The flag N is set if the MSB of the result is set, and cleared otherwise. Z is set if the result is zero, cleared otherwise. C is set to reflect the last bit rotated out of the operand. X is set according to the last bit rotated out of the operand.

Example: Add the four signed bytes in a long-word using the ROR instruction. The four signed bytes are in register D0, and the result should be placed in D1 [white group].

We will solve this problem in two different ways: Both techniques employ the ROR instruction. The first solution is given in the following:

```
CLR.L    D3       ; clear the temporary holding register
CLR.L    D1       ; clear the register to hold the result
MOVE.B   D0,D1    ; place the first byte into D1
ROR.L    #8, D0   ; byte 2 goes to the least significant byte
                  ; of D0, byte 3 to the second least
                  ; significant byte of D0, byte 4 goes to
```

```
                          ; the third least significant byte of D0
                          ; and byte 1 goes to most significant
                          ; byte of D0.
MOVE.B      D0, D3        ; Send the low byte of D0 (i.e., byte 2) to D3
ADD         D3, D1        ; add byte 2 to byte 1, D1=byte 2+byte 1.
ROR.L       #8, D0        ; Byte 3 moves one byte to the left,
                          ; and successively the other bytes move
                          ; one byte to the right.
MOVE.B      D0, D3        ; Move the low byte to D3
ADD         D3, D1        ; add the third byte to D1, D1=B3+B2+B1.
ROR.L       #8, D0        ; Byte 4 moves to the least significant
                          ; byte of D0, and the other bytes move into
                          ; D0 accordingly.
```

The next solution is based on SWAP, and the resultant program is shown:

```
CLR.L       D3          ; clear the temporary register, D3
CLR.L       D1          ; clear D1
MOVE.B      D0, D1      ; the least significant byte moves to D1
SWAP        D0          ; interchange the least significant word
                        ; with the most significant word of D0
MOVE.B      D0, D3      ; move the original third most significant
                        ; byte into D3,
ADD.W       D3, D1      ; add. The result is D1= Byte 3 + Byte 1.
ROR.L       #8, D0      ; the original byte 4 goes to the byte 1 position and
                        ; all other bytes get moved into D0 in a
                        ; counterclockwise direction.
MOVE.B      D0, D3      ; Get the original byte 4 of D0
ADD         D3, D1      ; get the sum in D3, now D1=B4+B3+B1.
ROR.L       #8, D0      ; B1 goes to least significant byte of
                        ; D0, B2 to the second least significant
                        ; byte of D0, B3 to the third least
                        ; significant byte of D0, and B4 to the
                        ; most significant byte of D0.
```

3.5.5 Bit Manipulation Operations

The bit manipulation instructions allow the MC68020 to perform operations on one or more specific bits in a register or in a memory location. The bit can be set, reset, or changed during the execution of the instruction. The bit manipulation instructions are shown in Figure 3.18. They are useful for memory management, graphics and communication applications.

The MC68020 supports variable-length bit field operations with fields of up to 32 bits. Bit field instructions are shown in Figure 3.19. We will now define a number of terms to explain bit field operations.

Instruction	Operand Syntax	Operand Size	Operation
BCHG	Dn, <ea>	8, 32	~ (< bit number > of destination) → Z → bit of destination
	#<data>, <ea>	8, 32	
BCLR	Dn, <ea>	8, 32	~ (< bit number > of destination) → Z
	#<data>, <ea>	8, 32	0 → bit of destination
BSET	Dn, <ea>	8, 32	~ (< bit number > of destination) → Z
	#<data>, <ea>	8, 32	1 → bit of destination
BTST	Dn, <ea>	8, 32	~ (< bit number > of destination) → Z
	#<data>, <ea>	8, 32	

FIGURE 3.18. Bit processing instructions. (Courtesy Motorola Inc.).

Instruction	Operand Syntax	Operand Size	Operation
BFCHG	<ea> {offset : width}	1—32	~ Field → Field
BFCLR	<ea> {offset : width}	1—32	0's → Field
BFEXTS	<ea> {offset : width}, Dn	1—32	Field → Dn; Sign Extended
BFEXTU	<ea> {offset : width}, Dn	1—32	Field → Dn; Zero Extended
BFFFO	<ea> {offset : width}, Dn	1—32	Scan for First Bit Set in Field; Offset → Dn
BFINS	Dn, <ea> {offset : width}	1—32	Dn → Field
BFSET	<ea> {offset : width}	1—32	1's → Field
BFTST	<ea> {offset : width}	1—32	Field MSB → N; — (OR of all bits in field) → Z

N.B. All bit field instructions set the N and Z bits.

FIGURE 3.19. Bit field processing instructions. (Courtesy Motorola Inc.).

Bit field: A string of 1 to 32 consecutive bits within a bit array.
Bit array: A collection of bits that can be addressed in increments of bits.
Base address: The base address is the address of the byte that contains bit [0] of the array.
Bit field number: This specifies the relative offset from the most significant bit of the base address.
Bit field offset: This identifies the bit field number of the most significant bit of the field. It has a range of from $-(2^{31})$ to $(2^{31} - 1)$ bits in the memory.
Bit field width: The bit field width indicates the number of bits in the field, which ranges from 1 to 32.

These terms are illustrated through the following example:

The base address points to a memory location. The bit offset of -20 means that we have to count 20 locations before the starting location of the base address. If we specify that the bit field width is 4, then we need to count 4 bits from the offset.

3.5.6 Program Control Operations

The program control operations give us the capability to perform conditional branch, unconditional branch, and return instructions. The following testing and branching conditions are supported by the MC68020 processor.

CC—carry
CS—carry set
EQ—equal
F—false
GE—greater or equal
GT—greater than
HI—high
LE—less or equal
LS—low or same
LT—less than
MI—minus
NE—not equal to
PL—plus
T—always true
VC—overflow clear
VS—overflow set

During conditional operations, the CPU checks the Condition Code Register (CCR) before making a decision. In such cases, the CCR bits may get set or cleared as a result of the change in program flow. The program control instructions are shown in Figure 3.20.

3.5.7 System Control Operations

The system control instructions are summarized in Figure 3.21. They fall into three categories, namely: privileged instructions, trap-generating instructions, and status register instructions. They can be executed under the privileged mode. Any attempt to execute privileged instructions from the user mode will initiate an exception processing.

Conditional Instruction	Operand Syntax	Operand Size	Operation
Bcc	<label>	8, 16, 32	If condition is true, then PC + d → PC
DBcc	Dn, <label>	16	if condition false, then Dn − 1 → Dn if Dn ≠ −1, then PC + d → PC
Scc	<ea>	8	if condition is true, then 1's → destination; else 0's → destination

Unconditional Instruction	Operand Syntax	Operand Size	Operation
BRA	< label >	8, 16, 32	PC + d → PC
BSR	< label >	8, 16, 32	SP − 4 → SP; PC → (SP); PC + d → PC
CALLM	#< data >, < ea >	none	Save processor status in Stack Frame; Load new status from destination
JMP	< ea >	none	destination → PC
JSR	< ea >	none	SP − 4 → SP; PC → (SP); destination → PC
NOP	none	none	PC + 2 → PC

Return Instruction	Operand Syntax	Operand Size	Operation
RTD	#< d >	16	(SP) → PC; SP + 4 + d → SP
RTM	Rn	none	reload saved module state from stack frame; place module data area pointer in Rn
RTR	none	none	(SP) → CCR; SP + 2 → SP; (SP) → PC; SP + 4 → SP
RTS	none	none	(SP) → PC; SP + 4 → SP

FIGURE 3.20. Program control instructions. (Courtesy Motorola Inc.).

3.5.7.1 The Privileged Instructions

RESET is an important privileged instruction. It is generally used to recover from a catastrophic system failure. It causes the RESET pin of the MC68020 to be asserted low for a 512-clock period. This pin is generally wired to the "reset" lines of all external devices, and will cause them to "RESET." The execution of the RESET instruction does not alter the contents of the internal registers of the CPU. Rather, it generates an external reset pulse for external devices. At the completion of the RESET instruction, the processor will execute the next instruction from the instruction sequence.

The STOP (stop program execution) instruction causes the MC68020 to cease fetching and executing instructions. The CPU will not resume program execution until it receives a trace, interrupt, or externally generated reset signal. The CPU

Privileged Instruction	Operand Syntax	Operand Size	Operation
ANDI	#< data >, SR	16	immediate data ∧ SR → SR
EORI	#< data >, SR	16	immediate data ⊕ SR → SR
MOVE	< ea >, SR SR, < ea >	16 16	source → SR SR → destination
MOVEC	Rc, Rn Rn, Rc	32 32	Rc → Rn Rn → Rc
MOVES	Rn, < ea > < ea >, Rn	8, 16, 32	Rn → destination using DFC source using SFC → Rn
ORI	#< data >, SR	16	immediate data V SR → SR
RESET	none	none	assert RESET line
RTE	none	none	(SP) → SR; SP + 2 → SP; (SP) → PC; SP + 4 → SP restore stack according to format
STOP	#< data >	16	immediate data → SR; STOP

Trap Generating Instruction	Operand Syntax	Operand Size	Operation
BKPT	#< data >	none	if breakpoint cycle acknowledged, then execute returned operation word, else trap as illegal instruction.
CHK	< ea >, Dn	16, 32	if Dn < 0 or Dn > (ea), then CHK exception
CHK2	< ea >, Rn	8, 16, 32	if Rn < lower bound or Rn > upper bound, then CHK exception
ILLEGAL	none	none	SSP - 2 → SSP; Vector Offset → (SSP); SSP - 4 → SSP; PC → (SSP); SSP - 2 → SSP; SR → (SSP); illegal instruction Vector address → PC
TRAP	#< data >	none	SSP - 2 → SSP; Format and Vector Offset → (SSP); SSP - 4 → SSP; PC → (SSP); SSP - 2 → SSP; SR → (SSP); Vector address → PC
TRAPcc	none #< data >	none 16, 32	if cc true, then TRAP exception
TRAPV	none	none	if V then take overflow TRAP exception

Condition Code Register Instruction	Operand Syntax	Operand Size	Operation
ANDI	#< data >, CCR	8	immediate data ∧ CCR → CCR
EORI	#< data >, CCR	8	immediate data ⊕ CCR → CCR
MOVE	< ea >, CCR CCR, < ea >	16 16	Source → CCR CCR → destination
ORI	#< data >, CCR	8	immediate data V CCR → CCR

FIGURE 3.21. System control operations. (Courtesy Motorola Inc.).

does not release control of the system bus after the execution of the STOP instruction. A trace exception will occur if the trace was activated before the execution of the STOP instruction. The STOP instruction is extremely valuable when the system must wait for a specific 'event' to occur. In such a case, the designer will set the proper interrupt level, disable the trace bit, and execute a STOP instruction.

RTE (return from exception) allows the CPU to return from exception processing in an organized manner. During exception processing, the MC68020 takes a snapshot of its internal registers, SR and PC, and automatically saves them in the supervisory stack. The RTE pulls those values from the supervisory stack upon completion of the exception service routine. Exception processing will be covered in Chapter 7.

The MOVE, MOVEC, and logical instructions must be initiated under the supervisory mode when operated on protected resources like USP, SR, Rc, and Rn. Otherwise, the CPU will enter the exception processing mode.

Trap generation instructions (TRAP, TRAPV, and TRAPcc) cause the PC to be loaded from several designated program memory locations. The TRAPs are generated automatically by the CPU under certain error conditions, or can also be triggered via software control.

The CPU initiates exception processing as a result of the execution of a TRAP instruction. A vector number is generated by the CPU to reference the exception vector specified by the low-order 4 bits of the instruction. Thus, TRAP can be used to generate any of 16 different software interrupts. The trap instruction is simply written as

TRAP #n

where 'n' represents the trap vector number that is to be used by the CPU to locate the starting point of the exception processing routine. In fact, TRAP #n turns out to be a powerful systems programming tool for enlarging the instruction repertoire.

The TRAPV (trap on overflow) instruction tests the overflow (V) flag in the condition code register, and traps to a specific memory location "01C hex" if V is clear. If V is clear, execution continues with the next sequential instruction.

The CHK (check register against bounds) is a conditional trap-generation instruction. It jumps to memory location "018" hex if the register's value is less than zero or greater than an addressed 'upper bound' operand. This is useful in checking the length of an array. Each time we need to access the array, we can execute the CHK instruction to ensure that the array's bounds have not been violated.

3.5.8 Binary Coded Decimal Operations

The MC68020's arithmetic instructions assume that numbers are based on two's complement binary representation. The two's complement number is arrived at by first complementing the original binary number and than adding 1 to it. Such assumptions cannot be made for business applications. Here, the binary representation of numbers is inadequate to implement the floating-point numbers because accuracy to two decimal places is required. Today, many high-level languages support the decimal representation of numbers, to meet the business needs. Binary Coded Decimal (BCD) instructions are designed to facilitate the construction of a library of routines for decimal arithmetics.

A BCD digit has a range of 0 through 9, which is represented by four bits. A BCD digit is expressed as

$$D = B3(2^3) + B2(2^2) + B1(2^1) + B0(2^0)$$

where Bi is a binary digit of 0 or 1.

Since a BCD digit requires four bits, two BCD digits can be grouped and stored as a single byte. This form is called the packed BCD representation. In an unpacked BCD number, a single BCD digit is stored as a byte where the high-order four bits are stored as 0s. Algorithms for the multiplication and division of BCD numbers typically use the unpacked representation.

The five instructions to support operations on BCD numbers are shown in Figure 3.22.

ABCD (add decimal with extend)
NBCD (negate decimal with extend)
SBCD (subtract decimal with extend)
PACK
UNPACK

Three of the five instructions are used to operate on BCD values. The rest facilitate the conversion of byte-encoded numerical data, such as ASCII or EBCDIC strings to BCD and vice versa. These instructions include the X bit in the operand, and only change the Z bit if a nonzero result is generated. For this reason, the programmer must initialize X=0 and Z=1 before the use of the first BCD instruction in their programs. This can be achieved via MOVE #4, CCR.

ABCD (add decimal with extend) utilizes BCD arithmetic to add a source operand to the destination operand, including the extend bit, and stores the result in the destination address. The assembler syntax is:

```
ABCD.B  Dm, Dn
ABCD.B  -(Am), -(An)
```

Instruction	Operand Syntax	Operand Size	Operation
ABCD	Dn, Dn −(An),−(An)	8 8	$source_{10}$ + $destination_{10}$ + X → destination
NBCD	< ea >	8	0 − $destination_{10}$ − X → destination
PACK	−(An), −(An) #< data > Dn, Dn, #< data >	16 → 8 16 → 8	unpacked source + immediate data → packed destination
SBCD	Dn, Dn −(An),−(An)	8 8	$destination_{10}$ − $source_{10}$ − X → destination
UNPK	−(An), −(An) #< data > Dn, Dn, #< data >	8 → 16 8 → 16	packed source → unpacked source unpacked source + immediate data → unpacked destination

FIGURE 3.22. Binary coded decimal instructions. (Courtesy Motorola Inc.).

The packed BCD operands are interpreted in two ways: a) When two registers are involved, then the operands are added in the destination data register along with the value of the X flag; and b) When an operation involves memory to memory, transfer addresses are first decremented, the source operand byte is added to the destination operand, along with the X flag, and then the sum is placed on the destination address.

Example: Write a program that takes the current sales figure from a memory location pointed to by A0, and calculates the running total of sales, and save the result using A1 as the pointer. The program should trigger an error routine if it finds that the running total exceeds 999999 (because we are a small shop).

```
        MOVE.W   #4, CCR        ; set flags X=0 and Z=1
        MOVEQ    #2, D0         ; set the iteration limit
        ADDQ.W   #3, A0         ; load the string size for current
                                ; sales figure
        ADDQ.W   #3, A1         ; load the string size of the
                                ; running total for the shop
LOOP:   ABCD     -(A0), -(A1)   ; calculate the current total
        DBF      D0, LOOP       ; complete the addition
        BCS      ERROR          ; total exceeds the possible limit
        :
        :
ERROR:  :                       ; initiate error processing
```

NBCD (negate BCD with extend) subtracts both the destination operand and the value of the extend (X) flag from zero, and saves the result at the destination operand. NBCD produces a ten's complement of the destination if the extend bit is zero, or a nine's complement if the extend bit is one. The syntax is:

NBCD.B <EA>

where the condition codes reflect the result as:

X = Set the same as the carry bit
N = Undefined
Z = Cleared if the result is nonzero, unchanged otherwise
V = Undefined
C = Set if a decimal borrow occurs, cleared otherwise

SBCD (subtract decimal with extend) subtracts the contents of the source operand, including the value of the X bit, from the contents of the destination operand and saves the result at the destination operand. The syntax is:

SBCD.B Dm, Dn
SBCD.B -(Am), -(An)

The condition codes are set as follows:

X = Set the same as the carry bit
N = Undefined
Z = Cleared if the result is nonzero, unchanged otherwise
V = Undefined
C = Set if a decimal borrow occurs, cleared otherwise

The instruction has two modes:

1. Data register to data register mode: The data registers specified in the instruction contain the operands.
2. Memory to memory mode: The address registers specified in the instruction access the operands from memory, utilizing the predecrement addressing mode.

The memory to memory mode facilitates multibyte BCD subtraction, since a string of BCD digits representing a decimal number would normally be stored with the least significant digits as right-justified at the higher memory address.

Example: Assume that a store needs to issue a refund to a customer. The amount to be refunded is passed to the program via register D0, and the total through register D1. Before issuing a refund, we must check whether there is enough money in the cash register to cover the refund. Refunds are to be issued if cash is available, otherwise notify the manager.

```
        MOVE.W   #4, CCR     ; set Z=1 and X=0
        SBCD.B   D0, D1      ; subtract the refund
        BCS      NEG         ; if C=1, not enough funds
        BRA      REST        ; go ahead issue a refund
NEG:    :                    ; call manager
        :
REST:   <rest of the program> ; open the till and display
```

Example: Subtract two positive BCD bytes (within the range of 00 through 99). The first can be found in D0, and the other in D1. If the result of the subtract operation is found as positive, then save the result in D0 and place "+" (ASCII #$2B) at location A0. Otherwise, place the "-" (ASCII #$2D) at location A0.

```
MOVE.W   #4, CCR       ; initialize X=0 and Z=1
SBCD.B   D1, D0        ; D0=D0-D1
BCS      MINUS         ; result is negative
MOVE.B   #$2B, (A0)    ; result is positive, write "+"
BRA      NEXT          ; continue on to the next part
```

```
                                        ; of the program
MINUS:  ANDI      #$EF, CCR             ; make sure that X is reset
        NBCD.B    D0                    ; change D0-D1 in D0 to D1-D0
        MOVE.B    #$2D, (A0)            ; put negative sign "-" at (A0)
```

PACK: PACK adjusts and packs the low four bits of each of two bytes into a single byte. It is designed to facilitate the conversion of ASCII numbers to BCD. The assembler syntax is

```
PACK   -(Am), -(An), #<adjustment>
PACK   Dm, Dn, #<adjustment>
```

When operands are found in the data registers, then the adjustment is added to the content of source register. Bits [11:8] and [3:0] of the intermediate result are concatenated and placed in bit [7:0] of the destination register. The remainder of the destination register is unaltered.

With the predecrement addressing mode, the two bytes from the source are fetched and concatenated, along with the adjustment word. It then writes the second and the four hex digits (bits [11:8] and [3:0]) of the result to the destination byte.

Example: Develop a program that receives a string of ASCII characters from a buffer ending at 6000H, and puts the extracted digits at a buffer ending at location 7000H. The size of the ASCII buffer is 24.

```
        MOVE.L    #24, D0                ; set up the digit counter
        MOVEA.L   #$6000, A0             ; ending of ASCII character
                                         ; buffer
        MOVEA.L   #$7000, A1             ; ending of the digit buffer
LOOP:   PACK      -(A0), -(A1), #$CFD0   ; convert from ASCII
        DBF       D0, LOOP               ; continue until D0 contains -1
```

UNPK: The unpack instruction places the two BCD digits from the source operand byte into the lower nibbles of two bytes, and places zero bits in the most significant nibbles of both bytes. It then adds the adjustment to it before writing the word to the destination address.

The assembler syntax is

```
UNPK   Dm, Dn, #<adjustment>
UNPK   -(Am), -(An), #<adjustment>
```

In the case where both operands are in data registers, the instruction unpacks the source register contents, adds the extension word, and places the result in the destination register without modifying the high word.

In the second case, the specified addresses are decremented first, the operand is fetched, the adjustment word is added to the operand, and the result is written into the decremented addressed location.

3.6 MULTIPROCESSOR OPERATIONS

Communications between the MC68020 and other processors in the system is accomplished by executing a special set of instructions. The instructions fall in two categories: the multiprocessor instructions; and the coprocessor instructions.

The multiprocessor instructions are designed to enable the MC68020 to communicate with another CPU in the system. They are shown in Figure 3.23. TAS, CAS and CAS2 are created to facilitate multiprocessing capabilities. Using these, the MC68020 does not release the bus mastership, because they initiate indivisible read-modify-write bus cycles.

The TAS (Test and Set) instruction is designed to implement a "semaphore," a mechanism for programs (or tasks) that are executing independently to synchronize their activities. The basic concept of the semaphore is that a resource (for

Read-Modify-Write Instruction	Operand Syntax	Operand Size	Operation
CAS	Dc, Du, < ea >	8, 16, 32	destination — Dc → CC; if Z then Du → destination else destination → Dc
CAS2	Dc1 : Dc2, Du1 : Du2 (Rn) : (Rn)	16, 32	dual operand CAS
TAS	< ea >	8	destination — 0; set condition codes; 1 → destination(7)

Coprocessor Instruction	Operand Syntax	Operand Size	Operation
cpBcc	< label >	16, 32	if cpcc true then PC + d → PC
cpDBcc	< label >, Dn	16	if cpcc false then Dn - 1 → Dn if Dn ≠ -1, then PC + d → PC
cpGEN	User defined	User defined	operand → coprocessor
cpRESTORE	< ea >	none	restore coprocessor state from < ea >
cpSAVE	< ea >	none	save coprocessor state at < ea >
cpScc	< ea >	8	if cpcc true, then 1's → destination; else 0's → destination
cpTRAPcc	none #< data >	none 16, 32	if cpcc true then TRAPcc exception

FIGURE 3.23. Multiprocessor operations. (Courtesy Motorola Inc.).

example, a file, a block of memory, an I/O device, etc.) can be shared by various programs or processors. Various flags, priority, or queueing schemes are used to determine who gets what resources and for how long. Typically, a program examines the availability of the resource, and grabs it if it is available, by setting an agreed value in an assigned status bit or byte. The resource is eventually relinquished by clearing the status flag, so that other programs can access it.

The TAS instruction helps system designers to implement such a scheme. Without this instruction, two tasks working independently might get into the following situation during printing. Task A may test the printer status flag and find it to be zero (i.e., the printer is idle), and then get interrupted before setting it to 1 (i.e., the printer is taken). Task B may also test the same flag, find it to be zero, and set it to 1. Task A would then proceed as if no flag has been set, at the end of the interrupt operation. Now both tasks will think that they have the printer. The TAS instruction guards against such situations, where the MC68020 will not respond to an interrupt or bus request from external devices during the execution of the instruction.

The TAS first tests a value in memory or a register, using the usual addressing modes, and compares that value against '0'. The current value of the operand determines whether the CPU will set the N and Z flags in the CCR or not. If the operand value is '0', then Z is set to 1. If the most significant bit (bit-7) of the operand is found as 1, then N is set to 1. Finally, the TAS unconditionally sets the destination operand's sign bit (i.e., bit-7) to 1, forcing the byte to be negative. During these 'read-modify-write' activities, the processor will not release the system bus. This is known as an 'indivisible read-modify-write' operation. For example, let us assume that a byte at address $2FFF signals the following event to all user programs:

($2FFF) = 00 means that the printer is free
($2FFF) = not zero means that it is busy printing

One can implement a print driver via the TAS instruction, where a program will test the printer status and will do something else if it is busy, otherwise it will set the memory location $2FFF to '$80' and occupy the printer. This is shown below:

```
WAIT:    TAS   $2FFF      ; Test and set the CCR. Then
                          ; set it to $80
         BNE   WAIT       ; printer is busy do some other thing.
         :                ; Else, the printer is free, use it
         :
RELPR:   CLR.B  $2FFF     ; release the printer
         :                ; by setting the value to $00.
         :
         :
```

In practice, a print driver will be more complex. System designers will create virtual printers and require the programs to spool their printing job into them. The print driver will accept jobs from the virtual printers via a print scheduler, and complete them. At the completion of printing, the print driver will signal the virtual printer through the print scheduler that the printing has been completed. The print driver will then accept the next job from the print scheduler.

CAS and CAS2 (Compare And Swap) instructions also guard against resource contention. As explained, TAS supports a simple guarding mechanism. CAS extends the concept a step further, and CAS2 goes two steps further. The format for the instructions is:

CAS Dc, Du, <EA>

where

Dc—Identifies the data register that contains the value to be compared against the operand in the memory.
Du—Specifies the data register whose content will be written to the memory operand location if the comparison is successful.

CAS compares Dc and the operand found at the location pointed by "EA." If they are the same, then Du replaces the operand, otherwise it will try again.

CAS2 Dc1:Dc2, Du1:Du2, (Rn1):(Rn2)

where

Dc1 and Dc2 fields—Identify the data registers that contain 'test values' to be compared against the two memory operands. If Dc1 and Dc2 happen to be the same data register, and the comparison fails, then the content of the memory location (operand 1) will be put into the data register Dc.
Du1 and Du2 fields—Specify the data registers that contain the updated values to be written to the first and second memory operand locations in the event of a successful comparison.
Rn1 and Rn2 fields—Identify the numbers of the registers that contain the addresses of the first and second memory operands, respectively. If the operands overlap in the memory, the results of memory update are undefined.

CAS2 performs two comparisons and, in the event of any mismatch, memory operands will not be updated. This instruction only allows register (address or data) indirect modes. The use of the instructions are explained in the following example.

Example: Two tasks want to submit jobs to a printer in a multitasking environment. The access to the printer is achieved through a global variable 'pr—queue', and print buffers get assigned based on the value of the variable 'pr—value'. For example, let us assume that Task A loads the current value of pr—queue into D0, and then performs its update operations in D1. The time slice of Task A expires immediately. Task B then comes along and updates the pr—queue, and gets the same buffer (that Task A should have received) and writes into it. At the end of the Task B's time slice, Task A returns and writes into the same buffer written to by Task B. Such a situation can be avoided using the CAS instruction, which is illustrated in the following:

```
        MOVE.L  pr_queue, D0        ; get the buffer number
LOOP:   MOVE.L  D0, D1              ; maintain a copy
        ADDQ.L  #1, D1
        CAS.L   D0, D1, pr_queue    ; check whether someone
                                    ; grabbed the buffer.
        BNE.S   LOOP                ; if yes, the buffer is lost
        :
        :
        :                           ; continue
```

The update is performed only if D0 remains unchanged during the update operation. Otherwise, the update is not performed. The current value of pr_queue is then loaded in D0 and another attempt is made.

The CAS2 instruction enhances the capability of the CAS instruction. CAS2 performs two comparisons and updates two variables when the results of the comparisons are equal. CAS2 copies new values into the destination addresses, if the results of both comparisons are equal; otherwise, the contents of the two destination addresses are copied into the compare operands.

The following example [MC68030 1989] illustrates the use of CAS2, where it deletes an element from a linked list.

```
        LEA     HEAD, A0            ; Get the starting address of the link list
        MOVE.L  (A0), D0            ; keep a copy in D0 for comparison
LOOP:   TST.L   D0                  ; check whether the list is empty
        BEQ     EMPTY               ; if empty, do not remove anything
        LEA     (NEXT,D0), A1       ; get the address of the forward link
        MOVE.L  (A1), D1            ; copy the forward link value into D1
        CAS2    D0:D1,              ; if nobody has deleted
                D1:D1, (A0):(A1)
```

```
                                        ; the entry, then update the HEAD and
                                        ; forward pointers
        BNE        LOOP                 ; failed to delete the entry nothing to delete
EMPTY:
```

The algorithm works as follows: The LEA loads the effective address of HEAD into A0. The next instruction, MOVE, loads the address of the pointer HEAD into D0. The TST then checks for an empty list. If the list is found empty, then the program branches to label it EMPTY. Otherwise, it executes the next instruction from the instruction flow. The second LEA loads the address of the forward link into A0 from the NEXT pointer into the newest element on the list. The MOVE instruction then loads the pointer's newest element of the list into A1. The next MOVE then loads the contents of the pointer into D1. CAS2 then compares the address in A1. If no element has been inserted or deleted to another program while this ''program'' has been executing, then the results of these comparisons are equal, and the CAS2 instruction stores the new value into the location HEAD. If an element has been inserted or deleted, the CAS2 instruction loads the new address from the location HEAD into D0, and BNE branches the execution flow to instruction TST, in order to initiate another try.

The coprocessor instructions shown in Figure 3.23 are used in support of a coprocessor protocol. The protocol and the use of these instructions are explained in detail in Chapter 9.

Chapter 4

Hardware Design for the MC68020 Microprocessor

The MC68020 is the fourth processor to be added to the MC68000 family. The original 16-bit architecture is enhanced to include support for virtual memory and virtual machines, as well as coprocessor support. This chapter will illustrate the pins and their usage in designing the hardware interface. The block diagram of Figure 4.1 describes ten groups of signal pins, based on their functionalities. They are: address bus, data bus, transfer size bus, asynchronous bus control, function codes, cache control, interrupt control, bus arbitration control, bus exception control, and excitation control.

It is through these signal lines that the MC68020 is interfaced to external circuitry such as memory, peripheral devices, and the coprocessor. It is important to note that the MC68020 does not have any dedicated bus for the support of synchronous peripherals.

4.1 PROCESSOR STATUS PINS

The function control bus is composed of three signal lines: FC0, FC1, and FC2. These three state outputs identify the address space and the current bus cycle. They signal to the external circuitry which type of bus cycle is in progress. That is, whether data or a program is being accessed, and whether the CPU is in the user state or supervisory state, or in CPU space. The terms state and mode are used interchangeably throughout this book.

Table 4.1 lists the function code assignments. The address spaces 0 and 4 are reserved by Motorola for future use. Hardware designers should not use them, to maintain future compatibility. The address space 3 is reserved for the user,

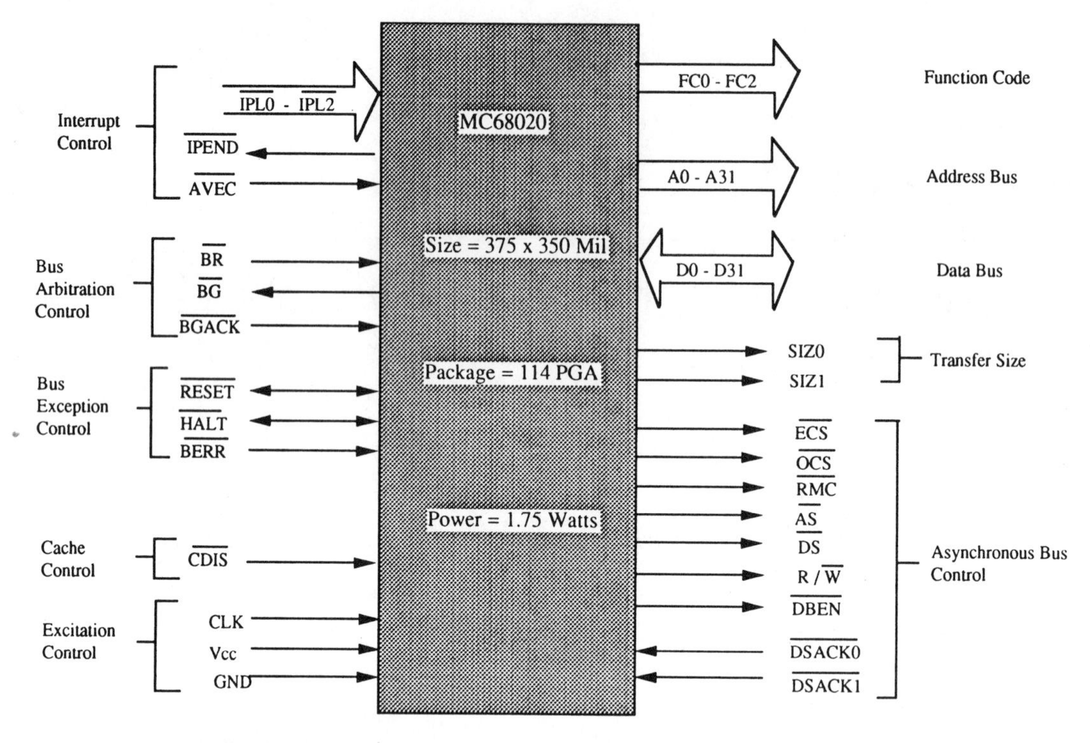

FIGURE 4.1. Functional signal of the MC68020, including its pins.

which means that Motorola will never use it in their future product enhancements. The hardware designer can use this address space in any way he/she sees appropriate. Notice that the function code 111 has a special name called the CPU space.

The MC68020 asserts the function code signals at the beginning of each bus access, and they remain valid until the beginning of the next read or write cycle. The function code signals FC1 = 1 and FC0 = 1 indicate that program space is being accessed where the program counter contains the source address, or that

TABLE 4.1. Function Code Assignments.

FC2	FC1	FC0	Cycle Type
0	0	0	(Undefined—Reserved for Motorola)
0	0	1	User data space
0	1	0	User program space
0	1	1	(Undefined—Reserved for user)
1	0	0	(Undefined—Reserved for Motorola)
1	0	1	Supervisory data space
1	1	0	Supervisory program space
1	1	1	CPU space

TABLE 4.2. Address Space of the MC68020 during CPU Space Cycles.

Address of Function	Function Description	Encoded Value in the Address Lines A19–A16
0	Breakpoint acknowledgment	0000
1	Access level control	0001
2	Coprocessor communication	0010
3	Reserved	0011
4	Reserved	0100
5	Reserved	0101
6	Reserved	0110
6	Reserved	0110
14	Reserved	0110
15	Interrupt acknowledgment	1111

reset vectors are being fetched by the CPU. FC1 = 0 and FC0 = 1 indicate that the data space is being accessed. In the data space, most of the operands are read (i.e., the PC is not the originator of the address) or written into, or vectors other than the RESET are being fetched.

A new address space has been added to the MC68020 called the CPU space. The processor puts 111 on the function code lines as it enters into this space. Four of the sixteen available functions are defined within the CPU space. These are the: interrupt acknowledge, breakpoint acknowledge, access level control (module call and return), and coprocessor communication cycles. The rest are reserved by Motorola for future use. The functions are listed in Table 4.2, along with their encoded values, which appear on the address lines A19 through A16.

4.2 REQUIREMENTS FOR INTERFACING MEMORY AND PERIPHERAL DEVICES

The control of the MC68020's bus is asynchronous. That is, once a bus cycle is initiated by the CPU to read (input) or write (output) instructions or data over the bus, it is not completed until a response is received from the memory or peripheral subsystems. This response may come as an acknowledge signal, which indicates the device's size and transfer status, i.e., complete or incomplete. Based on this information, the processor decides whether to complete its current bus cycle or not. This allows flexibility for interfacing with slower memory or peripheral devices.

No low-level protection mechanisms are provided by the MC68020 for the physical memory or peripheral devices. As a result, the designer must incorporate them externally. The function code signals can be used with the address bus signals to write-protect certain areas of the address space. Such schemes are used by the coprocessor to ensure that certain operations are performed during the desired CPU state. The processor validates the address on the address bus by asserting $\overline{AS}$. Thus $\overline{AS}$, FC2, FC1, FC0, and the address lines are used to generate the device or memory select signals. This allows the designer to create several address spaces. One such scheme is shown in Figures 4.2a and 4.2b, where the address space from $00000000 through $7FFFFFFE is reserved for supervisory software (the operating system), and the space from $80000000 to $FFFFFFFF is kept for user software. The extra hardware guarantees that the user software will never be able to write over the data in the supervisory space, even by mistake.

4.2.1 Synchronization

The MC68020 interacts with external memory/devices in an asynchronous manner. But it must synchronize these signals internally because the processor's internal operation is completely synchronized. The 'sync delay' is the total time required for the processor to sample an external asynchronous signal, determine the signal's state, and synchronize the signal's state with respect to the internal clock. This is shown in Figure 4.3.

A sample window exists during which the processor latches the level of the asynchronous input signal. All input signals must be held stable during this window. The CPU samples the input signals at the falling edge of the clock, validates

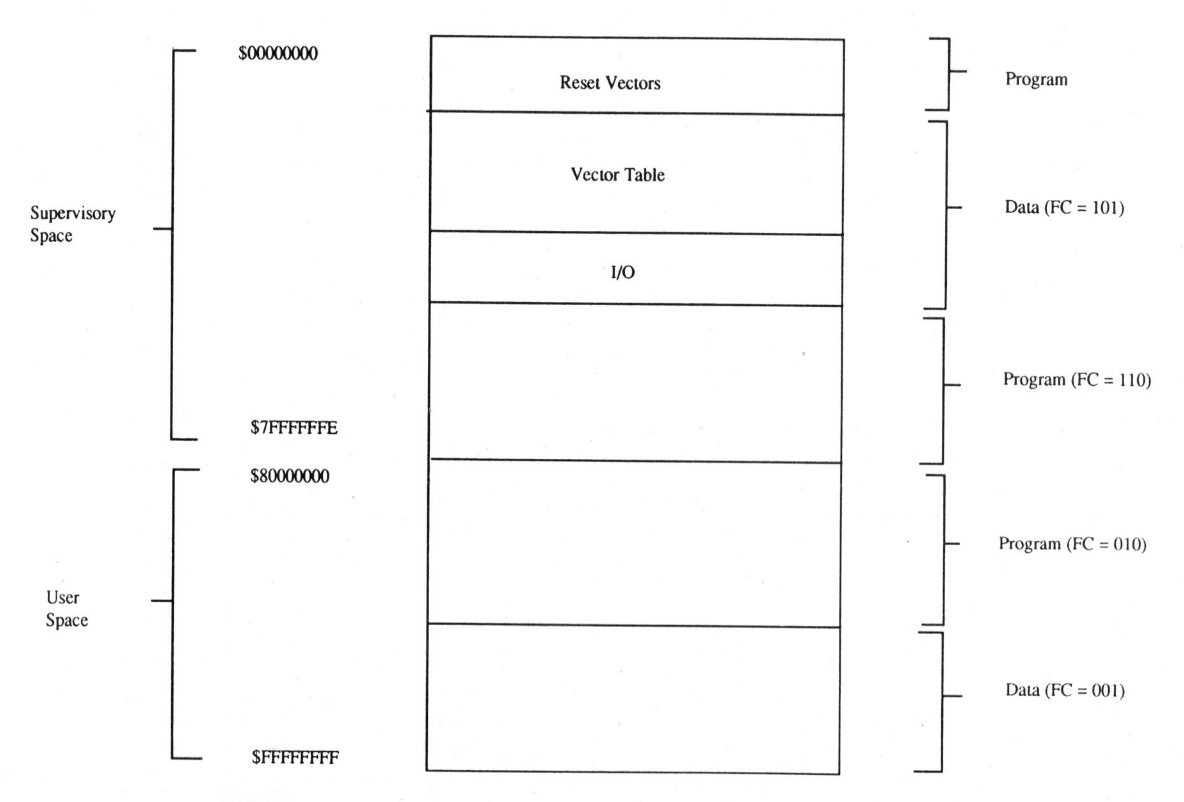

FIGURE 4.2a. Memory map utilizing features of the function control pins.

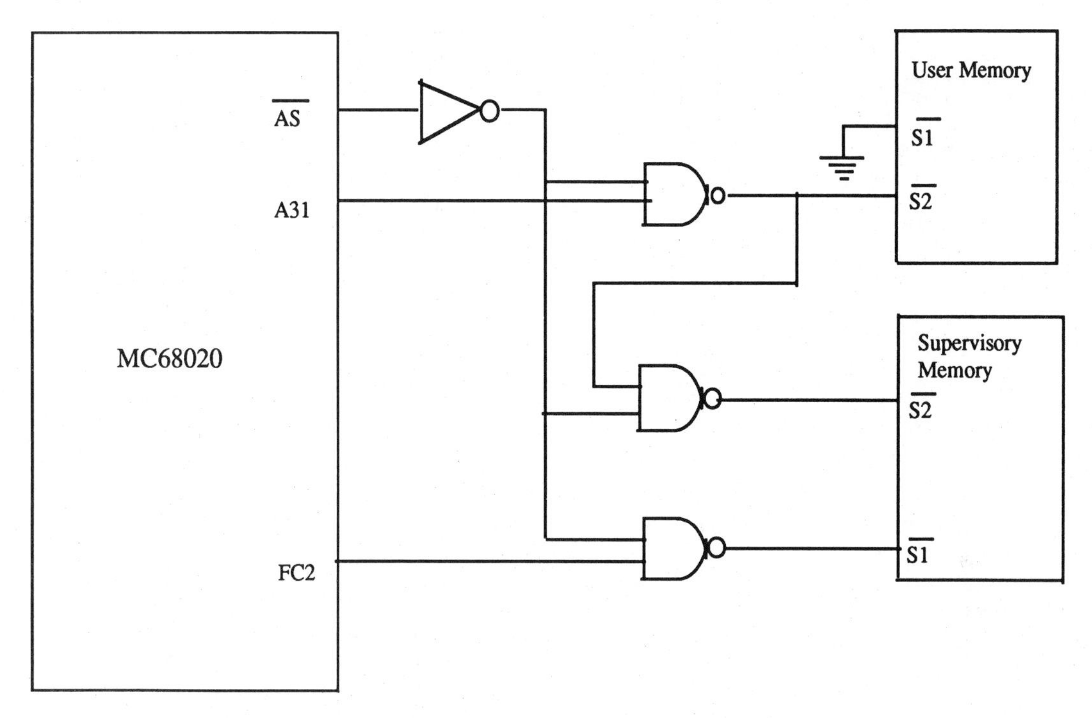

FIGURE 4.2b. Memory protection logic utilizing external logic.

them internally on the leading edge of the internal clock, and acts on them on the falling edge of the clock. If an input signal makes a transition during the sample window, the level recognized by the CPU is not predictable. However, the CPU will always resolve the latched level internally to a state of 1 or 0 before acting on signals.

4.2.2 Memory and Device Organization

In most MC68020-based systems, memory is organized as words (16 bits) or long-words (32 bits), and rarely as bytes (8 bits). Under byte organization, the memory is treated as a sequence of 8 bits, and reading/writing refers to a single byte, where multiple bytes will require several bus references. In word organization, the basic data-type movement involves 16 bits, and memory can still be byte addressable. Under long-word organization, memory is also byte addressable, but the data-type movement involves 32 bits. For efficient data transfer, words and long-words should be aligned on even addresses in memory. If they are not, then the CPU will complete the transfer over several bus accesses. As far as the high-level language programmer is concerned, this underlying protocol is automatically performed by the system. However, the assembler language programmer must be aware of the protocol and the implications of misalignment on the system's performance.

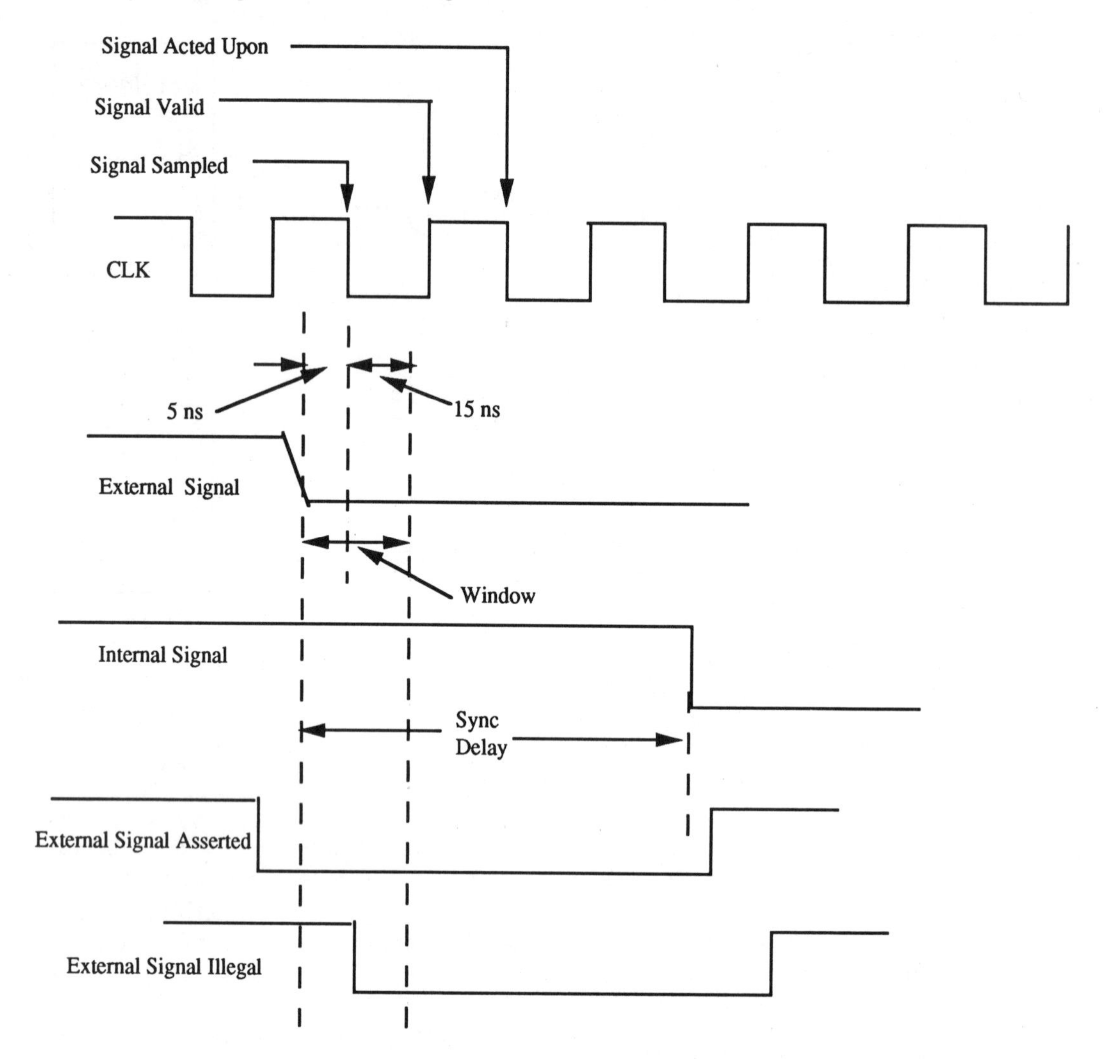

FIGURE 4.3. Synchronization of asynchronous input signals and their recognition.

Using the MC68020, the system designer must follow a definite convention in order to connect peripheral devices and memory to build a system, where peripheral devices are memory mapped. This means that memory and peripherals do not have separate address spaces. Instead, the designer allocates a part of the memory address space to the peripheral devices. Therefore, both memory and peripherals are accessed in same way through the asynchronous bus interface. Thus, we will use a common term "PORT" to identify a peripheral or memory location. A port must be connected to a particular section of the data bus. This is illustrated in Figure 4.4. A 32-bit port is tied to data pins D31 through D0, a 16-bit port is connected to data pins D31 and D16, and an 8-bit port is always attached to data pins D31 through D24. This left-justification of the port is re-

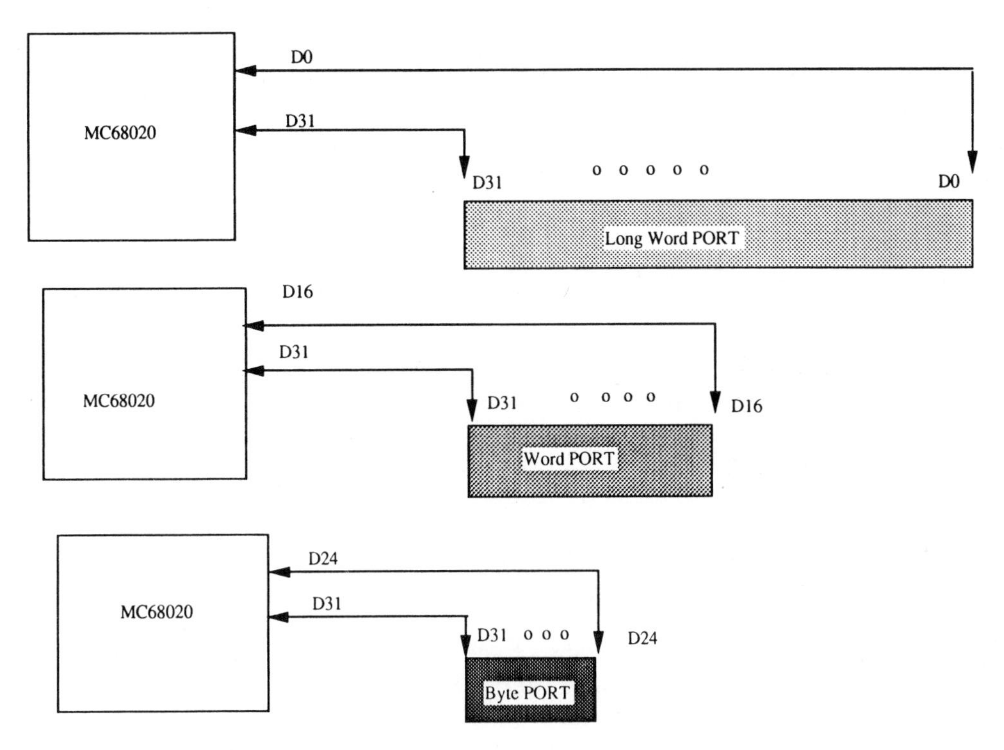

FIGURE 4.4. PORT connections to the data bus.

quired, because the most significant byte of an operand remains at the lowest address, and is accessed first. The least significant part of the operand is resident at the highest address, and is accessed last. During a bus cycle, the MC68020 always attempts to transfer 32 bits of data over the bus by assuming a long-word size port at the beginning of a bus cycle. The above scheme helps the processor to minimize the number of bus cycles required to transfer data to the 8-, 16-, and 32-bit ports.

4.2.3 Dynamic Bus Sizing

The MC68020 allows operand transfers to or from 8-, 16-, or 32-bit ports by dynamically determining the port size during each bus cycle. When the MC68020 begins a bus cycle, it does not have any prior knowledge of the port size of the slave device. During the asynchronous bus transfer, the slave follows fairly simple rules. The BYTE port always asserts $\overline{\text{DSACK0}}$, the WORD port asserts $\overline{\text{DSACK1}}$, and the LONG-WORD port asserts both $\overline{\text{DSACK0}}$ and $\overline{\text{DSACK1}}$. As long as both $\overline{\text{DSACK0}}$ and $\overline{\text{DSACK1}}$ are high, the CPU continues to increment its internal clock cycle and enters into the 'software wait cycle'. The MC68020 has no internal timeout mechanism, and will keep on inserting wait cycles endlessly.

During a bus cycle, the MC68020 declares how many bytes it wants to transfer

TABLE 4.3. Size Pin Assignments.

SIZ1	SIZ0	Operand Size
0	1	1 byte
1	0	2 bytes or word
1	1	2 bytes
0	0	4 bytes or long-word

by encoding that information on size pins (SIZ0 and SIZ1 signal lines) whose encoding is listed in Table 4.3. However, an actual transfer will depend on the port width and operand alignment. For example, if the CPU wants to transfer a long-word to a word-size PORT, the SIZx will contain '00' (i.e., 4 bytes are to be transferred). In reality, however, only 2 bytes will be transferred, because the PORT size is 16 bits. During the second bus cycle, the CPU will encode '10' to SIZx, and the last 2 bytes will be transferred. In order to interface with the MC68020, external enable circuitry should be designed to generate corresponding enable signals, using size pins and address pins A0 and A1, so that the correct port gets activated during the data transfer. For example, 4-byte enable signals need to be generated for a long-word device. A 2-byte enable signal would be generated for a word-size device.

During an operand transfer, the responding PORT declares its size, utilizing the $\overline{\text{DSACKx}}$ pins. The status of these pins also acts as an acknowledge signal, which tells the MC68020 that it should complete its current bus cycle. The encodings of the $\overline{\text{DSACKx}}$ pins to signal the current port size are listed in Table 4.4.

4.2.4 Byte Addressing

The routing and duplicating MUX (multiplexer) internal to the MC68020, in conjunction with the size pins(SIZ1 and SIZ0) and address lines (A0 and A1), determines where to put the individual operand bytes on the CPU's data pins during a bus cycle. This is illustrated in Figure 4.5. Assuming that the address pins A31–A2 are pointing to a 4-byte block of a long-word PORT, then A1 and A0 represent the most significant byte to be accessed from the base reference address. The offset encoding is listed in Table 4.5. The size pin indicates the number of bytes to be accessed starting at that base address. For example, when A1 = 1, A0 = 0.

TABLE 4.4. $\overline{\text{DSACKx}}$ Pin Assignments.

$\overline{\text{DSACK1}}$	$\overline{\text{DSACK0}}$	Encoded Response
1	1	No response—Insert software wait to current bus cycle
1	0	8-bit size data port—Complete the bus cycle
0	1	16-bit size data port—Complete the bus cycle
0	0	32-bit size data port—Complete the bus cycle

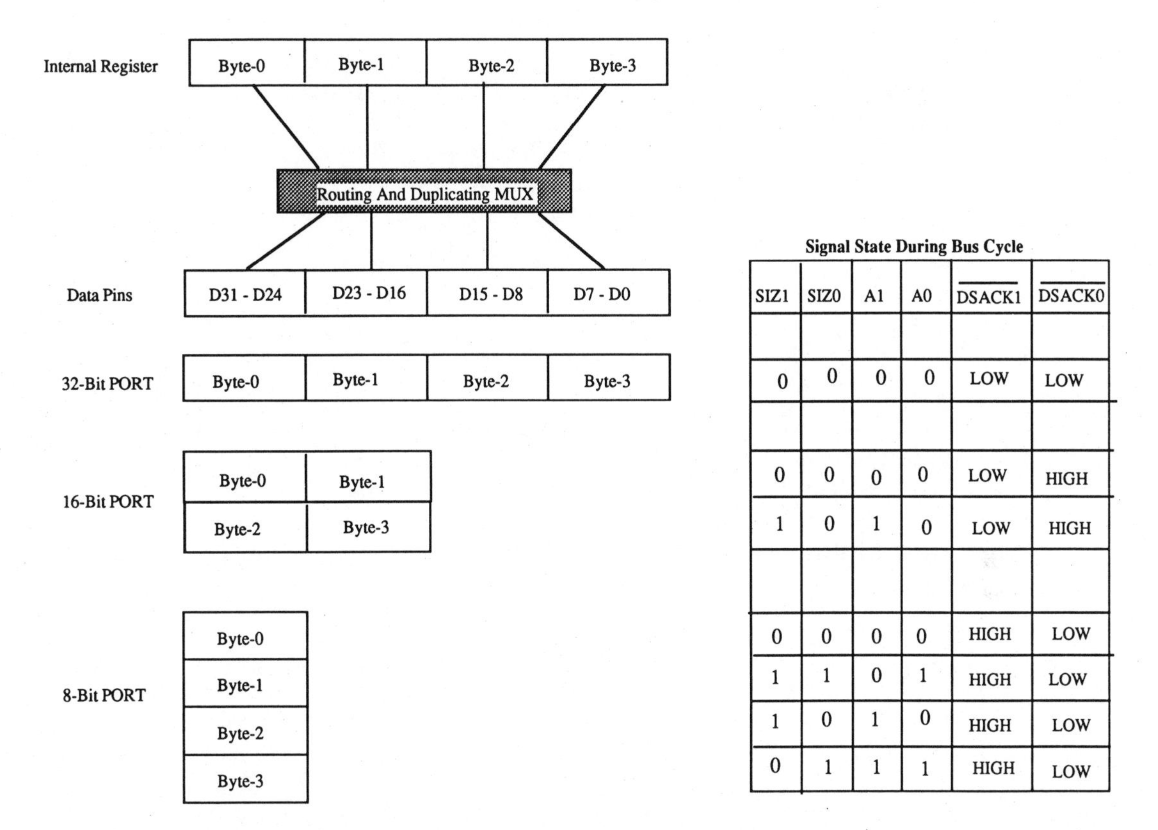

FIGURE 4.5. Aligned long-word transfer from various sizes of PORTs.

Then, byte-2 of the 4-byte block is the most significant byte of the operand to be accessed. If the size pins are zero, then 4 bytes are to be accessed, starting at Byte-2. Thus, Byte-2, Byte-3, Byte-4, and Byte-5 will be transferred. Two bus cycles will be required to access all 4 bytes.

For word size port, address lines A31–A1 point to a 2-byte block, where A0 represents the block offset. This offset points to the most significant byte of the operand to be accessed. The size pin indicates the number of bytes to be accessed, starting with the most significant byte. In the case of A0 = 1, byte-1 of the 2-byte block is the most significant byte to be accessed. If the size pins are zero, then 4

TABLE 4.5. Offset Coding.

A1	A0	Offset from the Base Address
0	0	0 bytes offset from base
0	1	1 byte offset from base
1	0	2 bytes offset from base
1	1	3 bytes offset from base

bytes will be accessed, starting at byte-1. Thus, byte-1, byte-2, byte-3, and byte-4 will be transferred. Three bus cycles will be required to transfer all 4 bytes.

Byte addressing of a bit port is straightforward. The address line A31–A0 points to the most significant byte to be accessed. The size pin indicates the number of bytes to be transferred from the byte-block. Using this convention, one should be able to deduce which byte is to be accessed for any combination of size and address values.

4.2.5 Normal Data Transfer Operation

Using the information on the size and the address pins, one can deduce where the routing and duplicating MUX places the information to and from the internal registers to the external CPU data pins. This is illustrated in Figure 4.6. It should be noted that the byte operand always comes from the low-order byte (i.e., byte-3) of the internal register. Likewise, the word operand comes from the low-order word (that is, byte-2 and byte-3) of the processor's internal registers. However,

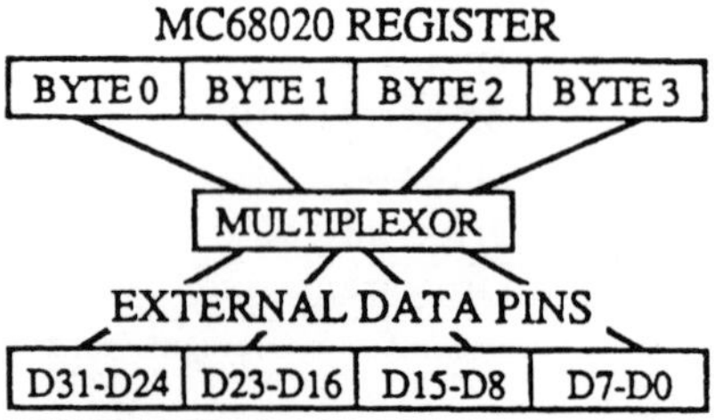

MC68020 REGISTER

BYTE 0	BYTE 1	BYTE 2	BYTE 3

MULTIPLEXOR

EXTERNAL DATA PINS

TRANSFER SIZE	SIZ1	SIZ0	A1	A0	D31-D24	D23-D16	D15-D8	D7-D0
BYTE	0	1	X	X	BYTE 3	BYTE 3	BYTE 3	BYTE 3
WORD	1	0	X	0	BYTE 2	BYTE 3	BYTE 2	BYTE 3
WORD	1	0	X	1	BYTE 2	BYTE 2	BYTE 3	BYTE 2
3 BYTE	1	1	0	0	BYTE 1	BYTE 2	BYTE 3	BYTE 0*
3 BYTE	1	1	0	1	BYTE 1	BYTE 1	BYTE 2	BYTE 3
3 BYTE	1	1	1	0	BYTE 1	BYTE 2	BYTE 1	BYTE 2
3 BYTE	1	1	1	1	BYTE 1	BYTE 1	BYTE 2*	BYTE 1
LONGWORD	0	0	0	0	BYTE 0	BYTE 1	BYTE 2	BYTE 3
LONGWORD	0	0	0	1	BYTE 0	BYTE 0	BYTE 1	BYTE 2
LONGWORD	0	0	1	0	BYTE 0	BYTE 1	BYTE 0	BYTE 1
LONGWORD	0	0	1	1	BYTE 0	BYTE 0	BYTE 1*	BYTE 0

X = don't care
* = byte ignored on read, this byte output on write

FIGURE 4.6. Internal-to-external data routing performed by the MC68020. (Courtesy Motorola Inc.).

during a byte transfer, the MUX places byte-3 of the internal register to all 4 bytes of the external data bus. No matter what size port responds, byte-3 is already in the correct position. For word transfer, the MUX places byte-2 and byte-3 of the internal register on the external data bus according to the value of A0, which represents an offset to a 2-byte block. Even if A0 = 1, byte-2 and byte-3 are already in the correct position for any port size. That is, byte-2 will be transferred on D31-D24 if a byte port responds. Byte-3 on D23–D16 will also be transferred if a word port responds.

Consider the case where the MC68020 wants to transfer 4 bytes with A1 = 0 and A0 = 1. Byte-0 is on D31–D24 in case a byte port responds. Byte–0 is on D23–D16, byte-1 is on D15–D8, and byte-2 on D7–D0, in case a long-word port responds. Thus, the MUX has placed the content of an internal register to the proper external data pins with no assumption of the size of any data port. During the first bus cycle, the CPU positioned byte-3 on the data pins D31–D24. During this bus cycle, if a byte port responds, then byte-3 will be transferred. Byte-3 will also be transferred if a long-word port responds.

4.2.6 Aligned Transfer

The aligned transfer comes into the picture when the operand resides on the word boundary of the address space. The processor achieves maximum efficiency if it can perform its operations in this mode. An aligned transfer depends somewhat on the port size. During a data transfer, data from the external data pins is transferred into and out of the internal registers via the data and routing multiplexer hardware, as shown in Figure 4.5. This MUX routes the data to and from any registers to the CPU's external data pins. The MC68020 always tries to transfer the maximum amount of data during a bus access.

In the case of a long-word transfer, the MC68020 asserts 0 on pins SIZ0 and SIZ1. It also asserts 0 on pins A0 and A1 to indicate an aligned long-word transfer. During data transfer, the MUX routes byte-0 to D31–D24, byte-1 to D23–D16, byte-2 to D15–D8, and byte-3 to D7–D0, respectively. The 32-bit port responds by asserting both $\overline{\text{DSACK1}}$ and $\overline{\text{DSACK0}}$. This completes the bus transfer, because all 4 bytes are transferred.

For a word port transfer, all signals start out the same as before, with the same bytes to the same data lines. When the MC68020 detects that only $\overline{\text{DSACK1}}$ is asserted, it completes the bus cycle, having determined that only 2 bytes have been transferred. The CPU increments the address by 2 and decrements the size pin encoding by 2. Now the size pins indicate that the CPU intends to transfer two more bytes. The routing and duplicating MUX routes byte-2 to D31–D24 and byte-3 to D23–D16. During the second bus cycle, the CPU also makes no assumption as to the size of the port. In this case, the processor receives $\overline{\text{DSACK1}}$ and determines that the remaining 2 bytes have been transferred.

Let us consider what happens when the CPU transfers the same 32 bits to an 8-bit port. The first bus cycle starts the same way as before, when the byte port responds with /DSACK0, the processor determines that only one byte is transferred. The CPU then increments the address byte by 1, decrements the size pin encoding by 1. The next cycle is at one address higher and the size pins indicate that 3 bytes remain to be transferred. The routing MUX places byte-1 on D31–D24. Again $\overline{\text{DSACK0}}$ is asserted and CPU completes another bus cycle. For the last bus cycle, the size pin encoding indicates that one byte remains to be transferred. The routing MUX places byte-3 on D31–D24. When $\overline{\text{DSACK0}}$ is asserted, the processor determines that the last byte has been transferred.

4.2.7 Misaligned Long-Word Transfer

The MC68020 does not enforce any data alignment restrictions during operand transfer. Transfer may occur on any address boundary. For example, transferring a long-word to a 32-bit port with A0 = 1 or A1 = 1 is possible. Transferring a word to a word port or long-word port with A0 = 1 is also feasible. During a misaligned transfer, some performance degradation will occur, due to the multiple bus access that the CPU will make to transfer all bytes of an operand. However, the MC68020 requires the instruction stream to fall on a word boundary. As a result, if the program counter gets loaded with a value of A0 = 1, the processing of an illegal address exception will be initiated by the CPU.

Now, let us look at the events that may occur during a misaligned long-word transfer to a 32-bit port. This is illustrated in Figure 4.7. During the first bus cycle, the size pin indicates a transfer of 4 bytes. The address pin A0 causes the MUX to transfer byte-0 through byte-2 to be placed on external data pins D23 through D0. When the 32-bit port responds by asserting $\overline{\text{DSACK1}}$ and $\overline{\text{DSACK0}}$ as 0, the processor determines that three bytes are transferred. The address is then incremented by 3 and the size encoding is decremented by 3. The second bus cycle indicates an address 3 bytes higher and the size pins indicate that one byte needs to be transferred. When the size byte indicates 1 byte, the MUX routes byte-3 to all bytes of the data pins. Since A1 and A0 indicate the offset, byte-3 should only be written into D31–D24 of the slave. When the slave responds by asserting $\overline{\text{DSACKx}}$, the CPU determines that the byte has been transferred.

Again, it is interesting to observe what happens if the same misaligned long-word needs to be transferred to a 16-bit port. The first bus cycle is identical to the first bus cycle for the 32-bit port transfer. When the 16-bit slave responds by asserting $\overline{\text{DSACK0}} = 0$, the processor determines that only one byte has been transferred. The address is incremented by 1 and the size encoding is decremented by 1. The second bus cycle points at an address one byte higher and the size pins indicate 3 bytes to be transferred. As soon as the 16-bit port responds by asserting $\overline{\text{DSACK0}}$, the MC68020 determines that the next two bytes were transferred. The address is incremented by 2 and size is decremented by 2. The size encoding indicates that 1 byte is to be transferred, and the MUX routes byte-3 to all bytes of the data bus. The offset of 0 indicates that byte-3 should only be written to D31–D24 of the slave. Note that the port asserts the appropriate $\overline{\text{DSACKx}}$ to indicate its port size, independent of the transfer size encoding on the size pins.

During a misaligned transfer, care should be taken so that bytes are not overwritten. That is, during the long-word transfer to a 32-bit port, only 3 bytes are to be written during the first bus cycle, and only one byte during the next bus cycle. Similarly, some bytes must not be written to the 16-bit slave. Therefore, external hardware must be used to create 4 separate byte enable signals, one for

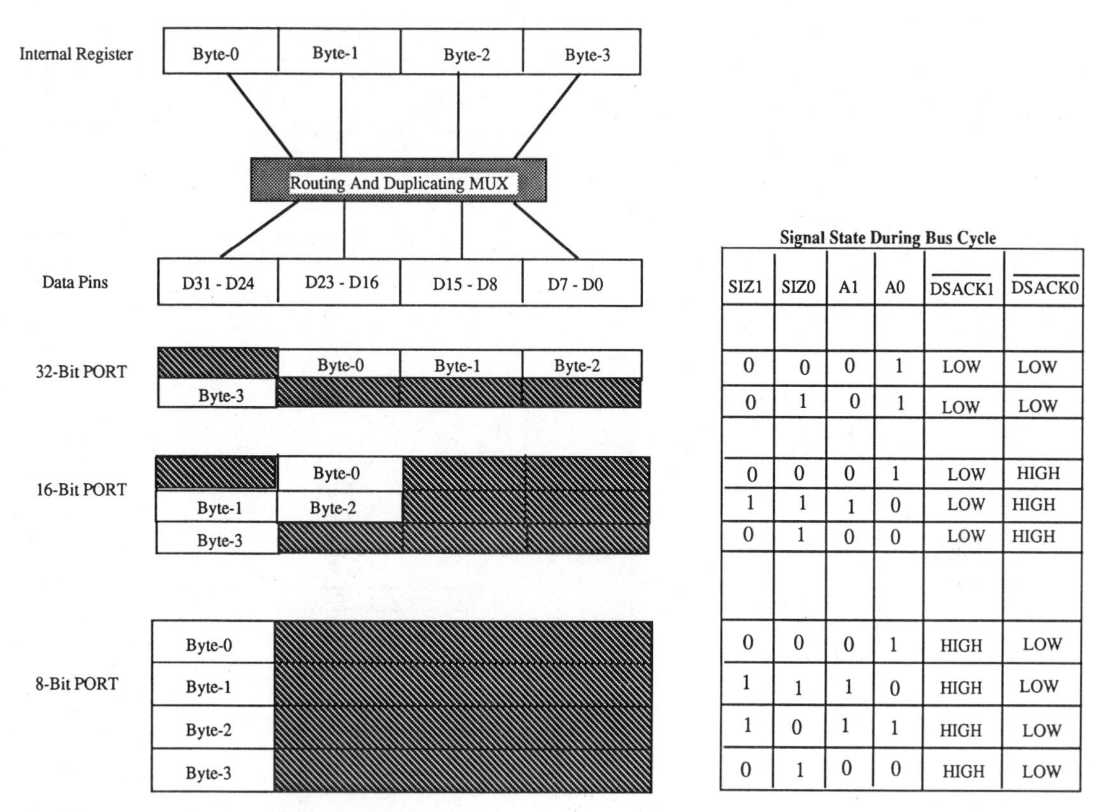

SIZ1	SIZ0	A1	A0	$\overline{DSACK1}$	$\overline{DSACK0}$
0	0	0	1	LOW	LOW
0	1	0	1	LOW	LOW
0	0	0	1	LOW	HIGH
1	1	1	0	LOW	HIGH
0	1	0	0	LOW	HIGH
0	0	0	1	HIGH	LOW
1	1	1	0	HIGH	LOW
1	0	1	1	HIGH	LOW
0	1	0	0	HIGH	LOW

FIGURE 4.7. A misaligned long-word transfer.

each byte of the 32-bit port. This will enable the MC68020 to determine which byte is to be enabled during a bus transfer. Two more byte enables should be created for the 16-bit port, if 16-bit devices are to be interfaced with the MC68020.

Figure 4.8 illustrates the interface enable logic for a word-size port, and the truth table used in the design of the circuit. Figure 4.9 shows the interface logic for a long-word port, along with the truth table. The enable signals are functions of the size pins and address lines A0 and A1.

Correct external interpretation of bus control signals are critical to ensure valid data transfer operations. The MC68020 system designers designed the circuit so that the correct byte data strobe signals are generated to route the correct byte to the proper byte location in the memory map. The MC68020 always drives all external data lines during the write operation. This requires careful control of the enable signals for the independent bytes of a data port.

4.2.8 Bus Cycle Timing Diagram

All timing of this section can be found in Appendix C [Motorola 85]. The timing diagram shown in Figure 4.10 is valid for the MC68020, operating at 16.67 MHz.

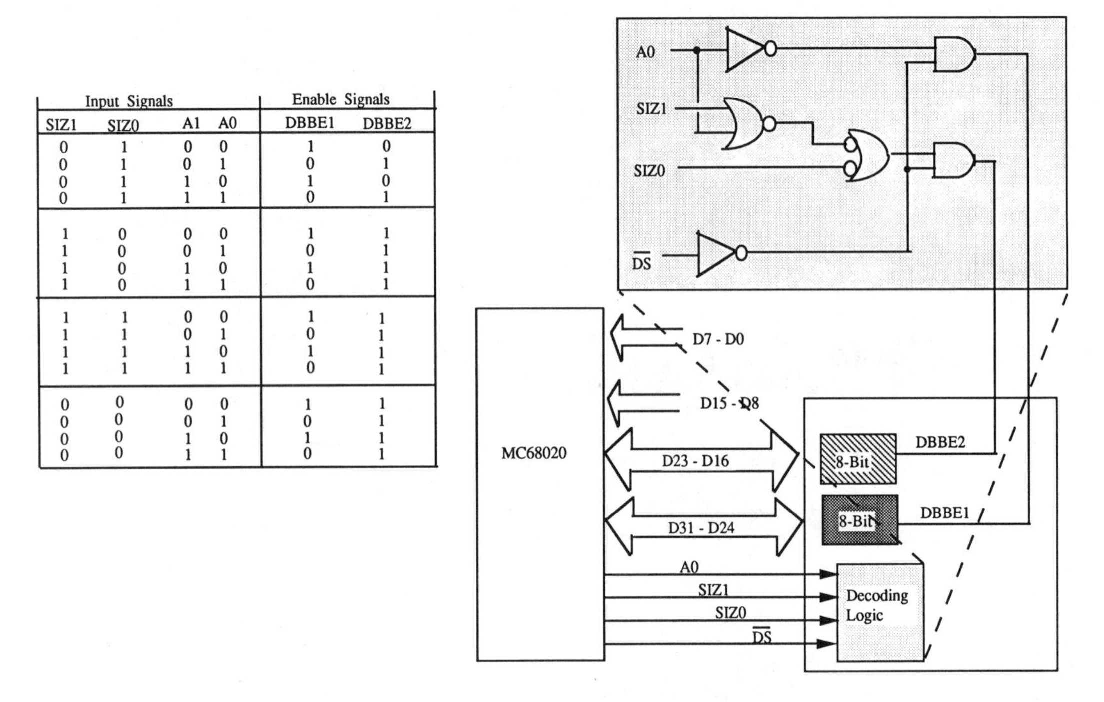

Input Signals				Enable Signals	
SIZ1	SIZ0	A1	A0	DBBE1	DBBE2
0	1	0	0	1	0
0	1	0	1	0	1
0	1	1	0	1	0
0	1	1	1	0	1
1	0	0	0	1	1
1	0	0	1	0	1
1	0	1	0	1	1
1	0	1	1	0	1
1	1	0	0	1	1
1	1	0	1	0	1
1	1	1	0	1	1
1	1	1	1	0	1
0	0	0	0	1	1
0	0	0	1	0	1
0	0	1	0	1	1
0	0	1	1	0	1

FIGURE 4.8. Interface enable logic for a word-size PORT.

The processor's bus cycle requires a minimum of 3 clock cycles or 6 states. The bus cycle begins with the rising edge of a clock cycle (which is also the beginning of state-0), and causes the address information, function codes, and size information to propagate to the respective pins. Bubble number 6 indicates that this information will become valid on the pins from 0 to 30 nsec after the rising edge of clock state S0. Under normal operation, the address, function code, and size information will be clocked out to pins during the state S0 of the next bus cycle. However, if the processor relinquishes the control of the bus, then bubble number 7 indicates how long it will take for these signals to reach the high-impedance state after the end of state S5.

$\overline{\text{ECS}}$ and $\overline{\text{OCS}}$ have the same timing characteristics, except that $\overline{\text{OCS}}$ is asserted only during the first bus cycle of an operand. $\overline{\text{OCS}}/\overline{\text{ECS}}$ are asserted by the MC68020, and bubble number 6a indicates that they will become valid from 0 to 20 nsec after the beginning of the state S0, and remain valid for less than half the clock period. These signals are negated at the end of state S0. Bubble number 12a indicates that it takes 0 to 30 nsec for this signal to get to a high-impedance state after the end of state S0. Both $\overline{\text{AS}}$ and $\overline{\text{DS}}$ have identical timing during the read bus cycle. They are asserted at the falling edge of state S0. Bubble number 9 indicates that they will become asserted from 3 to 30 nsec after the beginning of state S1. However, bubble number 9a indicates that $\overline{\text{DS}}$ may become asserted 15 nsec before or after $\overline{\text{AS}}$ is asserted, and depends on the load.

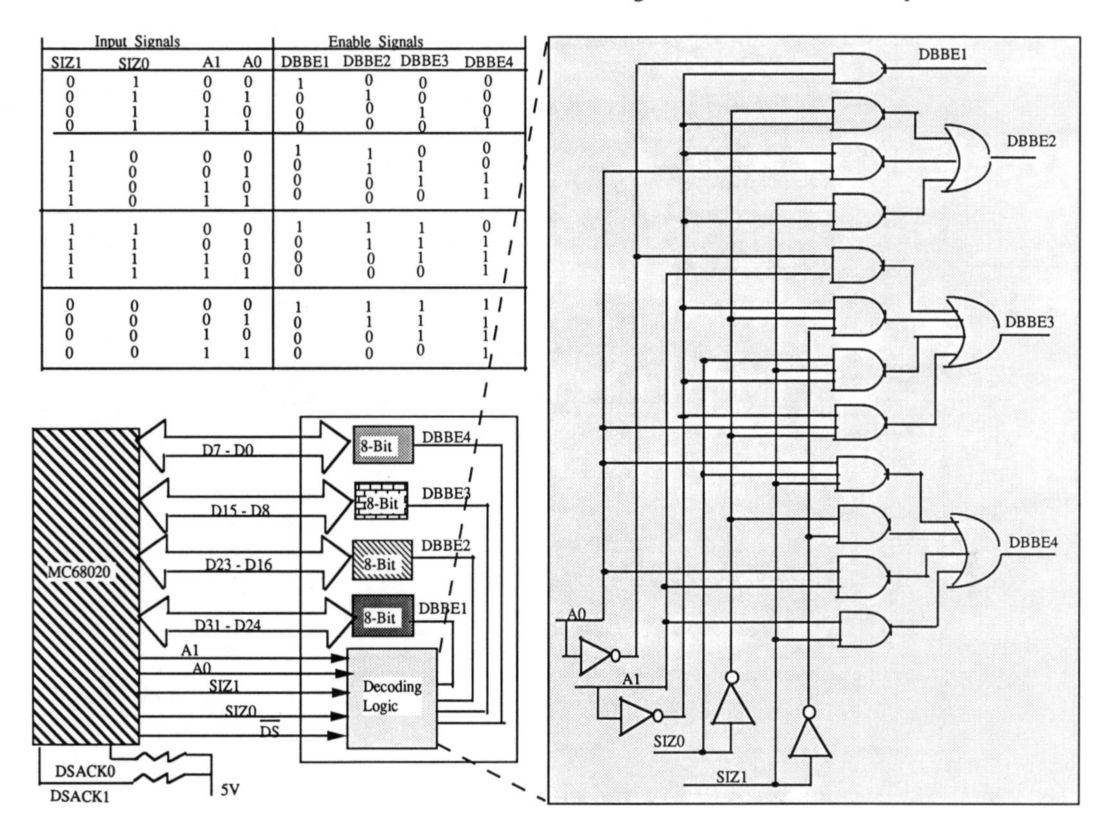

Input Signals				Enable Signals			
SIZ1	SIZ0	A1	A0	DBBE1	DBBE2	DBBE3	DBBE4
0	1	0	0	1	0	0	0
0	1	0	1	0	1	0	0
0	1	1	0	0	0	1	0
0	1	1	1	0	0	0	1
1	0	0	0	1	1	0	0
1	0	0	1	0	1	1	0
1	0	1	0	0	0	1	1
1	0	1	1	0	0	0	1
1	1	0	0	1	1	1	0
1	1	0	1	0	1	1	1
1	1	1	0	0	0	1	1
1	1	1	1	0	0	0	1
0	0	0	0	1	1	1	1
0	0	0	1	0	1	1	1
0	0	1	0	0	0	1	1
0	0	1	1	0	0	0	1

FIGURE 4.9. Interface enable logic for a long-word size PORT. (Courtesy Motorola Inc.).

Bubble number 12 indicates that $\overline{AS}$ and $\overline{DS}$ will be negated by the processor at any time between 0 to 30 nsec after state S4.

During the write bus cycle, $\overline{DS}$ gets asserted at any time from 3 to 30 nsec after the end of state S2, as indicated in bubble number 9b, which is one clock cycle later than the assertion of $\overline{DS}$ during the read bus cycle. The $R/\overline{W}$ line becomes redefined for a new bus cycle at the rising edge of S0. Bubble number 20 indicates that the value of $R/\overline{W}$ will become valid for a write operation from 0 to 30 nsec after state S0. In the case of a consecutive write bus cycle, this line will remain low for the entire duration of the bus cycle, and will remain high during the consecutive read bus cycle.

The MC68020 places data on the data bus at the beginning of S2 during a write operation. Bubble number 23 indicates that the data will become valid on the data pins no later than 30 nsec after the beginning of state S2. Bubble number 53 indicates that the data are held over on the data bus until the end of state S5.

The asynchronous input signals $\overline{HALT}$, $\overline{DSACKx}$, autovectors, cache disable, $\overline{RESET}$, $\overline{BERR}$, $\overline{BR}$, $\overline{BGACK}$, and interrupt lines are sampled on the falling edge

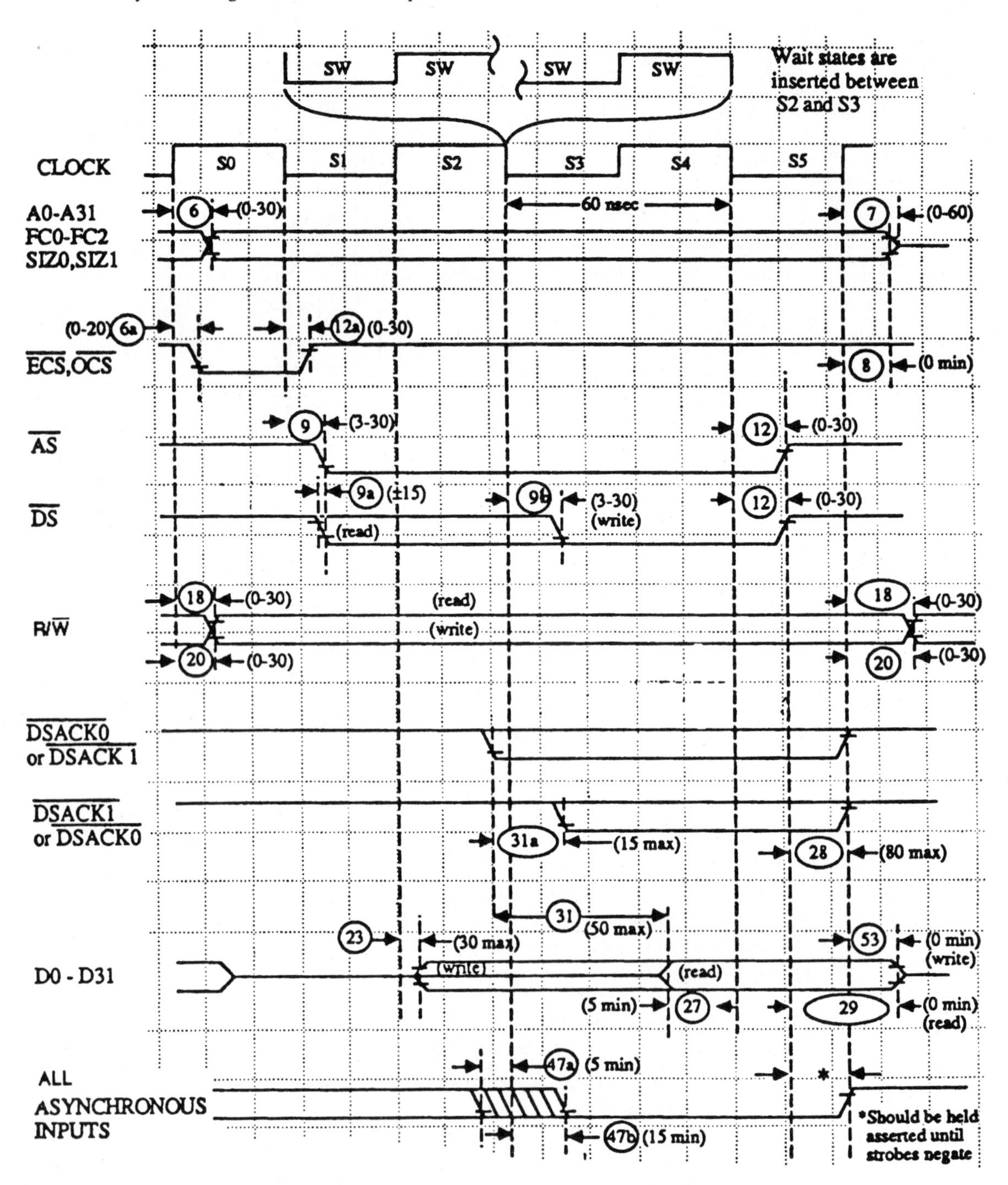

FIGURE 4.10. Bus cycle timing diagram. (Courtesy Motorola Inc.).

of the clock. If these signals are found to be asserted, they are acted upon during the falling edge of the next clock cycle. They must meet both the asynchronous signal setup time, identified in bubble number 47a, and the hold time, shown in bubble number 47b. These signals must be asserted on the pins at least for 5 nsec before the falling edge of the clock, and must remain asserted for a minimum of 15 nsec after the falling edge of the clock.

The sampling of the asynchronous inputs takes place any time during this window. As a result, signals must remain asserted throughout this period, if recognition is necessary on a specific clock edge. If an asynchronous signal does not meet the setup or hold time requirements at one edge, but meets them at the other edge, then the input will be recognized on the next clock cycle. When the $\overline{\text{DSACKx}}$ are not recognized on the falling edge of the state S2, the CPU enters into a software wait cycle and samples the $\overline{\text{DSACKx}}$ again on the next falling edge of the clock. The samplings are repeated until the $\overline{\text{DSACKx}}$ are recognized. Once they are recognized, the processor continues with states S3, S4, and S5 and completes the bus cycle.

The proper asynchronous protocol requires that any asynchronous input should be held constant until the $\overline{\text{AS}}$ and $\overline{\text{DS}}$ are negated. Bubble number 31a indicates that both $\overline{\text{DSACKx}}$ should be asserted within 15 nsec, in order to ensure their recognization by the CPU. In reality, it does not matter which $\overline{\text{DSACK}}$ comes first, as long as both are asserted within 15 nsec of each other. Bubble number 47a indicates that if one $\overline{\text{DSACK}}$ meets the asynchronous input setup time and the other $\overline{\text{DSACK}}$ is asserted within 15 nsec of the first, then both will be recognized as asserted on that falling edge. However, if both $\overline{\text{DSACKx}}$ do not meet the input setup time, but are asserted 15 nsec of each other and remain asserted, then they will not be recognized until the next falling edge of the input clock. If a read bus cycle is in progress, and if the CPU recognizes $\overline{\text{DSACKx}}$ as asserted, then the processor will latch the data at the end of state S4.

Bubble number 27 indicates that the data must be present on the CPU's pins at least 5 nsec before the falling edge of the clock at the end of state S4. Bubble number 31 indicates that, if the data is present no later than 50 nsec after the assertion of the first $\overline{\text{DSACK}}$, then the data will be read correctly by the CPU. This allows the designer to ignore the asynchronous input setup time and data setup time requirements, and to generate these signals without the knowledge of the CPU's clock edges.

4.3 SPEED CALCULATION

The determination of the instruction execution time of the MC68020 and subsequent processors is difficult. This is because the exact execution time of an instruction or operation depends on various factors, namely:

- Instructions in cache and/or cache enabled
- Operand alignment
- Memory port size
- Memory speed
- Instruction data size
- Addressing mode used, memory indirection modes, including the number of extension words utilized.
- Prefetch sequence, even or odd word alignment of Op-word
- Instruction overlap and instruction sequences.

The timing guidelines presented in this section are designed to help the software designer/programmer to predict the execution speed of a program. Timing information outlined in this section is for situations when the instruction in question resides in the cache, and for when (on the average) it is not in the cache. In addition, the timings for exception processing, context switching, and interrupt processing are included, so that designers of multitasking and/or real-time systems can predict the task switching overhead, worst-case interrupt latency, and other time-critical parameters.

The MC68020 can be viewed as composed of five independently scheduled resources (Figure 2.10). These are: the instruction cache, the bus controller, the microsequencer (including control logic), the execution unit, and instruction pipes. Since very little of the scheduling is directly attributable to the boundaries of the instruction, it is as a result difficult to make accurate estimates of the time required to execute a particular instruction without knowledge of the context within which the instruction is executing.

The microsequencer is either executing microinstruction or awaiting the completion of accesses that are necessary to continue the execution. The control logic of the microsequencer directs the bus controller, instruction execution, and pipeline operations, to calculate the effective address and set the condition codes. The microsequencer initiates instruction prefetch and generates the necessary control signals to validate the instruction word in the pipe.

As discussed earlier, three-word instruction pipes decode the instruction opcodes. Each unit has a status bit to indicate whether the word in the stage was loaded with data from a bus cycle that was terminated prematurely. Stages of the pipe are only filled in response to specific prefetch requests issued by the microsequencer.

The instruction cache feeds the instruction prefetch portion of the microsequencer. The prefetch of an instruction that matches the on-chip instruction cache causes no delay in the instruction execution, since the CPU does not initiate an external bus cycle. The instruction cache also operates synchronously with the external bus during instruction cache fills following an instruction cache miss.

The bus controller unit consists of micro-bus controller and instruction fetch pending buffer, which carry out all reads in the event of cache miss. The microbus controller performs the bus cycles issued to the bus controller by the rest of the CPU. It also carries out the dynamic bus sizing. It always initiates a longword read operation during prefetch. The CPU reads two words, which may load two instructions at once, or be two words of a multiword instruction, into the cache holding register.

4.3.1 Concurrency

The bus controller and the sequencer may operate on an instruction concurrently. The bus controller can perform read or write operations, while the sequencer controls an effective address calculation or sets condition codes. The sequencer may also request a bus cycle that the bus controller cannot perform immediately. In such a situation, the bus cycle is queued and the bus controller runs the cycle at the completion of current cycle. Thus, the concurrent operations between the sequencer and bus controller introduces ambiguity into the instruction time calculation, due to the potential overlap of instruction execution.

4.3.2 Overlap and the Best Case

Overlap is the time, measured in clock periods, that an instruction executes concurrently with another instruction. Figure 4.11 shows three instructions, A, B, and C, including their overlap period.

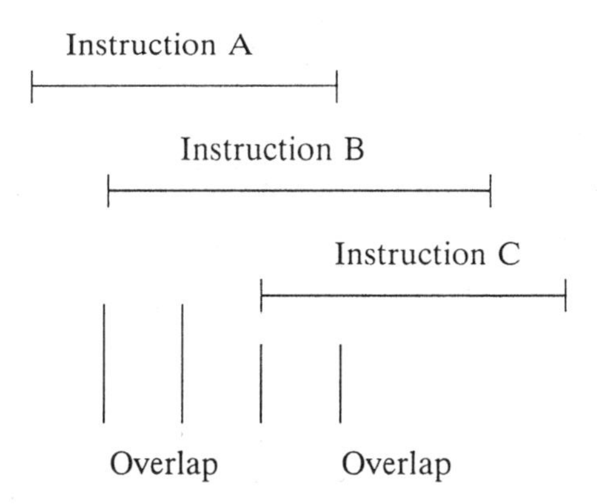

FIGURE 4.11 Instruction overlap.

Such overlap reduces the overall execution time of the three instructions. The first overlap time decreases the overall execution time for instructions A and B. The second overlap decreases the overall execution time for instructions B and C. Thus, the total execution time for instructions A, B, and C will be much lower than the sum of their individual execution times. It is possible that the execution time of an instruction may be completely absorbed by overlap with a previous instruction. This would result, for this instruction, in a net execution time of zero clock.

4.3.3 The Instruction Timing Table

The following assumptions apply to the instruction times in Appendix D [Motorola 85]. They are:

- Memory operands, including the system stack, are long-word aligned.
- The data transfer between the MC68020 and the memory takes place over a 32-bit bus.
- All memory access requires 3 bus cycles, i.e., no wait state is involved.
- No exception takes place during the instruction execution.

Three values are given for each instruction and effective addressing mode, and they are:

- The BC (Best Case) assumes that the instruction resides in the cache, the cache is enabled, and maximum overlap occurs with other instructions.
- The CC (Cache Case) also assumes that the instruction is in the cache, the cache is enabled, but no overlap takes place with other instructions.
- The WC (Worst Case) assumes a disabled cache or the instruction is not in the cache, and no overlap occurs with other instructions.

Entries in the instruction timing table contain four numbers, three of which are within the parentheses.

- The outer number represents the total number of clock cycles utilized by the processor to execute the instruction.
- The first number inside the parentheses tells the number of read bus cycles utilized by the instruction.
- The second value represents the maximum number of instruction access cycles performed by the instruction, including all of the prefetch bus cycles needed to keep the instruction pipe filled.
- The last value within the parentheses indicates the number of write bus cycles performed by the instruction.

For example:

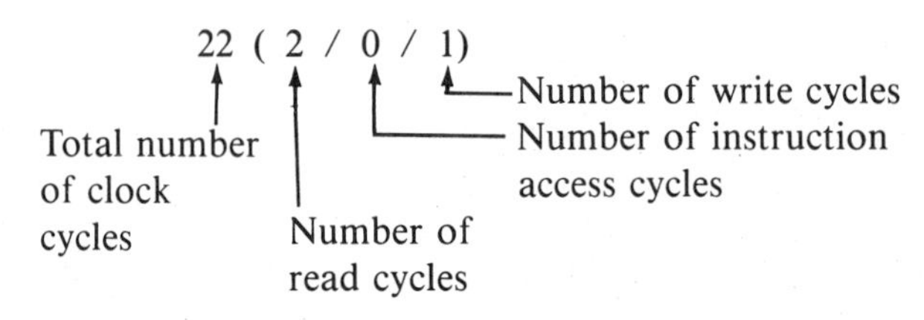

The total number of bus activity clocks for the above example is derived in the following way:

(2 reads * 3 clocks/read) +
(0 instruction access * 3 clocks/access) +
(1 write * 3 clocks/write)
= 9 clocks of bus activity

22 total clocks − 9 clocks (bus activity) = 13 internal clock cycles

The example used here was taken from the best-case fetch-effective addressing mode for a MOVE instruction. The source-effective addressing mode is [(d32, B), 1, d32] with a destination mode (xxx).L. The timing is also valid where the source addressing mode is (d31, B) and the destination-effective addressing mode is ([d16, B], 1) and also for the best-case timing mode. The MOVE instruction with same addressing modes for the 'cache case' timing number is 25(2/0/1).

The instruction timing tables are utilized to calculate 'best case' and 'worst case' bounds for some target instruction stream. The determination of the exact timing from the tables is very difficult, because the instructions may not be mutually exclusive. Thus, the only way to calculate the exact execution time is to count the number of clock cycles consumed during the execution of the instructions.

4.3.4 Instruction Timing Example

Consider the ADD instruction below and calculate its execution time.

ADD.L ([$2000, A3, D3.W * 2], $8000), D6

The instruction uses memory indirect with preindexing, with both base and outer displacement for the source-effective addressing mode, and a data register for the destination. We will also assume that the MC68020 executes the instruction under the following constraints:

- Due to memory, the operand read bus cycle requires two wait cycles.
- The prefetch requires three clock cycles of wait state.
- The operand write bus cycle requires three clock cycles of wait state.
- The memory is organized in 32-bit ports.
- The processor is operating at 16.67 MHz.

In this example, the base and outer displacement are all 32 bits. We will determine the best-case and worst-case timing for this instruction. The instruction execution timing table in Appendix D [Motorola 85] indicates that the fetch effective address time must be added. The table does not show the index register scaling, because scaling does not take additional time. We are, however, required to calculate the timing for the 16-bit base displacement, a base register, and memory indirection with a 32-bit outer displacement.

The best-case timing for the source-effective address fetch is 13 clocks, including 2 read with no prefetch and no write. The worst-case timing is 23 clocks, including 2 reads and 2 prefetch with no wait. The number of clock cycles for the wait state must be added. That is, for 2 clock cycles prefetch, we must add 3 clock cycles. For the best case, there are no prefetch, no writes, and only operand read. This gives a total of:

$$13 \text{ clocks} + 4 \text{ clocks} = 17 \text{ clocks}$$

For the worst case, there are operand read and prefetch bus cycles, which give a total

$$23 \text{ clocks} + 4 \text{ clocks} + 9 \text{ clocks} = 36 \text{ clocks}.$$

Using an operating speed of 16.67 MHz, the best-case timing is found to be 1.02 microsec., and the worst-case timing is 2.16 microsec.

4.4 BUS ARBITRATION CONTROL

Program-controlled data transfer requires a significant amount of the processor's time to transmit a small amount of data per unit time. Here, the processor cannot engage in other processing functions during such data transfer operations. The interrupt mechanism provides an enhancement over the previous technique, and increases the data transfer rate. It requires less software overhead and allows for concurrent processing. However, for many applications, the interrupt technique cannot meet performance goals.

The Direct Memory Access (DMA) technique overcomes these limitations by providing an ability to request the system bus for the direct transfer of data between an I/O device and memory without the intervention of the processor. The processor completes the current bus cycle, if one is in progress, relinquishes the bus, and then grants control of the bus to the requesting device. The data does not flow through the processor and, as a result, no program intervention is required during the transfer.

4.4.1 DMA Techniques

Direct Memory Access (DMA) is commonly used for three types of data transfer: burst, cycle stealing, and transparent modes [Nikitas 84]. The burst transfer mode is the simplest and most often used. Here, the operation of the processor is suspended while the data transfer occurs, into the memory. A block of data is transferred while the processor is locked out. Therefore, either the processor or the device gets to use the bus at any given time. The maximum data transfer rate in burst mode is limited by the read or write access time of the memory, including the operating speed of the requesting device. Burst transfer mode is frequently used in transferring data to or from a floppy or hard disk. This mode is also used in the design of multiprocessors.

Under the cycle stealing mode, data is transferred concurrently, while other processing is carried out by the processor. The steps are similar to burst transfer mode, except that one unit of data is transferred between two bus accesses. The process continues until all data is transferred. The requesting device steals cycles from the microprocessor, during which it transfers data. As a result, processing carried out by the CPU is slowed down accordingly.

Under the transparent mode, memory cycles are stolen by requesting devices while the processor is performing some internal operation and is not using the system bus. The requesting device gains an immediate access to the memory, during which it transfers data. This continues until the requisite data are transferred. This technique requires external logic to detect when the address and data buses are not used. This DMA technique is transparent to the processor, in that it does not interfere with or slow down the normal rate of processor execution.

4.4.2 Multiprocessor Support

The MC68020 has extended the basic DMA transfer capability, to support multiprocessor design. It provides a mechanism for a shared system bus and shared memory environment. A common protocol is implemented to support DMA and the multiprocessor, and they are implemented with the help of $\overline{BR}$, $\overline{BG}$, and $\overline{BGACK}$ signals. The protocol also includes an arbitration mechanism to support a single bus master at any one time. The bus controller within the MC68020 manages the bus arbitration signals and allows either the processor or an external device to become a bus master at any instant, where the processor is assigned the lowest priority. Systems that include several potential bus masters will require external circuitry to resolve priorities among contending devices. The multiprocessor support for the MC68020 will be discussed in more detail in Chapter 6.

In its simplest form, the bus access protocol consists of three steps, which are: the bus request, the bus grant, and the bus grant acknowledgment. The steps are described in the following subsections.

4.4.3 Bus Request

Those external devices capable of becoming the bus master can request the bus by asserting the Bus Request ($\overline{BR}$) signal.

All contending devices' bus request lines must be wired ORed to the $\overline{BR}$ pin of the CPU, signaling that an external device requires the system bus. If no acknowledge is received by the CPU while the bus request signal is active, the CPU reclaim the bus as soon as the bus request is negated. This eliminates unnecessary confusion, because the arbitration logic may inadvertently respond to noise, or an external device may decide that it no longer requires the bus before it is granted the bus mastership.

4.4.4 Bus Grant

The CPU asserts the Bus Grant ($\overline{BG}$) signal to announce that the bus will become available at the end of the current bus cycle. The $\overline{BR}$ may arrive at any time during a bus cycle or between cycles. The $\overline{BG}$ is only asserted in response to a $\overline{BR}$. It is usually asserted as soon as the $\overline{BR}$ has been synchronized and recognized, except when the MC68020 has made an internal decision to execute a bus cycle. Again, the $\overline{BG}$ is not asserted until the end of a read-modify-write cycle (i.e., $\overline{RMC}$ is negated). If more than one device contends for the bus, then the external arbitration logic must select the current bus master, upon detecting a valid $\overline{BG}$ signal.

It is quite possible that a system may route $\overline{BG}$ signals through a daisy-chained network or through a specific priority-encoded network. In such a situation, these interfaces will not alter the processor's operation.

4.4.5 Bus Grant Acknowledgment

The requesting device asserts the Bus Grant Acknowledge ($\overline{BGACK}$) signal, to indicate that it has assumed bus mastership. The would-be bus master must fulfill the following criteria through the normal bus arbitration procedure, before acquiring the bus.

a. It must receive the $\overline{BG}$ through an arbitration process.
b. The Address Strobe ($\overline{AS}$) must be negated, indicating that no bus cycle is in progress, and the external device must ensure that all appropriate processor signals have been placed in the high impedance state.
c. The termination signal $\overline{DSACKx}$ for the most recent cycle must have become inactive, indicating that all external devices have relinquished the bus.
d. $\overline{BGACK}$ must be inactive, indicating that no bus master has claimed ownership of the bus.

It is extremely important to understand that the MC68020 will recognize the bus arbitration request during normal processing, $\overline{RESET}$ and $\overline{HALT}$ assertions, and even when the CPU has halted due to a double bus fault.

4.4.6 Bus Arbitration Control

The bus arbitration control unit in the MC68020 is implemented with a state machine, shown in Figure 4.12. The input signals labeled R and A are internally synchronized versions of the bus request and bus grant acknowledge, respectively. All asynchronous inputs to the MC68020 are internally synchronized within two clock cycles. The bus grant is internally labeled as G, and the internal high-impedance control signal is labeled as T. T becomes true only when the address, data, and control buses float (high-impatience state) after the next rising edge following the negation of the address strobe ($\overline{AS}$) and the read-modify-write ($\overline{RMC}$) signal.

Changes to the states occur in the state machine on the next rising edge of the clock after the internal signal is valid. At State-0, G and T are both negated. The CPU remains bus master during this state and the internal arbitration mechanism remains in the *idle* condition. The arbitrator remains in this state as long as re-

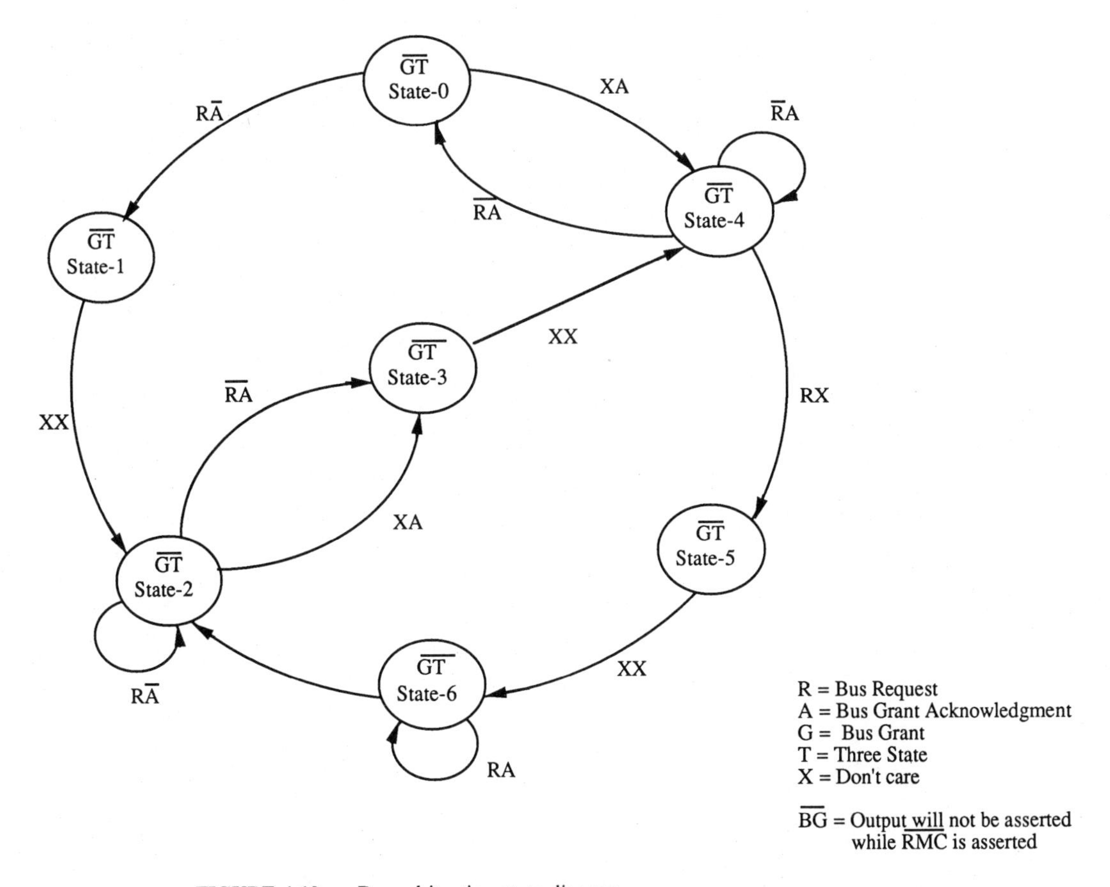

FIGURE 4.12. Bus arbitration state diagram.

quest (R) and acknowledge (A) are negated. The machine makes a transition to State-1 upon receiving a request. At this state, both G and T are asserted.

The next clock takes the machine to State-2, where G and T are held. The bus arbiter remains at this state until A is asserted or R is negated (i.e., the request is withdrawn by the requesting device). The transition to State-3 occurs if one of the conditions described in Section 4.4.5 is fulfilled. The grant G is negated during State-3.

The next clock takes the arbiter to State-4, where grant G remains negated and signal T remains asserted. The arbiter asserts acknowledge A and remains here until A is negated or request R is asserted. The machine returns to State-0 if it receives A. Upon entering into State-0, the arbiter negates T.

The machine enters into State-5 as it receives R, and then makes a transition to State-6 on the clock pulse.

The read-modify-write sequence is normally indivisible, to support the semaphore and synchronization for multiprocessing. The MC68020 asserts the $\overline{RMC}$ signal, and causes the bus arbitration state machine to ignore the R signal that occurs after the first read cycle of the read-modify-write sequence, by not granting G.

Figure 4.13 shows the bus arbitration timing diagram. The timings are calculated for a CPU running at 16.67 MHz. Bubble number 35 indicates that the MC68020 takes a maximum of 3.5 clock cycles to generate the $\overline{BG}$ signal after recognizing $\overline{BR}$ as asserted. Bubble number 33 specifies that the CPU will take a maximum of 30 nsec to generate $\overline{BG}$ from the falling edge of the clock. Bubble number 37 indicates that the MC68020 will wait between 1.5 clock cycles to 3.5 clock cycles to recognize $\overline{BGACK}$ as asserted, before it decides to take away the system bus. Bubble numbers 39 and 39a indicate that the CPU will negate $\overline{BG}$ for a minimum of 90 nsec, and will stay asserted for a minimum of 90 nsec. The system designer should follow these guidelines before designing the external interface logic.

4.5 INTERRUPT CONTROL

The program-driven I/O technique has some inherent flaws, namely:

a. It cannot handle asynchronous, unpredictably occurring external events efficiently.
b. It fails to provide an acceptable solution to the most important criteria for real-time applications, which require the CPU to respond instantly to external events.

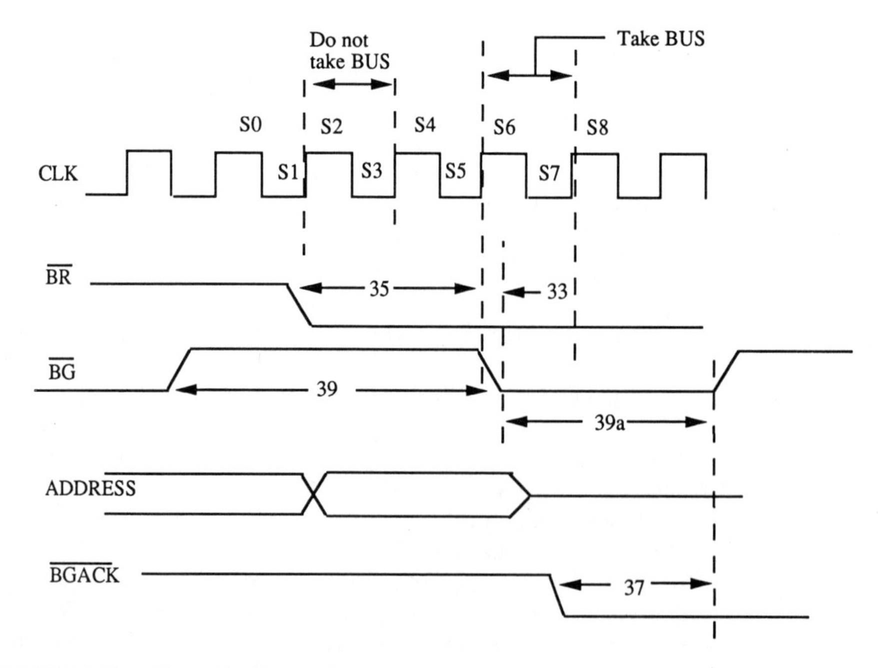

FIGURE 4.13. Bus arbitration timing diagram.

c. It does not optimize the CPU's processing power, because it may become I/O-bound in inquiring on the status of all devices, and thus has little free time to do useful work.

To overcome these problems, the designers of the MC68000 family of processors implemented an *interrupt* mechanism. Here, the data transfer is initiated by an interrupt signal issued by an external device, requesting an information exchange with the CPU. Upon recognizing an interrupt signal, the CPU will suspend execution of any program, and it will enter into supervisory mode (which will be discussed later), identify the type of interrupt, save the CPU state in the interrupt stack, and finally transfer control to the corresponding interrupt-handling routine to service the device. When the routine finishes servicing the interrupting device, the saved CPU state is restored and the normal program execution resumes at the completion of the interrupt service.

4.5.1 Initialization of the Interrupt Mask

During the power-up reset, the MC68020 sets the interrupt mask bits I0, I1, and I2 in the status register to level 7. This disables all of the incoming interrupts. The mask must be modified to a lower priority level through software in order to activate the interrupts. For example, the reset initialization routine may set the mask bits to "000," which would enable all interrupts.

4.5.2 General Interrupt Processing Sequence

MC68020 architecture supports two types of interrupts, namely *vectored* and *autovectored* interrupts. We have already seen in Chapter 2 that the MC68020 provides three signal lines, $\overline{\text{IPL0}}$, $\overline{\text{IPL1}}$, and $\overline{\text{IPL2}}$, for interrupts. The internal interrupt-pending flag is brought to the outside world via the $\overline{\text{IPEND}}$ pin (Figure 4.14). The pin associated with autovectoring is known as $\overline{\text{AVEC}}$. The autovector signal is the only remnant of the synchronous bus that existed in the previous generation of MC68000 processors. The signal is only used during the interrupt acknowledgment bus cycle.

When the MC68020 receives an interrupt via the interrupt pins, it compares this incoming level against the mask bits I0, I1, and I2 of the status register, to determine whether to process the interrupt. The CPU will not recognize an interrupt until the following conditions are met:

1. The incoming interrupt level must be higher than the current values of the interrupt mask bits in the status register. That is, if the SR mask value is 100, then the CPU masks out all incoming interrupts from levels 2 down to 0.
2. The current interrupt request level over the three interrupt control lines must be held until the processor acknowledges the interrupt, by initiating an interrupt acknowledge bus cycle (IACK).

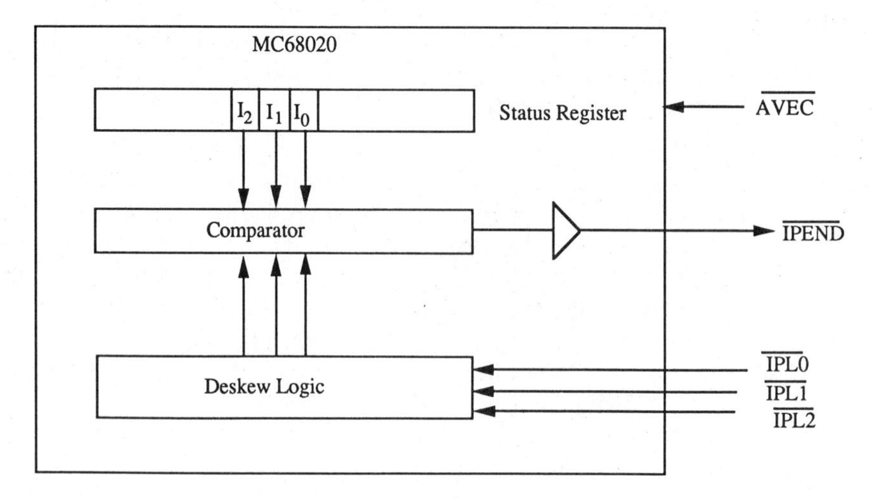

FIGURE 4.14. Interrupt control signals.

These rules guarantee that the interrupt will be processed. Table 4.6 illustrates the seven priority levels, 1 through 7, and the corresponding signal levels on pins $\overline{IPL0}$, $\overline{IPL1}$, and $\overline{IPL2}$, along with the interrupt mask levels required for the recognition of the requested interrupt. The interrupt levels 1 through 6 are level sensitive, whereas level 7 is edge triggered.

The MC68020 indicates that it needs the bus to service an interrupting device by asserting $\overline{IPEND}$. The current bus master may relinquish the bus momentarily to allow the CPU to service the interrupt. $\overline{IPEND}$ is asserted whenever the deskewed interrupt level is higher than the mask level found in the SR register. $\overline{IPEND}$ negates momentarily whenever the interrupt level changes, until the new level has been deskewed. Deskewed logic will not consider the interrupt level as valid, and present that to the internal comparator, until it has seen the same level for two consecutive falling edges of the clock. Thus it is possible for an interrupt request which is held for as short a period as two clock cycles to be recognized.

TABLE 4.6. Priority Levels and Status of the Interrupt Pins.

Requested Interrupt Level	Control Line Status			Level of Masked Interrupt
	$\overline{IPL2}$	$\overline{IPL1}$	$\overline{IPL0}$	
0	High	High	High	No request received
1	High	High	Low	0
2	High	Low	High	0 through 1
3	High	Low	Low	0 through 2
4	Low	High	High	0 through 3
5	Low	High	Low	0 through 4
6	Low	Low	High	0 through 5
7	Low	Low	Low	0 through 6

$\overline{\text{IPEND}}$ is also negated when the CPU, as a part of exception processing, raises the mask of the interrupt level about to be acknowledged.

A recognized interrupt may not initiate an immediate exception processing, rather it is only made pending. It is done to signal the external devices that the MC68020 has recorded an interrupt as pending. The processor completes the current instruction before taking any action on the interrupt. This wait period may be equal to the time required to execute the most complex instruction. Again, the exception processing starts at the next instruction boundary, provided that a higher priority exception is not waiting to be serviced.

4.5.3 Interrupt Acknowledgment Cycle

To correctly manage an interrupt request, the processor must determine the starting location of the service routine corresponding to the request. All members of the MC68000 family generate an interrupt acknowledgment bus cycle. They procure the interrupt vector number by generating it internally, or by receiving it from an external source. The vector number is specifically used to determine which vector table entry is to be loaded into the program counter.

The interrupt acknowledgment sequence is shown in Figure 4.15. During the IACK bus cycle, the processor outputs the interrupt level being serviced on the address lines A1 through A3. This asserts control signals $R/\overline{W} = 1$, $\overline{\text{AS}} = 0$, $\overline{\text{DS}} = 0$, $\overline{\text{SIZ0}} = \overline{1}$, and $\overline{\text{SIZ1}} = 0$. The processor also sets the function codes FC0–FC2 to '111' and places the current level of the interrupt request over address lines A1–A3. The requesting device then either places an interrupt vector number on the data bus, asserts the appropriate value to $\overline{\text{DSACK0}}$ and $\overline{\text{DSACK1}}$ to declare its port size, or it asserts $\overline{\text{AVEC}}$ to request the processor to internally generate the vector number corresponding to requested interrupt level. Upon receiving $\overline{\text{DSACKx}}$, the CPU latches the vector number from the data bus, and negates $\overline{\text{DS}}$ and $\overline{\text{AS}}$. The interrupting device then negates $\overline{\text{DSACKx}}$ and releases the control of the data bus. This completes the IACK bus cycle.

During the response mode of the IACK bus cycle, if the CPU fails to receive any response in the form of $\overline{\text{AVEC}}$ or $\overline{\text{DSACKx}}$, it will automatically enter into a software 'wait cycle' and wait for a response from the interrupt requesting device. The MC68020 will keep on waiting indefinitely. In order to avoid such a situation, the system designer must provide a *watch dog timer*, which will assert $\overline{\text{BERR}}$ to terminate the interrupt acknowledge bus cycle.

4.5.4 Vectored Interrupt

This technique is used to interface external devices capable of responding with a vector number. Peripheral devices designed to work with the MC68000 family of processors has a vector number register implemented internally. This allows an I/O device to guide the CPU to its service routine during the interrupt acknowledgment cycle. The device places the contents of the vector number register on the least significant byte of the data port, and asserts the appropriate $\overline{\text{DSACKx}}$. The CPU first extracts the 8-bit vector number from the data bus, then multiplies the vector number by 4 (by shifting the value by two bit positions to the left). It

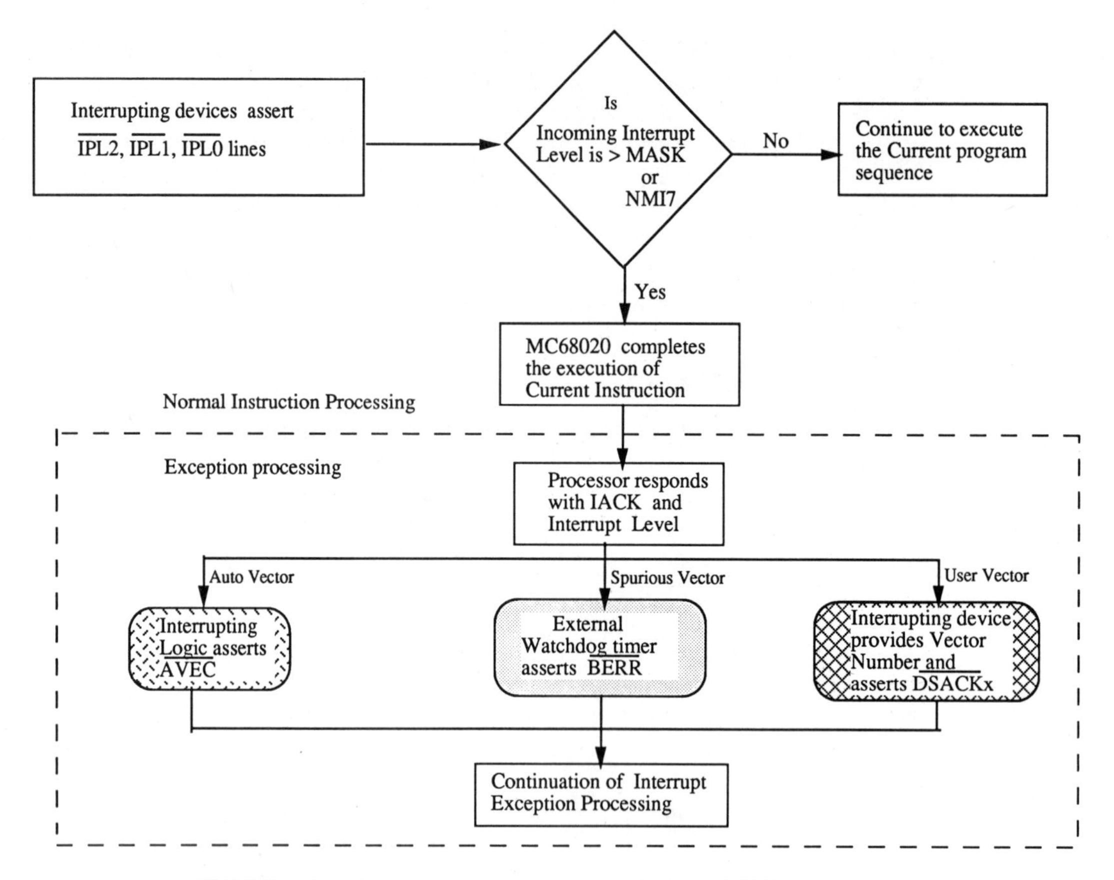

FIGURE 4.15. Interrupt processing sequence for the MC68020.

then fills the upper address bits with zeros, and creates the address that points to the vector to be loaded into the program counter of the CPU. Control is then transferred to the respective interrupt handling routine for the interrupting device. The diagram below shows how the processor reads an 8-bit data from the data bus and creates the address for the vector pointer table.

D31	D24 D23	D0
b7 b6 b5 b4 b3 b2 b1 b0	x x x x x	x x x x x x

A31	A9	A2 A1 A0
000000 . . .	0000b7 b6 b5 b4 b3 b2 b1 b0 0	

x = indicates do not care

4.5.5 Autovector Interrupt

The MC68020 supports a simpler interrupt mechanism for devices that are unable to provide the vector number during the interrupt acknowledgment cycle. This is called the autovector mode of operation. Here, external hardware is needed to recognize the IACK bus cycle and generate $\overline{\text{AVEC}}$. This allows the MC68020 to internally generate an interrupt vector number corresponding to the interrupt level. The CPU generates the vector address based on the encoded value of the $\overline{\text{IPL0}}$, $\overline{\text{IPL1}}$ signal lines, and fetches the address of the service routine from fixed memory locations 64 Hex—7C Hex. For example, if the vector level is '0', then the processor will fetch the address of the service routine from memory location $64. This set of default vector addresses are part of the MC68000 family architecture.

4.5.6 Spurious Interrupt

During the interrupt acknowledgment cycle, if no device responds by asserting $\overline{\text{DSACKx}}$ or $\overline{\text{AVEC}}$, then the CPU enters into a software wait cycle and continues to sample these lines. Since the MC68020 does not have any internal timeout circuitry, it is the responsibility of the system designer to generate a $\overline{\text{BERR}}$ to terminate the vector acquisition phase. The CPU distinguishes this error from other bus errors. It enters into exception processing by fetching the spurious interrupt vector, and proceeds to execute the service routine specifically designed to resolve this situation.

4.5.7 Interrupt Interface Logic

Interrupt interface logic, designed to connect intelligent and nonintelligent devices to the MC68020 microcomputer system, is illustrated in Figure 4.16. It interfaces with four different I/O devices. Three of the devices can supply interrupt vector numbers, and the bottom one in the diagram will request the processor to internally generate the vector. Interrupting devices capable of providing vector numbers do so by placing the value on the least significant byte of the data port. Since an I/O device with an 8-bit port is connected to the D31 through D24 CPU data pins, its vector number will be placed on those pins. Again, the least significant byte of a 16-bit port is connected to the D23–D16 CPU data pins, and thus the vector number will be placed on those pins. The least significant byte of an I/O device with a 32-bit port is connected to the D7–D0 CPU data pins and, as a result, the vector number will be placed on those pins. That is, the CPU must first know the port size of the interrupting device before being able to determine which pins will have the vector number.

Let us look at the sequence of events that take place if all four interrupting devices happen to interrupt the CPU at the same time. That is, when interrupt request levels 3, 5, 6, and 7 are all asserted simultaneously. The 3-bit priority encoder chip 74LS148 will resolve the priority of the incoming requests, and will present the highest interrupt request level to the CPU. In this case, it will assert level 7. This request level is nonmaskable, so that at the completion of the current

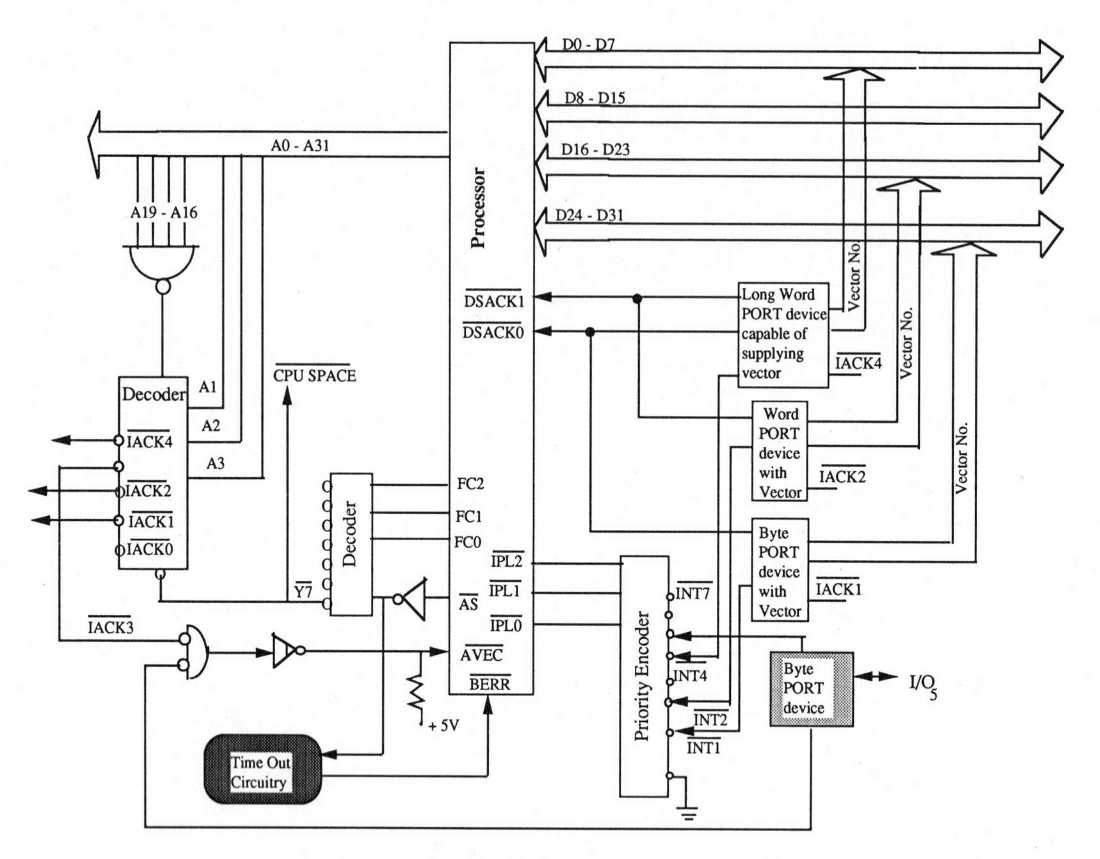

FIGURE 4.16. Interrupt interface logic example.

instruction execution, the CPU will initiate the exception processing, raising its internal mask bit to level 7, and generating an interrupt bus cycle for level 7.

The MC68020's function codes (FC0–FC2) are input to the 74LS138 decoder. As mentioned earlier, the function codes are all 1's, causing the $\overline{Y7}$ line to assert low, which, in turn, enables the 74LS138 on the extreme left. During interrupt acknowledgment, the address lines A19–A16 must also be decoded, along with the function code signal lines to determine the CPU space type. The address lines A19–A16 appear at the inputs of 4-wire NAND gates. When A19–A16 are all 1s, it fully defines the bus cycle as an interrupt acknowledgment. Now, the other input of the second 74LS138 is enabled. This decoder monitors the address lines A1–A3, which define the interrupt level being recognized. Since level 4 is recognized first, it causes the bottom output, called $\overline{ICAK4}$, to be asserted. This signal is fed back to the requesting device that it is connected to as a 32-bit I/O port. In response, the 32-bit I/O port places its vector number on data pins D7–D0, and asserts both $\overline{DSACK0}$ and $\overline{DSACK1}$. The MC68020 receives the vector, multiplies it by 4, and adds it to the value of the Vector Base Register (VBR). This resultant effective address is then used to fetch the 32-bit starting address of the service routine. The service routine is then accessed from the memory and the program execution begins.

At the completion of the service routine, the level 4 interrupt is removed. The last instruction of the service routine, RTE, restores the context that existed prior to the level 4 interrupt request. A similar sequence of events will follow for the level 2 interrupt. The interface logic will generate $\overline{IACK2}$, and the 16-bit I/O device will place its vector number on the D23–D16 pins of the data bus, and will only assert $\overline{DSACK1}$. Upon completion of the level 2 interrupt, the CPU will initiate the processing of interrupt level 1. A similar sequence of events will take place during the service of this interrupt. When $\overline{IACK1}$ is asserted, the 8-bit I/O device will place its vector number on the D31–D24 pins of the CPU'S data bus, and will only assert $\overline{DSACK0}$. The interrupt request 3 will be recognized, assuming that it arrived now. When $\overline{IACK3}$ is generated, the 74LS02 gate will assert $\overline{AVEC}$. In response, the CPU will ignore the data bus, and internally generate a vector number corresponding to the interrupt request. In this case, the autovector entry level 3 will be selected. The processor will fetch the memory location $68 for the address of the interrupt service routine.

In the example, if (for some reason) an interrupt level 4 is asserted due to a noise glitch, and remains long enough for the processor to deskew and recognize it, the CPU would generate $\overline{IACK4}$ for level 4. No I/O device would respond with the vector number and $\overline{DSACKx}$ or $\overline{AVEC}$ would not be asserted. As a result, the CPU would continuously insert a 'wait state' until the bus timeout circuitry causes a bus error. In this case, the processor would enter into a spurious interrupt service routine.

4.5.8 Daisy-Chain Configuration

A system designer may decide to utilize a single-line vector interrupt system. In such a case, a daisy-chain configuration is generally used for implementing the priority arbitration logic. The external devices are placed in a chain, where any number of devices may request an interrupt simultaneously. The device that is physically closest to the MC68020 has the highest priority and is serviced first, at the end of a proper interrupt signal acknowledgment. All lower-priority devices down the chain are inhibited from interrupting the CPU, while the higher-priority device's interrupt is being serviced.

4.6 SYSTEM CONTROL

As outlined earlier, the system control bus allows the designer to place the MC68020 in a known state.

4.6.1 RESET ($\overline{RESET}$)

The $\overline{RESET}$ is a bidirectional signal line. Using this, an external entity resets the system, or the processor resets external devices by executing the RESET instruction. When power is applied to an MC68020-based system, external circuitry must assert the $\overline{\text{RESET}}$ line, and hold that for a minimum of 100 msec. Figure 4.17 illustrates the timing diagram for a power-up reset operation. It highlights the relationship among $\overline{\text{RESET}}$, Vcc, and bus signals.

Upon detecting an externally generated reset, the MC68020 responds by completing the active bus cycle in an orderly fashion, and then reads the contents of vector table entry 0 from address $00000000, and loads the value into the Interrupt Stack Pointer (ISP). The CPU then reads vector table entry 1 from address $00000004, and loads that into the Program Counter (PC). The MC68020 also initializes the interrupt mask level in the SR to seven, clears the T1, T0, and M bits, and sets the S bit. The vector base register is also initialized to $00000000. The processor enables the cache control bit in the cache control register, invalidating the cache entries. The processor does not alter other internal registers.

When a RESET instruction is executed, the CPU drives the $\overline{\text{RESET}}$ signal for 512 clock cycles. In this case, the MC68020 resets the external devices connected to the system, and the internal registers remain unchanged. The external devices connected to the $\overline{\text{RESET}}$ pin get reset at the completion of the reset instruction. An external $\overline{\text{RESET}}$ signal that is asserted to the processor during the execution of a RESET instruction must extend beyond the reset period of the instruction by at least 8 clock cycles in order to reset the processor.

Figure 4.18 shows an example of reset logic. The logic device contains the multivibrator MC3456 (one can use an equivalent device), which generates a $\overline{\text{RESET}}$ signal of 100 msec width during the powerup reset operation. The bottom part of the logic is designed to provide for a push-button or single-shot reset.

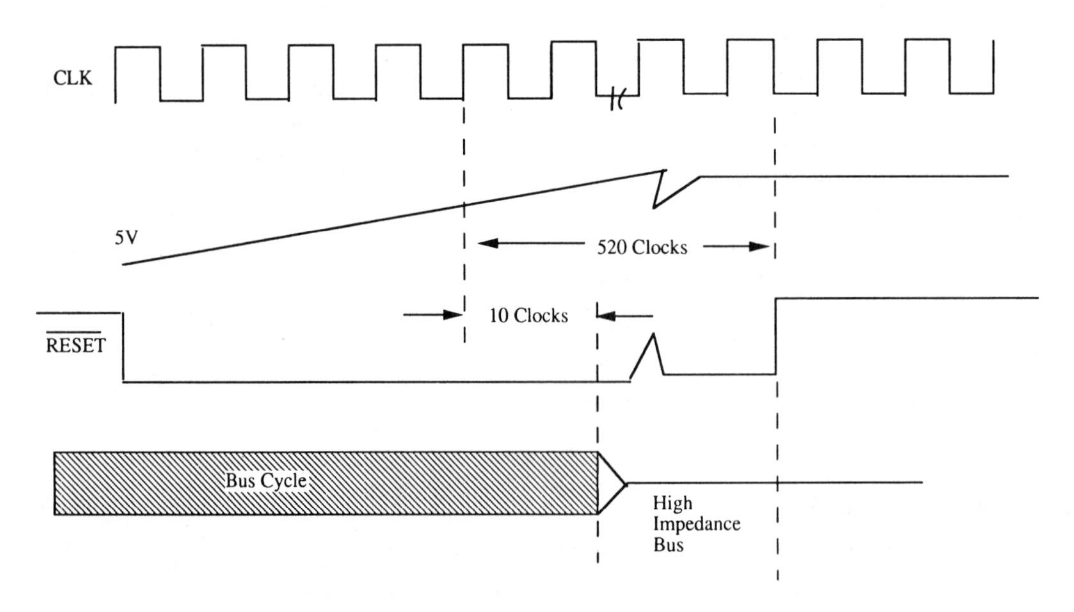

FIGURE 4.17. Initial reset timing operation.

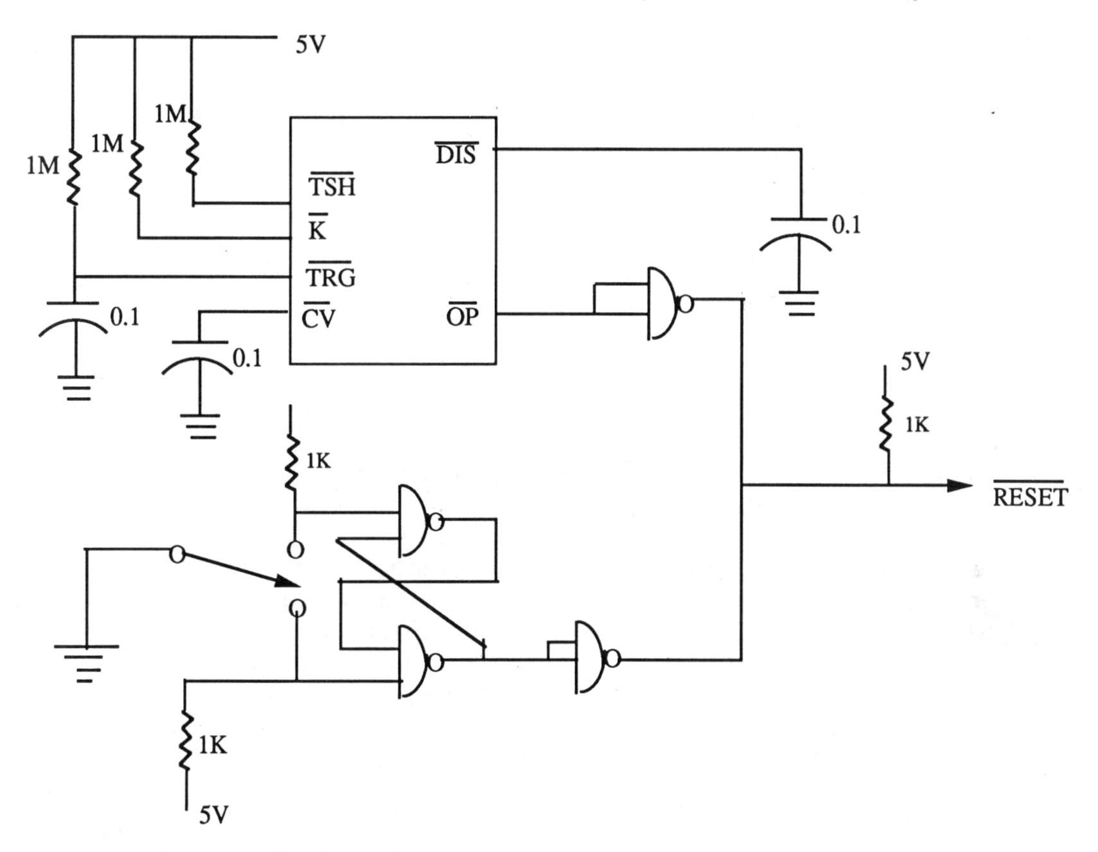

FIGURE 4.18. An example of single-pulse reset logic.

4.6.2 Halt ($\overline{\text{HALT}}$)

An input to this pin tells the processor that it is to suspend bus activities. $\overline{\text{HALT}}$ becomes an output when the MC68020 determines that the system is unusable. The processor labels the system as unusable if, during the processing of an exception, it cannot save its state in a supervisory stack, or cannot read a table entry that points to a service routine.

When the $\overline{\text{HALT}}$ signal is asserted, and $\overline{\text{BERR}}$ is not asserted, the MC68020 halts all external bus activities at the next cycle boundary. The processor does not terminate a bus cycle. Rather, it will continue the execution until the necessary operands are not found in the internal cache memory.

Upon recognizing $\overline{\text{HALT}}$, the CPU places the data bus into a high-impedance state, drives the control bus into an inactive state (not high-impatience state), and retains the previous states of the address bus, function code, and size pins, along with the R/$\overline{\text{W}}$ pin. The halt operation does not interfere with bus arbitration. The CPU also does not service interrupt requests while it is halted, but asserts the $\overline{\text{IPEND}}$ signal when appropriate.

By negating and reasserting $\overline{HALT}$, one can create a single-step (i.e., bus-cycle by bus-cycle) operation. Such a mode allows the user to debug or step through one bus cycle at a time. It is important to realize that a bus error during a halt assertion will initiate a retry operation by the CPU. Such single-step operations and software trace capabilities allow the system debugger to trace bus cycles, execute instructions when required, and initiate change to the program flow. These capabilities, when coupled with software debugging tools, create a powerful debugging environment.

4.6.3 Bus Error ($\overline{BERR}$)

It is quite feasible that the MC68020 may get into a software wait cycle during asynchronous transfer. This is due to the nonavailability of the $\overline{DSACKx}$ signal. If this happens, then the execution of the current instruction would not be completed. Instead, the processor would hang-up at the instruction and this is known as a *bus error*. The bus error is not detected by the MC68020, but instead must be detected by external circuitry, and signaled to the processor. External logic would assert $\overline{BERR}$ to indicate a bus error. In response, the processor aborts the current bus cycle, negates all strobes, and initiates an exception processing. The address of the exception routine is defined by vector 2. This routine can be written in such a way that it would attempt to correct the problem, to display the bus error condition, or print the 'address' that caused the bus error.

The prefetch capability and pipeline architecture of the MC68020 require that the CPU process bus errors in different ways. If the bus error occurs during operand read or write, then the CPU initiates exception processing immediately. If the bus error occurs during instruction prefetch, however, then the CPU defers exception processing until the instruction is actually needed, the flag in the pipeline unit indicates that it contains unusable data, or the cache entry is found to be invalid.

The $\overline{BERR}$ signal in also used to implement the memory protection and virtual memory system. In a protected memory scheme, a memory management unit maps and confines users within the areas assigned to them. If a user tries to access an area of memory previously defined, the MMU unit detects that as faulted access and asserts the $\overline{BERR}$.

In a virtual memory system, if a faulted access is detected, then it can mean that the required page is not in the memory and must be copied from the disk, so that the program can continue its execution. Again, during the access of a RAM memory module, if an error is detected and the correction logic has determined that the data error cannot be corrected, then the correction logic may assert $\overline{BERR}$. In such situations, the bus error service routine is required to be such that it can identify and/or resolve the errors.

4.6.4 Retry Operation

When both $\overline{BERR}$ and $\overline{HALT}$ are asserted by an external device the CPU will retry the bus cycle. Here, the MC68020 terminates the bus cycle, places the control signals in their inactive state, and does not begin another bus cycle until the HALT signal is negated by the external logic. It retries the previous bus cycle, using the same access information, after an appropriate synchronization delay. The external device must negate the BERR signal before S2 of the read cycle.

4.6.5 Double Bus Fault

When a bus error or address error occurs, the processor initiates an exception processing. During this exception processing, if the CPU fails to stack the necessary information into the stack so that it can recover from the error condition, this second error is then called a *double bus fault.* A bus error during an externally generated reset is also called a double bus fault. After reset, the CPU fetches the address of the ISP and PC to restart program execution. If a bus error or address error occurs during the reading of the vector table, the processor will consider that to be a double bus fault and will halt its operation.

However, a retried bus cycle does not constitute a bus error or contribute to a double bus fault. The CPU continues to retry the same bus cycle as long as the external hardware requires it. Only an external reset operation can restart a halted processor, and a bus arbitration can still occur.

4.6.6 The DTACKx, BERR, and HALT Relationship

The external hardware can request the CPU to re-execute or rerun an existing bus cycle by asserting both the HALT and BERR signals simultaneously. The BERR and HALT should be held asserted until the address strobe and data strobe are negated. BERR should then be negated before or with the HALT signal. Any type of bus cycle may be rerun, including the read or write bus cycle of the indivisible read-modify-write operation. This is possible because the read modify control signal remains asserted during the entire sequence. Another variation of the retry is called relinquish and retry, where the external hardware can request that the CPU relinquish control of the bus, and then rerun its last bus cycle when control of the bus has been returned.

To request this operation, bus error (BERR), halt (HALT), and bus request (BR) must all be asserted at the same time. When the CPU ends the bus cycle and negate its strobes, the external hardware should negate the bus error before or with the halt. It declares the owning of the bus by asserting a bus grant acknowledgment and then negating the bus request. Bus grant acknowledgment should be held asserted until it is desired to let the CPU back on to the bus. As soon as the bus grant acknowledgment is negated, the CPU will regain ownership of the bus and rerun the last bus cycle. The page memory management unit utilizes the relinquish and retry operation during the execution of a table-walking algorithm. System designers who use the relinquish and retry operation must pay close attention in situations where read-modify control is asserted, since the CPU will not relinquish the bus during these indivisible operations. Any device that requires that the CPU give up the bus and retry the bus cycle during the read-modify bus cycle must assert a bus error. However, a bus request and a halt must not be included. The bus error-handler software should examine the information placed on the stack during bus error exception processing, and take appropriate action to resolve the fault.

4.6.7 System Control Timing

The system control timing shown in Figure 4.19 provides an understanding of the timing relationships among $\overline{DSACKx}$, $\overline{HALT}$, and $\overline{BERR}$. All timings for this diagram are taken from the timing table of the MC68020, operating at 16.67 MHz. Bubble number 47a specifies the asynchronous input setup time required for asynchronous input signals, and includes the bus error signal if it is asserted in the absence of $\overline{DSACKx}$. In order to guarantee that the bus error will be recognized on an individual falling edge of the clock, it must be asserted at least 5 nsec beforehand. Bubble number 47b and bubble number 47a together specify the

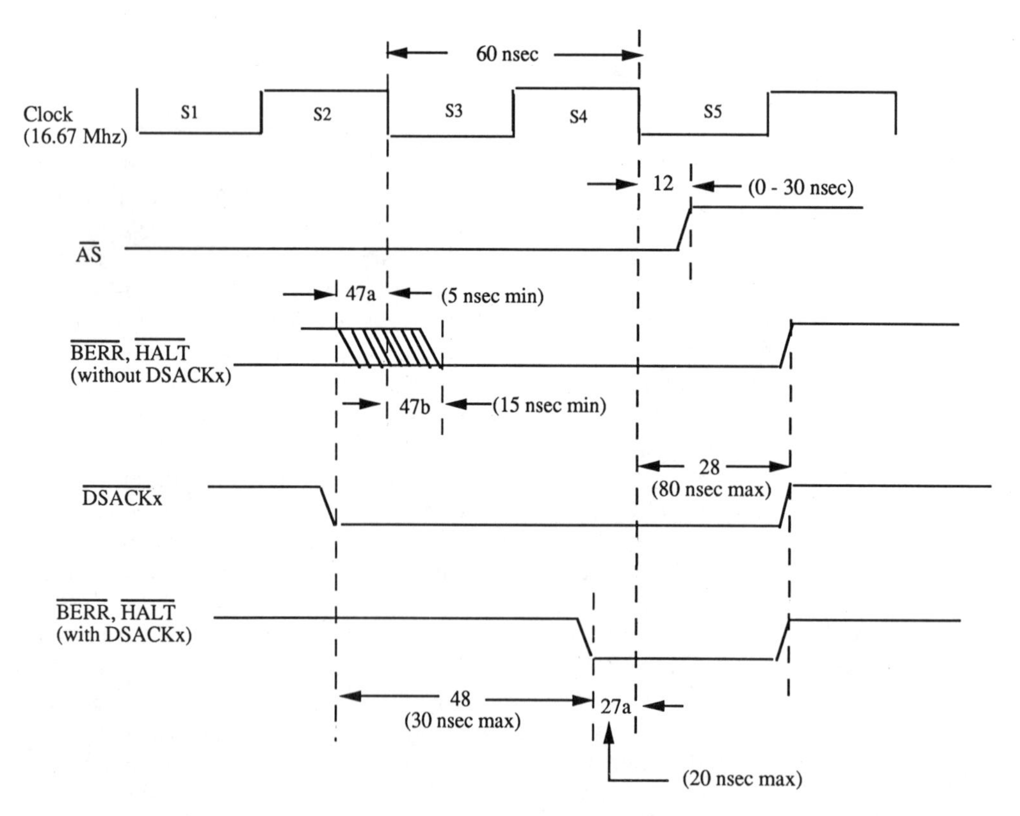

FIGURE 4.19. The timing relationships among system control signals. (Courtesy Motorola Inc.).

sampling window. The asynchronous input is sampled between 5 nsec and 15 nsec from the falling edge of the clock. This means that all asynchronous input must be held asserted within the sampling window, in order to be recognized as asserted. However, $\overline{\mathrm{BERR}}$ can be asserted up to one clock cycle late after the $\overline{\mathrm{DSACKx}}$ is being asserted and still be recognized [MTTA2 87].

If $\overline{\mathrm{DSACKx}}$ meets the sample and hold-time requirements, as specified by bubble numbers 47a and 47b on a specific falling edge of the input clock, then the $\overline{\mathrm{BERR}}$ signal must also meet the setup time requirement of bubble number 27a on the next falling edge of the clock, in order to be recognized. Furthermore, $\overline{\mathrm{BERR}}$ must be held asserted throughout the sample window on that clock edge. Bubble number 48 indicates that, if a bus error is asserted, then the bus error will still be recognized regardless of when these asynchronous inputs were asserted relative to the CPU clock. This allows designers to issue a late $\overline{\mathrm{BERR}}$ signal without the knowledge of when the transition to the CPU clock has happened.

In order to properly control the termination of the bus cycle, the $\overline{\mathrm{BERR}}$ and $\overline{\mathrm{HALT}}$ should be asserted or negated on the rising edge of the clock. This ensures that when the two signals are asserted simultaneously, the required setup time bubble number 47a will be met in the same states for both signals. A rerun request can also be issued late, up to a clock cycle after the $\overline{\mathrm{DSACKx}}$ being asserted, and still be recognized by the CPU. If the $\overline{\mathrm{DSACKx}}$ meets the setup time requirement of bubble number 27a on the next falling edge of the clock, then a late bus error and late rerun will have the same timing.

Chapter 5

Programming Techniques

The software capabilities of the MC68020 were introduced in Chapter 3, in terms of instructions and addressing modes. Programming examples were used to illustrate data movement, and perform calculations and various functions. In this chapter, we will explore various programming techniques that are useful for solving complex problems. The emphasis is on the implementation and the utilization of the high-level syntax generally found in modern programming languages.

We will introduce the structured programming technique in this chapter and its implementation, utilizing the instructions supported by the MC68000 family of processors. Structured programming relies heavily on the support of data structures like array, link list, tables, etc. We will highlight their creation and manipulation via examples in this chapter. However, only a few of the many topics of concern in structured programming will be addressed here.

At the end of the chapter, we will outline program design techniques and develop guidelines and how to apply them to design complex programs. We will also design a speech synthesis program for the MC68020 as an example of complex structured programing.

5.1 MC68020 PROGRAMMING–A GLOBAL VIEW

An assembly language program written for the MC68000 family of processors is divided into four sections. In many cases, these sections can be mixed to provide for a better design of the algorithm. We have seen all of the elements in Chapter 3. Now we will formalize them. The sections are:

Assembly directives: These are provided by the user to the assembler for defining data and symbols, setting the assembler and linking conditions, and specifying output format. The directives do not produce machine code.

Assembly language instruction: These are the MC68000 family instructions, including those defined with labels.

Data storage directives: These allocate data storage locations containing initialized or uninitialized data.

END directive: This is the last statement in an assembly language source code and causes termination of the assembly process.

5.1.1 Assembly Language Program Body

Each line of the MC68020 program, excluding special constructs, is made up of four distinct fields, where not all may be present at all times. The order of the fields are:

- Label
- Opcode
- Operand/operands
- Comment field

The *label* is a symbolic name for a memory address, and may be defined only once in the program. The assembler vendors generally publish a set of guideline on their usage. In general, it begins with a character (A-Z), and may be followed by letters, digits, or special symbols. The length of the label will vary from one compiler to another. However, none will allow a blank character within the label field. We will always use the following convention: A label will begin in column 1 and will terminate with a colon (:), and will never be greater than eight characters long.

The *opcode field* contains symbolic names for the machine instructions and assembler directives. The assembler supports preformat instructions.

The *operand field* specifies the source and destination for the read and write operations of the instruction. We will always follow Motorola's conventions of assembly Language programming throughout this book.

The *comment field* is the last field on a line, and provides room for program documentation. This field will begin with Semicolon (;). For a comment field starting at column 1, our assembler utilizes "/" to start and terminate it. The comment is ignored by the assembler, but is included in any listing of the source program. Comments are a valuable documentation tool. The user should use them liberally. Other assemblers may use slightly different conventions.

5.1.2 Storage Definition

The MC68020 assembler supports three storage allocation directives to reserve data space. They are:

5.1.2.1 Define Constant (DC)

The syntax is:

DC.<size> <constant>

This allows for both the allocation and initialization of data space with constant values. The size parameters may be 'B' (Byte), 'W' (Word), and 'L' (Long-word). The constants can be numbers or characters enclosed by single quotation marks.

5.1.2.2 Data Storage (DS)

The syntax is:

DS.<size> <count>

This directive only reserves memory locations. The contents in the locations are not initialized. The size parameters may be 'B' (Byte), 'W' (Word), or 'L' (Long-word). The count identifies the number of memory location to be set aside by the assembler.

5.1.2.3 Data Constant Block (DCB)

This has the following syntax:

DCB.<size> <length>, <value>

This tells the assembler to reserve a block of initialized memory locations. The size of each element of the block is specified by the *size* field. The *length* identifies the number of items in the block, and the data item is initialized to *value.*

5.2 IMPLEMENTATION OF HIGH-LEVEL LANGUAGE CONSTRUCTS

Today's high-level languages provide constructs for conditional code execution and loops. They are essential to the implementation of any algorithm. These constructions get translated into test and branch instructions supported by the processor. The MC68020 not only implemented the basic test and branch instructions, but also enhanced them in such a way that programmers or compiler writers can take advantage of the enhanced features to generate compact codes. In this section, we will examine the high-level constructs and their equivalent assembly language implementation syntax, with examples.

5.2.1 If . . . Then Construct

This is a fundamental construct of any high-level language. The syntax is:

```
IF <boolean expression>
  THEN
    statement-1
```

```
        ELSE
          statement-2
      ENDIF
```

Here, the CPU first evaluates the boolean expression and, if the result yields a TRUE, then it executes 'statement-1'; otherwise, it executes 'statement-2'.

The assembly language implementation is:

```
         <set CCR bits>
         Bcc <ELSE>
              statement-1
            :
            :
       ELSE:  statement-2
            :
```

5.2.2 If-Then-Else Construct

Very often, a program has to make an 'either-or' choice. If the condition is TRUE, then it is required to execute 'statement-1', otherwise it executes 'statement-2'. The syntax in the high-level language is:

```
IF   <boolean expression>
  THEN
    statement-1
  ELSE
    statement-2
    next instruction
```

The equivalent assembler syntax is:

```
         <set CCR bits>
         Bcc <ELSE>
            statement-1
            BRA <NEXT>
       ELSE:  statement-2
       NEXT:  statement-3
              :
              :
```

A conditional branch will direct the program control to 'statement-2', and the instructions in sequence. In the case where 'statement-1' is executed, an unconditional BRA branch transfers control to 'statement-2'.

5.2.3 While . . . Do Construct

A program loop is a natural extension of a simple conditional construct. The most common form of a loop is given by the WHILE . . . DO construct, whose syntax is given by:

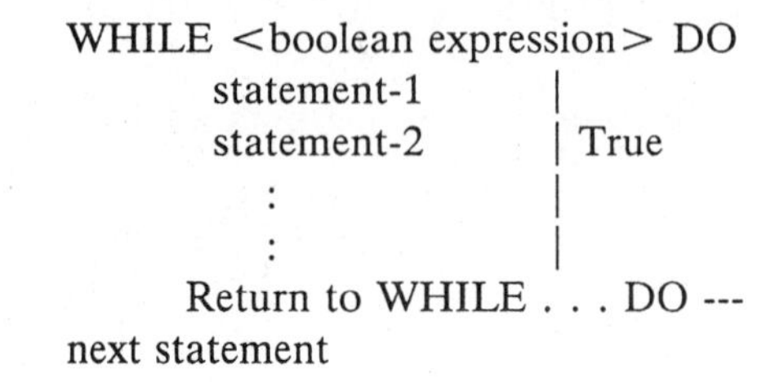

WHILE <boolean expression> DO
statement-1
statement-2
True
:
:
Return to WHILE . . . DO ---
next statement

The most common form of assembly language implementation is:

```
        <set CCR bits>
WHILE:  Bcc <NEXT>
        statement-1
        statement-2
           :
           :
        BRA <WHILE>
NEXT:      next statement
```

The value of the expression is first stored in a data register. The boolean expression is tested by a conditional branch. The return is implemented by an unconditional BRA branch.

5.2.4 Loop . . . Exit Construct

Today's programming languages implement loop constructs in numerous ways. A popular choice is to unify these constructs in a single, all-purpose LOOP . . . EXIT statement. The 'exit' condition can be tested at any appropriate point in the statement sequence. If the result is TRUE, an exit from the loop is performed. The syntax for this implementation is:

```
-> LOOP
      statement-1
      statement-2
FALSE| EXIT WHEN condition
      statement-3        |
         :               | TRUE
    END LOOP
    next statement
```

The equivalent assembly language syntax is:

```
-> LOOP
  |      statement-1
  |      evaluate condition
FALSE|  Bcc <NEXT>             ----|
  |      statement-3       |
  |         :           | TRUE
  |-- END LOOP             |
   NEXT: next statement <-------|
```

In a loop control situation, as described above, the object is, in many cases, to get out of the loop if the condition code is met. This is illustrated in the following example.

Example: Check the validity of the positive numbers in a table. If you find a negative number, then stop checking. Assume that A0 points to the start of the table in memory, and the table has a maximum of 50 entries that are word size. If the program fails to find a negative number, then make D1 = −1 and D0 = 0. On the other hand, if the program finds a negative number, then copy that into D0 and record the entry number in D1.

```
        CLR.W   D0          ; initialize D0
        MOVE.M $49, D1      ; set up the counter to scan the
                            ; table
SCAN:   TST.W (A0)+         ; simply compare operands to 0 and
                            ; set the Z flag if 0, or else set N=1
        DBMI D1, SCAN       ; decrement the counter and scan the
                            ; table until all entries are
                            ; checked or the first negative number
                            ; is found.
        TST.W   D1          ; Test D1 to check for 0 or negative
        BMI     NONE        ; D1 = -1, then no negative number found
                            ; if so, then go to NONE
        MOVE.W              ; D1 not equal to -1. The program
        -2(A0), D0
                            ; will fall through, because the 'MI'
                            ; condition was met during the test,
                            ; i.e., a negative number is found.
                            ; retrieve the entry number from the
                            ; table. The last iteration decrements the
                            ; counter by 2 and places the result in D0.
        SUBI.W #50, D1      ; Calculate D1-50, where D1 has been
                            ; decremented n times from its
                            ; initial value of 49. This entry
                            ; contains the negative value.
```

```
        NEG.W   D1              ; Make it a positive number, and
NONE:   next statement          ; continue with the program because
                                ; no negative number is found.
```

It is important to note that the DBcc instruction can be utilized to extract all kinds of numbers from the table. For example:

```
SCAN:   CMPI.W #$0D, (A0)+   ;
        DBEQ, SCAN           ; Keep looking until you find a
                             ; carriage return in the table
                             ; or no entry is left.
```

Example: Find the last 0 entry in a table of any length. Assume that (A0) points to the base of the table, whose first word holds the number of signed words that follow. Record the last 0's entry position in D1. If no 0's are found, then set D1 = −1. If the table is empty, then set D1 = 0, or else make D0 = 0.

```
        MOVE.W (A0)+, D1    ; Load D1 with the length of the
                            ; table. A0 now points to the first
                            ; entry of the table.
        SEQ   D0            ; Set if equal. If D1=0 then the table
                            ; is empty. If so, the Z flag would be
                            ; set. In that case, set D0=FF, or
                            ; else 00. Thus, D0=FF signifies an
                            ; empty table, while D0=00 signifies a
                            ; nonempty table.
        BEQ   EMPTY         ; D1 was 0. Branch to empty. Nothing
                            ; needs to be done or searched.
        MOVE.W  D1, D2      ; Place the number of words in the table into
                            ; D2 as well.
        ASL.W  #1, D2       ; D2 now contains the number of bytes in the
                            ; table.
        ADDA.W  D2, A0      ; A0 will now point to one word
                            ; beyond the end of the table.
SCAN:   TST.W  -(A0)        ; Test the last word entry in the table.
                            ; Note the predecrement on A0
                            ; in view of ADDA.W D2, A0 above.
        DBEQ D1, SCAN       ; As long as it is not equal to 0, and D1 is
                            ; not equal to -1, continue.
        TST.W  D1           ; We are here because either a 0
                            ; element was found, or because none
                            ; was found and D1 = -1. We are asking
                            : whether
                            ; D1 is negative, i.e., if no 0 entry was
                            ; found in the table.
        BMI   NONE          ; D1 is negative. Table
```

```
                          ; unsuccessfully searched. Branch to
                          ; NONE with D1 = -1.
ADDQ.W #1, D1             ; D1 = D1 + 1. This compensates for the
                          ; SUBQ.W  #1,D1 in the previous
                          ; instruction. D1 now has the actual
                          ; position of the last 0 element.
NONE:   next instruction
                          ;
EMPTY:  continue
                          ; We reached here with D1 = 0 and
                          ; D0 = FF, indicating an empty table.
```

5.2.5 The For Loop Construct

This is a special form of the WHILE . . . DO construct, where a fixed number of iterations are performed. When a FOR instruction is entered, its counter variable is assigned the initial value. The counter value is incremented each time that the FOR statement's action is carried out. When the counter variable represents the final value, the LOOP iterates one last time and the program moves on to the next statement. The syntax for this statement is:

```
FOR counter = Initial_Value TO Final_Value
  DO
    statement-1
    statement-2
ENDFOR.
```

The assembly language implementation for this is given via an example.

Example: Design a program that will take a positive number and generate a sum total according to the following formula:

$$SUM = 1+2+3. \; . \; . \; . \; +N$$

This program calls two subroutines, HEXIN and HEXOUT. The first routine will extract the positive number, and the second routine will output the sum.

```
START:   JSR HEXIN
         MOVE.W  D0, D2     ; Make a copy of N.
         CLR.W   D0         ; Clear D0 to hold the sum.
         MOVEQ   #1, D1     ; Initialize the counter
LOOP:    CMP.W   D2, D1     ; completed the number of iterations.
         BGT     FINISH     ; Exit loop if D1 > N.
         ADD.W   D1, D2     ; Calculate the sum.
         ADDI.W  #1, D1     ; Increment the counter.
         BRA     LOOP       ; Go through the loop.
FINISH:  JSR     HEXOUT     ; Output the sum.
```

```
        JSR     STOP
        END
```

5.3 NEW ENHANCED INSTRUCTIONS

One of the goals of the MC68020 designers was to enhance the richness of the instructions available in the MC68000 processor. They have added several new instructions, and enhanced the capabilities of various old instructions. The new instructions provide more flexibility, and we will describe the usage of several of them through examples in the following subsections.

5.3.1 Bit Field Instruction

In programs written in high-level languages, variables that will not assume very large values are often packed in the memory through the use of bit fields. For example, a 6-bit, a 3-bit, and two 3-bit variables can be packed into a 16-bit integer, rather than using an entire integer for each variable, thereby saving memory space. Without the bit field instruction, one would have to design programs like the following:

```
FETCH:  MOVE.W  #$0380, D0     ; load Mask.
        ADD.W   [EA], D0       ; Get field.
        LSR.W   #7, D0         ; Justify.
STORE:  LSL.W   #7, D0         ; Shift field.
        MOVE.W  #$FC7F, D1     ; Load mask.
        AND.W   [EA], D1       ; Get word.
        OR.W    D1, D1         ; Add field.
        MOVE.W  D1, [EA]       ; Store it.
```

Such implementation requires a significant amount of memory, and does not make good use of the register set to hold the temporary result.

The bit field operations of the MC68020 are designed to implement such operations with a single instruction. These instructions are very flexible and fast. The general format for a bit field instruction is:

```
BFnn  <ea> {offset, width},Dn
where;
        'nn'     is the bit field specifier
        <ea>     is any general addressing mode
        'offset' is the number of bits from the
                 effective address where the field starts
        'width'  is how wide the bit field is.
```

Both offset and width can have constant values specified in the instruction stream, or variable values contained in data registers. The offset has a range from

-2^{31} to $2^{31}-1$, and the width of the field can be from 1 to 32 bits. The instructions BFTST, BFSET, BFCLR, and BFCHG do not use a destination data register, since they simply read the specified bit field, set the condition codes according to the source value, and write the modified value as all 1's, all 0's, or the complement of the fields for the BFSET, BFCLR, and BFCHG instructions. Finally, the BFEXTS, BFEXTU, and BFINS instructions are used to move bit fields in and out of data registers. For example, the above routines can be replaced by two instructions:

```
FETCH:   BFEXTU   [EA] {#6, #3}, D0
STORE:   BFINS   D0, [EA] {#6, #3}
```

The offset calculation of a bit field starts from the most significant bit of the byte specified by the effective address to the most significant bit of the bit field and that bit field may span over any byte, word, or long-word boundary in an arbitrary fashion.

The bit field instructions are also intended for hardware control functions and complex data manipulations. For example, one might use a bit field data type to map variables onto hardware control registers of a device that a driver is being written for. Now, the bit field manipulation instructions, such as BFTST, BFSET, or BFCLR, can be used to toggle control registers in the device or to check status fields easily. Thus, the bit field instructions are very useful in developing fast executable device drivers.

5.3.2 CHK Instructions

The designers of the MC68020 have also enhanced the CHK instructions. Now, a 32-bit value in a data register can be compared to the lower bound of zero and the upper bound value operand found in the effective address. The syntax is:

```
CHK <ea>, Dn
```

If the data register value is less than zero or greater than the upper bound, an exception processing will occur and instruction execution resumes in the 'check service routine.' The 'N' is the only defined condition code bit when the check instruction completes. When $N=1$, the register content was less than zero. For $N=0$, the register value exceeded the upper bound. If CHK found that the data register value is within the bounds, then all condition codes are undefined, and the instruction execution continues to the next instruction in the instruction stream. The use of CHK is illustrated in the following example:

```
CHK.L   (A1), D1
```

In this case, D1 contains a value of $A002 and the memory location pointed by A1 contains $A000. If that is so,

0 < D1.L < $A000

The contents of D1 are compared against the operand pointed to by A1, which is $A000. Since the register value is greater than the upper bound, an exception processing is initiated.

The CHK2 has a syntax of;

CHK2 <ea>, Rn

The CHK2 instruction will set the carry bit if the data is out of bounds, and clear the carry bit if it is within bounds. The zero bit is set if the registers are within either bound, and cleared otherwise. This is illustrated in the following example:

CHK2.L (A1), D1

In this case, D1 contains $2000, and the memory locations pointed to by A1 contain $1000 and $2200, respectively, as illustrated below:

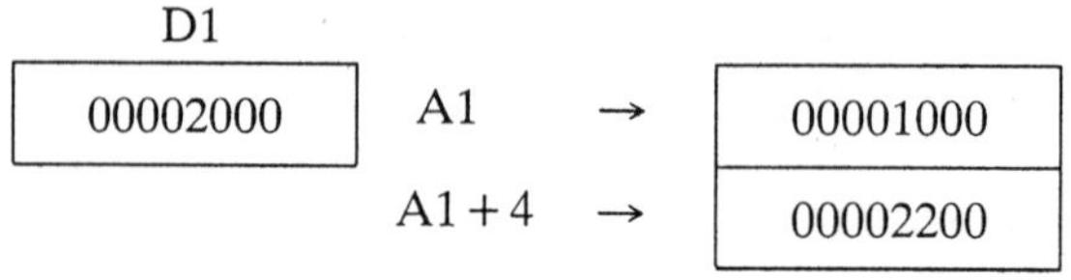

D1			
00002000	A1	→	00001000
	A1+4	→	00002200

In this case, $1000 < D2.L < $2200, no exception processing is initiated, and the processor will execute the next instruction in the instruction stream.

5.3.3 CMP2 Instructions

CMP2 has the syntax:

CMP2 <ea>, Rn

It compares the value in Rn against the bounds pair at the effective address location, and sets the condition codes accordingly. The upper bound and lower bound values must be in memory. The lower bound resides at the effective address location, and the upper bound resides immediately following, at the higher address. The size of the data to be compared and the bounds to be used may be specified as B/W/L in the instruction itself.

If the compared register is a data register and the operand size is byte or word, then only the appropriate low-order part of the data register is checked. If the checked register is an address register and the operand size is byte and word, then the bound operand is sign-extended to 32 bits, and the resultant operand is compared against the full 32 bits of the address register.

The CMP2 instruction will set the C (Carry) bit if the data is out of bounds, and clear the carry bit if it is within bounds. The Z (Zero) bit is set if the registers are within either bound, and cleared otherwise. The difference between CHK2 and CMP2 are that, if the value in the registers are out of bounds, then CHK2 causes an exception to occur. On the other hand, CMP2 does not. In the event of exception processing, CHK2 uses the same exception vector as the CHK instruction.

The CMP instruction performs either signed or unsigned bounds. The MC68020 automatically evaluates the relationship between the two bounds to determine which type of comparison is appropriate. If the programmer desires the bound value to be evaluated as a signed value, then the processor uses the arithmetically smaller value to be the lower bound. If the bounds are to be evaluated as unsigned values, then the logically smaller value should be the lower bound. This is shown in the following:

EA	→	Lower Bound
EA + size	→	Upper Bound

The operation of the CMP2 instruction is illustrated below:

Example 1: CMP2.W (A1), D1
where

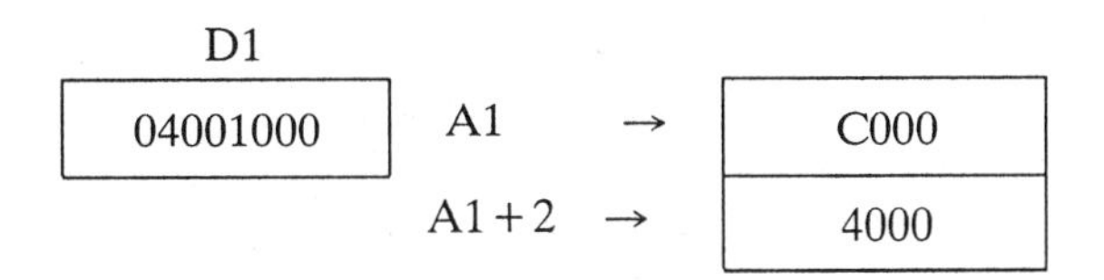

First Case: Signed comparison:
$-\$4000 < \text{D1.W} < +\$4000$, therefore: $C=0$
$-\$4000 < \text{D1.W} = +\$4000$, therefore $Z=0$
Second Case: Unsigned comparison:
$\$4000 < \text{D1.W} < +\$C000$, therefore: $C=1$
$\$4000 < \text{D1.W} = +\$C000$, therefore $Z=0$

Both CMP2 instruction examples are identical. The differences between the two are in how the upper bound and the lower bound are arranged. In the first case, the word value of $C000 resides in memory as the lower bound, and $4000 as the upper bound. With this arrangement, the programmer directs the CPU to interpret the bound values as the signed numbers. The 2's complement arithmetic $C000 has an actual value of −$4000. Thus, the instruction determines that the word contained in data register D1 is $1000, which is greater than −$4000 and less than $4000. As a result, the carry bit is then clear. The word in data register D1 is not equal to −$4000 and $4000, so that Z=0.

In the second case, the value $4000 is the specified lower bound, and $C000 as the upper bound. With this, the programmer directs the MC68020 to interpret the numbers as unsigned numbers. Now the values represent magnitudes with no sign involved. Thus, the instruction determines whether the word contained in D1, $1000, is greater than or equal to $4000, and less than or equal to $C000. The value in D1 is less than $4000. Thus, the carry bit is set to 1, indicating that it is out of bounds, since the word that D1 contains is not equal to either bound. The Z bit is cleared, N and V are undefined, and the X-bit remains undefined.

5.3.4 TRAPcc Instructions

The designers of the MC68020 have added a new trap instruction called TRAPcc, with the following syntax:

```
TRAPcc.L #<data>
```

This allows a conditional trap on any of the sixteen different conditions. The conditions are identical to the condition codes associated with the set-on-condition and decrement branch instructions. The trap-on-condition instruction evaluates the status of the condition code bit, and initiates an exception processing. The vector number is generated to reference the TRAPcc exception vector. The trap-on-condition instruction may have an additional word or two words specified, which follow the op-word. This word or long-word value consists of a parameter used by the trap condition handler to further define the requested resources that it should perform. The MC68020 makes no interpretation of these values, but must know how many words to follow the op-word, so that it can adjust the program counter to the beginning of the following instruction.

5.4 PROGRAMMING EXAMPLES

The fundamental data types supported by the MC68020 are signed or unsigned integers, BCD Integers, and Boolean variables. They are considered to be fundamental data types because processor instructions are available to process them directly. The MC68020 has extended these fundamental data types with the help of a coprocessor. The coprocessor can support any type of data type and structure like floating-point arithmetic, complex variables, scientific data types, etc.

The definition of new data types and the logical relationships defining their organization leads to the study of the *data structure.* The implementation of various data structures are beyond the scope of this book. However, because of their importance in realizing efficient programs, we will study several basic data types.

5.4.1 Adding an Entry to a List of Elements

A *list* is an example of a basic data structure. It consists of a set of elements of a single data type stored in contiguous locations in memory. We will write a program that will simply search the data elements in a list. The search will be performed element by element for any occurrence of the value that our program wants to add to the list. The element to be added to the list contains the register D7. If the element is already present in the list, the program will return $AA in D6 without duplicating it into the list. Otherwise, the program will insert the element into the list and will return <$00 (NULL). The starting address of the list is found in A0. The first element of the list contains the length of the list. One solution of this problem is given in the following:

```
ADDLIST:  MOVEA.L A0, A1       ; Make a copy of the list pointer
          MOVE.W  (A1)+, D1    ; Extract the length of the list from the first element of the list.
          SUBQ  #1, D1         ; Subtract -1 from D1.
NITEM:    CMP  (A1)+, D7       ; Match found? If yes,
          BEQ.S  ITIS          ; it is in the list.
          DBF  D1, NITEM       ; Match not found, continue.
          MOVE.W  D7, (A1)     ; Add the number to the list.
          ADDQ  #1, (A0)       ; Update the number of the element
                               ; count. This is the first element of the list.
          MOVE.W  #$00, D7     ; Insert null into D7.
ITIS:     MOVE.W  #$AA, D6     ; Duplicate item, copy AA in D6.
          STOP
          END
```

5.4.2 Deleting an Element

We will now develop a program that will delete an element from the same list. The example program is given below:

```
DELLIST:  MOVEA.L A0, A1       ; Keep a copy of the starting
                               ; address of the list.
          MOVE.W (A1)+, D1     ; Extract the length of
                               ; the list from the first
                               ; element of the list.
          SUBQ  #1, D1         ; Subtract -1 from D1.
NITEM:    CMP  (A1)+, D7       ; Match found? If
          BEQ.S  ITIS          ; yes, go and delete from list.
          DBF  D1, NITEM       ; Match not found, continue.
          BRA.S DONE           ; All done.
```

```
ITIS:     MOVE.W (A1)+, -4(A1)   ; Move a word up the list.
          DBF  D1, ITIS          ; Have all elements been moved?
          SUBQ #1, (A0)          ; Yes, subtract 1 from length.
          MOVE.W #$00, D7        ; Insert null into D7.
          JMP  FINISH            ; Job is done.
DONE:     MOVE.W #$AA, D6        ; Duplicate item, copy AA in D6.
FINISH:   STOP
          END
```

This program is very similar to the previous one. The NITEM loop compares each element in the list against the value in D7. If a match occurs, then the program branches to the ITIS loop, which moves all subsequent elements up one word location, and then decrements the element length counter, which is the first element of the list.

5.4.3 Finding the Minimum and Maximum Value from a List

We will solve this problem by picking the first element from the list, and assuming that the element has the minimum and maximum values. We will then compare each remaining element in the list to that minimum value. If the program finds a value that is less than the minimum value, then that element becomes the new minimum value. Likewise, if the program encounters a value that is greater than the maximum value, then that element becomes the new maximum.

```
MIN:      DS.W   1               ; Reserve a location for the minimum value.
MAX:      DS.W   1               ; Reserve a location for the maximum value.
LIST:     MOVE.W (A0)+, D1       ; Copy starting location of the list.
          SUBQ #1, D1            ; Decrement D1.
          MOVE.W (A0), MIN       ; Initially make it first.
          MOVE.W (A0)+, MAX      ; Make the next element as the maximum.
CMIN:     MOVE.W (A0)+, D7       ; Load the next element for comparison.
          CMP    MIN, D7         ; If yes, update minimum value.
          BEQ    COUNT
          BCC.S  CMAX
          MOVE.W  D7, MIN        ; Found the minimum, update minimum.
          BRA    COUNT
CMAX:     CMP    MAX, D7         ; Is this a new maximum value?
          BLS    COUNT
          MOVE.W  D7, MAX        ; New maximum, update the value.
COUNT:    DBF    D1, CMIN        ; Have we checked the entire list?
          STOP                   ; Job done.
          END
```

The algorithm works as follows: First, we reserve the data space for MIN and MAX. We then load the element count, and decrement the counter register D1 by 1. We then assume the second element as the current minimum, and the third element as the current maximum. We then load the next element from the array into D7, and then compare it against MIN. Now, we may encounter one of the three situations:

1. If D7 = MIN (result due to CMP instruction), the program branches to the instruction labeled by COUNT, and checks whether all elements in the list have been processed.
2. If D7 is greater than MIN, then the program branches to CMAX. Now D7 is compared against current MAX element.
3. If the value in D7 is less than the current MIN, the program executes the next instruction in the sequence and updates MIN with the contents of D7.

The execution continues until the entire list is searched. It should be realized that this program will only work if all elements in the array are positive numbers.

5.4.4 Linked List

We have seen in the previous section that elements are ordered successively and occupy contiguous blocks of memory. In order to manipulate elements, we need a more efficient data structure. One such data structure is the *link list,* which does not require contiguous storage of its elements.

The diagram in Figure 5.1 represents a simple link list. Each item in the list contains two pieces of data, namely, the element and the pointer. The pointer indicates where other pieces of the item are located. The item also contains a special pointer to indicate the first member of the list. In the diagram, the last item contains a special symbol to indicate the end of the list. One can build a more complex data structure based on this simple concept.

Implementing a link list is fairly simple, because of the powerful addressing mode supported by the MC68000 processor family. The following example shows a subroutine that creates a link list in a free area of memory with the beginning address LLIST. The items in the list are each ten bytes in length, and the list will be initialized to contain 16 entries. The segment of codes starting at the label LINK computes the address of the next item of the list, and then stores the address as the pointer to that item. The last pointer value is set to NULL.

```
                        ; Create a Link List

         NULL      EQU 0
LNLIST:  MOVEM.L D0-D1/A0-A1, -(SP)     ; Save the register on stack.
         LEA      LLIST, A0             :
         MOVE.L   A0, FIRST             ; Set first to point to LIST.
         MOVE.L   #$A, D0               ; Set up bytes/item.
         MOVE.L   #$10 , D1             ; Set up the number of items in
                                        ; list
```

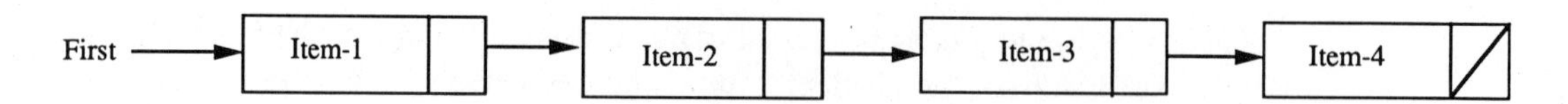

(a) A simple LINK list

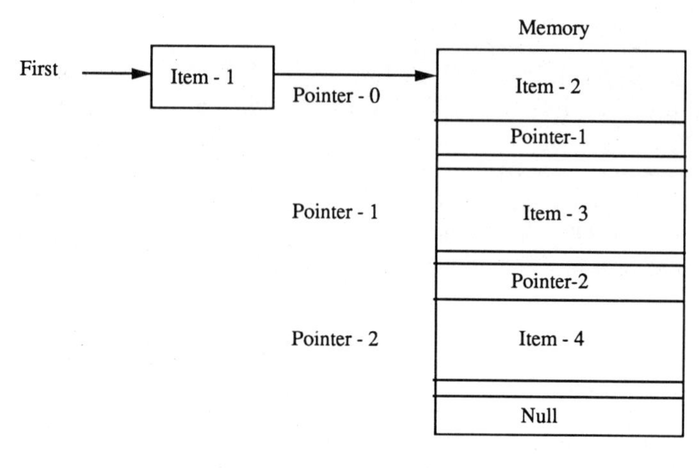

(b) Memory Allocation

FIGURE 5.1. A LINK list and memory allocation.

```
LINK:   MOVE.L A0, A1                    ; Item address into A1.
        ADD.L  D0, A0                    ; Compute the next item.
        SUBQ.L #1, D1                    ; Decrement the item number.
        BNE    LINK                      ; Continue until D1=0.
        MOVE.L #NULL, (A1)               ; Rewrite last pointer value.
        MOVEM.L (SP)+, D0–D1/A0–A1       ; Restore registers from stack.
        RTS
        FIRST  DS.L   1                  ; Pointer to the first item
        LLIST  DS.L   160                ; memory block reserved
                                         ; for link list.
        END
```

5.4.5 Data Structure for Tables

An array is a list of equal-sized data elements stored in consecutive memory locations. A table is an array of arrays specified by rows and columns. Tables are generally implemented as two-dimensional arrays. The concepts can be extended to implement multidimensional arrays, in which elements can be accessed by three or more indices.

Elements in a table are generally accessed as an offset to its starting address. An access function takes the row and column indices of an element and returns the address of the desired element. The access function for the table requires several parameters, namely:

S = The starting address of the table in the memory.
I = The number of bytes required to keep a row of elements.
J = The number of bytes required to store a single element.

If we assume A to be a table, then the address of element

A[row][column]
is given by: S + (I * row) + (J * column)

From implementation point, it is always preferable to create a table by creating a descriptor, as shown below:

Row	Column	Row Size	Element Size	Row	Column
Upper Bound	Upper Bound	I	J	Index	Index

This descriptor can be implemented as a record. The descriptor can then be passed to the called routine by the calling routine. We will utilize such a technique to develop a program where the fields of the descriptors are defined as offsets to the register. An address register will be used as a pointer to such a descriptor. A subroutine PTABLE will print the contents of the table by rows. The program is shown in the following:

```
ROWBD   EQU   0        ; offset for row bound
COLBD   EQU   2        ; offset for column bound
RSIZE   EQU   4        ; offset for row size
CSIZE   EQU   6        ; offset for column size
ROW     EQU   8        ; offset for row pointer
COL     EQU   10       ; offset for column pointer
TADRES  EQU   28       ; offset to table on the stack
TDESC   EQU   24       ; offset to table descriptor on the stack
SIZE    EQU   8        ; two long-word parameters on the stack

PTABLE:  MOVEM.L D0–D2/A0–A1, -(SP)   ; save into the stack
         MOVEA.L  TADRES(SP), A0     ; save the address of the table
         MOVEA.L  TDESC(SP), A1      ; save the table descriptor
         MOVE  #0, ROW(A1)           ; set the lower bound of the row
PR0:     MOVE  ROWBD(A1), D1
         CMP.W  ROW(A1), D1          ; compare the row and row bound
         BLT    PRT3                 ; if rowbd < row,print done
         MOVE   #0, COL(A1)          ; set column = 0 for row print
```

```
PR1:     MOVE COLBD(A1), D2
         CMP.W COL(A1), D2                ; compare column and column bound
         BLT PRT2                         ; if colbd < column row print

         /********************************************/
              MOVE   ROW(A1), D1
              MULU   RSIZE(A1), D1  ; D1= I*ROW
              MOVE   COL(A1), D2
              MULU   CSIZE(A1), D2  ; D2= J*COL
              ADD.W D2, D1          ; D1=access offset
         /********************************************/

         MOVE  0(A0,D1.W), D0             ; use direct access to element
         JSR     DECOUT
         ADDQ.W  #1, COL(A1)              ; column = column + 1
         BRA     PR1                      ; go back for the next column element
PR2:     JSR     NLINE                    ; end of row. Print new line
         ADDQ.W  #1, ROW(A1)              ; ROW = row +1
         BRA PR0                          ; go to next row
PR3:     MOVEM.L (SP)+, D0-D2/A0-A1
         MOVE.L  (SP), SIZE(SP)           ; reset the stack
         ADDQ.L  #SIZE, SP                ;
         RTS
             END
/ ************************************************************
  The main routine MAIN calls PTABLE to list the contents of the table.
  The main routine stores the address of the table and its
descriptor on the stack as input parameters for the subroutine
PTABLE.
************************************************************/

START:   PEA TABLE                        ; store of the address of the table
                                          ; on to the stack
         PEA TDESC                        ; push the table descriptor onto the stack
         JSR PTABLE                       ; call the subroutine to get the
                                          ; elements from the table
         STOP
         END

TABLE:   DC.W  20, 30, 40, 50, 60, 70
         DC.W  80, 90, 10, 21, 31, 41
         DC.W  42, 43, 44, 45, 46, 47
         DC.W  51, 52, 53, 54, 55, 56
TDESC:   DC.W  3, 5, 8, 8                 ; example of descriptor fields
         DS.W  3                          ; example of index fields
```

5.5 BUILT-IN SUPPORT FOR STRUCTURED PROGRAMMING IN THE MC68020

As mentioned in the previous section, the concept of subroutines or procedures is vital to the creation of structured programs. In order to separate subroutines, the data structure and algorithm should be confined to the subroutine itself. The only way that a subroutine should transfer information is through parameter passing during the calling sequence. This separates the implementation from the calling sequences. The MC68000 family of processors supports a number of methods for passing parameters. The method is selected when the program is designed, and this choice constitutes an important part of the program design.

The information needed by the subroutine/procedure is defined in terms of parameters which allow the subroutine to handle general cases rather than operate on specific values. A subroutine written in high level language is taken for illustration purposes

SUBR (A, B, C, D, E)

This routine receives parameters A, B, C, and D, and puts the result in parameter E. This routine can be called by any program with various arguments, as long as the arguments are of some data type.

5.5.1 Reentrant Program

A program that may be called to be executed by means of an interrupt or subroutine call, and be reentered and operated properly, is known as a reentrant program. In assembly language, a program may be called in a number of times by different programs, or by the program itself to do a job. A good example is a program that calculates the factorial of a number. In order to calculate the value of factorial 4, the program calls itself four times. In such an event, the processor must separate the value of the variable during each calling sequence. The MC68000 family of processors supports reentrant calling sequences. This support mechanism is built-in within the architecture of the machine.

In this processor family, the subroutines are referenced by the instructions:

```
JSR <SUBR>    ; Jump to the subroutine
BSR <SUBR>    ; Branch to the subroutine
```

The way that the processor returns control to the calling routine is through the execution of the instructions:

```
RTS   ; Return from subroutine (used during user mode)
RTE   ; Return from exception (used during supervisory mode)
```

The JSR/BSR instruction causes the return address within the calling program to be saved onto the system stack. The control is transferred to the subroutine at address <SUBR>. The address may be specified by any of the control addressing modes. In general, subroutines need to receive various kinds of information that allow it to handle general cases, rather than operate on specific values.

The mechanisms of defining parameters and transmitting arguments to subroutines are different in assembly language from high-level languages. The MC68000 processor family provides a number of techniques for passing the arguments:

- Register transfer
- Mail box
- In-line code
- Stack frame

5.5.2 Register Transfer

The simplest method of passing arguments from one routine to another is by using a register from a predetermined set of registers, when called. The results generated by a subroutine are placed in predefined registers before the return instruction is executed. In general, data values are passed to any of the eight data registers D0–D7. Similarly, the address registers are used to pass addresses that may point to data values or may contain the starting address of a data table.

The advantages of this scheme are:

- It is simple.
- It is efficient in its execution.
- It has little memory overhead.

The main disadvantage is that the number of arguments that can be passed is limited to the number of available registers in the processor. In the MC68000 family, this is limited to 15 registers. For example:

```
MOVE.W   VAL, D7      ; argument value
MOVE.L   TABLE, A5    ; starting address of the table
LEA      HEAD, A6     ; address of head
JSR      SUBR
```

The first three statements pass the arguments to the registers. The first statement will set up a 16-bit value in D7, an address pointer in location TABLE in A5, and the address HEAD in A6.

The subroutine SUBR can now access the values in the registers directly to perform its operation.

5.5.3 Mail Box

When a large number of parameters are to be passed, a mailbox can be set up in the memory to facilitate this activity. The mailbox contains the values or addresses in a predetermined sequence. The arguments can be accessed by the subroutine after it has been passed the starting address of the area. The same mailbox could be used by several subroutines requiring different parameters, as long as the area is large enough to hold the maximum number of arguments.

The main advantages are;

- The registers are not used up.
- A dynamic set up is possible.

The disadvantages are that the access mechanism is slow; and it is non-reentrant

Example: The calling routine could set up a mailbox in the following way:

```
MOVE   VAL1, MBOX       ; store data
MOVE   VAL2, MBOX + 2   ; store second argument
MOVE   VAL5, MBOX + 8   ; fifth argument
LEA    MBOX, AO         ; put the address of the mailbox
JSR    SUBR             ; call subroutine
MBOX     DS.W 10        ; reserve area for mail box
```

In this example, 10 words are reserved for the mailbox, to which 5 parameters are supplied by the calling routine. The subroutine SUBR could access the values using indirect addressing with displacement. The instruction

```
MOVE 6(A0), D4
```

transfers the fourth parameter to data register D4. A number of variations are possible to define the parameter area in memory.

5.5.4 In-Line Coding

In in-line coding, the values to a subroutine are passed by coding the values following the call to the subroutine. This method defines argument values as constant, and they do not change after assembly. These values are defined by DC directives following the call.

Example:

```
JSR     SUBR
DC.W     ARG   ; in-line argument
```

The return address is pushed onto the stack by the processor on subroutine call. This address points to locations of the "arguments" in the instruction sequence that must be restored. The following statements can be executed by the called subroutine to load the argument into the least significant word of D1 and point the return address on the stack to the word beyond the value:

```
MOVEA.L   (A7), A0     ; get pc value
MOVE      (A0)+, D1    ; get first argument
MOVEA.L   A0, (A7)     ; push return address
:
:
RTS
```

The first instruction loads the contents of the PC into A0 from the stack. Register A0 then acts as a pointer. The second instruction loads the first argument into D1. After A0 is incremented, it points to the next instruction in the calling program following the in-line argument. The next instruction in the subroutine call pushes the correct return address on the stack, overwriting the value saved by the JSR instruction. The RTS instruction is used to restore the PC value and return control to the calling program.

The main advantages of this technique are:

- CPU registers are not used.
- The codes are position independent.

The disadvantages are:

- Slower access
- Static
- Return address must be adjusted

5.5.5 Stack Frame

One of the fundamental issues in the design of a subroutine involves the concept of transparency. The concept of the stack frame assures that the details of the subroutine operation are transparent to the calling program. The stack can be used as temporary storage of the register's contents by each subroutine and for each return address. This concept can be extended to the stack frame.

The stack frame is a block of memory in the stack that is used for return addresses, input parameters, output parameters, and local variables. Such a facility supports multiprogramming and recursive calling of the subroutine.

The MC68000 processor family provides two instructions to allocate and deallocate a data area called a frame in the stack part of the memory. The FRAME can be used for local storage of parameters or other data. The instructions are:

LINK— Link and allocate
UNLK— Unlink or deallocate

The general format is:

LINK An, d

The instruction saves the contents of the specified address register onto the stack. After the push, the address register is loaded with the value from the updated stack pointer. Finally, the 16-bit sign-extended displacement is added to the stack pointer. Now, the local variables can be stored in this area, as illustrated in Figure 5.2. For example:

Instruction		*Operations*
LINK	A1, #-$C	1. A1 ---> - (SP)
		2. SP ---> A1
		3. SP + (-$C) --> SP

The execution of this instruction causes the current contents of A1 to be pushed onto the stack. The stack pointer value will then be decremented by two words. The updated value of the stack pointer will then be loaded into A1. The stack pointer will be decremented by C bytes to create work space. The updated value of the stack pointer will point to the current top of the stack.

Once the input arguments are processed, and the outputs are stored on the stack, the prior data frame must be restored just before returning to the calling program from the subroutine.

The UNLK instruction is used to retrieve information from the stack-frame area. The general format is: UNLK An. An example of its use follows:

Instruction	*Operation*
UNLK A1	1. A1 ---> SP
	2. (SP)+ ---> A1

The execution of UNLK involves two steps, and is shown in Figure 5.3. First, the CPU loads the value of the frame pointer A1 into the stack pointer (which points to the old value of A1 saved on the stack by the LINK instruction). Then,

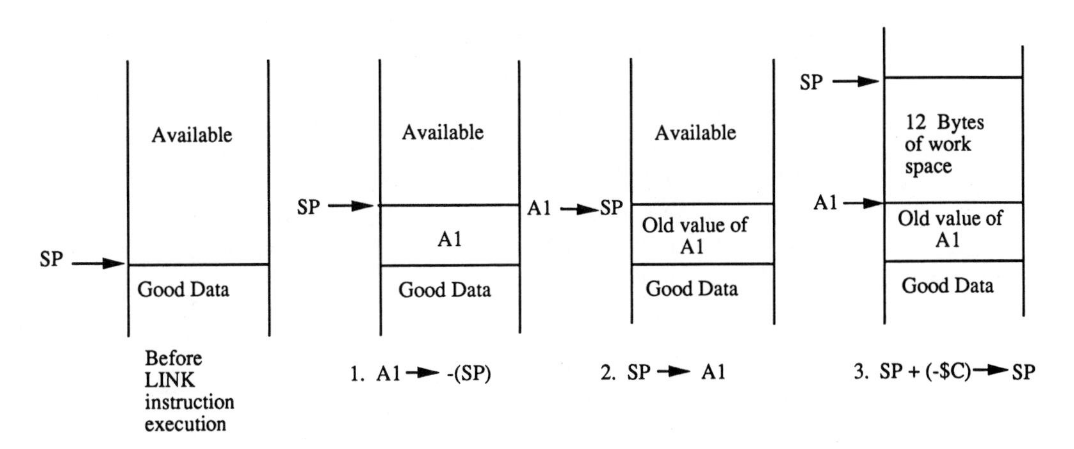

FIGURE 5.2. Execution of a LINK operation.

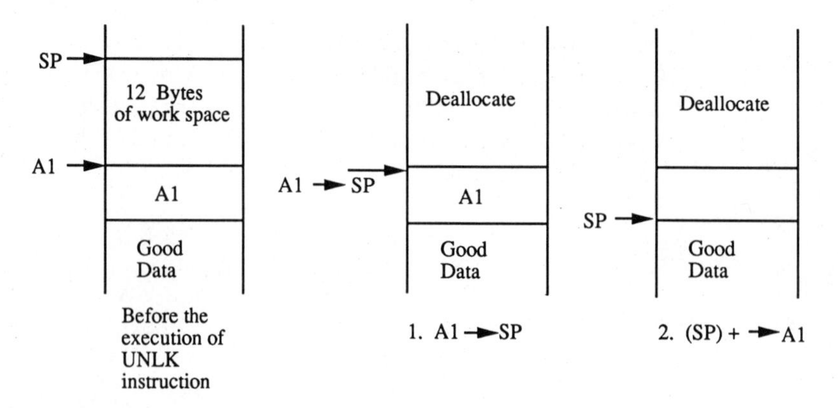

FIGURE 5.3. Execution of an UNLK instruction.

A1 is restored to its previous value, using the autoincrement mode of addressing, so that SP can point to the return address.

The next example illustrates a possible sequence in the calling program and the operation of the subroutine. The accompanied diagram, Figure 5.4, shows the status of the stack as the processor executes the instruction.

Example:

```
ORG     $1000
M       EQU     8
M       EQU     8
/*********************************************/
        ADD.L   #-N, SP             ; output area
        MOVE.L  ARG, -(SP)          ; input argument
        PEA     X                   ; input address X
        JSR     SUBR                ; call the subroutine
        ADD.L   #8,SP               ; skip over inputs
        MOVE.L  (SP)+, D1           ; read input
        MOVE.L  (SP)+, D2           ; read input
        ARG     DC.L  $01234567     ; argument to pass
X       DS.B    200                 ; table whose address is passed
SUBR:   LINK    A1, #-M             ; save old frame pointer
        MOVE.L  LOC1, -4(A1)        ; save local variable 1
        MOVE.L  LOC2, -8(A1)        ; save local variable 2
        :
        :
        ADD.L   #1, -4(A1)          ; change local variable
        MOVE.L  8(A1), A2           ; get x
        :
        :
```

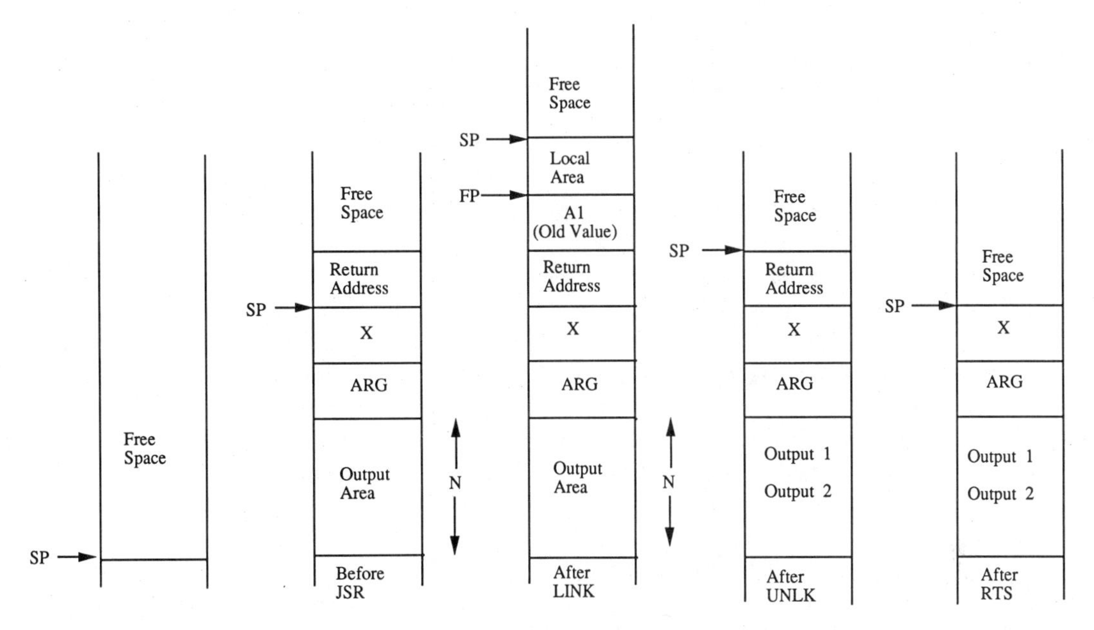

FIGURE 5.4. Examples of LINK, UNLK, and subroutine calls.

```
MOVE.L   OUT1, 16(A1)        ; push an output
UNLK     A1                  ; restore
RTS
OUT1     DC.L   $ABCDEF98
OUT2     DC.L
```

The subroutine first executes the LINK instruction to create a stack frame and define the Frame Pointer (FP). The instruction saves the value of A1 on the stack, and replaces A1 with the value of the Stack Pointer (SP). The FP now points to the bottom of the local area. The displacement is then added to the SP so that it points N bytes further down the memory. Local variables are stored in this area and accessed by displacements from the value in the FP.

Once the input arguments are processed and the outputs are stored on the stack, the UNLK instruction is executed. This instruction releases the local area and restores the stack pointer contents so that it points to the return address.

5.5.6 Module Instruction

The module instruction of the MC68020 Call Module (CALLM) and Return from Module (RTM) supports a superior coordinated and flexible mechanism for calling and returning from a program. The branch to subroutine and return from subroutine instructions provide the simplest form of this mechanism. The combination of parameter passing, locating the variable storage register usage, and removal of parameters are undefined in a conventional subroutine mechanism. The MC68020 extended the fundamental concepts associated with TRAP0 through TRAP15, including the return from exception instructions. The enhanced features provide a calling and returning mechanism with the status register automatically saved and restored. Furthermore, they raise the privilege level of the user to the supervisory level, and lower the privilege back down upon return.

During the execution of the call module and the return from module instruction, the processor can generate CPU space Type-1 bus cycles. These are called access level controls, to which the external hardware responds. Motorola's *page memory management unit* has this hardware built-in. In addition, the module instruction supports run-time relocation, allowing one copy of the library routine to be shared by many programs. Programs can dynamically link into executable library functions at run time, rather than link to their own copies of the library at compile time. This reduces disk space and memory utilization. This run-time relocation is possible because the call module instruction does not call the module directly, rather it indirectly calls the module via the module descriptor, whose contents are maintained by the operating system.

The module instruction automatically coordinates and updates the data pointer for variable storage, pass parameters, and saving and restoring of the condition code. In addition, the module instruction supports a hierarchical protection mechanism with finer granularity than the familiar supervisory and user levels. Up to 256 privilege levels are supported by this mechanism. This hierarchical protection mechanism does require external hardware to perform access level checking of the MC68020.

5.5.6.1 Module Descriptor

This is shown in Figure 5.5. The first long-word contains control information used during the execution of the CALLM instruction. The rest of the locations store data, which may be loaded into processor registers by the CALLM instruction. The OPT field indicates how many arguments are to be passed to the called

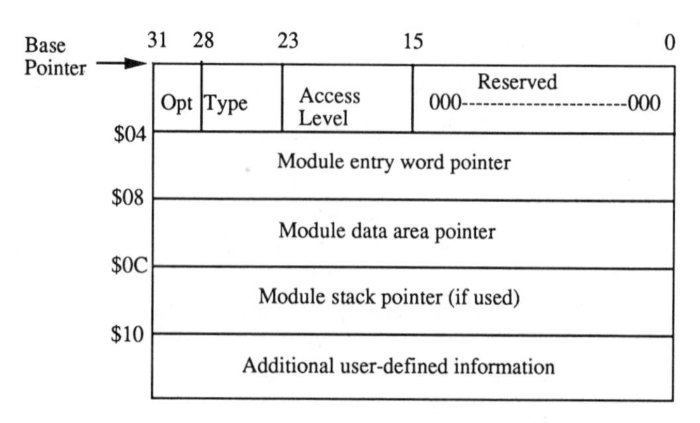

FIGURE 5.5. Module descriptor format.

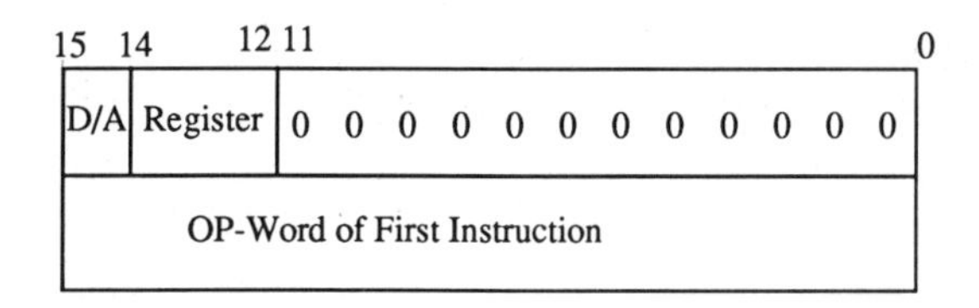

FIGURE 5.6. Module entry word.

module. The values '000' and '100' are recognized, and the rest will initiate an exception processing.

The value '000' tells the CPU that the called module expects to find arguments from the calling module on the stack below the module stack frame. The MC68020 will automatically copy the old stack to a new stack in the event of a stack pointer update during the call.

The value '100' tells the CPU that the called module will access the arguments from the calling module through an indirect pointer in the stack of the calling module. Thus, the MC68020 does not copy the arguments, but rather, puts the value of the stack pointer from the calling module in the module stack frame.

The TYPE descriptors $00 and $01 are the only ones supported. Type $00 defines a module for which there is no change in the access right, and the called module builds its stack from the top of the stack used by the calling module. Type $01 defines a module for which an access right may change. The access level control bus register is shown in Figure 5.6. The Current Access Level (CAL) register contains the access level right of the currently executing module. The Increase Access Level (IAL) is the register through which the processor requests an access right increase. The Decrease Access Level (DAL) register is the register used to request an access right decrease. The access status register allows the MC68020 to query the external hardware as to the validity of the intended access level transition. Table 5.1 lists the valid values of the access status register.

In case of a change in the access right, the called module may have a separate stack area from that of the calling module. If the access change requires a change of stack pointer, then the old value is saved in the module stack frame and the new value is taken from the module descriptor stack pointer field. The module descriptor types $10 through $1F are user definable. The user can disable any module by setting a single bit in its descriptor, without the loss of any descriptor

TABLE 5.1. Valid Access Rights.

Value	Validity	Action Taken by MC68020
00	Invalid	Format error
01	Valid	Access rights remain intact
02–03	Valid	Change access rights with no change of rights
04–07	Valid	Change access rights and change stack pointer
Other	Undefined	Take format error exception

information. The processor will generate a format error exception when it sees a user-defined module descriptor.

5.5.6.2 Module Stack Frame

The description of the module stack frame is shown in Figure 5.7. The processor creates it during the CALLM instruction and removes it during an RTM instruction. The first two 32-bit areas contain control information passed by CALLM to the RTM instruction. The module descriptor pointer contains information to be restored on return to the calling module. The saved program counter area holds the address of the next immediate instruction after the CALLM instruction. The OPT and TYPE fields specify the argument options and type module stack frame, and are copied to the frame from the module descriptor by the CALLM instruction. The RTM instruction will return a format error in case of an invalid OPT and TYPE field. The access level is saved in the *saved access level* area. The *saved stack pointer* area contains the value of the old stack pointer before the CALLM instruction is executed, and it is restored during the execution of RTM.

5.5.6.3 Module Instruction Usage

During the execution of a call module instruction, an immediate data value must be specified to indicate the number of bytes of parameters to be passed to the called module. An effective address pointing to a module descriptor must also be included. The program containing the call module instruction must define a RAM storage area for the module descriptor. The location of the module descriptor is known to the operating system at run time. The operating system loads the starting address of the library routine to be called into the module descriptor.

	15		12					7	0
SP →	Opt			Type					Saved access level
	0	0	0	0	0	0	0	0	Condition codes
	0	0	0	0	0	0	0	0	Argument byte count
	Reserved								
+$08	Module descriptor pointer								
+$0C	Saved program counter								
+$10	Saved module data area pointer								
+$14	Saved stack pointer								
+$18	Arguments (optional)								

FIGURE 5.7. Module call stack frame. (Courtesy Motorola Inc.).

The diagram in Figure 5.8 gives an overview of the call module (CALLM) and return from module (RTM) operation. When the call module instruction executes, it saves the current module description into the stack and loads the new module stack from the module descriptor. The start address fetch from the module descriptor points to the library routine. The last instruction of the library routine is the return from the module instruction. It reloads the saved module state from the stack frame, and the user program execution continues with the next instruction.

5.5.6.4 Module Component

The module description, module stack frame, and the called module get invoked during the execution of CALLM and RTM instructions. The module descriptor location is generally defined by the programmer with the description declaration, which must be known to the operating system. The contents of the descriptor are filled in by the operating system at run time. The CALLM instruction execution, shown in Figure 5.9, inspects the information in the module descriptor to determine whether an access level change will be made to the location of the called module, the data area, and (optionally) another stack area. The module stack frame is created by the called module. The type of the module call, the address of the called module, and the contents of the register used for the data points are saved. Optionally, the old stack pointer value may be saved. The first word of the called module defines a register to be used as the data area pointer. The CALLM instruction has to read this word to determine which register contents to save on the stack before loading it with the data pointer value found in the module descriptor. The second word of the CALLM is the first op-word. The last instruction of the module is the RTM instruction, which loads the register

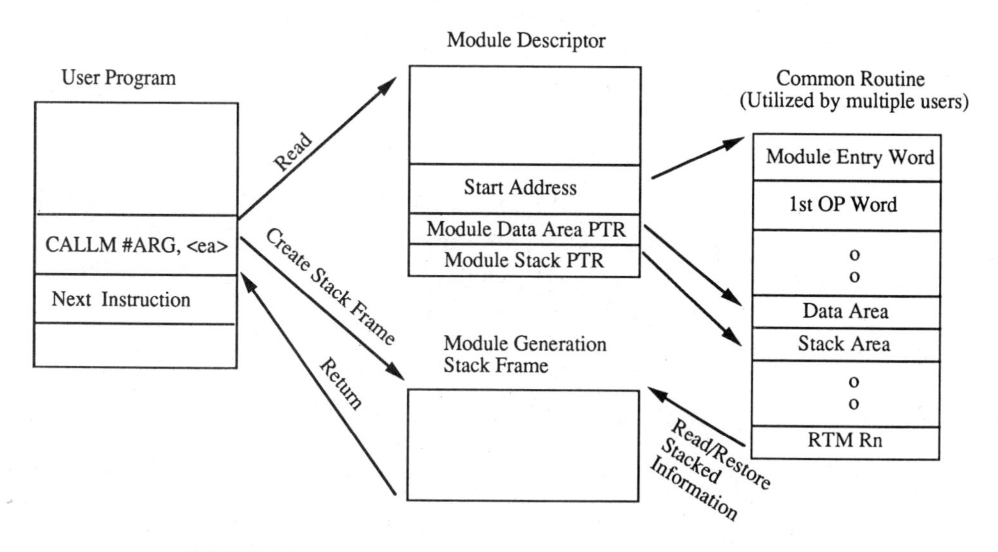

FIGURE 5.8. CALLM and RTM examples.

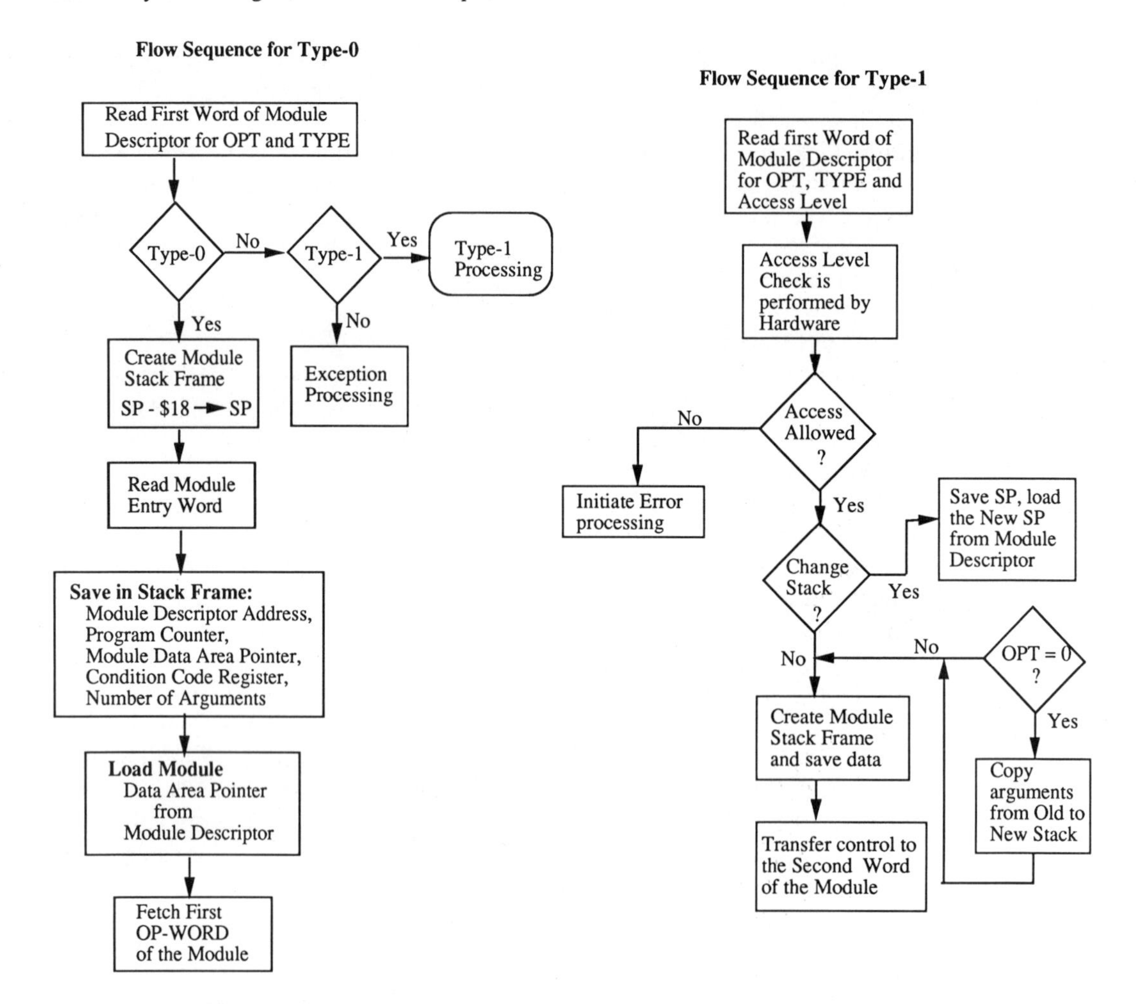

FIGURE 5.9. Execution flow sequence for CALLM instruction. (Courtesy Motorola Inc.).

specified with the stack module data area pointer value and restores the program counter, and (optionally) the stack pointer RTM also reads the argument byte count and adds this to the stack pointer, which removes the arguments from the stack.

5.6 POSITION-INDEPENDENT CODE

In all of the examples, the program-occupied fixed locations in memory, and the starting address, were defined by the origin (ORG) directive of the assembler.

If these programs were to be moved to another area in memory, reassembly with a new origin would be required.

This situation is not at all acceptable for Read-Only Memory (ROM) based programs. In these cases, the programs must be able to be relocated after they are assembled.

Most programs in ROM are strictly position independent, since the starting address of the ROM program is defined by the system designer, based on the requirements of the system. A ROM-based floating-point routine may have a starting address of $2000 in one system, and start at location $1000 in another.

Position independence is supported in the MC68020 through several addressing modes. All program counter relative addresses are position independent. Address register indirect with index or displacement addressing can be used to create position-independent code in a scheme called *base register addressing.* The memory reference indirect addressing mode is valuable in creating position-independent codes.

Today's multiuser systems employ memory management systems that perform the relocation, so that position-independent coding is becoming less important. However, many embedded systems do not utilize the Memory Management Unit (MMU), and do utilize monitor-based hardware. A monitor supports a minimum set of functions, such as: the ability to examine and modify memory, load codes, execute programs, set breakpoints, examine registers, etc. Thus, it is a common practice to implement monitors in a position-independent fashion.

Two types of position independent coding techniques exist in the MC68000 processor family, and these are *static* and *dynamic.* A program is strictly position independent if it can be loaded as a whole at a different starting address and then executed to conclusion with no change in result. Dynamic position independent implies that the program can be interrupted at any point, moved to another location, and restarted with no change in result. We are going to concentrate our attention on static position-independent coding through examples.

```
MOVE.W   #64, D7      ; position independent
MOVEA.L  #BUFF, A6    ; not position independent
```

5.6.1 PC Relative Addressing

When a location referenced in a program is at a fixed distance from the instruction making the reference, the PC relative addressing mode can be used to create position-independent code. As long as the relative displacement is not changed, the program will execute correctly anywhere in memory. For example:

```
LEA  POINT(PC), A6
```

5.6.2 Address Register Indirect Addressing

This can only be utilized if the address register has been loaded in a position-independent manner. This is illustrated by:

```
LEA      POINT(PC), A6
ADD.L    (A6), D6
```

5.7 MULTIPROCESSOR OPERATIONS

Multiprocessor operations allow the MC68020 to communicate with other processors in the system. These operations fall into two categories:

- Mulitprocessor operations
- Coprocessor operations

Multiprocessor operations are designed to enable the MC68020 to communicate with another CPU in the system. The instructions TAS, CAS, and CAS2 are created to facilitate this capability. The MC68020 does not release the bus mastership during the execution of these instructions, because they initiate indivisible read-modify-write bus cycles.

The coprocessor operations and the associated protocol are described in Chapters 9, 10, and 11.

5.7.1 Test and Set (TAS)

The TAS instruction is designed to implement a mechanism for programs that are executing independently, to synchronize their activities. Without this instruction, two programs working independently might find themselves in the following situation. One program might test a flag and find it to be zero, and then get interrupted before setting it. The other program might also test the same flag, find it to be zero, and set it. The first program would proceed as if no flag has been set at the end of the interrupt operation. To guard against such a situation, the MC68020 will not respond to an interrupt or bus request from external devices during the execution of the TAS instruction. This guarantees that the entire instruction will execute without the possibility of a destination operand being changed during the execution of the operation.

The TAS first tests a value in memory or a register, using the usual addressing modes, and compares that value against zero. The current value of the operand determines whether the CPU will set the N and Z flags of condition code register or not. If the operand value is zero, then Z is set to 1. If the most significant bit (bit-7) of the operand is found to be 1, then N gets set to 1. Finally, TAS unconditionally sets the destination operand's sign bit 7 to 1, forcing the byte to be negative.

5.7.2 Semaphore Implementation

The basic concept of a semaphore is that a resource, for example, a file, a block of memory, an I/O device, etc., can be shared by various programs or CPUs. Various flags, priority, semaphores, or queueing schemes are used to determine who gets what resources, including their duration. Typically, a program examines the availability of a resource, and grabs it if available by setting an agreed-upon value in an assigned status bit or byte. The resource is eventually relinquished by clearing the status flag, or status bit, so that other programs are then free to access it. For example, let us assume that a byte at address $2FFF signals the following event to all user programs:

($2FFF) = 00 means that the printer is free
($2FFF) = not zero means that it is busy printing

We can implement a mechanism using a semaphore, where a program will test the printer status and will do something else if it is busy, otherwise it will set the memory location and occupy the printer.

```
WAIT:       TAS $2FFF      ; Test and set the CCR. Then
                           ; set it to $80.
            BNE WAIT       ; If the printer is busy, do some
                           ; other thing.
            :              ; If the printer is free, use it.
            :
            CLR.B $2FFF    ; Release the printer
            :              ; by setting the value to $00.
            :
            :
```

In the real world, such a simple printer-locking technique will be more complex. Users will create a virtual printer and spool their job into it. The printer program will accept the jobs from the virtual printer queue according to certain priority scheme, and print that job. At the completion of the printing, the printer program will signal another program to inform the requesting program that its job has been completed.

5.7.3 The Compare and Swap Instructions

CAS and CAS2 (Compare and Swap) instructions guard against multiuser accesses to the same resource. As explained in the past, TAS supports the simplest technique of guarding against the problem of simultaneous access to the same resource, where a *flag* is set up at a well-defined place known to all the users, and can only be set by one user. Such a mutual agreement is a necessary precondition for implementing multiuser safeguards. CAS goes one step further than TAS, and CAS2 goes two steps further in implementing such safeguards. The instruction formats are:

CAS Dc, Du, <EA>
where,
Dc field identifies the data register that contains
the value to be compared against the

operand in the memory.

Du field specifies the data register whose contents will be written to the memory operand location if the comparison is successful.

CAS compares Dc and the operand found at the location pointed to by EA. If they are the same, then Du replaces the operand. Otherwise, the CAS will try again.

CAS2 Dc1:Dc2, Du1:Du2, (Rn1):(Rn2)

where,

Dc1 and Dc2 fields identify the data registers that contain 'test values' to be compared against the two memory operands. If Dc1 and Dc2 happen to be the same data register, and the comparison fails, then the content of the memory location (operand 1) will be put into the data register Dc.

Du1 and Du2 fields specify the data registers that contain the updated values to be written to the first and second memory operand locations in the event that the comparison is successful.

Rn1 and Rn2 fields identify the numbers of the registers that contain the addresses of the first and second memory operands, respectively. If the operands overlap in the memory, the results of any memory update are undefined.

CAS2 performs two comparisons and, in the event of any mismatch, memory operands will not be updated. This instruction only allows register (address or data) indirect modes.

Example: Two processes want to submit jobs to a printer in a multitasking environment. The access to the printer is achieved through a global variable pr_queue, and print buffers get assigned based on the pr_value. For example, let us assume that process-1 loads the current value of pr_queue into D0, and then performs its update operations in D1. The time slice of process-1 expires immediately. Process-2 then comes along and updates the pr_queue and gets the same buffer (which process-1 should have received) and writes into that. At the end of process-2's time slice, process-1 gets back and writes into the same buffer written to by process-2. Such a situation can be avoided by using the CAS instruction, which is illustrated in the following example.

```
        MOVE.L   PR_QUEUE, D0        ; Get the buffer number.
LOOP:   MOVE.L   D0, D1              ; Maintain a copy.
        ADDQ.L   #1, D1
        CAS.L    D0, D1, PR_QUEUE    ; Check whether someone
                                     ; grabbed the buffer.
        BNE.S    LOOP                ; If yes, the buffer is lost.
        :
        :
        :                            ; Continue.
```

The update is performed only if D0 remains unchanged during the update operation. Otherwise, the update is not performed, and the current value of PR_QUEUE is loaded into D0 and another try is initiated.

The CAS2 instruction extends the CAS instruction's capability another step ahead, where it performs two comparisons and updates two variables whenever the results of the comparisons are equal. The CAS2 copies the new values into the destination addresses, if the results of both comparisons are equal. Otherwise, the contents from the two destination addresses are copied into the compare operands.

The following example [MC68030 1989 Motorola Inc.] illustrates the use of the CAS2, where it deletes an element from a linked list.

```
        LEA      HEAD, A0                   ; Get the starting address of the link list.
        MOVE.L   (A0), D0                   ; Keep a copy into D0 for comparison.
LOOP:   TST.L    D0                         ; Check whether the list is empty. If
        BEQ      EMPTY                      ; empty, do not remove anything.
        LEA      (NEXT, D0), A1             ; Get the address of the
                                            ; forward link.
        MOVE.L   (A1), D1                   ; Copy the forward link value into D1.
        CAS2     D0:D1, D1:D1, (A0):(A1)    ; If nobody has deleted
                                            ; the entry, then update the HEAD and
                                            ; forward pointers
        BNE      LOOP                       ; failed to delete the entry.
EMPTY:                                      ; Nothing to delete.
```

The algorithm works as follows: the LEA loads the effective address of HEAD into A0, and the next instruction, MOVE, loads the address in pointer HEAD into D0. TST then checks for an empty list. If the list is found to be empty, then the program branches to EMPTY, otherwise it executes the next instruction in the flow. The second LEA loads the address of the NEXT pointer in the newest element on the list into A1. The MOVE instruction then loads the pointer in the newest element on the list into A1, and the following MOVE loads the contents of the pointer into D1. CAS2 then compares the address in A1. If no element has been inserted or deleted to another program while this ‘‘program’’ has been executing, then the results of these comparisons are equal and the CAS2 instruction stores the new value into location HEAD. If an element has been inserted or deleted, the CAS2 instruction loads the new address from location HEAD into D0, and BNE branches the execution flow to instruction TST in order to initiate another try.

5.8 PROGRAMMING EXAMPLE

The Linear Predictive Coding (LPC) method is one of the most popular approaches for processing speech. Two issues influence the quality of digitized speech. They are windowing and preemphasis. They are important in the accurate determination of the speech parameters.

The popularity of LPC stems from the fact that it employs a model of speech production that works amazingly well. The parametric representation of speech through the LPC also facilitates the use of very low bit rates for speech coding, down to about 2.4 kbits/sec. Another very important characteristic of the LPC parameters is that they preserve essentially all of the intelligibility information of the speech signal, and so they can be (and have been) used for speech recognition. In this way, since the same parameters can be used for coding or recognition, the task of building a system with both functions becomes somewhat easier.

The speech production model is shown in Figure 5.10. For software implementation, this model can be transformed into a model, shown in Figure 5.11. This model assumes that incoming speech is sampled within a sampling window. The digitized samples are then passed through a Durban filter. The output is then analyzed through an analyzer, and signals are then compensated for errors. The resultant outputs are then synthesized to reconstruct the original sample points. The synthesizer requires adjustment during the synthesis process. Interested readers are encouraged to look to the references for a more detailed explanation.

The program is implemented using four modules, which are called:

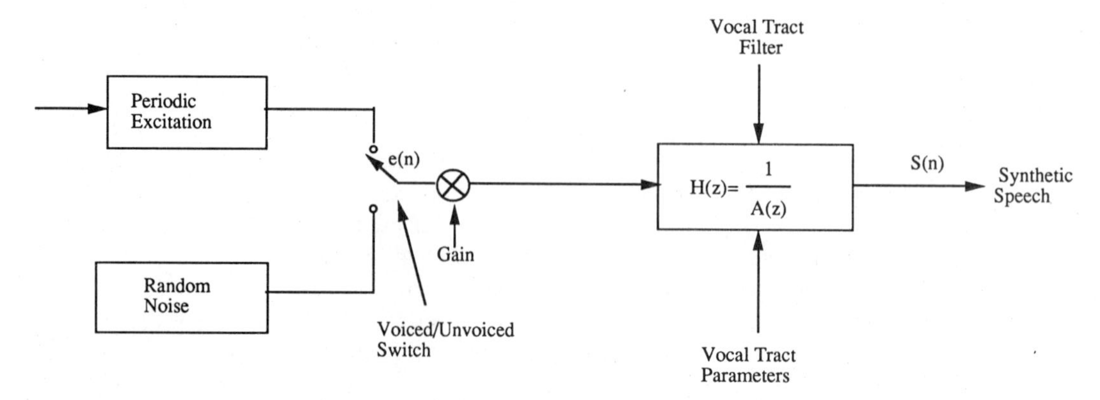

FIGURE 5.10. Speech production model for LPC.

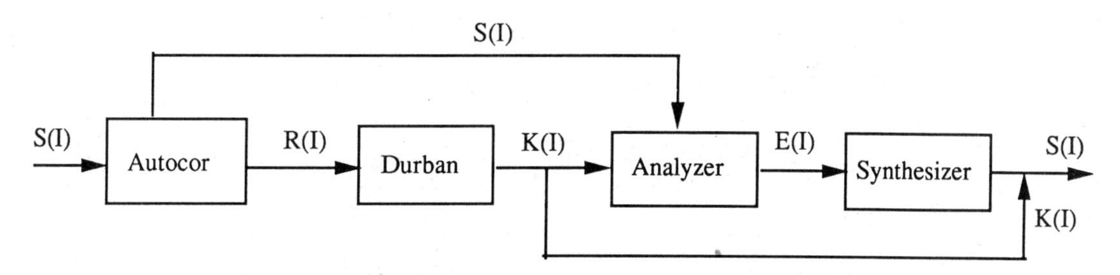

FIGURE 5.11. Software implementation of the model.

- Durban
- Analyzer
- Synthesizer
- Autocorrelation

The Durban routine uses the Durban algorithm to transform the autocorrelation coefficients R(1) − R(I + 1) to the reflection coefficients K(1) − K(I) and to the filter coefficient A(1) − A(I + 1) with A(1) = 1.

The autocorrelation is given by the formula:

R(I) = R(−I) = Sum of the elements S(N)S(N + 1)
where N = 0 to N = N − I − 1 and N = sampling window.

The program listing follows:

```
/*
/*  Simple Durban recursion for lattice reflection coefficient
/*
    XREF ARRAY_R
    XREF M
    XDEF ARRAY_K
    XDEF DBN
/*
/*      Register used
        D7  Index I (2, 4, 6, 8, . . .)
        D6  E(0), E(1), E(2), . . .
        D1  K(I)
        A6  ARRAY_R POINTER
        A5  ARRAY_A POINTER
        A4  ARRAY_K POINTER
/*
/*      PROGRAM EQUATES
/*
   N  EQU 7
```

```
                /* This program uses fixed-point arithmetic with the radix
                   point
                /* assumed to be in the middle of the word. For example,
                   $12.34
                /* is represented by $1234.
                /*

DBN:
  MOVE.L  #M, D2                              ; radix compensation register
  MOVE.W  #2, D7
  LEA     ARRAY_R, A6
  LEA     ARRAY_A, A5
  LEA     ARRAY_K, A4
/*
/* perform first iteration
/*
  MOVE.W      (A6), D6                        ; set E(0) = r(0)
  MOVEQ       #0, D1                          ; zero out upper part for divide
  MOVE.W      2(A6), D1                       ; get r(1)
  EXT.L       D1                              ; sign-extend
  LSL.L       D2, D1                          ; per shift for divide
  DIVS        D6, D1                          ; calculate r(0)/r(1)
  NEG.W       D1                              ; form -r(0)/r(1)
  MOVE.W      D1, (A4)+                       ; store k(1) away
  MOVE.W      #$100, (A5)                     ; A(0)=1 in this radix arithmetic
  MOVE.W      D1, 2(A5)                       ; A(1)=K(1)
  MULS        D1, D1                          ; square K(1)
  LSR.L       D2, D1                          ; compensate for radix representation
  NEG.W       D1                              ; form (1-K(I))^2
  ADD.W       #$100, D1                       ; add 1 to radix arithmetic
  MULS        D1, D6                          ; form E(1)
  LSR.L       D2, D6                          ; compensate for radix representation
/*
/* Perform the remaining iteration
/*
LOOP1:
/* First calculate the numerator of K(I)
  MOVE.W      D7, D5                          ; calculate the LOOP2 index
  LEA         4(A6,D5.W), A3                  ; ARRAY_R working register
  LSR.W       #1, D5
  MOVEA.L     A5, A2
  MOVEQ       #0, D1                          ; initialize SUM
LOOP2:
  MOVE.W      -(A3), D0                       ; get R(I)
  MULS        (A2)+, D0                       ; compensate for radix representation
  LSR.L       D2, D0
```

```
  SUB.W       D0, D1
  DBF         D5, LOOP2
/*
/* calculate k(i)
/*
  EXT.L       D1                   ; sign-extend D1 for divide
  LSL.L       D2, D1               ; preshift for divide
  DIVS        D6, D1               ; K(I) in register D1
  MOVE.W      D1, (A4)+            ; save K(I) in its array
  MOVE.W      D1, 2(A5,D7.W)       ; A(i) = K(i)
  MOVE.W      D7, D0               ; counter index
  SUBQ        #1, D0               ;
  LSR.W       #2, D0
  LEA         2(A5,D7.W), A3       ; A(I-1) in A3
  LEA         2(A5), A2            ; A(1) in A2
LOOP3:
  MOVE.W      D1, D4               ; K(I) is in D4
  MOVE.W      D1, D5               ; K(I) is in D5
  MULS        -(A3), D5            ; A(I-1) * K(I)
  LSR.L       D2, D5               ; compensate for radix representation
  ADD.W       (A2), D5             ; A(1)+A(I-1)*K(I)
  ADD.W       (A3), D4             ; A(I-1)+A(1)*K(I)
  MOVE.W      D4, (A3)             ; new A(I-1)
  MOVE.W      D5, (A2)+            ; new A(1)
  DBF         D0, LOOP3            ; end LOOP3
/*
/* Calculate the new E(I)
/* K(I) is still in D1
/*
  MULS        D1, D1               ; K(I)^2
  LSR.L       D2, D1               ; compensate for radix representation
  NEG.W       D1
  ADD.W       #$100, D1            ; 1-K(I)^2 in the radix arithmetic
  MULS        D1, D6               ; E(I-1)*(1-K(I)^2)
  LSR.L       D2, D6               ; compensate for radix representation
  ADDQ.W      #2, D7               ; ordinary arithmetic okay here
  CMP.W       #N*2, D7             ;
  BNE         LOOP1
  RTS
      DS.W    32                   ; buffer
  ARRAY_A DS.W    32
  ARRAY_K DS.W    32
  END
```

```
/*
/* Sample circular autocorrelation function generator
```

```
        /*
          M   EQU 8
          N   EQU 8
        ACR:
        /* First loop through the data, shifting it left one bit
        /* so it becomes radix 512. It is entered as radix 256
        /*

  MOVE.W     #7,D0              ; loop counter
  LEA        ARRAY_S, A0        ; get the data array
LOOPX:
  LSL        (A0)+              ; shift each datum left
                                  once
  DBF        D0, LOOPX          ; end shift loop
  MOVE.L     #M, D0             ; radix compensation

           /*
           /* LOOP1 steps through the autocorrelation shift index, I
           /* LOOP2 steps through the data values calculating the cir-
              cular
           /* autocorrelation for the lag given in LOOP1
           /*

  MOVEQ      #0, D7
  LEA        ARRAY_S, A6        ; pointer to data array
  LEA        ARRAY_R, A5        ; autocorrelation array
                                  pointer
           LOOP1:
           /* initialize for the autocorrelation LOOP2
  MOVE.Q     #0, D6             ; initialize index N
  MOVE.W     D7, D5             ; initialize index M
  MOVEQ      #0, D4             ; initialize SUM register
  MOVEQ      #0, D3             ; clear upper bits of D3
LOOP2:
/* perform the multiply S(N)*S(M)   ; where M=N+I MOD
                                      N
  MOVE.W     (A6,D6.W), D3      ; set S(N)
  MULS       (A6,D5.W), D3      ; perform S(N)*S(M)
  ADDQ.W     #2, D5             ; increment index M(mod
                                  2N)
  CMP.W      #(N*2), D5         ; wrap M to zero
  BNE        AC001              ; B-NO
  MOVEQ      #0, D5
AC001:
  ADDQ.W     #2, D6             ; increment index N
```

```
        CMP.W       #(N*2), D6
        BNE         LOOP2
/*
/* end LOOP2
/*
        MOVE.W      D4, (A5,D7.W)       ; save the autocorrelations
        ADDQ.W      #2, D7              ; increment index I
        CMP.W       #(N*2), D7
        BNE         LOOP1
        RTS
        ARRAY_R     DS.W  32
        ARRAY_S     DC.W  $0300
                    DC.W  $0300
                    DC.W  $0300
                    DC.W  $0300
                    DC.W  $0
                    DC.W  $0
                    DC.W  $0
                    DC.W  $0
        END
```

```
/*
/* Sample lattice filter (analyzer) program
/*
   XREF   ARRAY_K
   XREF   ARRAY_S
   XREF   M
   XREF   ARRAY_E
   XREF   ANAL
/*
/* The following equates define functions provided
/* by the MC68020 debugger via the TRAP function
/*

INCHNP    EQU    0     ; input character in D0 register/no parity
OUTCH     EQU    1     ; output character from a register
PDATA1    EQU    2     ; output string
PDATA     EQU    3     ; output CR/LF, then a string
OUT2HS    EQU    4     ; output two Hex and a space
OUT4HS    EQU    5     ; output 4 Hex and a space
OUT8HS    EQU    12    ; output 8 Hex and a space
PCRLF     EQU    6     ; output CR/LF
SPACE     EQU    7     ; output a space
MONITR    EQU    8     ; enter a 68020 debugger
VCTRSW    EQU    9     ; vector examine/switch
```

```
BRPKT     EQU   10          ; user program breakpoint
PAUSE     EQU   11          ; task pause function
NUMFUN    EQU   13          ; number of available functions


        /* Note:
        /* 1. Arithmetic is done with the radix point.
        /* 2. The output of this filter is the error signal, which is
        /*                 f(7).
        /* 3. The optimal reflection coefficients are calculated in the
        /*                 DURBAN program
        /* 4. The program loops through all N-1 stages of the lattice
        /* filter (inner loop), and then through all N values of the
        /*                 data (outer loop).
        /*
        /* Register Usages:
        /* A6    POINTER TO ARRY_K
        /* A5    POINTER TO ARRY_F
        /* A6    POINTER TO ARRY_B
        /* A6    POINTER TO ARRY_S
        /* A6    POINTER TO ARRY_E
        /* D7  INNER LOOP COUNTER N-2 TO 0
        /* D6  OUTER LOOP COUNTER N-1 TO 0
        /* D3  RADIX COMPENSATION REGISTER
        /* D2  HOLD NEW B(I-1) VALUE BEING CALCULATED
        /* D1  USED AS ACCUMULATOR FOR NEW B(I)
        /* D0  USED AS ACCUMULATOR FOR NEW F(I)
        /*

        N  EQU 8                          ; length of data sequence
       /* The program starts here
       /*
       ANAL:
        MOVE.W    #(N-1), D6              ; outer loop index
        LEA       ARRY_S, A3
        LEA       ARRY_E, A2
        MOVE.L    #M, D3                  ; setup for radix point
                                            compensation.

          /*
          /* outer loop starts
          /*
          LOOP1:
          /*
          /* Initial calculations and register initialization for inner
          /* loop.
```

```
/*
 LEA            ARRY—K, A6
 LEA            ARRY—F, A5
 LEA            ARRY—B, A4
/*
/* f0 =     Sn
/*

MOVE.W   (A3)+, D2              ; Sn --> D2 as new B0
MOVE.W   D2, (A5)               ; Sn --> f0 as new
MOVE.W   #(N-2), D7             ; inner loop index

         /* inner loop starts
         LOOP2:
         /* Calculate new fi, i=1, 2, 3,. . . . 7
         /* A3 points to new f(i-1)
         /* A4 points to old b(i-1)
         /* A6 points to Ki
         /* D0 is the accumulator
         /*

MOVE.W   (A6), D0               ; get K(i)
MULS     (A4), D0               ; form K(i)*b(i-1)
LSR.L    D3, D0                 ; radix arithmetic adjustment
                                ; form f(0)+K(i)*B(i-1)

         /*
         /* Calculate new b(i), i=1,2,3 . . .
         /* D1 is the accumulator
         /*
         /*

MOVE.W   (A6)+, D1              ; get K(i)
MULS     (A5)+, D1              ; form K(i)*f(i-1)
LSR.L    D3, D2                 ; radix arithmetic adjustment
ADD.W    (A4), D1               ; form b(i-1)+K(i)*f(i-1)

/*
/* Store calculated values
/* f(i) is updated immediately for use in subsequent calculations.
/* b(i) is held briefly till the next iteration so old values
/* are used. D2 is the b(i) holding register
/*
```

```
        MOVE.W      D0, (A5)            ; save f(i)
        MOVE.W      D2, (A4)+           ; save b(i-1)
        MOVE.W      D1, D2              ; HOLD B(I) over until
                                          the next pass
                    /*
                    /* Inner loop end and for update
       DBF          D7, LOOP2
                    /*
                    /* cleanup and miscellaneous for each
                     datum through the lattice
                    /*
        MOVE.W      D2, (A4)            ; last b(i) is saved
     MOVE.W         (A5), (A2)+         ; save error for synthe-
                                          sizer
                    /*
                    /* Outer Loop Ends
                    /*
        DBF         D6, LOOP1
        RTS
/*
        DS.W        16                  ; buffer
                    ARRY_F:
                    DC.W  0
                    DC.W  0
                    DC.W  0
                    DC.W  0
                    DC.W  0
                    DC.W  0
                    DC.W  0
                    DC.W  0
                    DC.W  0
                    DC.W  0
                    DC.W  0
                    DC.W  16
ARRY_B:
                    DC.W  0
                    DC.W  0
                    DC.W  0
                    DC.W  0
                    DC.W  0
                    DC.W  0
                    DC.W  0
                    DC.W  0
                    DC.W  0
                    DC.W  0
                    DC.W  0
```

```
                  DC.W  16
ARRY_E            DS.W    32
   END
/*
/* Lattice program synthesizer
/*
   XREF  ARRY_K
   XREF  ARRY_E
   XREF  M
   XREF  SYNT
   XDEF  ARRY_QQ
/*
/* Global module equates
/*
   N  EQU    8
/* Program starts here
SYNT:
/* First register
  MOVE.W      #(N-1), D6              ; Outer loop index
  LEA         ARRY_E, A3
  LEA         ARRY_QQ, A2
  MOVE.L      #M, D3
/* Outer loop starts
LOOP1:
/*
/* Initial calculation and register initiali-
  zation for inner loop
/*
  LEA         ARRY_K+2*(N-1), A6
                                      ; point to past last item
  LEA         ARRY_F+2*(N-2), A5      ; point to last item
  LEA         ARRY_B+2*N, A4          ; point to past last item
/* f7=En
  MOVE.W      (A3)+, (A5)             ; En --> new f7
  MOVE.W      #(N-2), D7              ; inner loop index
              /* Inner loop starts
              LOOP2:
              /*
              /* Calculate f '(i-1), i=7, 6, . . . 1
              /* A5 points to new f(i)
              /* A4 points to old b(i)
              /* A6 points to K(i)
              /* D0 is the accumulator
              /*
  MOVE.W      -(A6), D0               ; get K(i)
  MULS        -4(A4), D0              ; form K(i)*b(i-1)
  LSR.L       D3, D0                  ; radix arithmetic adjust-
```

```
                                                    ment
    NEG.W       D0                                  ; form -K*b
    ADD.W       (A5), D0                            ; form f'(i)-K(i)*b(i-1)
/*
/* Calculate new b(i) i=7, 6 . . . 1                ; where b(0) = f(0)
/* D1 IS THE ACCUMULATOR
/* A6 POINTS TO k_i
/* A5 points to f_i
/* A4 points to past b_i
    MOVE.W      (A6), D1                            ; get K(i)
    MULS        D0, D1                              ; form K(i)*f'(i-1)
    LSR.L       D3, D1                              ; radix arithmetic adjust
    ADD.W       -4(A4), D1                          ; form b(i-1)+K(i)*f'(i-1)

/*
/* Store calculated values
/* f(i-1) is updated immediately for use in subsequent calculations
/* b(i) is updated immediately
/*

    MOVE.W      D0, -(A5)                           ; save f'(i-1)
    MOVE.W      D1, -(A4)                           ; save b'(i)

/*
/* Inner loop end for updates
/*
    DBF     D7,LOOP2
/*
/* cleanup and miscellaneous for each datum through the lattice
/*

        MOVE.W      D0, -(A4)       ; make b(0)=f(0)
        MOVE.W      D0, (A2)+       ; save the output signal
    /* Outer loop ending
        DBF         D6, LOOP1
        RTS
        DS.W        16              ; buffer
    ARRY—F:
                    DC.W  0
                    DC.W  0
                    DC.W  0
                    DC.W  0
                    DC.W  0
                    DC.W  0
```

```
                DC.W  0
                DC.W  0
                DC.W  0
                DC.W  0
                DC.W  0
                DS.W  16
ARRY—B:
                DC.W  0
                DC.W  0
                DC.W  0
                DC.W  0
                DC.W  0
                DC.W  0
                DC.W  0
                DC.W  0
                DC.W  0
                DC.W  0
                DC.W  0
                DS.W  32
ARRY_QQ         DS.W32
 END
```

5.9 MODULAR PROGRAMMING TECHNIQUES

Problem solution, leading to the development of executable programs, is a complex discipline, because we must limit ourselves to programs that we can fully understand and manage intellectually. As a result, it is important that we develop a strategy to design software.

This section summarizes the structured program design concepts that have major impact at the code level, the module level, and the system level. It introduces three major program design methodologies: functional decomposition, data flow design, and data structure design. It then outlines a guideline of how programs should be designed.

5.9.1 Motivation

The major motivation for looking at program design methodologies is the desire to reduce the cost of producing and maintaining software. Statements appearing in industry literature shed some light on the goals and tenets of structured programming.

a) The purpose of structured programming is to control complexity.
b) Structured programming allows verification of the correctness of all steps in the design process, and thus automatically leads to self-explicable and easily modifiable codes.

c) The task of organizing one's thoughts in a way that leads, in a reasonable time, to an understandable expression of a computing task, has come to be called structured programming.

These statements indicate that a major objective of structured programming is the simplification of the program development process. One way to harness complexity is to break a problem into small, easily understood pieces. One of the fundamental problems is how to break down a problem, as far as software is concerned.

5.9.2 Functional Decomposition

Functional decomposition is simply the divide-and-conquer technique applied to programming, as shown in Figure 5.12a.

The design process can be divided into the following steps:

- Clearly state the intended function.
- Divide, connect, and check the intended functions by expressing them as an equivalent structure of properly connected subfunctions, each solving a part of the problem (Figure 5.12b).
- Divide, connect, and check each subfunction far enough to feel comfortable.

In following this procedure, the key to successful program design is rewriting, followed by more rewriting. The designer should make every effort at each step to conceive and evaluate alternate designs.

A useful mind set is to pretend that there exists a language powerful enough to solve the problem in only a handful of commands. In level-1 decomposition, one should write down a handful of commands, and together they form a complete program. In level-2 decomposition, refinement should be made to the level-1 instructions into a set of less powerful instructions. By continuing to successively refine each instruction, one gets to a program that can be executed by a real computer.

There are several problems involved in applying this technique. First, the

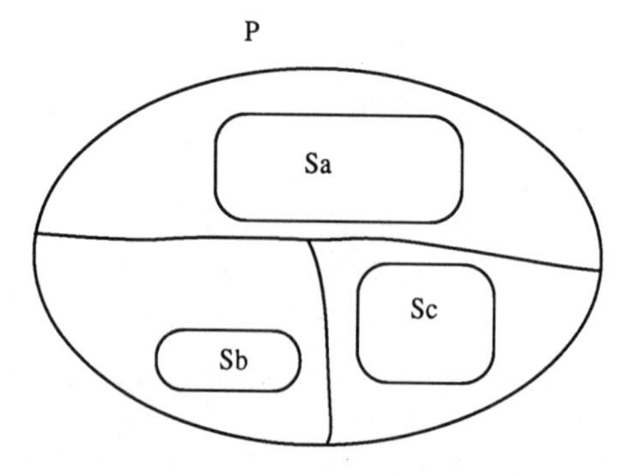

FIGURE 5.12a. Divide and conquer.

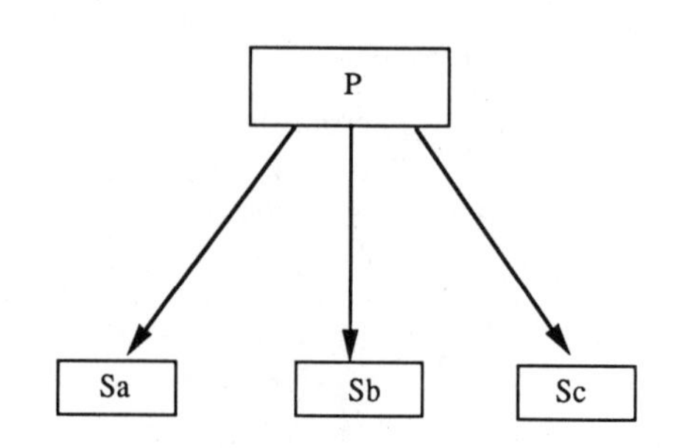

FIGURE 5.12b. Stepwise refinement.

method specifies that a functional decomposition be performed, but it does not say what, with respect to the decomposition, should be conducted. One can decompose with respect to time order, data flow, logical groupings, access to a common resource, control flow, or some other criterion.

If you decide to decompose with respect to time, you get modules like initialize, process, and terminate, and you have a structure with temporal cohesion. If you cluster functions that access a shared data base, you have made a start toward defining abstract data types and will get communicational cohesion. If you decompose using a data flowchart, you may end up with sequential cohesion. If you decompose around a flowchart, you will often end up with logical cohesion. The choice of "what to decompose with respect to" has a major effect on the appropriateness of the resulting program.

The major advantage of functional decomposition is its general applicability. The disadvantages are its unpredictability and variability. The chance of two people independently solving a given problem in the same way are practically nil.

5.9.3 Data Flow Design

The data flow design method was first proposed by Larry Constantine, and then extended by Ed Yourdan and Glen Myers. In its simplest form, it resembles a functional decomposition with respect to data flow. Each block of the structure chart is obtained by the successive assignment of functionalities to black boxes that transform an input data stream into an output data stream.

The first step in using the data flow design method is to draw a data flow graph, as shown in Figure 5.13. This graph is a model of the problem environment, which is transformed into the program structure. While the modules in functional decomposition often tend to be attached by a "USES" relationship, the bubbles in a data flow graph could be labeled BECOMES. That is, data input A "becomes" data output B. Data B becomes C, C becomes D, etc. The only shortcoming of this decomposition is that it tends to produce a network of programs—not a hierarchy of programs.

Given the data flow graph in Figure 5.13, the modules of the structure chart can be generated very quickly, as shown in Figure 5.14, and defined as Get A, Get B, and Get C. Also defined are modules S_a, which transforms A into B; S_b, which processes B into C; and S_c, for generating D from C. The output module is illustrated by the Put D module.

Thus, the data flow design method can be broken into the following four basic steps:

a) Model the program as a data flow graph.
b) Identify input, output, and the purpose of the central transform elements.
c) Factor the input, output, and central transform branches to form a hierarchical program structure.
d) Refine and optimize.

These steps can usually be performed and verified independently, separating concerns by partitioning both the design process and the problem solution.

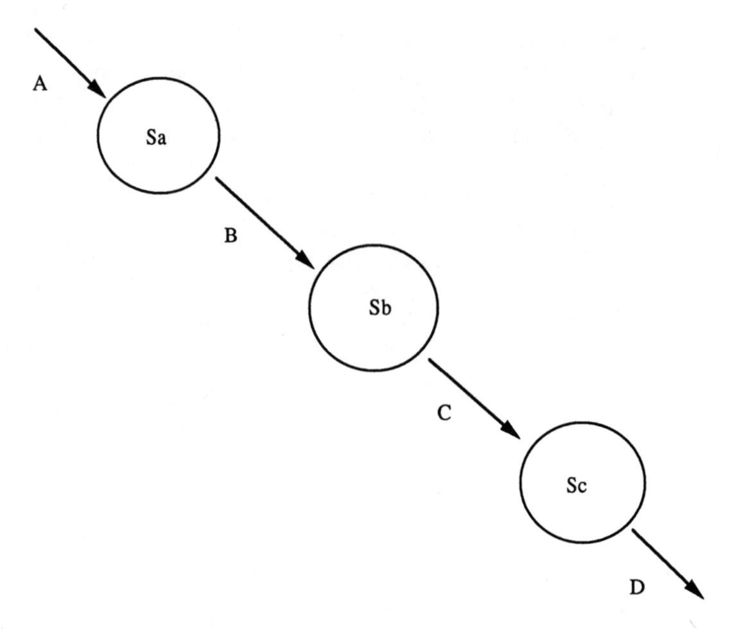

FIGURE 5.13. Data flow diagram.

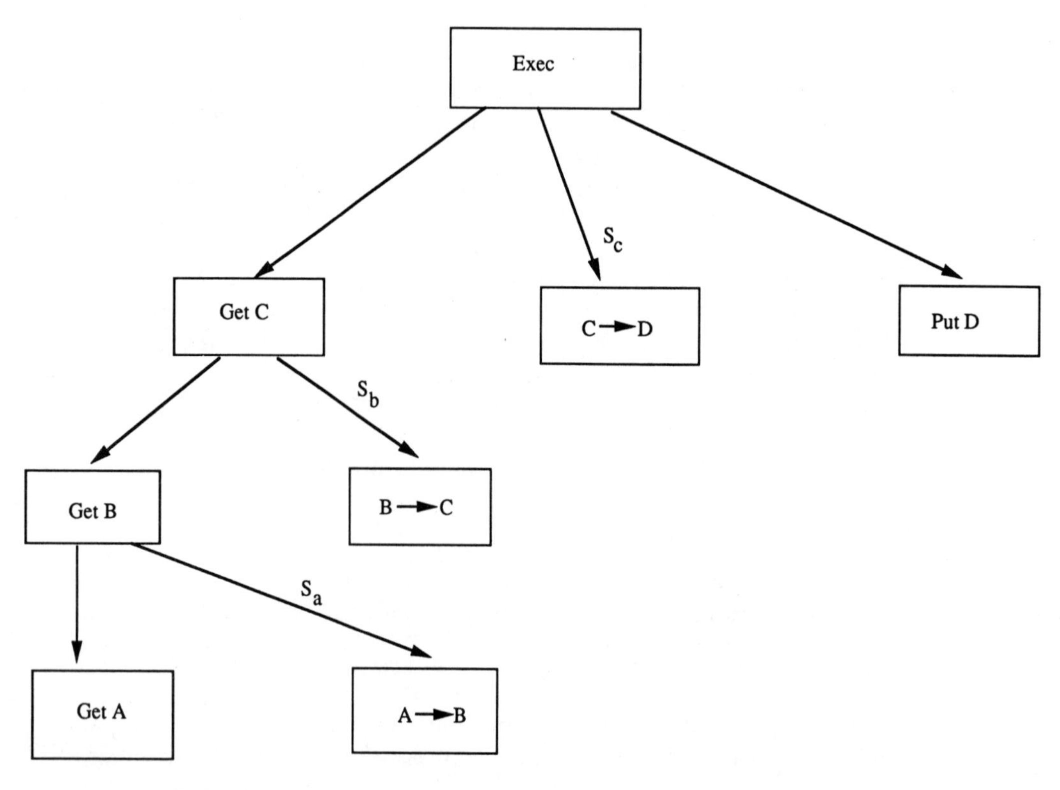

FIGURE 5.14. Structure chart.

5.9.4 Data Structure Design

The basic premise in data structure design methodology is that a program views the world through its data structures. Therefore, a correct model of the data structures can be transformed into a program that incorporates a correct model of the world.

When the program structure is derived from the data structure, the relationship between different levels of each resulting hierarchy tends to be an "is composed of" relationship. For example, an output report is composed of a header, followed by a report body, followed by a report summary. Such a static relationship does not change during the execution of the program, thus forming a base for modeling the problem.

The design process involves the following steps:

a) Form a system network diagram that models the problem environment.
b) Define and verify the data-stream structures.
c) Derive and verify the program structures.
d) Derive and allocate the elementary operations.

These steps can usually be performed and verified independently.

The system network diagram for a simple program is represented schematically in the upper portion of Figure 5.15a. In its simplest form, it represents a network of functions that consume, transform, and produce sequential data structures.

Since the program structure models the data structures, and since most operations are performed on data elements, one can list and allocate executable operations to each component of the program structure. These elementary operations

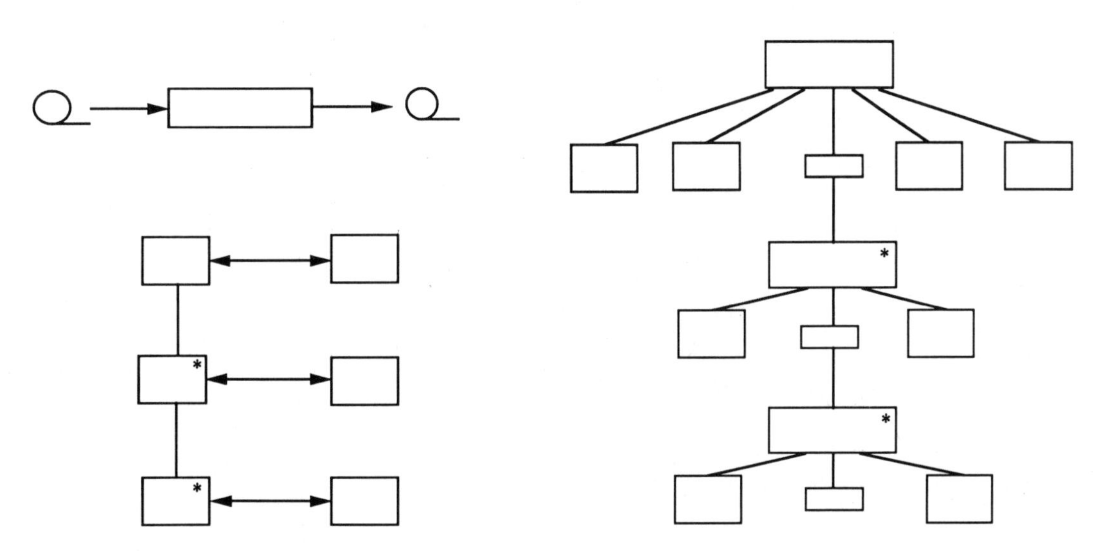

FIGURE 5.15a. System network/data structure/operations.

FIGURE 5.15b. Structure diagram.

are denoted by the small square in the structure diagram, and are shown in the lower portion of Figure 5.15a and in Figure 5.15b.

The major drawback to the data structure design methodology is that it is being developed from the bottom up. That is, although it is clear at this point how to apply it to small problems, the correct method for extending it to large system problems is not understood clearly.

5.9.5 An Interim Procedure

The following guideline can be used as an interim procedure, until we formulate the appropriate method for designing the structure of software.

- Construct a data flow model of the original problem.
- Construct data structure diagrams that correspond to each data flow path.
- Cluster and combine bubbles that can be treated by one simple program.
- Use the data structure design method to combine simple programs and reduce the number of intermediate files.
- Assign definite functionality to each block.
- Apply the functional decomposition technique to each block and create modules that can be implemented as simple subroutines or procedures.
- Now implement each block as a concurrent, asynchronous task, if you are operating under a suitable programming environment.
- Develop an executive program, which will implement total control, will initiate a call to each task, and will pass data from one task to another task.

The above methodology has been used to design a speech processing program.

Chapter 6

System Design Using the MC68020

Today, microprocessors are incorporated into computer systems to control their operation. The processor directs system activities by executing programs and performing input/output operations. The processor communicates with other hardware modules through a system bus, over which the transfer of control signals, addressing, and data movements take place. Although the performance depends on speed of the processor, the overall throughput is influenced by the size of the system bus, the access time of the memory, and the various interconnected devices.

The features of the MC68020 that help meet system, software, and interfacing requirements for designing complex systems are illustrated in this chapter. During the design of a system, the designers need to decide whether to design application-specific hardware based on a proprietary bus, or to adopt a commercially available bus to design the system. This chapter also describes the relevant issues via examples.

6.1 THREE VIEWS OF SYSTEM DESIGN

In the 20 years since the introduction of the first microprocessor, a remarkable diversity of processors has emerged. In addition to traditional microprocessors, the single-chip microcontroller, digital signal processors, special-purpose processor, graphics processors, etc. have enriched that diversity. Within each category, similar devices are available from different manufacturers, all with various bells and whistles.

In theory at least, this diversity presents the designer with many choices—the opportunity to select a processor that is optimized for a specific application. In reality, the vast number of processors makes it difficult to evaluate all of them and to make "the most appropriate" selection. The selection process gets complicated because of the need to consider various issues beyond the processor itself—business and support issues as well. Thus, the evaluation task can be simplified by evaluating the processor from the views of [Harman85]:

- System design
- Hardware design
- Software design

It is essential to understand that the system designer, software designer, and hardware designer all focus on different aspects of the processor, where the views may coincide or even conflict in some instances. For example, the software and system designers may have similar goals, that is, debugging a prototype system, but their approach to achieve the goals may differ considerably. Figures 6.1–6.3 summarize the various criteria used during the selection process by the three different design regimens.

6.1.1 The System Designer's View

The system designer is concerned with the overall operation of the system, including the architecture, performance, reliability, and cost over the entire life cycle (i.e., concept to field service and enhancements) of the system. The system de-

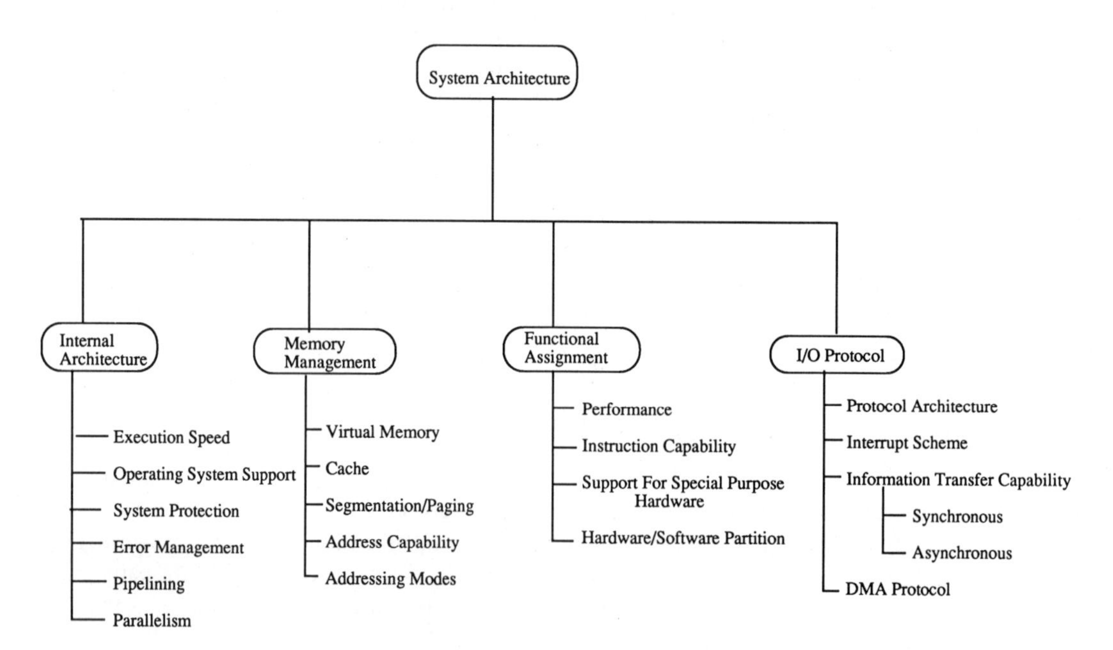

FIGURE 6.1. Functional separation of system architecture.

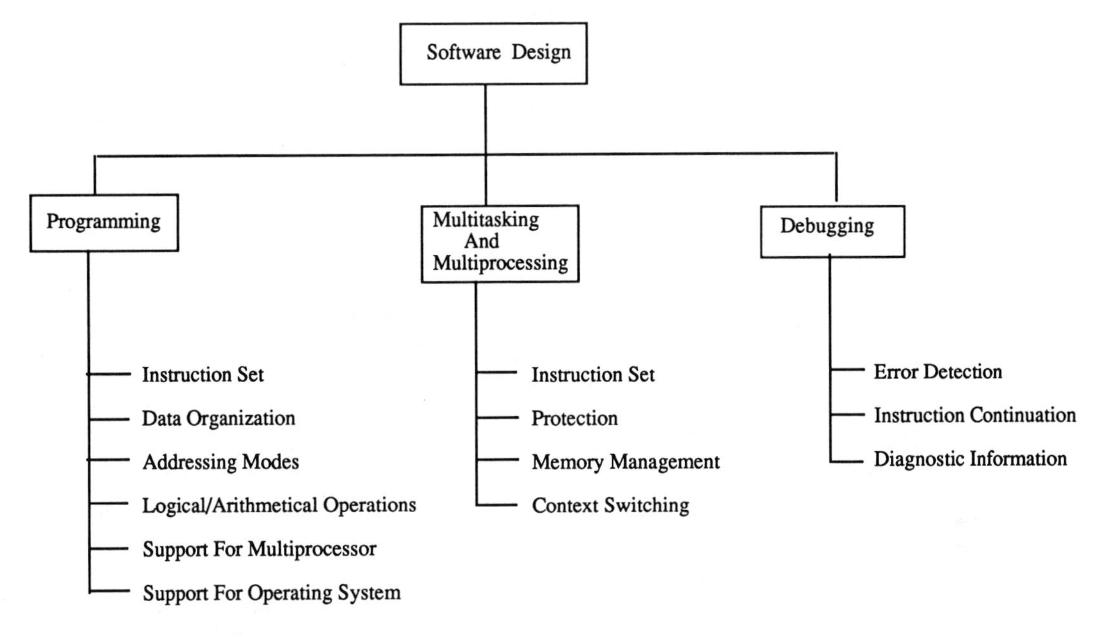

FIGURE 6.2. The software designer's concerns.

signer is also concerned with the memory size, its allocation to the operating system, and the selection/development of application software. The I/O protocol is also of concern, because of its critical importance in directing the coordination between software and hardware during data transfer. The concerns include:

- Price.
- Performance.
- System chip counts.

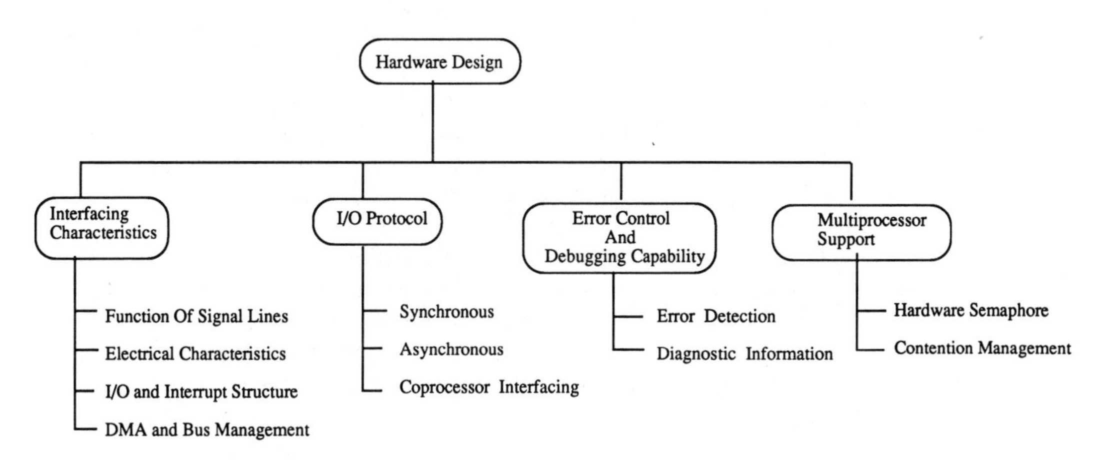

FIGURE 6.3. The hardware designer's concerns.

- Number of sources for the processor.
- Range of devices within the family.
- Prospects for future enhancements.
- Software development tools, including assemblers, compilers, and debuggers.
- Availability of operating system and application software.
- Hardware development tools, including in-circuit emulators and logic analyzers with a disassembler.
- Reputation of processor and development tool vendors.

The relative importance of these factors varies greatly depending on the situation. For a processor that is expected to become part of a very high volume product, price and second sourcing become extremely important. For a product with complex software and a tight schedule, on the other hand, quality of development tools become the dominant concern.

Embedded applications generally depend on third-party software, but the availability of a real-time operating system is also important. A real-time operating system is available for every processor, but more selection exists for the MC68000 family.

Evaluating the performance of the processor for an application is a difficult task. Although MIPS (Million of Instructions/Second) and MFLOPS (Millions of Floating-point instructions/Second) are used as general criteria, the true performance will depend on such things as:

- The instruction sets.
- The addressing capability.
- The speed of the memory.
- The architecture of the processor (i.e., the number of instruction pipelines, the on-chip cache, internal parallel operation, etc.).
- The cache and its implementation.
- Functional assignments.

By standardizing on an architecture, one can gain the maximum leverage from development tools, hardware, and software design experience. Different price performance can be achieved by using faster clock rates, enhanced functionality, and memory-system designs, so a family of different processors may play an important role in the selection of a processor. The system designer should also consider how the architecture will evolve over the time. The MC68000 architecture has been in place for over two decades, and has evolved over that period and many products have been successfully launched, based on this architecture.

The I/O is a critical distinguishable characteristic of the microprocessor. Some microprocessors are designed to work smoothly with vendor-supplied peripheral devices, but otherwise don't work smoothly. Some are oriented towards supporting one technology (i.e., TTL, CMOS, ECL, etc.), while others support more than one technology. The number of interrupt lines, and the time and technique used to respond to each are also important.

The system designer's job is to sort through this maze and make the most cost effective solution for his application.

6.1.2 The Software Designer's View

The various concerns of software designers are shown in Figure 6.2. Programming support can have a major impact on the selection of the processor. If the decision is made to implement the software in assembly language, then the quality of the compiler won't matter much. However, if (like most designers) we plan to develop software in a high-level language, then a poor compiler can throw away much of the processor's raw speed.

The efficiency of the compiler will depend on the instruction set, addressing modes, data organizations, arithmetic and logical operations, etc. supported by the processor. We have already seen that the MC68020 has a powerful instruction set and rich addressing modes. Various special instructions are also available to support multitasking and multiprocessing operations. In addition to these, the MC68020 incorporates several features to aid in the debugging and testing of programs.

The MC68020's architecture also facilitates the implementation of the operating system. To prevent errors in application programs from affecting the overall operation of the system, the processor provides two modes of execution. *Supervisory mode* is designed to support the operating system and to have full control of the processor and system functions. The *user mode* is more suitable for running application software. In addition to the basic levels of protection, the MC68020 supports an additional 256 levels of protection without any support from the memory management unit.

Various hooks are provided by the MC68020 for the smooth transition from one protection level to the next. For example, when the TRAP instruction is executed in a user mode program, control is returned to supervisory mode. This is a convenient way of procuring operation system resources from the application programs. The instruction TAS allows several processors to share a common memory area in multitasking/multiprocessing applications. We have already explored the power of the processor and its rich instruction sets, including these special features, in Chapter 5. Many of the remaining chapters are devoted to exploring the software capabilities and remaining concepts in more detail.

6.1.3 The Hardware Designer's View

Hardware designers are concerned with the implementation of the system using off-the-shelf and/or custom devices. The processor interacts with other hardware elements of the system via the system bus, where the transfer of addresses and data occurs under the watchful eye of the processor. The capability and flexibility of the processor in this regard is determined by the functions of the signal lines from the processor. Sophisticated I/O protocols and interrupt capabilities eliminate the requirements for a great deal of special hardware during the design of the system. In addition to address, data, and control signals, the MC68020 provides the state of the processor and other relevant information to external circuits, a feature that simplifies the hardware design.

Price is always an important issue in hardware design. However, what really matters is not what the microprocessor costs, but the cost of the complete system. A processor without an on-chip cache may seem like a bargain, but if hundreds of dollars worth of SRAMs (Static Random Access Memory) must be added to gain the desired performance level, then a more expensive processor with some on-chip cache might be a wiser selection. The MC68020, with its on-chip cache, favorably resolves this issue.

Various types of special-purpose processors are required today to achieve the desired performance. The MC68020 supports a special interface called the coprocessor interfacing protocol, to help hardware designers to interface any special-purpose hardware. Motorola has created various devices to operate as coprocessors, making the family functionally rich.

In order to enhance the performance of systems, designers are adopting multiprocessor technology. The MC68020 provides a set of features that enables designers to implement that technology very efficiently.

The peripherals available for a microprocessor can also make a big difference in chip count and system cost. The availability of a DMA controller, interrupt controller, and a programmable chip-select logic device can virtually eliminate the need for hardware design, other than memory and application-specific I/O. Today, we can get MC68000 family support chips from various vendors. It is important to understand that cost isn't limited to the price of the chips, it is also affected by the board space, power requirements, and testing time, all of which contribute to the final price of the system.

The hardware designers' concerns are illustrated in Figure 6.3. We will herein highlight the interfacing capability and design memory, I/O system, and multiprocessor system using the MC68020.

6.2 MEMORY SYSTEM DESIGN

The first thing any hardware designer does is to create an architecture from the functional specification. The memory map for the system is then created. The processor architecture then dictates the organization of the memory and how the I/Os are to be connected. The steps involved in memory system design are:

- Determine the memory organization.
- Estimate the RAM and ROM (Read-Only Memory) required for the system.
- Determine the address boundaries for each type of memory.
- Select the memory devices.
- Determine the I/O organization.
- Determine the arrangement of memory devices necessary to provide the required word length and memory size.
- Draw the detailed memory map.
- Determine the address decoding logic.
- Determine the buffering requirements.
- Determine the required speed for the memory devices.

In the MC68000 family of processors, I/O devices are always memory mapped. This means that memory and I/O do not have separate address spaces. Instead, the designer must allocate a part of the memory locations to I/O devices. Therefore, both memory and I/O are accessed in the same way, through the asynchronous protocol. Again, the architecture of the MC68020 requires that the memory locations between the address range $00000000 through $000003FF must be set aside for the exception vector table. The first four words must reside in nonvolatile memory (i.e., in some form of ROM). The remainder of the memory can be general purpose.

The MC68020 supports the division of the memory system into program and data memory. The program memory contains the op-codes of the instructions in the program, the direct addresses of operands, and data for immediate source operands. The data storage memory contains data operands that are to be processed by the instructions. Data operands contain variables, vectors, stacks, queues, strings, table lists, and various types of data found separate from the instructions and fixed operands.

During bus access, the processor outputs the status code over function code lines (FC2, FC1, and FC0), to indicate whether it is accessing program or data memory. One use of the function code lines is to facilitate memory partition via hardware. This is done by decoding the function codes by external logic to produce enable signals for the user program segment, user data segment, supervisory program segment, and supervisory segment.

One implementation is illustrated in Figure 6.4. A simple decoder 74S153 is used to partition the user memory segment and a supervisory memory segment. The output of the decoder generates individual chip select signals, and is utilized to select various memory banks within the segments. The designer can create various forms of memory organizations by extending this concept.

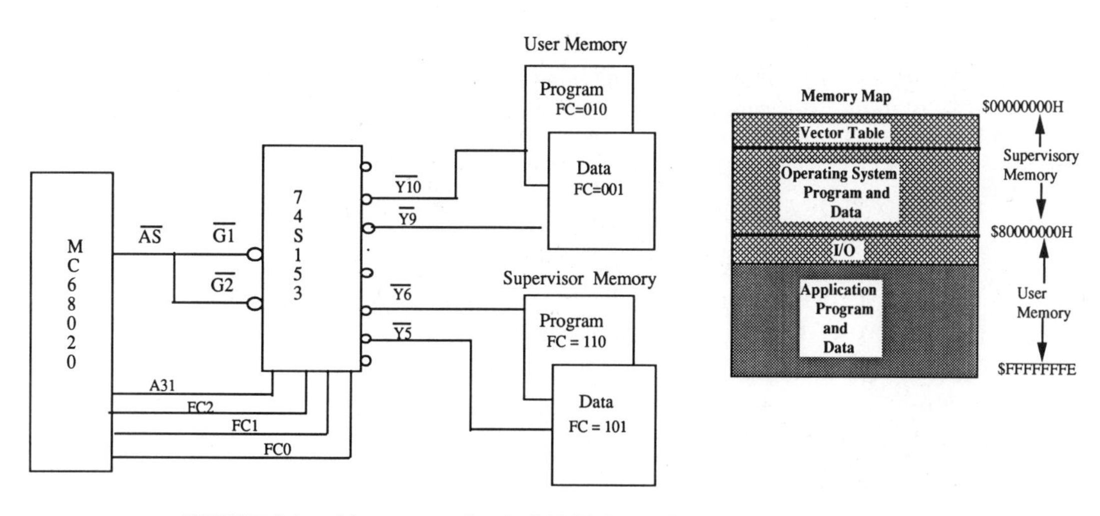

FIGURE 6.4. Memory map for the MC68020-based system.

6.2.1 Selection of Memory Devices

The cost of the memory system for the MC68020 can become a major factor. If we need a few kbytes of memory for programs and data, then it won't matter much if we use fast SRAM. However, if we need several megabytes of memory, the cost differences between SRAM and DRAM—or even between slow DRAM and fast DRAM—can be significant. We should be careful about systems that depend on separate instruction memory and data memory to attain high performance. We may be surprised by the cost and design complexity of the memory system of such design.

Most memory systems for the MC68020 are designed around DRAM devices. DRAMs are very compact, and are cheaper than SRAM. However, DRAMs need to be refreshed periodically in order to retain their contents. Programmable dynamic RAM controllers are used to interface DRAM and the processor.

We will examine two design examples in this section. The first design uses memory without wait states, and uses DRAM devices. The other design is based on SRAM.

6.2.2 Memory System Design for the MC68020 with a Zero Wait State

This design consists of National Semiconductor's DP8422 Dynamic RAM Controller, two PALs (Programmable Array Logic 16R4D and 16L8D), and a page detector (ALS631). The design accommodates two banks of DRAM, each bank being 32 bits in width, giving a maximum memory capacity of either 8 Mbytes, using 1 M × 1-DRAMS, or 32 Mbytes, using 4 M × 1-DRAM chips. This design is based on 1 M × 1 DRAM operating at a 16 Mhz clock speed. The schematic diagram for interfacing the DP8422A to the MC68020 is shown in Figure 6.5. The DRAM controller is operating in mode 1. An access cycle begins when the MC68020 places a valid address on the address bus and asserts the $\overline{\text{AS}}$ line, if a refresh is not in progress. The proper $\overline{\text{RAS}}$ and $\overline{\text{CAS}}$ will be asserted, respectively, depending upon programming bits C6, C5, and C4 for the $\overline{\text{RAS}}$ and $\overline{\text{CAS}}$ configuration, after guaranteeing the programmed value of row address hold time and column address setup time.

The High-Speed Access ($\overline{\text{HSA}}$) output signal of the page detector indicates whether the current access is in the same page as the previous access or not. $\overline{\text{ADS}}$ and $\overline{\text{AREQ}}$ is kept low if the current access is in the page, otherwise $\overline{\text{ADS}}$ and $\overline{\text{AREQ}}$ will be forced to go high to terminate the burst access. Internal refresh logic automatically generates a refresh request every 15 nsec.

Two PALs are used in the design. The PAL1 output signals are:

$\overline{\text{CS}}$ = The output signal is chip-select.
$\overline{\text{CSD}}$ = This output signal is chip-select delayed by one clock.
$\overline{\text{ASD}}$ = This output signal is address-strobed (also used as an access request, $\overline{\text{AREQ}}$ to DP8422A).
$\overline{\text{DTACKDx}}$ = This output signal is data-transfer delayed by one clock.

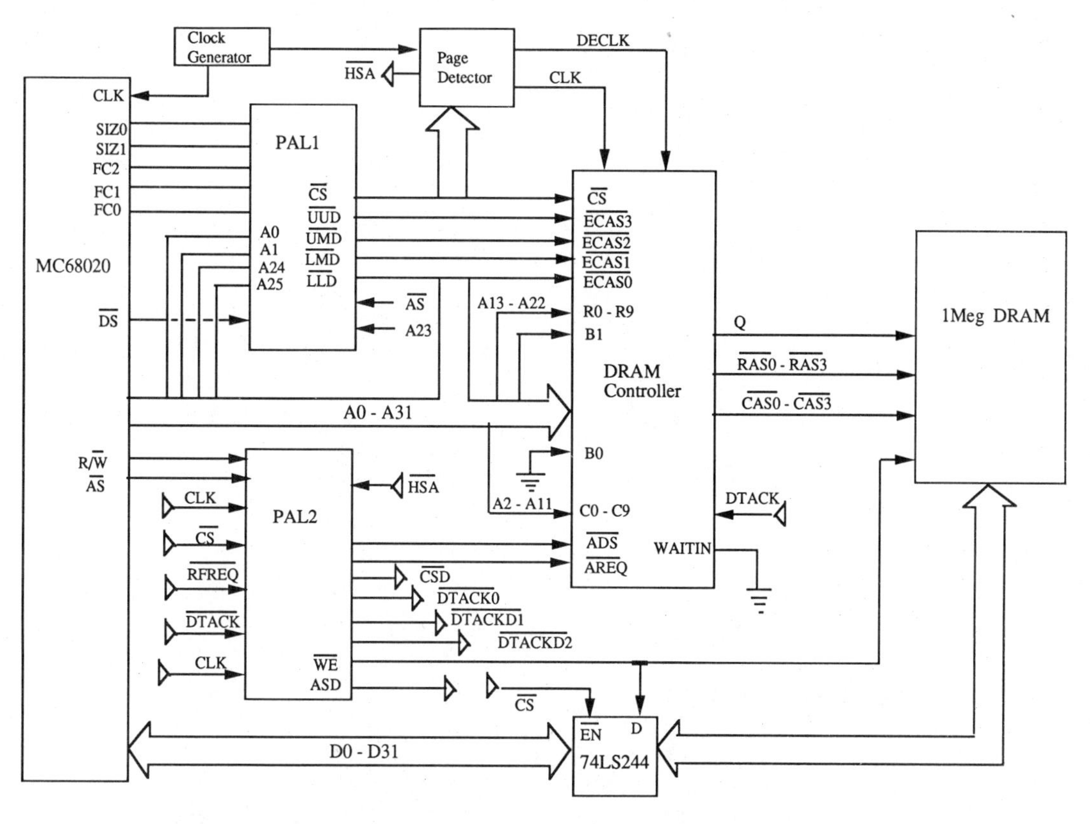

FIGURE 6.5. A memory system using DRAM and a DRAM controller.

$\overline{DTACK1}$ = This output signal is data-transfer delayed by two clocks.
$\overline{DTACK2}$ = This output signal is data-transfer delayed by three clocks.
$\overline{WE}$ = This output signal is write-enabled to DRAM
$\overline{DSACK}$ = This output signal is data-transfer and size acknowledged.
$\overline{UUD}$ = Select the upper upper byte.
$\overline{UMD}$ = Select the upper middle byte.
$\overline{LMD}$ = Select the lower middle byte.
$\overline{LLD}$ = Select the lower lower byte.
$\overline{IACK}$ = This output signal is interrupt acknowledged.

PAL1(16L8D) equations are:

Inputs are: A0, A1, A23, A24, A25, FC2, FC1, FC0, SIZ0, SIZ1, $\overline{DS}$, $\overline{AS}$
PAL1 Output equations are:

$$\overline{IACK} = \overline{FC2} + \overline{FC1} + \overline{FC0}$$

$$\overline{CS} = \overline{A23} * \overline{A24} * \overline{A25} * \overline{FC2} * \overline{FC1} * FC0 +$$

$\overline{A23}*\overline{A24}*\overline{A25}*\overline{FC2}*FC1*\overline{FC0}$ +
$\overline{A23}*\overline{A24}*\overline{A25}*FC2*\overline{FC1}*FC0$ +
$\overline{A23}*\overline{A24}*\overline{A25}*FC2*FC1*\overline{FC0}$

$\overline{UUD} = \overline{A0}*\overline{A1}*\overline{DS}*\overline{AS}$

$\overline{UMD} = \overline{SIZ0}*\overline{A1}*\overline{DS}*\overline{AS}*$ +
$A0*\overline{A1}*\overline{DS}*\overline{AS}$ +
$SIZ1*\overline{A1}*\overline{DS}*\overline{AS}$

$\overline{LMD} = \overline{A0}*\overline{A1}*\overline{DS}*\overline{AS}$ +
$\overline{A1}*\overline{SIZ0}*\overline{SIZ1}*\overline{DS}*\overline{AS}$ +
$SIZ1*SIZ0*\overline{A1}*\overline{DS}*\overline{AS}$ +
$\overline{SIZ0}*\overline{A1}*A0*\overline{DS}*\overline{AS}$

$\overline{LLD} = AO*SIZ0*SIZ1*\overline{DS}*\overline{AS}$ +
$\overline{SIZ0}*\overline{SIZ1}*\overline{DS}*\overline{AS}$ +
$A0*A1*\overline{DS}*\overline{AS}$ +
$A1*SIZ1*\overline{DS}*\overline{AS}$

The PAL2(16R6D) equations are:

Inputs are: CLK2, $\overline{CS}$, $\overline{HSA}$, $\overline{AS}$,
RFRQ, $\overline{DTACK}$, CLK, R/$\overline{W}$
$\overline{ADS} = \overline{HSA} * \overline{ASD}$
$\overline{DSACK} = \overline{DTACK} * \overline{DTACKD}$
$\overline{DTACKD} := \overline{DTACKD} * \overline{CLK} + \overline{DTACKD} * CLK$
$\overline{DTACKD1} := \overline{DTACKD} * \overline{CLK} + \overline{DTACKD1} * CLK$
$\overline{DTACKD2} := \overline{DTACKD1} * \overline{CLK} + \overline{DTACKD2} * CLK$
$\overline{ASD} := \overline{CSD} * CLK * \overline{AS}$ +
$\overline{CSD} * \overline{HSA} * \overline{AS}$ +
$\overline{CSD} * \overline{RFRQ} * \overline{AS}$ +
$\overline{RFRQ} * CSD * \overline{AS}$ +
$\overline{RFRQ} * CLK * \overline{AS}$ +
$\overline{RFRQ} * \overline{HSA} * \overline{AS}$ +
$\overline{CSD} * CLK * RFRQ$ +
$\overline{CSD} * \overline{HSA} * RFRQ$

$WE := R/\overline{W} * AS * CLK + \overline{WE} * \overline{CLK}$

$CSD := \overline{CS} * \overline{CLK} + \overline{CSD} * CLK$

The boolean operators used in the above equations are:

':=' Replaced by (after clock)
'=' Equality
'*' AND
'+' OR
'/' Complement

The timing paths are critical for the design of memory system. The type of device that is interfaced to the MC68020 determines exactly which paths are most critical. The address-to-data paths are typically the critical paths for static devices, since there is no penalty for initiating a cycle to these devices and later validating that access with the appropriate bus control signal. Conversely, the address-strobe-to-data-valid path is often most critical for dynamic devices, since the cycle must be validated before an access can be initiated.

For devices that signal termination of a bus cycle before data is validated (e.g., a hardware error or external cache), the critical path, in order to improve the performance, may be from the address to the assertion of $\overline{\text{BERR}}$. Finally, the address-valid-to-$\overline{\text{DSACKx}}$ asserted path is most critical for very fast devices and external caches (if any). This is because the time interval between the valid address to when $\overline{\text{DSACKx}}$ must be asserted to terminate the bus cycle is minimal.

6.2.3 Memory System Design Using SRAM

When the MC68020 is required to operate at a high frequency, a no-wait-state external memory system without any disruption will most likely be composed. This section discusses the design of such a memory system, which requires a higher level of performance and bus utilization than the previous design.

The MC68020 attains its highest performance when the external memory system can support a three-clock asynchronous bus protocol. Figure 6.6 shows the design of a memory bank and its interface to the processor. The devices used are:

- 16 K × 4 SRAMs MCM6290 with a 35nsec access time.
- 74F244 buffers.
- 74F32 OR gates.
- PAL-16L8D or its equivalent.

The design can be divided into three subsections:

a) The byte selection and address decoding are implemented via the PAL-16L8D.
b) The main memory.
c) The buffer section.

The first section consists of two 74F32, a 74F74D flip-flop, and a PAL-16L8D. The PAL generates six memory-mapped signals and four byte-select signals for write operations, buffer control signals, and the cycle termination signal. The byte-select signals are asserted during write operations, when the processor is addressing the 64 K bytes contained in the memory bank. Then, only the appropriate byte is being written into, as indicated by the SIZ0, SIZ1, A0, and A1 signals. The four signals, $\overline{\text{UUD}}$, $\overline{\text{UMD}}$, $\overline{\text{LMD}}$, and $\overline{\text{LLD}}$, control data bytes over data lines D31–D24, D23–D16, D15–D8, and D7–D0, respectively. $\overline{\text{AS}}$ is used to qualify the byte-select signals to avoid spurious writes to memory before the address is valid.

The second section contains the memory devices. The current design uses four

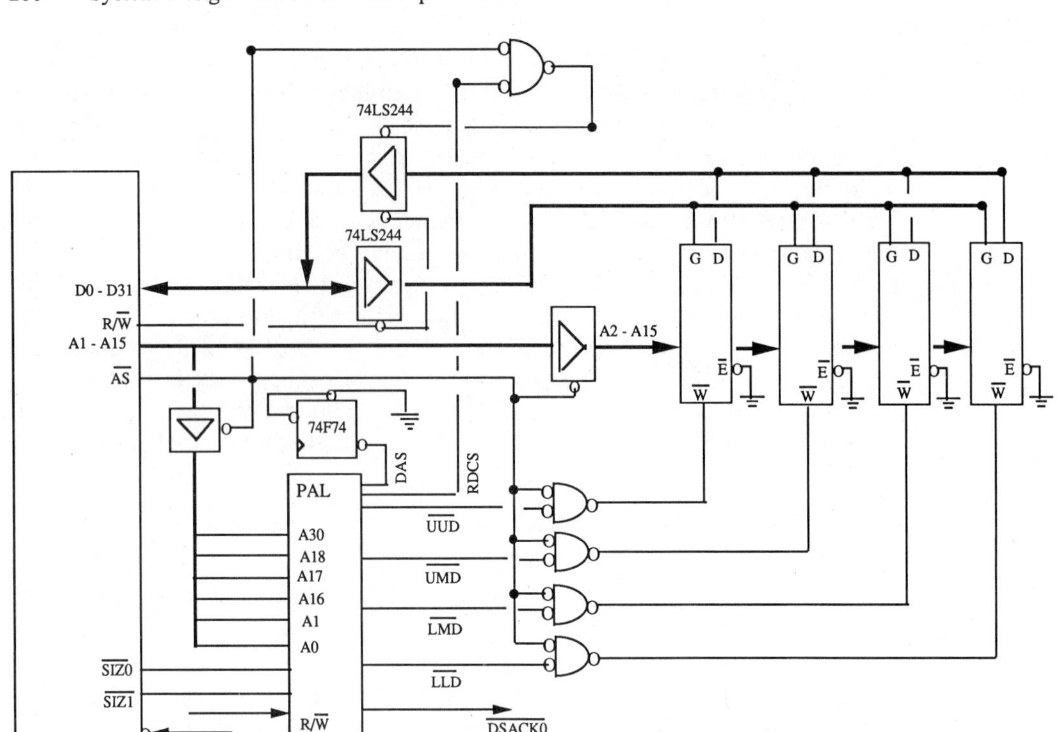

FIGURE 6.6. A memory system design using SRAMs.

of the MCM6290 SRAM devices. The important features of these devices are: $\overline{W}$ = write enable, $\overline{E}$ = chip enable, $\overline{G}$ = output enable, and multiplexed data pins. The enable pins on the SRAMs are grounded for both simplicity and to speed up the memory access. If the designer wishes to include some type of enable circuitry to take advantage of the low bus utilization to save power consumption, then the timing in this design will be preserved if the memory's $\overline{E}$ signal is asserted before the falling edge of clock. The $\overline{G}$ provides increased system flexibility and eliminates the bus contention problem.

Data buffers are required to meet the drive requirements for memory devices. 74F144 are used as data buffers. However, 74ALS244 can also be used. The $\overline{RDCS}$ signal qualified with $\overline{AS}$ controls the data buffers during read operations.

Bus contention may occur in such a design. To eliminate the possibility of contention, we have used two buffers and utilized $\overline{G}$ pins to enable the MCM6290 during read operations. Another popular technique to reduce bus contention is to put a current-limiting series resistor on each bus line. The series resistor does not eliminate bus contention, but it helps to reduce the large transient currents associated with bus contention. However, the series register will increase the access time. In general, the value of these resistors should not exceed 100 ohms.

The PAL equations are given below:

$$\mathrm{UUD} = \overline{\mathrm{A0}}*\overline{\mathrm{A1}} + \overline{\mathrm{RW}}*\overline{\mathrm{A16}}*\overline{\mathrm{A17}}*\overline{\mathrm{A18}}*\mathrm{A30}$$

$$\begin{aligned}\mathrm{UMD} = {} & \mathrm{A0}*\overline{\mathrm{A1}}*\overline{\mathrm{RW}}*\overline{\mathrm{A16}}*\overline{\mathrm{A17}}*\overline{\mathrm{A18}}*\mathrm{A30} + \\ & \overline{\mathrm{A1}}*\overline{\mathrm{SIZ0}}*\overline{\mathrm{RW}}*\overline{\mathrm{A16}}*\overline{\mathrm{A17}}*\overline{\mathrm{A18}}*\mathrm{A30} + \\ & \overline{\mathrm{A1}}*\mathrm{SIZ1}*\overline{\mathrm{RW}}*\overline{\mathrm{A16}}*\overline{\mathrm{A17}}*\overline{\mathrm{A18}}*\mathrm{A30}\end{aligned}$$

$$\begin{aligned}\mathrm{LMD} = {} & \overline{\mathrm{A0}}*\mathrm{A1}*\overline{\mathrm{RW}}*\overline{\mathrm{A16}}*\overline{\mathrm{A17}}*\overline{\mathrm{A18}}*\mathrm{A30} + \\ & \overline{\mathrm{A1}}*\overline{\mathrm{SIZ0}}*\overline{\mathrm{SIZ1}}*\overline{\mathrm{RW}}*\overline{\mathrm{A16}}*\overline{\mathrm{A17}}*\overline{\mathrm{A18}}*\mathrm{A30} + \\ & \overline{\mathrm{A1}}*\mathrm{SIZ0}*\mathrm{SIZ1}*\overline{\mathrm{RW}}*\overline{\mathrm{A16}}*\overline{\mathrm{A17}}*\overline{\mathrm{A18}}*\mathrm{A30} + \\ & \mathrm{A1}*\mathrm{SIZ1}*\overline{\mathrm{RW}}*\overline{\mathrm{A16}}*\overline{\mathrm{A17A18}}*\mathrm{A30}\end{aligned}$$

$$\begin{aligned}\mathrm{RDCS} = {} & \overline{\mathrm{A16}}*\overline{\mathrm{A17}}*\overline{\mathrm{A18}}*\mathrm{A30}*\mathrm{RW} + \\ & \overline{\mathrm{A15}}*\overline{\mathrm{A17}}*\overline{\mathrm{A18}}*\mathrm{A30}*\overline{\mathrm{RW}}*\mathrm{DAS}\end{aligned}$$

6.3 INTERFACING PROGRAM-DRIVEN I/O DEVICES

This section covers a technique for interfacing external devices to implement program-driven input/output operations. For the MC68020, interfacing I/O devices is similar to interfacing the memory devices. An I/O device responds to a bus cycle initiated by the processor, and can also request service from the processor by generating an interrupt request.

I/O devices may be classified into three general categories:

a) Human-oriented device, such as keyboards, LEDs, CRT displays, etc.
b) Computer-oriented devices like floppy drives, hard-drives, communication controllers, etc.
c) External environment-oriented devices, designed to provide interface for devices like A/D and D/A converters, stepping motors, etc.

Most off-the-shelf I/O devices available for the MC68000 family are program driven, where data transfer takes place between the processor and device over one or more ports under the control of the program executed by the processor. A port may occupy one or more memory locations within the address space of the processor (i.e., the devices are memory mapped). Ports also act as buffers that facilitate the interfacing of various speed devices.

6.3.1 Interfacing Parallel Devices

In the MC68020-based system, parallel devices can be interfaced using the MC68230 or equivalent. The MC68230, also called a parallel interface/timer (PIT), provides double-buffered parallel interfaces and a system-oriented timer. The interfaces can be configured to operate in unidirectional or bidirectional modes, either 8 or 16 bits wide. We will use this device in designing an interface to a printer [Motorola AN-854].

6.3.1.1 Parallel Interface/Timer Devices

The configuration of the MC68230 is shown in Figure 6.7. The ports can be programmed to operate in unidirectional or bidirectional modes. In unidirectional modes, the configuration of each port pin is determined by the associated data directional register (i.e., whether they are input or output types). In bidirectional modes, the data direction registers are ignored, and the direction is determined by the state of the four handshake pins. This facilitates interfacing with a wide variety of low-, medium-, and high-speed peripherals. The timer may be clocked by the system clock or by an external clock in conjunction with the 5-bit prescaler. The timer is capable of generating periodic interrupts, square waves, or a single interrupt at the completion of the programmed time period. It can also be used as an indicator of elapsed time or as a watchdog timer for the system.

The MC68230, when interfaced, will occupy 32 bytes of memory locations within the address space of the processor. The internal registers are configured via the register select pins RS1–RS5 to set the ports, timer, and their mode of operation. The system bus interface allows for asynchronous transfer of data from the PIT to the MC68020 over the data lines D0 through D7. The Data

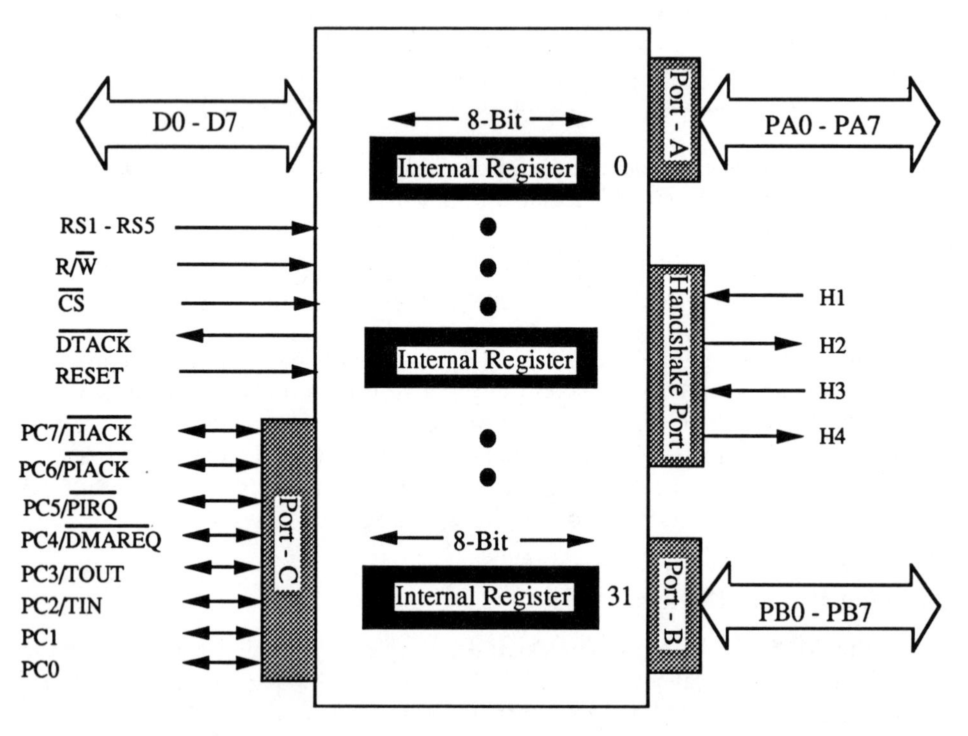

FIGURE 6.7. Architecture of the MC68230.

Transfer Acknowledgment ($\overline{\text{DTACK}}$), RS1–RS5, Timer Interrupt Acknowledgment ($\overline{\text{TIACK}}$), Read/Write line (R/$\overline{\text{W}}$), Chip Select ($\overline{\text{CS}}$), or Port Interrupt Acknowledgment ($\overline{\text{PIACK}}$) control data transfers between the processor and PIT.

6.3.1.2 Printer Interface Requirements

The printer interface requires 8 bits of data in parallel. The data transfer is achieved under control of two real-time handshake lines, $\overline{\text{DATA-STROBE}}$ and $\overline{\text{ACKNLG}}$. The setup and hold time with respect to $\overline{\text{DATA-STROBE}}$ input is 50 nsec. The printer provides three status signals, BUSY, PE (Printer Enable), and SLCT (select). The timing diagram for the interface signals is given in Figure 6.8. A HIGH on the select line means that the printer is on-line.

When the printer is ready to receive a byte of data, its PE output indicates a logic low and the SLCT (select) line indicates a logic level high. The PIT is required to output data plus a logic low-data strobe pulse upon recognizing this status. The printer responds by asserting a low pulse on the $\overline{\text{ACKNLG}}$ line, upon accepting a data byte. This indicates that the printer is ready for the next byte of data. However, if the printer is busy due to form-feed or printing activity, it responds with a HIGH BUSY signal and does not output the $\overline{\text{ACKNLG}}$ pulse until it is again available.

6.3.1.3 Design of the Interface Logic

Figure 6.9 illustrates the interconnection of the MC68020, MC68230 (PIT), and the printer interface. The PIT is configured to operate in unidirectional 8-bit

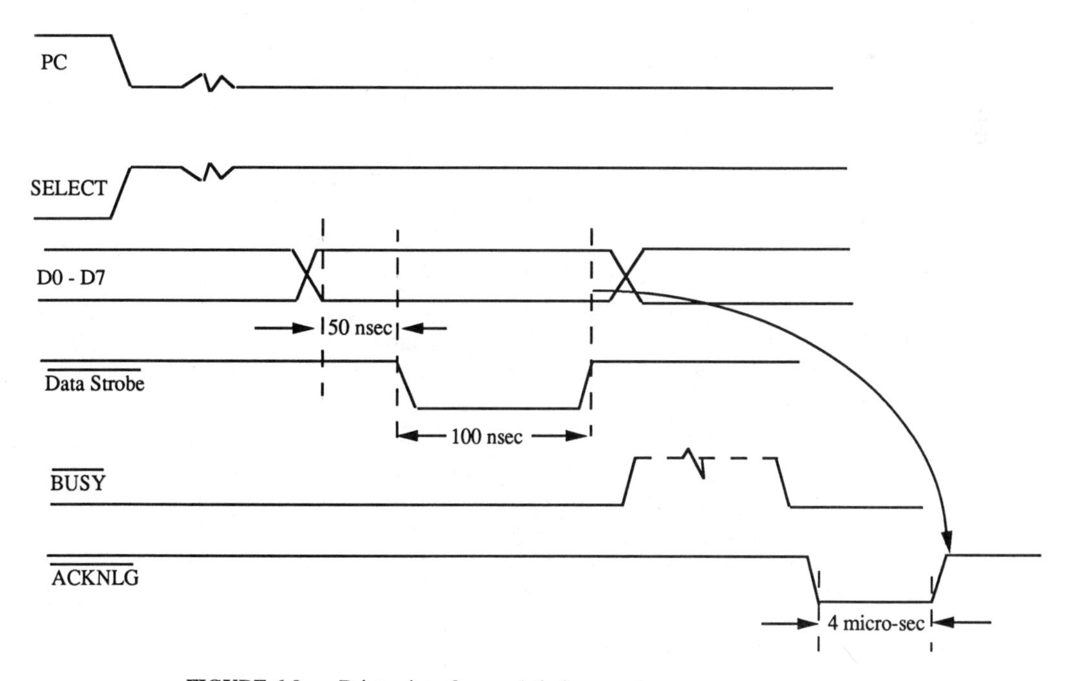

FIGURE 6.8. Printer interface and timing requirements.

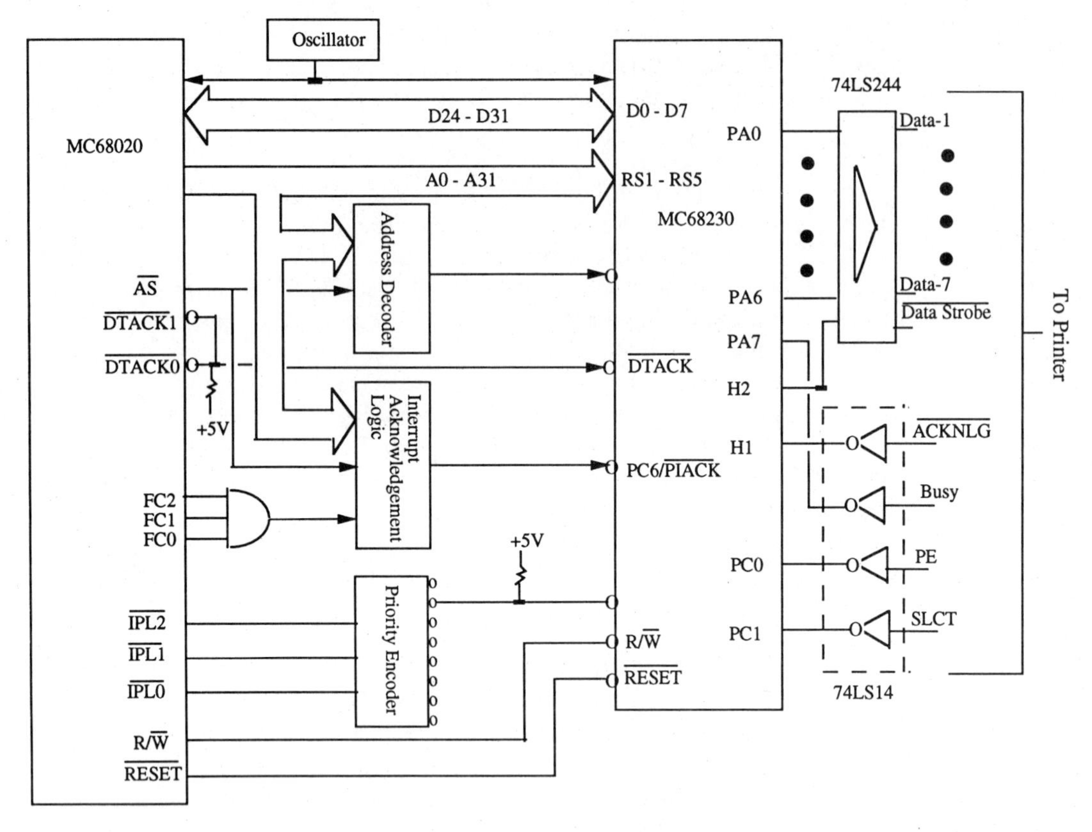

FIGURE 6.9. MC68020—PIT—Printer interface schematic diagram.

mode, with double-buffered output transfers chosen for Port A (PA7–PA0). The PA0–PA6 pins are connected to the printer via a 74LS244 driver. Since PA7 is not required for data output, it is used as the *Busy* input from the printer. This is done to illustrate the flexibility of PIT, and the designer may decide on another implementation. The H2 pin is configured to generate the pulse output required by the printer. This pin produces a minimum pulse of 180 nsec (3 clock cycles due to the write bus cycle) at 16.67 MHz, whenever new data appears at the Port A output pins. This pulse is received by the printer as its $\overline{\text{DATA-STROBE}}$ and is used to latch the new value. The low active $\overline{\text{ACKNLG}}$ output from the printer is then received by the 74LS14, and applied as an inverted input to the edge-sensitive H1 input pin of the PIT. An edge-triggered level change from high to low on H1 indicates that the printer is ready to accept new data. Again, H1 is designed to detect the trailing edge of the 4-microsecond $\overline{\text{ACKNLG}}$ pulse.

The PIT contains two full-static 8-bit latches in the output data path of each port. The designer can take advantage of this feature and design the I/O driver software in such a way that it checks H1 (Bit-0 of the port status register) after depositing each byte, to verify whether both buffers are full or not. The driver should return only when the buffers are full.

Since the MC68320 operates as a byte device, its data pins D0–D7 are connected to the D24–D31 data pins of the MC68020, and the $\overline{\text{DSACK1}}$ is tied to *high*. This will guarantee that the MC68020 will communicate with the device as a byte device.

6.3.1.4 Driver Software

The algorithm for the driver is implemented via four routines, namely:

1. LPOPEN—The initialization routine called by the server process during the power-up or reset operation. This routine will configure the MC68230 so that it operates according to the design.
2. Whenever the server needs to print, it repeatedly calls LPWRITE, and passes the buffer address, including the byte count. LPWRITE returns the status in register D0.
3. LPWRITE—This enables the interrupt after verifying the printer status. The PIT generates an interrupt as soon as interrupts are enabled.
4. LPINTR—This routine sends the data to the printer. As the printer receives data, it sends out an acknowledgment via $\overline{\text{ACKNLG}}$ to the PIT. This initiates the transfer of the next piece of data to the output lines, and new data to the double-buffered input Port A. This will eliminate the overhead of frequent interrupts.

The assembly language implementation is listed in the following:

```
        ORG $1000

    PGCR     EQU PIT+1     ; port general control register
    PSRR     EQU PIT+3     ; port general request register
    PADDR    EQU PIT+5     ; Port A data direction register
    PIVR     EQU PIT+$B    ; port interrupt vector register
    PACR     EQU PIT+$D    ; Port A control register
    PADR     EQU PIT+$11   ; Port A data register
    PCDR     EQU PIT+$19   ; Port C data register
    PSR      EQU PIT+$1B   ; port status register

        ; ********************************************
        ; LPOPEN—Called by the printer server routine for
        ;      initialization. It sets up the PIT for unidirectional
        ;      byte mode operation with Port A configured as an
        ;      output port, and H2 pulsed output for handshake
        ;      control.
        ; ********************************************

LPOPEN:     TST   FINFLAG          ; $FF = finished, idle
MOVE.B      #$7F, PADDR            ; 7 bits out and configure
MOVE.B      #$78, PACR             ; Port A submode 01,
```

```
MOVE.B     #$10, PGCR            ; Enable Port A in mode-0
MOVE.B     #$40, PIVR            ; interrupt vector
MOVE.B     #$18, PSRR            ; enable INPRPINS
RTS                              ; return from exception

           ;****************************************************
           ; LPWRITE—Trap handles set the parameters for data
           ;     transfer. D0=byte count, and A0=buffer address if the
           ;     printer is on-line. D0=return status.
           ;****************************************************

LPWRITE:   CLR.B   FINFLAG       ; just starting to execute
    MOVE.L D0, BYTECNT           ; save user parameters
    MOVE.L A0, BADDR
    BTST   #1, PCDR              ; on line?
    BEQ.S  LPWGO                 ;
NOGO:      CLR.L   D0            ; Set 0 to all bits
    NOT    D0                    ; set all to 1
    RTS                          ; return from exception
LPWGO:     BSET  #1, PACR        ; enable this interrupt
    TST.B  FINFLAG               ; wait for FINFLAG EQU $FF
    BEQ.S  LPW1                  ; OS buffering here
    CLR.B  D0                    ; normal return
    RTS                          ; return to caller.

        ;********************************************************
        ; LPINTR—Interrupt service routine. It gets the data from
        ;     the buffer and sends it to the PIT to be forwarded to the printer.
        ;********************************************************

LPINTR:    MOVE.L  A0, -(SP)     ; Save the value of A0
    MOVE.L BADDR, A0             ; get the buffer address
    TST.L  BYTECNT               ; any more data to print?
    NEQ.S  EMPTY                 ; Finished, go back
PRINTSM:   MOVE.B  (A0)+, PADR   ; move to PIT
    SUNQ.L #1, BYTECNT           ; decrement byte count
    BEW.S  EMPTY                 ; out of data stop.
    BTST   #0, PSR               ; See whether buffer is empty
    BNE.S  PRINTSM               ; yes, go back and continue
    BRA.S  NOTRDY                ; not ready, then save buffer address
EMPTY:     BCLR  $1, PACR        ; disable the interrupt.
    TST    FINFLAG               ; Set finished status
NOTRDY:    MOVE.L  A0, BADDR     ; save buffer address
    MOVE.L (SP)+, A0             ; restore A0
    RTE
```

```
BADDR       DC.L    0
BYTECNT     DC.L    0
FINFLAG     DC.B    0
PIT         DC.L    $20400         ; reserve space
            END
```

This example software will require modification to include it as a driver in a real-time operating system. A better approach is to utilize a semaphore designed to suspend the LPWRITE program. At the end of the completed data transfer, LPINTR will invoke LPWRITE, which will return for further data transfer.

6.3.2 Interfacing Through a Multifunction Peripheral

The MC68901 multifunction peripheral (MFP) provides the following resources to system/hardware designers:

- 8 programmable I/O pins with interrupt capability
- 16 source-interrupt controllers with individual source enable and masking
- 4 timers, 2 of which are multimode timers.
- Single-channel full-duplex Universal synchronous/Asynchronous Receiver/Transmitter (USART) and supports polynomial generator checker that supports byte synchronous formats.

The I/O lines and the address of the 24 registers that facilitate the MFP's configuration are shown in Figure 6.10. Programmers can utilize the directly ad-

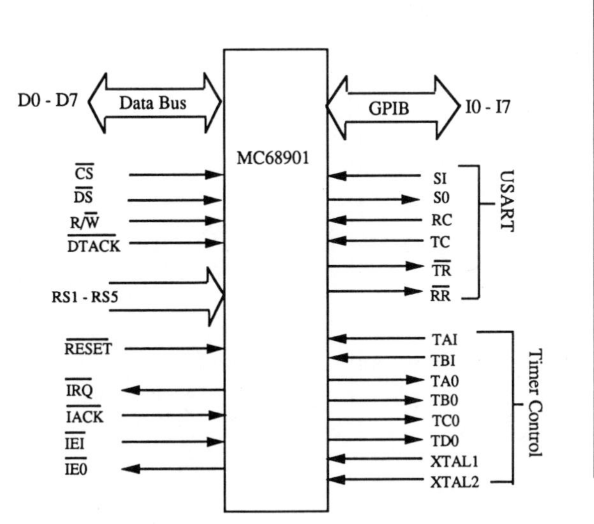

RS5	RS4	RS3	RS2	RS1	Register Name
0	0	0	0	0	General Purpose I/O Data Register
0	0	0	0	1	Active Edge Register
0	0	0	1	0	Data Direction Register
0	0	0	1	1	Interrupt Enable Register-A
0	0	1	0	0	Interrupt Enable Register-B
0	0	1	0	1	Interrupt Pending Register-A
0	0	1	1	0	Interrupt Pending Register-B
0	0	1	1	1	Interrupt In-Service Register-A
0	1	0	0	0	Interrupt In-Service Register-B
0	1	0	0	1	Interrupt Mask Register-A
0	1	0	1	0	Interrupt Mask Register-B
0	1	0	1	1	Vector Register
0	1	1	0	0	Timer A Control Register
0	1	1	0	1	Timer B Control Register
0	1	1	1	0	Timers C and D Control Register
0	1	1	1	1	Timer A Data Register
1	0	0	0	0	Timer B Data Register
1	0	0	0	1	Timer C Data Register
1	0	0	1	0	Timer D Data Register
1	0	0	1	1	Synchronous Character Register
1	0	1	0	0	USART Control Register
1	0	1	0	1	Receiver Status Register
1	0	1	1	0	Transmitter Status Register
1	0	1	1	1	USART Data Register

FIGURE 6.10. Block diagram of the MFP (MC68901), including register name and address.

dressable registers to control the operation of the device. The general-purpose I/O lines I0–I7 are 8-bit pin- programmable ports with interrupt capability. The data direction register bits individually configure them.

The USART is a single full-duplex serial channel with a double-buffered receiver and transmitter. Separate receive and transmit clocks are available, with separate receive and transmit status and data bytes. Each section has two interrupt channels: one for the normal condition, and the other for error conditions. All interrupt channels are edge-triggered.

The MFP contains four 8-bit timers capable of generating baud rate clocks for the USART. They can be configured to generate periodic interrupts, measure elapsed time, and count signal transitions. In addition to these capabilities, two timers are also capable of generating wave forms of any desired width. We will utilize this device to interface to a cassette recorder.

6.3.2.1 Hardware Design

The design shown in Figure 6.11 uses a MC68020 processor. The MC68901 interfaces the cassette with the processor, generates the interrupt, and supplies the vector number to the MC68020. The memory-mapped I/O technique is used. The registers of the MFP are mapped at memory locations starting at $2000.

The interfacing of the MFP is fairly simple. The signal lines $\overline{\text{RESET}}$, $\overline{\text{DS}}$, and R/$\overline{\text{W}}$ are connected directly to the corresponding pins of the MC68020. RS1–RS5 are connected to the A0–A4 registers of the processor. The data pins D0–

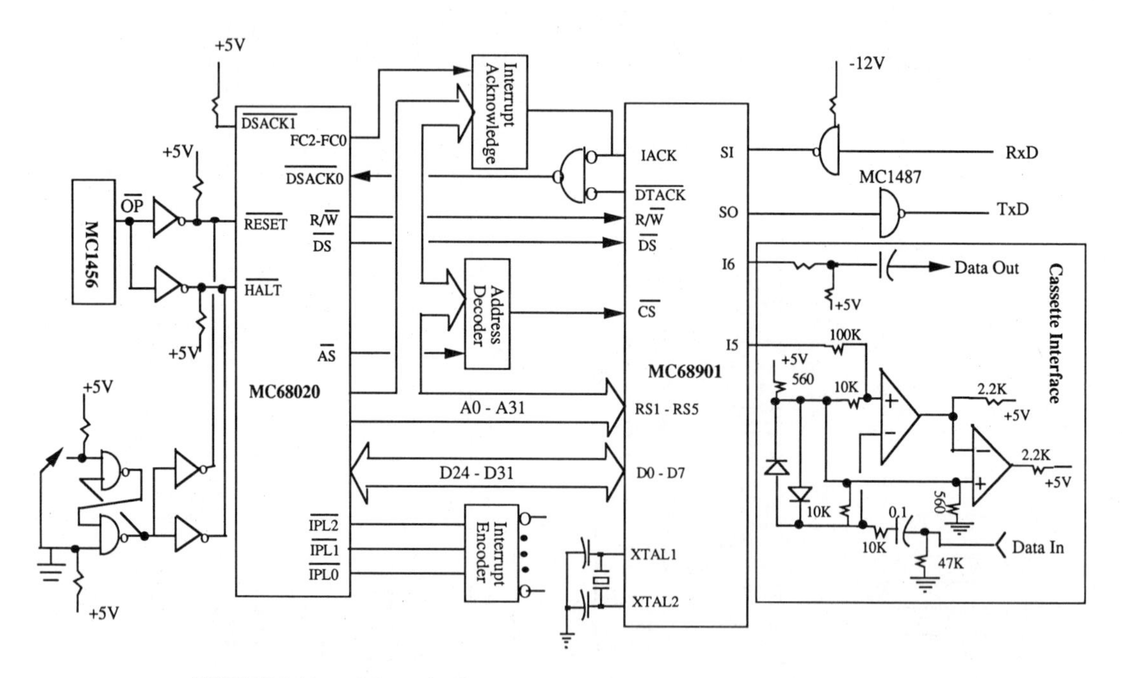

FIGURE 6.11. Schematic diagram for the hardware interconnection between an MC68020, an MC68902, and a cassette interface.

D7 of the MFP are connected to the data pins D24–D31 of the MC68020, which guarantees that the processor will transfer the data correctly in the internal registers. The $\overline{\text{DTACK}}$ is gated with the QD output of the 74LS164, and connected to the $\overline{\text{DSACK0}}$ pin of the processor, and $\overline{\text{DSACK1}}$ is tied to high to tell the processor that the MFP will provide a byte of data. The $\overline{\text{IRQ}}$ is connected to the interrupt encoder. The interrupt acknowledge logic shown in Figure 4.16 is utilized in this case. The $\overline{\text{IACK}}$ is generated when the processor enters into interrupt acknowledgment.

A 2.4576 MHz crystal is connected to the MFP to generate the desired baud rate. The timer C(TCO) is externally connected to the Receiver Clock (RC) and Timer D (TDO) is externally connected to the Transmitter Clock (TC). Although the software for this application assumes that the receiver and transmitter clocks operate at the same frequency, the MFP allows for separate clocks.

6.3.2.2 Cassette Interface

Two general-purpose I/O lines of MFP I5 and I6 are used for the cassette interface. Data is transmitted and received as square waves. The length of a single cycle of the square wave dictates whether a "1" or "0" is being transferred.

Data for the cassette interface is placed on I6 of the MFP. This output drives a resistor network, which divides the voltage by approximately 10. The cassette data output line is then connected to the microphone.

The input data from the recorder is transformed by a comparator and operational amplifier. Two IN914 diodes limit the voltage swing. The other comparator inverts the output of the previous comparator, which may or may not be required. The output of the first comparator is fed to the MFP via pin I5.

6.3.2.3 Software Design

The software is divided into five modules, namely:

- Initialization of the MFP
- Transmit character to the serial port
- Receive character from the serial port
- Transmit character from the cassette
- Receive character from the cassette.

The software is designed to reflect the hardware design. Embedded-system designers often design similar types of software.

6.3.2.4 MC68901 Initialization

Device initialization consists of the proper configuration of the device, starting the serial communication clock, and loading the USART control register. Timers C and D are used for the serial receiver and transmitter clocks, which operate at 9600 baud. The 2.4567 MHz clock is divided by loading $02 into both data registers C and D. The routine starts timers C and D in divide-by-4 mode. The USART control register is configured to operate in the divide-by-16 mode. In addition, the proper serial communication protocol is loaded into the USART control register. The USART is programmed to communicate in an asynchronous mode, with 1 start bit and 1.5 stop bit, with *odd* parity.

6.3.2.5 Serial I/O

Both receive and transmit routines look for the BREAK indication by reading a bit from the receiver status register. If a BREAK is detected at any time during serial communications, then a jump to a BREAK handler routine is made. The subroutine first transmits a message, and also checks for a CONTROL-W character and halts if one is received. Transmission is then resumed if any new data is received. The divide-by-16 mode should be used during serial communication, because it results in better noise rejection. In order to operate the USART in divide-by-1 mode, the receiver clock must be synchronized externally to the received data.

6.3.2.6 Cassette Driver

Communication is performed with the cassette via GPIP6 (bit 6 of the GPIP I/O port control register) and received through GPIP5. Data is recorded as a sequence of single-cycle square waves, with a 500 microsecond width representing *high* logic and a 1 millisecond width representing *low* logic. Timer A of the MFP is used for measuring the period in both the trasmit and receive routines. We assume that the first byte of any data stream will be a synchronization character 'S'. As a result, the receive routine assumes a synchronization character to also be an 'S'. In order to generate a bit stream, the receive routine measures the period length of all incoming square waves. After synchronization, data bytes are assembled from each successive 8 bits, and formed into a block. The implementation program is given in the following:

```
;*******************************************************************
; Initialization of the MC68901, transmit a character through the serial port,
; receive a character through the serial port, transmit a character to tape,
; receive a character from tape.
;*******************************************************************

        ORG $1000

            BASE    EQU $2000
            GPIP    EQU BASE + $01   ; general-purpose I/O
            AER     EQU BASE + $03   ; active edge
            DDR     EQU BASE + $05   ; data direction setup
            IERA    EQU BASE + $07   ; interrupt enable A
            IPRA    EQU BASE + $0B   ; interrupt pending A
            IMRA    EQU BASE + $13   ; interrupt mask A
            VR      EQU BASE + $17   ; vector
            TACR    EQU BASE + $19   ; timer A control
            TBCR    EQU BASE + $1B   ; timer B control
            TCDCR   EQU BASE + $1D   ; timer C/D control
```

```
TADR    EQU BASE + $1F    ; timer A data
TBDR    EQU BASE + $21    ; timer B data
TCDR    EQU BASE + $23    ; timer C data
TDDR    EQU BASE + $25    ; timer D data
UCR     EQU BASE + $29    ; USART control
RSR     EQU BASE + $2B    ; receiver status
TSR     EQU BASE + $2D    ; transmitter status
UDR     EQU BASE + $2F    ; USART data

; ********************************************************************
;
; Start the MC68901 transmitter and receiver clock for 9600 baud communication.
; Load the USART control register.
; ********************************************************************

INIT:   MOVE.B  #$02, TCDR          ; 1/4 transmitter clock
        MOVE.B  #$02, TDDR          ; 1/4 receiver clock
        MOVE.B  #$11, TCDCR         ; divide by 4
        MOVE.B  #$94, UCR           ; housekeeping work
        MOVE.B  #$01, RSR           ; start receiver clock
        MOVE.B  #$05, TSR           ; start transmitter clock
;
; Input character from serial port into D0
;
INCHNE: BTST.B  #3, RSR             ; echo ?
        BNE.S   BREAK               ; break, process it
        BTST.B  #7, RSR             ; check for incoming
                                    ;   data
        BEQ     INCHNE              ; continue checking
        MOVE.B  UDR, D0             ; get the received data
        RTS
;
; Send data through the serial port
;
OUTCH:  BSR     CHKBRK              ; check for break
        BTST.B  #7, TSR             ; buffer status
        BEQ     OUTCH               ; nothing to send
        MOVE.B  D0, UDR             ; buffer is full, send
                                    ;   data
; CHECK for CONTROL-W
;
        BTST.B  #7, RSR             ; read status
        BEQ     CTLW9               ; data not ready
        MOVE.B  UDR, D1             ; read data
        CMP.B   #CTLW, D1
        BNE     CTLW9               ; not CNTL-W
```

```
CTLWH:  BSR      CHKBRK          ; check for break
        BTST.B   #7, RSR         ; read status
        BEQ      CTLWH           ; wait for any new data
CTLW9:           RTS

;
; Check for break on serial line
;
CHKBRK:  BSTS.B  #3, RSR         ; read status
    RTS
;
; Process break as it arrives
;

BREAK:  BTST.B   #7, TSR         ; transmit ready ?
        BEQ      BREAK           ; wait until it happens
        MOVE.B   UDR, D0         ; read character
        BTST.B   #3, RSR         ; break reset
        BNE      BREAK
        RTS

;
; Transmit data in D2 to tape. A logic '0'
; is recorded as one square wave period of
; 1 milli-second duration. A logic '1' is
; recorded as one square wave period of the
; required duration.

TAPEO:   BSET.B   #6, DDR        ; set the GPIP6 pin as output
         BSET.B   #5, IERA       ; enable timer A interrupt
         MOVE.B   #1, D0         ; stop bit into D0
TAPE1:   ROL.B    #1, D2         ; data bit into D2
         BSR      TTST           ; wait until pulse done
         MOVE.B   #0, TACR       ; stop timer A
         MOVE.B   #10, D1        ; timer count for 1
         BTST.L   #0, D2         ; send !
         BNE      TAPE02         ; yes
         ADDI.L   #10, D1        ; no, timer count for 0
TAPE02:  MOVE.B   D1, TADR       ; set timer preload
         BSET.B   #6, GPIP       ; send 1 to tape
         MOVE.B   #$05, TACR     ; start timer A with 1/64
         BCLR.B   #6, GPIP       ; send 0 to the tape
         MOVE.B   #$05, TACR     ; start timer A with 1/64
         ALS.B    #1, D0         ; send 8 bits of data
         BNE      TAPE01         ; any more to send? If
                                 ; no, continue
         RST
```

```
        ;
        ;  Timer
           test
        ;
TTST:   CMP.B    #0, TACR      ; timer running
        BEQ      TTST1         ; no, return received
        BTST.B   #5, IPRA      ; time delay expired ?
        BEQ.S    TTST          ; no, still running
        BCLR.B   #5, IPRA      ; clear interrupt
TTST1:  RTS
        ;
        ; Receive data from tape
          into D0
        ;
TAPEIN: CLR.B    TACR          ; stop timer A
        CLR.B    D1            ; clear D1 register to accept data
T10:    BTST.B   #5, GPIP      ; wait for data to arrive
        BNE T10                ;
T20:    BTST.B   #5, GPIP      ; wait for the line to go high
        BEQ      T20
        ;
        ; Synchronize on 'S' data
        ;
TS:     ASL.B    #1, D1        ;
        BSR      T30           ; get bit from tape
        CMP.B    #'S', D1      ; found "S"
        BNE      TS            ; If no, continue
        MOVE.B   D1, (A6)+     ;
        ;
        ; Get data from tape
        ;
GC:     MOVEQ    $2, D1        ; send stop bit
GC10:   BSR      T30           ; get bit from tape
        ASL.B    #1, D1        ; stop in carry
        BCC      GC10          ; no
        BSR      T30           ; get the last bit
        RTS
T30:    MOVE.B   #$3B, TADR    ; load timer preload
        MOVE.B   #5, TACR      ; start timer in
        BTST.B   #5, GPIP      ; wait for low
        BNE      T40
        BTST.B   #5, GPIP      ; wait for high
        BNE      T50
        CLR.B    TACR          ; stop the timer
        MOVE.B   TADR, D3      ; store the value
        CMPI/B   #$1F, D3      ; logic 1?
```

```
        BLT      T60       ; no
        ADDQ.B   #1, D1    ; store 1
T60:    RTS
        END
```

6.4 MULTIPROCESSING CAPABILITIES OF THE MC68020

As processor costs comes down, it becomes economical to utilize several CPUs to design a system, to better distribute the work load for higher performance and efficiency at a reasonable cost. In order to design such a system, and get the CPUs to work together in a coordinated manner, the processors must have certain basic communication capabilities built into their architecture.

The MC68020's architecture supports two types of multiprocessor communication mechanisms, namely:

1. A bus arbitration scheme designed to allow several MC68020 bus masters to share the same bus. Three instructions are available to interlock bus cycles during interprocessor communications in a loosely coupled system.
2. Five instructions are available to support coprocessor interfacing, which can be used in a tightly coupled system.

Loosely coupled multiprocessors may be implemented in various ways. In such an architecture, each processor executes an instruction stream and manipulates data that are separate from each other. When interprocessor communication is needed, a shared memory area is used to pass messages, commands, and data between the various processors. Often, a dual-port RAM is used as a 'mailbox' for interprocess communication. The hardware designer of such a system must provide adequate hardware support for controlled access to the shared areas of RAM; in addition to a software mechanism for passing items through the 'mailbox'.

Another popular architecture is based on a time-shared bus, where processors that wish to transfer data must check the availability of the destination resource. It then informs the destination resource and initiates an actual data transfer once it gains bus mastership. In order to gain the bus mastership, each processor must go through an arbitration scheme.

In this section we will explore two designs based on the above concepts.

6.4.1 Multiprocessor Design Using Dual-Port RAMs

The asynchronous nature of the MC68020 makes it simple to utilize a Dual-Port RAM (DPR) to synchronize the activities of the processors in a multiprocessor system. A DPR has two independent ports for interfacing with the address/data/control buses of two separate processors. This allows two processors to access the same memory contents without interfering with each other. Thus, messages, instructions, data, etc. can be transferred from one processor to another very efficiently.

Access to the dual-port RAM is generally controlled by one or more semaphore registers. A semaphore register is simply a memory location set aside as a flag to indicate whether or not a dual-port RAM is currently in use. If the semaphore bit is set, one of the processors is using it. Other semaphore registers could be used to indicate the availability of messages, a change of contents, etc.

To reduce the memory overhead required for an individual semaphore, and the performance penalty associated with their management, the MC68020 provides three instructions for handling system variables. The first is TAS. It uses an indivisible read-modify-write bus cycle that locks other processors out of a multiport RAM during a semaphore operation. The other instructions are CAS and CAS2. These instructions also utilize the same indivisible read-modify-write bus cycle, and perform more complex operations that are tailored to the operations of counters, stack pointers, and queue pointers.

6.4.1.1 Hardware Considerations

The design of dual-port RAM-based multiprocessor is shown in Figure 6.12. Each processor has access to the dual-port RAM through its private bus. With asynchronous operation, it is difficult to predict when a MC68020 will initiate an access to the dual-port RAM. Thus, an arbitration circuit is required.

The arbitration logic is shown in Figure 6.13. Four D-latches form the basis

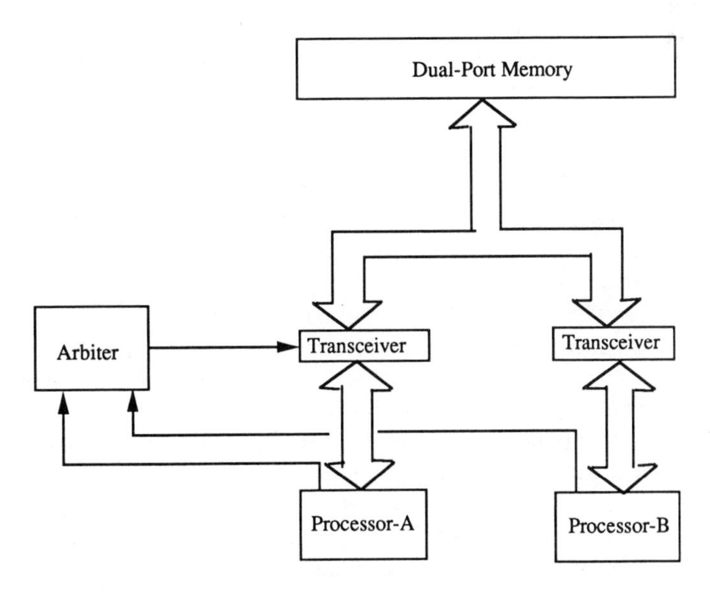

FIGURE 6.12. Multiport memory system configuration.

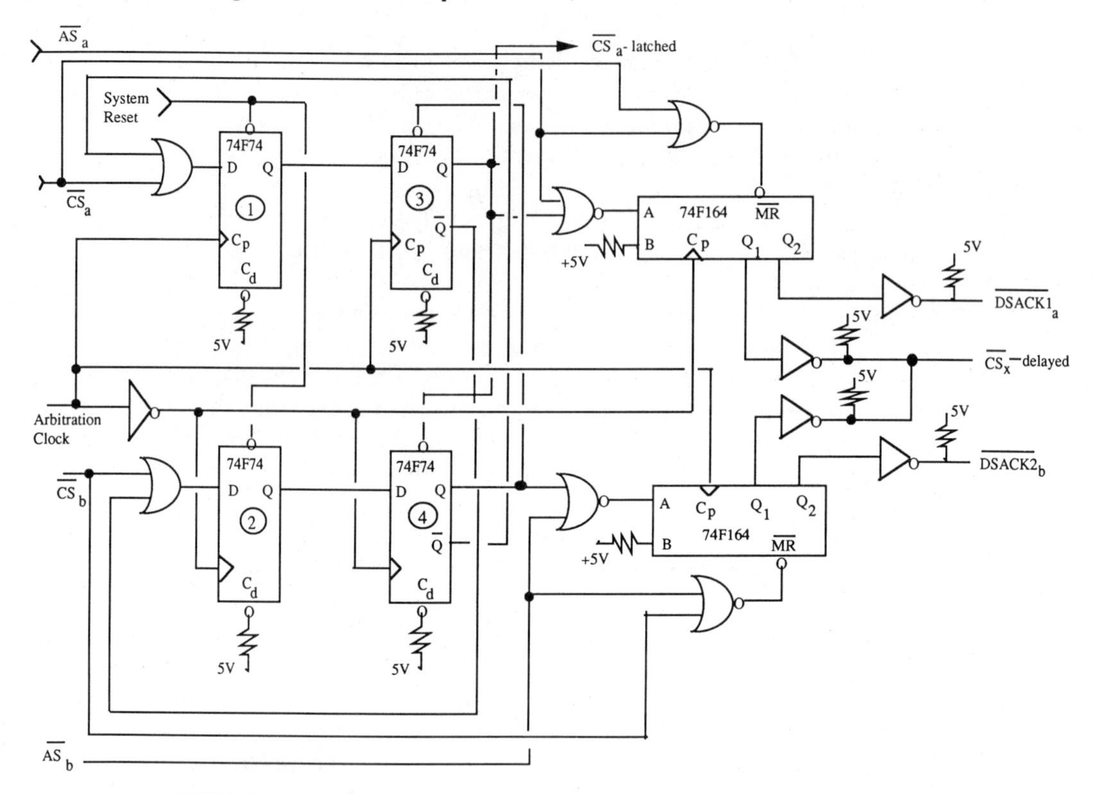

FIGURE 6.13. Dual-port RAM bus arbitration logic.

of this logic. Latches 1 and 2 are clocked on the opposite phases of the arbitration clock. At reset, the output $\overline{Q}$ of latches 3 and 4 are low. If two requests for the DPR arrives at the same time, then only one will go through latches 1 and 3, because they are clocked on opposite phases. The second set of latches, 3 and 4, provide a debouncing latch for the previous stage (latch pair 1 and 2). Latch pair 3 and 4 are also clocked by the arbitration clock. The feedback mechanism through the OR gate makes it possible to hold off the access of the other processor until the first has completed its access and releases its chip select ($\overline{CS_a}$, $\overline{CS_b}$) signal.

The latched request for a DPR signal ($\overline{CS_a}$ or $\overline{CS_b}$, to be called $\overline{CSx}$) is forwarded to the 74F164 shift register. One of the Shift Registers (SR) will be enabled because $\overline{AS}$ will be asserted with $\overline{CSx}$. Once the request signal is present at the serial input of the SR, it begins to propagate through on the positive transition of the clock. The SR outputs become $\overline{CSx}$ delays, and $\overline{DSACK1}$ for each processor. The delay becomes $\overline{DSACK1}$, and can be adjusted by tapping the appropriate output of the SR. The cycle completes when $\overline{CSx}$ or $\overline{ASx}$ are negated and clears the SR. The logic gates attached to the 74F164s allow the use of the TAS instruction. At the completion of the DPR access, the processor negates the $\overline{AS}$, which will clear the SR.

Figure 6.14 shows the static RAM and buffers required to allow two processors to access the same memory. MC68EC64 provides 8 Kwords of memory. The memory size can easily be changed using appropriate devices. Four 8-bit bidirectional buffers (74F245) provide buffering for the data bus. These buffers are controlled by the $\overline{CSx}$ data input ($\overline{CSx}$ data signals are the $\overline{CSx}$ delayed outputs of the 74F164 SR, delayed again by passing through the 74F244s). However, only one set of data buffers are activated at any time, as controlled by the arbitration circuitry. The data direction is controlled by the state of the $R/\overline{W}$ signal. The address lines are also buffered with $\overline{CSx}$ latched-signal enable inputs for the two 74F244s. The address setup time required for the MC68EC64 is met prior to the $\overline{CSx}$-delayed signal being presented to the chip select inputs of the MC68EC64 chips.

A 74F244 is used as a router for the $R/\overline{W}$, DBBE1, and DBBE2 signals from the two processors. The $\overline{CSx}$-latched signal from the arbitration logic determines which set of control signals are presented to the memory chips.

6.4.1.2 Control Software

The hardware enables the processors to read and write to all DPR locations. It is necessary to develop a driver routine that will allow each processor to initiate

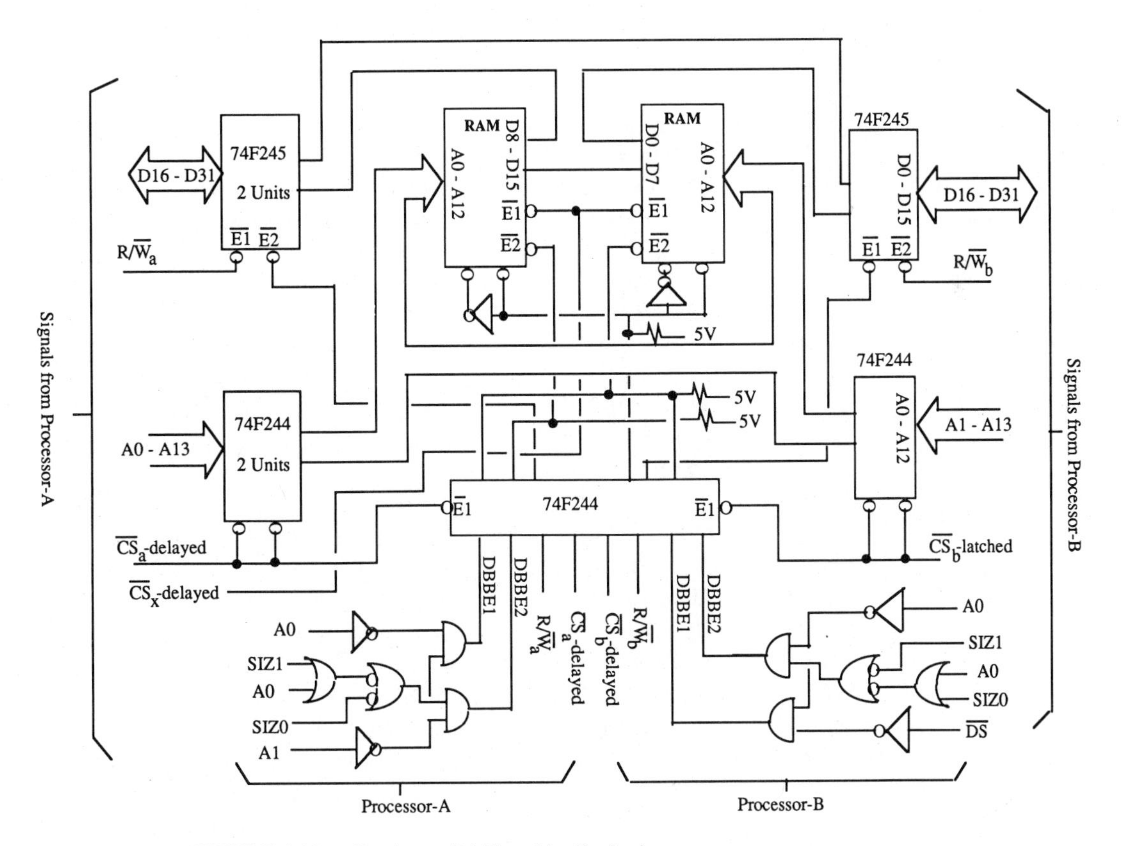

FIGURE 6.14. Dual-port RAM and buffer logic.

an access to the DPR. Two programs are included here as examples of message-passing techniques where a master sends a message to a slave processor, and vice versa. Here, a master takes characters sent from a terminal and stores them in a buffer until a carriage return is detected. The master then sets a semaphore flag and transfers the message from the buffer to the DPR. Upon relinquishing control of the DPR, the master sets a message flag in the low byte of a semaphore register to indicate that a message is waiting for the slave. The slave processor continuously checks this message flag to determine the presence of a message and, if found, prints the message out on its terminal and resets the semaphore.

```
;
; The routine sends a message
; from the master to the slave
; processor. The master routine
; reads characters from a terminal
; and puts them into MPR memory.
;
         ORG      $10000
         COUNT    DS.W     1              ; number of characters transferred
         CHBUFF   DS.W     100            ; buffer area reserved for characters
         CR  EQU  $0D                     ; carriage return
         SEMAFR   EQU      $20000         ; location of master's DPR
         MSFLAG   EQU      $SEMAFR+1      ; lower byte of semaphore word
START:            CLR.W    COUNT          ; clear the character count variable
                  MOVE.L   #CHBUFF, A3    ; set the pointer to the buffer area
INLOOP:           BSR      GETCHAR        ; call the routine to get characters
                  CMP.B    #CR, D0        ; end of string ?
                  BEW      ENDSTG         ; got the entire string
                  MOVE.B   D0, (A3) +     ; save characters in the buffer
                  ADDQ     #1, COUNT      ; increment the counter
                  BRA      INLOOP         ; get the next character
ENDSTG:           MOVE.B   #$0A, D0       ; line feed
TLOOP:            TAS.B    SEMAFR         ; dual-port RAM available
                  BMI      TLOOP          ; no, wait
                  MOVE.B   MSFLAG, D4     ; get the flag
                  CMP.B    #$FF, MSFLAG   ; flag already set by master
                  BNE      NOSET          ; no, not yet set
                  CLR.B    SEMAFR         ; reset semaphore
                  BSR      DELAY          ; enter into delay routine
                  BRA      TLOOP          ; try again
NOSET:            MOVE.B   #$FF, MSFLAG   ; set message flag
                  MOVE.L   #SEMAFR+2, A1  ; set pointer to DPR
                  MOVE.W   COUNT, (A1)+   ; store count
                  MOVE.L   #CHBUFF, A2    ; pointer to buffer
TXLOOP:           MOVE.B   (A2)+, (A1)+   ; copy the data from buffer
                  SUBQ     #1, COUNT      ; decrement counter
```

```
                BNE        TXLOOP           ; continue until finished
                CLR.B      SEMAFR           ; reset semaphore
                BSR        DELAY            ; wait
                BRA        START            ; go back and continue checking
;
; Delay routine
;
DELAY:          MOVE.W     #$5, D5
DLOOP:          NOP
                DBRA       D5, DLOOP        ; enter into the loop
                RTS
                END
;
; This routine is executed by the slave to
; receive the data from the DPR area
;
ORG $11000
    SEMA4       EQU        $21000           ; base line for slave's DPR
    MSFLAG      EQU        SEMA4 + 1        ; message flag is the lower
                                            ; byte of the semaphore word
    COUNT       EQU        SEMA4 + 2        ; number of bytes to be transferred
START:          TAS        SEMA4            ; DPR available
                BMI        START            ; not available
                MOVE.B     MSFLAG, D0       ; get the flag
                CMP.B      #$FF, D0         ; flag set ?
                BEQ        MSAVAL           ; yes, message available
                CLR.B      SEMA4            ; no, clear semaphore
                BSR        DELAY            ; wait around
                BRA        START            ; start the trial effort
MSAVAL:         CLR.B      MSFALG           ; reset the message flag
                LEA        SEMA4+4, A5      ; prepare for output
                LEA        SEMA4+4, A6      ; get the byte count
                ADDA       COUNT, A6        ; add the offset
                BSR        OUTRT            ; call the output routine
                                            ; to display the characters
                CLR        SEMA4            ; clear semaphore
                BSR        DELAY            ; call the delay routine
                BRA        START            ; go back to continue
;
; Delay routine for slave processor
;
DELAY:          MOVE.W     #$7, D5          ; set the counter
DLOOP:          NOP
                DBRA       D5, DLOOP        ; continue until D5=0
                RTS
                END
```

6.4.2 Multiprocessor Design Based on the VME Bus

The example multiprocessor system consists of eight MC68020s. A time-shared commercially available VME bus (see Subsection 6.6.2)has been used for inter-processor communication. The system also operates in the master/slave mode. One specific processor has been designated as the master, and has control over all operations of the system.

The slave processors execute jobs under the direct control of the master, and require its attention for I/O-related activities. Each processor has 32 Kbytes of private EPROM memory and 128 Kbytes of RAM, organized as a 32-bit wide memory for maximum system throughput. The total system has 3.5 Mbytes of global memory shared by all processors. The global memory occupies the address space from $10000 through $FFFFC. Figure 6.15 shows the memory map of the overall system. The master receives interrupts from the slaves and can interrupt any slave processor.

A daisy-chained arbitration scheme is used by the slave processors, which arbitrates the concurrent bus request. The architecture of the system is shown in Figure 6.16. The bus requester logic sends a request to the bus arbitrator over

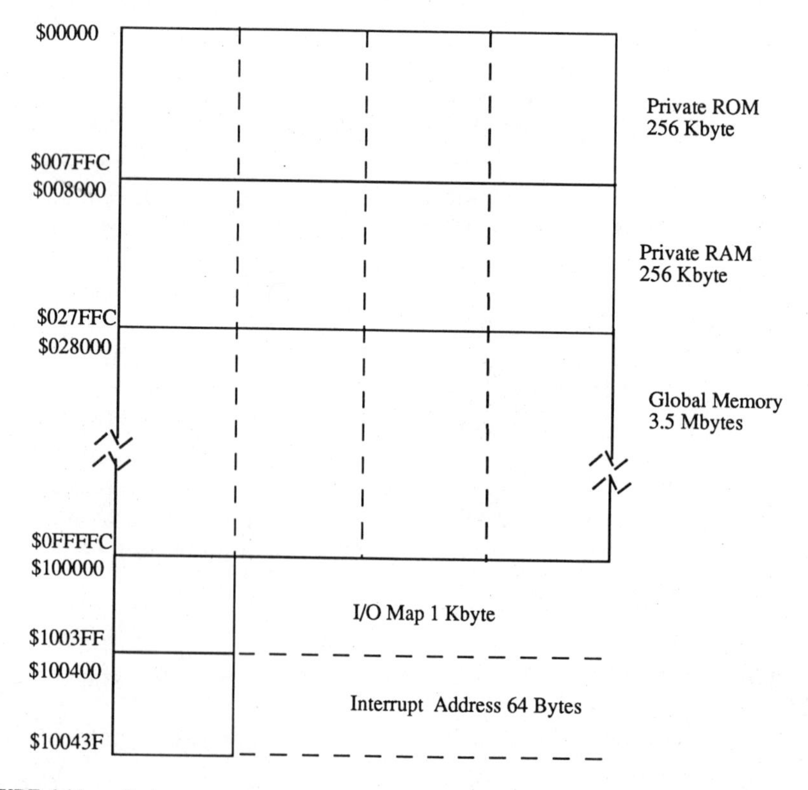

FIGURE 6.15. System memory-map.

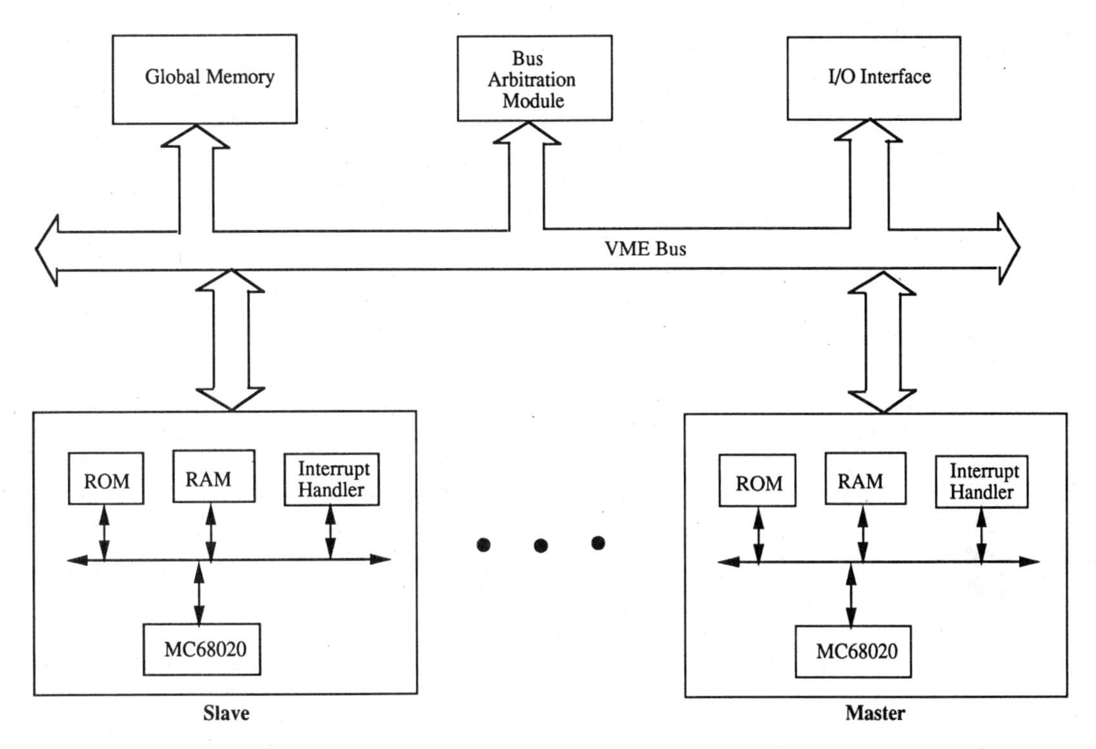

FIGURE 6.16. Block diagram of a multiprocessor system.

the VME bus. As the requester receives the bus grant acknowledgment signal, the buffer enable/disable logic allows the assertion of the global address, data, and control signals over the system bus (VME bus).

6.4.2.1 Design of the Processor Board

Figures 6.17a and 6.17b illustrate the design of the master processor board, along with the bus requester logic, buffer enable/disable logic, and the daisy-chain control logic. During the access of the private memory, the buffer enable/disable logic keeps the buffers disabled. In order to access any global memory location, the processor will assert address lines A17 through A19. To generate a bus request, address lines A17–A19 are logically NORed and connected to a D latch. The $\overline{AS}$ will clock this signal to the output of the D latch. The output of this D latch will drive the $\overline{BR}$ signal line of the system bus. This request is forwarded to the daisy-chain logic to indicate that a bus request is pending.

The daisy-chain logic receives the bus grant $\overline{BG}$ x IN from the bus arbitrator over the system. If a request is pending, then $\overline{BR}$ is low for this processor board. This, along with $\overline{BG}$ x IN (low in this case), signals the buffer enable/disable logic to enable the address, data, and control buffers. If no request is pending for the processor board, then it simply drives the $\overline{BG}$ x OUT line to transmit the grant signal to the next processor board in the chain.

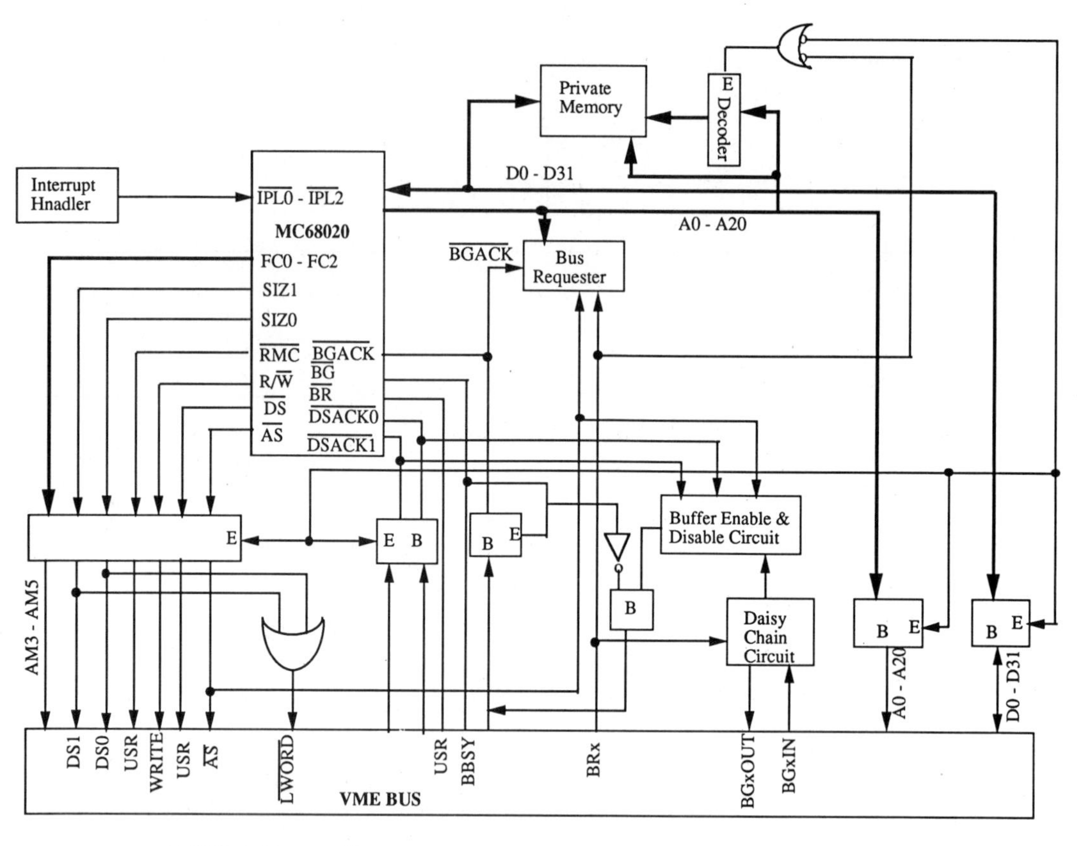

FIGURE 6.17a. The processor board.

The buffer enable/disable logic generates an output enable signal for address/data/control lines and the $\overline{\text{BGACK}}$ signal for the bus arbitrator. Since the arbiter removes the bus grant signal when it receives $\overline{\text{BGACK}}$ from the master requesting the bus, the output enable signal is the logical OR of the daisy-chain bus grant, and $\overline{\text{BGACK}}$ of each user. The $\overline{\text{BGACK}}$ is generated when the global device asserts the $\overline{\text{DSACKx}}$ of the bus master to which it is accessed. This will set the FF in this module and drive the output low. The complementary output of the FF is inverted to generate the $\overline{\text{BGACK}}$, which is now low and drives the BBSY line of the VME bus. The buffer is controlled by the on-board $\overline{\text{BG}}$ signal from the CPU, which is asserted when any other master requests the bus from this CPU when it has the mastership of the bus. The BBSY is a bidirectional line.

6.4.2.2 Design of the Bus Arbitration Module

Figure 6.18 shows the Bus Arbitration Module (BAM). MC68452 is used to realize this module. Since the VME bus allows only four bus request lines, the $\overline{\text{DBR7}}$–$\overline{\text{DBR4}}$ inputs are connected to $\overline{\text{BR3}}$–$\overline{\text{BR0}}$ of the VME bus to receive requests for bus arbitration. The bus grant line outputs $\overline{\text{BG7}}$–$\overline{\text{BG4}}$ are connected to $\overline{\text{BG3IN}}$–

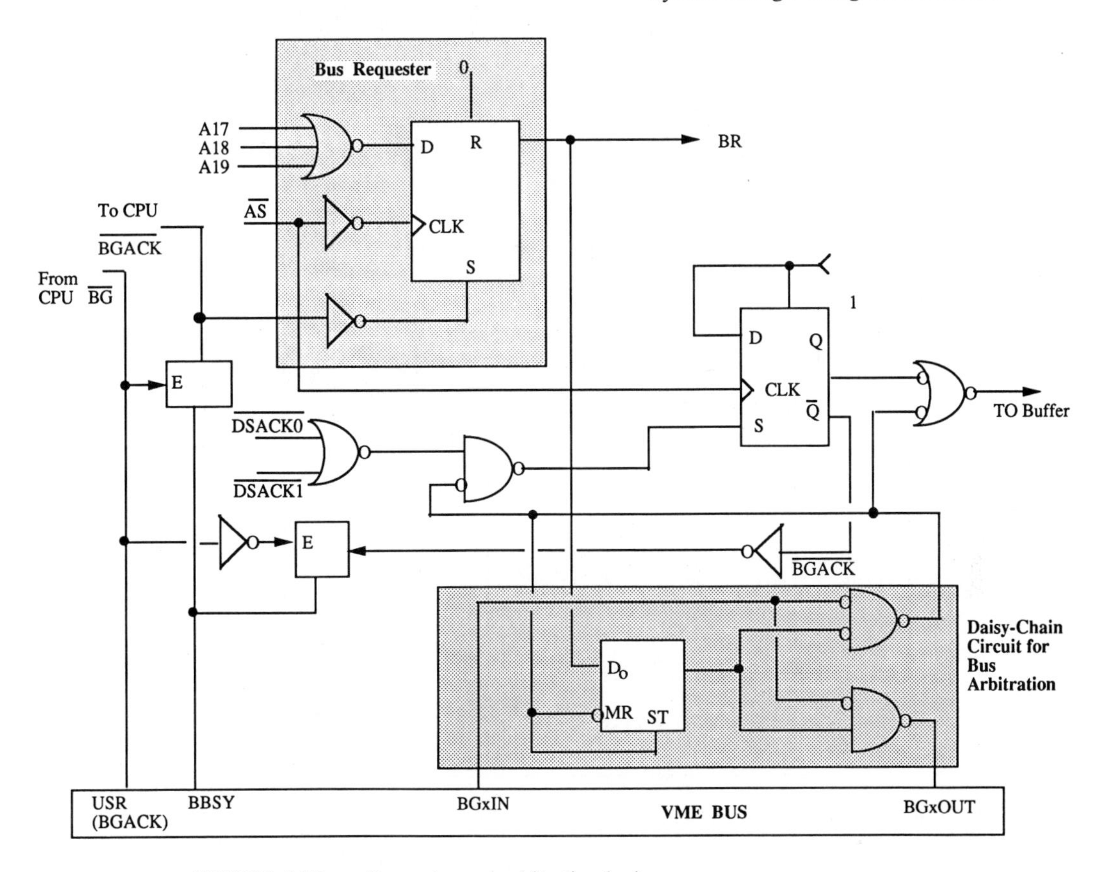

FIGURE 6.17b. Requester and arbitration logic.

$\overline{\text{BG0IN}}$ of the VME bus to issue grant signals to the requesters. Whenever a requester wants the bus mastership, it drives one of $\overline{\text{BRx}}$ lines on the VME bus. The specification of this bus allows more than one master to drive the same $\overline{\text{BRx}}$ line. The bus grant lines are daisy-chained, which allows the master closest to the arbiter to take control of the bus, if the request is pending from that processor.

The $\overline{\text{BBSY}}$ line from the VME bus is connected to the $\overline{\text{BGACK}}$ input of the BAM. This signal is asserted by the would-be master when it takes control of the bus and generates the $\overline{\text{BGACK}}$ signal. A VME bus line is used to connect the $\overline{\text{BG}}$ output of the CPUs when the BAM requests the current bus master to relinquish the bus. The current bus master will first signal the BAM that it has received the signal and will relinquish the bus at the end of the current bus cycle. The output signals $\overline{\text{BR}}$ and $\overline{\text{BCLR}}$ from the BAM are connected to the VME bus. $\overline{\text{BR}}$ is asserted by the BAM whenever there is a request (low) at the $\overline{\text{DBRx}}$ input. This line asserts the $\overline{\text{BR}}$ line of the current bus master. The $\overline{\text{BCLR}}$ line, when asserted, indicates that a higher priority request is pending. These two lines are buffered. The buffer is enabled by the $\overline{\text{RMC}}$ line of the MC68020. This scheme is utilized

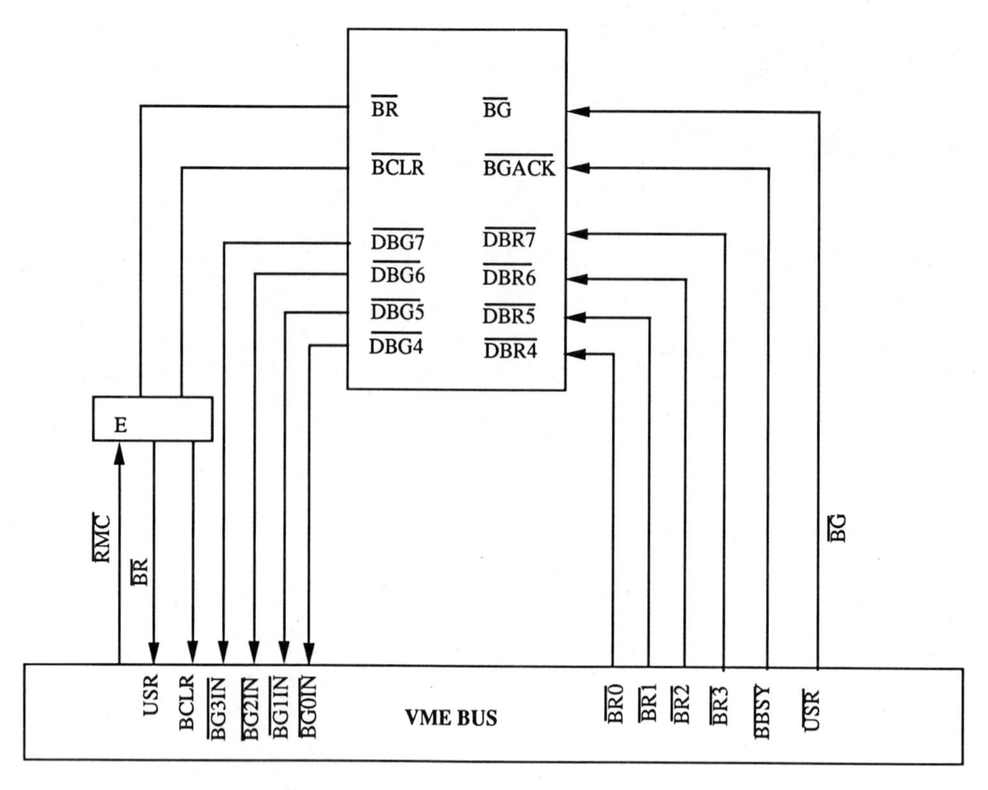

FIGURE 6.18. Bus arbitration module.

to provide the indivisible read-modify-write bus cycle for each processor. When the processor performs a read-modify-write cycle, all requests for the bus mastership by other processors are disabled, until the current bus master is through with the indivisible operation.

6.4.2.3 Design of the Local Memory

We have already seen that each processor has local memory and is divided into two sections. It is made of four 8 Kbyte devices. This memory contains software to boot the local CPU, initialize all I/O devices, and program the board-dependent functions, Maximum data throughput to the MC68020 is provided through the fast decoding logic and separate data transceivers supporting one-wait-state operations, if 100 nsec devices are installed (CPU clock speed of 16.67 MHz). This design is shown in Figure 6.19, which spans over 7FFFH–0000H.

The local SRAM is 128 Kbytes, and is installed on all local processor boards. Byte-wide SRAM, such as NEC's 32K × 8 bits, is used that retains its contents even if the power fails for a long time. The memory has an access time in the range of 85 to 150 nsec, and access times compatible with the EPROM. The design is shown in Figure 6.20 and spans over 27FFFH–8000H. This device is used because of its ability to interface to the memory bus. It has 15 address lines and

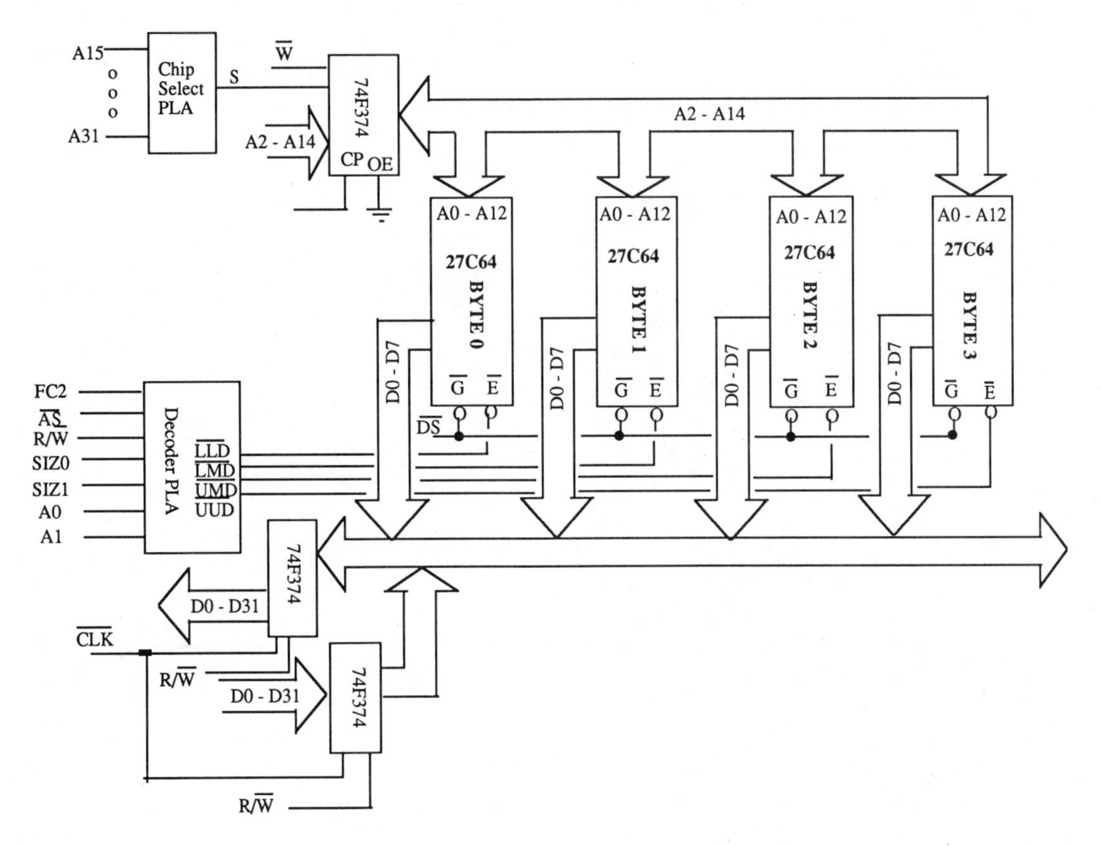

FIGURE 6.19. System EPROM—Boot and initialization (7FFF—0000H).

8 common I/O signals, an output enable ($\overline{OE}$) pin, and a write enable ($\overline{WE}$) pin. The chip select ($\overline{CS}$) pin controls the operation of the device. When the $\overline{CS}$ is high, the device is on standby and consumes little power. We minimized power consumption by enabling only the accessed devices in standby. The processor is interfaced with the local memory via a local bus. This allows the processor to access the local memory without having to go through the global bus and arbitration resolution mechanism.

6.4.2.4 The Global Memory Design

The global memory spans from $20000 through $0FFFFC. The organization is shown in Figure 6.21. It consists of 28 banks and each bank is 32 K × 32 bits. The address lines A0–A15 are connected to the RAM chips, while the address lines A15–A19 are decoded to select a specific bank. Two 4-by-16 line decoders (MC14514) are used for decoding each bank address. The decoder can select up to 32 banks. The address lines A15–A18 are used as select lines to the decoder and A19 selects one of the two decoders. The decoders are enabled by the $\overline{IACK}$ signal from the interrupt handler. This is used because when a processor is inter-

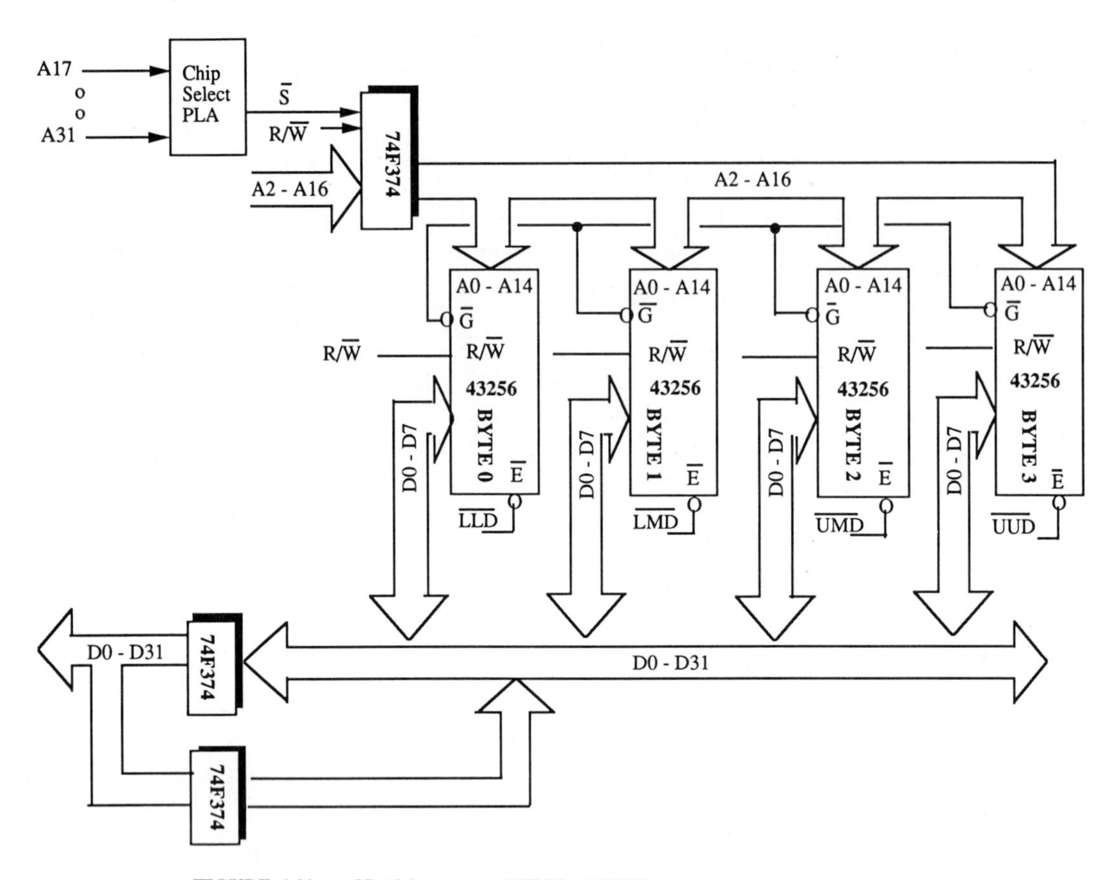

FIGURE 6.20. SRAM memory (27FFF—8000H).

rupted, it initiates an interrupt acknowledgment cycle and gains control of the global bus and drives all of the address lines to the Z-state. In his mode, the processor reads the vector number from the data bus and must not access the global memory. As a result, the global memory must be disabled during the IACK cycle of the processor. Thus, when the IACK line is asserted, both decoders are disabled simultaneously.

Figure 6.22 illustrates the simple $\overline{\text{DSACKx}}$ circuit. Since the memory is 32 bits wide, it should assert both $\overline{\text{DSACK0}}$ and $\overline{\text{DSACK1}}$ simultaneously. The circuit is realized using a D flip flop. The clock signal is tied to SYSCLK and the data input is connected to the $\overline{\text{AS}}$ pin from the VME bus. Whenever a valid address appears to the global memory, the set input of the D-FF is asserted and the output of the FF is low, which in turn drives $\overline{\text{DSACK0}}$ and $\overline{\text{DSACK1}}$ on the VME bus. At the end of the bus cycle, the $\overline{\text{AS}}$ is negated, which in turn negates the $\overline{\text{DSACKx}}$. It is important to note that $\overline{\text{DSACKx}}$ lines are user-defined on the VME bus.

6.4.2.5 The Interrupt Handler

The system operates in a master slave mode, where the master processor has the capability to interrupt any one of the slave processors at any time. The slave

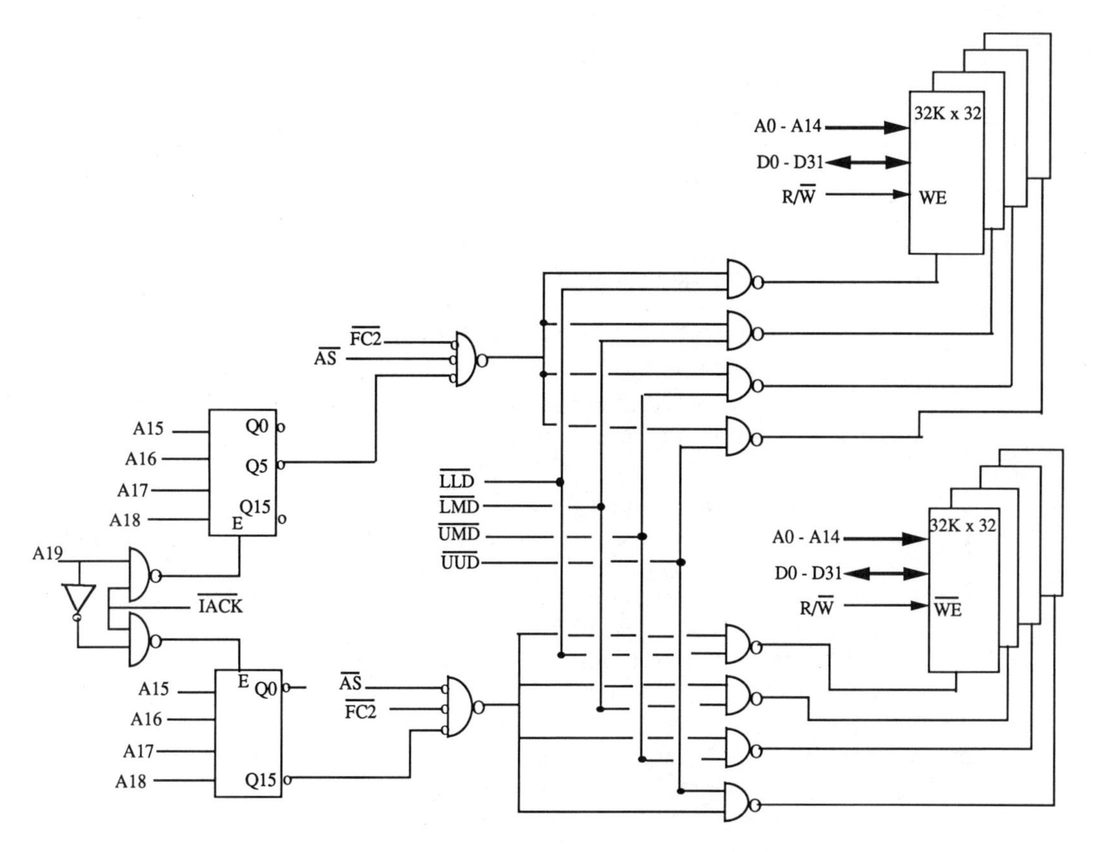

FIGURE 6.21. Global memory.

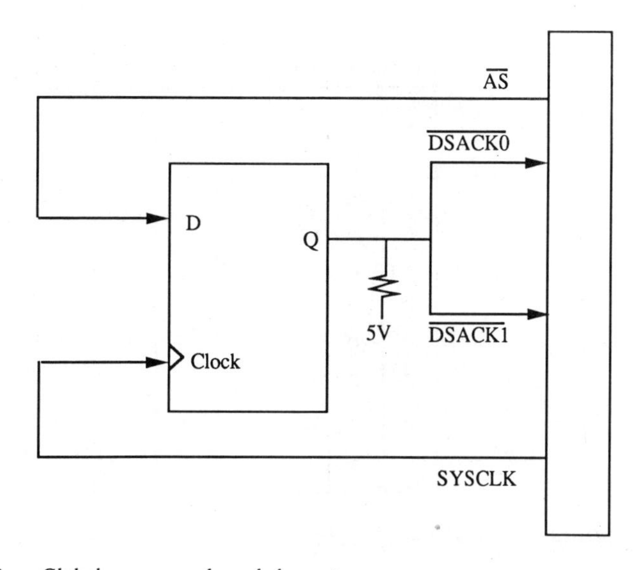

FIGURE 6.22. Global memory acknowledgment.

processors can also interrupt the master processor using a predetermined priority level. The master processor uses a special memory map technique to interrupt the slave processor. The system memory map in Figure 6.15 illustrates the global memory, I/O space, and the interrupt space. The 64 bytes of memory are utilized for the interrupt address. Figure 6.23 shows the address of the interrupt schemes.

The design of the interrupt circuit for master processor is illustrated in Figure 6.24. Whenever the master wants to interrupt a particular slave, it simply writes a byte into the reserved location for that slave. In practice, the byte is latched to an 8-bit buffer. When the master writes at this location, the address lines A10 and A20 are NANDed with $\overline{AS}$ to enable this buffer. The low output of the NAND gate asserts the $\overline{IRQ7}$ line on the VME bus. The address broadcast over the global bus is decoded at each memory board to determine which board and interrupt was sent by the master processor.

6.4.2.6 *The Interrupt Circuit for the Slave Processor*

The design of the interrupt circuit for slave processor is illustrated in Figure 6.25. The address lines A3–A5 identify the destination of the slave processor board, and the address lines A0–A2 determine the level of interrupt. Address lines A5–A3, including A10, A20, and $\overline{IRQ7}$, activate the designated slave processor's decoder. This decoder then decodes the address lines A0–A2 to determine the level of interrupt. The decoder output is fed to the priority interrupt encoder, and 74F348 encodes the interrupt level to the IPL0–IPL7 pins of the slave.

Upon receiving the interrupt, the slave compares the interrupt level against the mask level of its SR register, and initiates the IACK cycle. The slave processor's FC lines, including A16–A19, are logically ANDed to generate an IACK signal. This IACK line is connected to the user line on the VME bus, and ends up at the master processor board. Upon receiving the IACK signal, the master processor

Board Location			Interrupt Level			Remarks
A5	A4	A3	A2	A1	A0	
0	0	0	X	X	X	Master
0	0	1	0	0	0	Slave Processor's 7 Levels of Interrupt
0	0	1	0	0	1	
0	0	1	0	1	0	
0	0	1	0	1	1	
0	0	1	1	0	0	
0	0	1	1	0	1	
0	0	1	1	1	0	
0	0	1	1	1	1	

FIGURE 6.23. Processor board location and interrupt level.

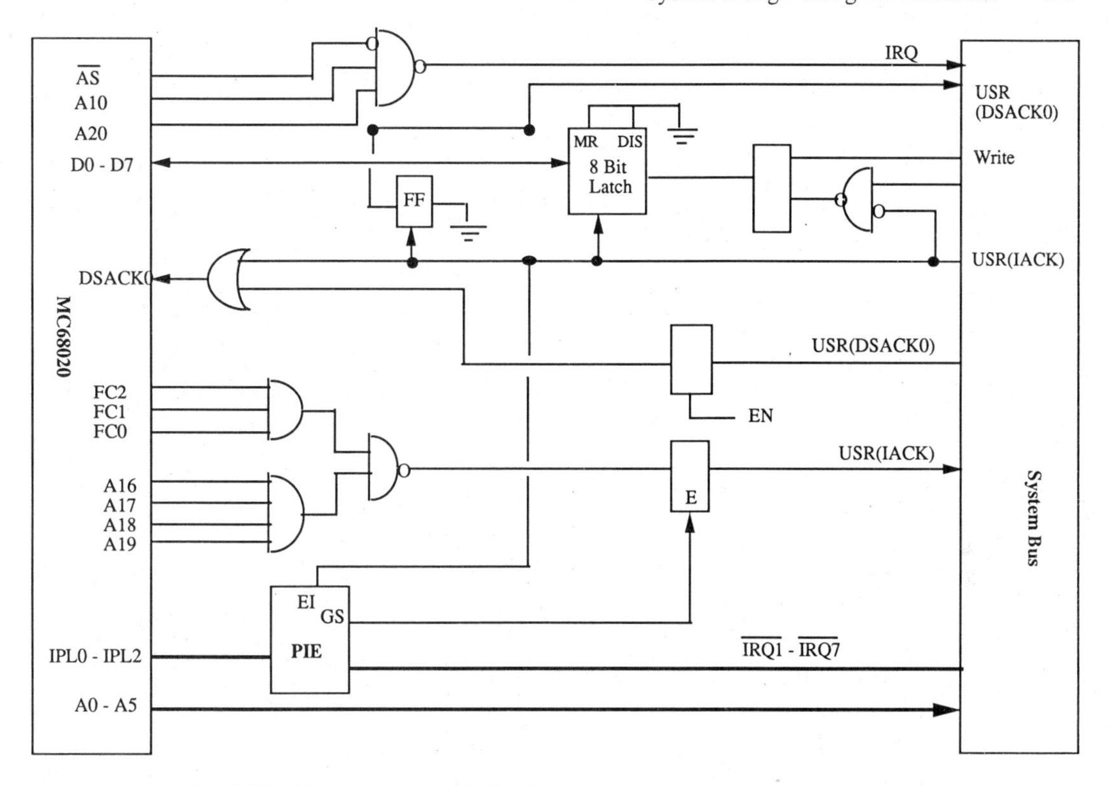

FIGURE 6.24. Interrupt logic for the master processor.

writes the vector number on the slave processor's latch register and asserts the $\overline{\text{DSACK0}}$ of the slave processor. The master then relinquishes control of the system bus, and the interrupted slave gains control of the bus. It receives the $\overline{\text{DSACK0}}$ from the master, and then reads the vector number from the master boards. The R/$\overline{\text{W}}$ line of the slave enables the output buffer of the latched vector number. After reading the vector number, the IACK cycle is terminated and it then jumps to the service routine.

6.5 DESIGN TRADE-OFFS

So far, we have designed systems with the ROM occupying the lower address and RAM occupying the higher address in the memory map. The I/O map started immediately after the boundary of the memory. In many embedded systems, the designers will select another strategy to design the system. The memory map for such a design is shown in Figure 6.26, where the *system area* (implemented using RAM) resides at the lower address, followed by the *user memory* (implemented using RAM chips), then the System ROM and the I/O interface registers, at the highest address of the system.

This architecture will allow the system designer to create a virtual machine by

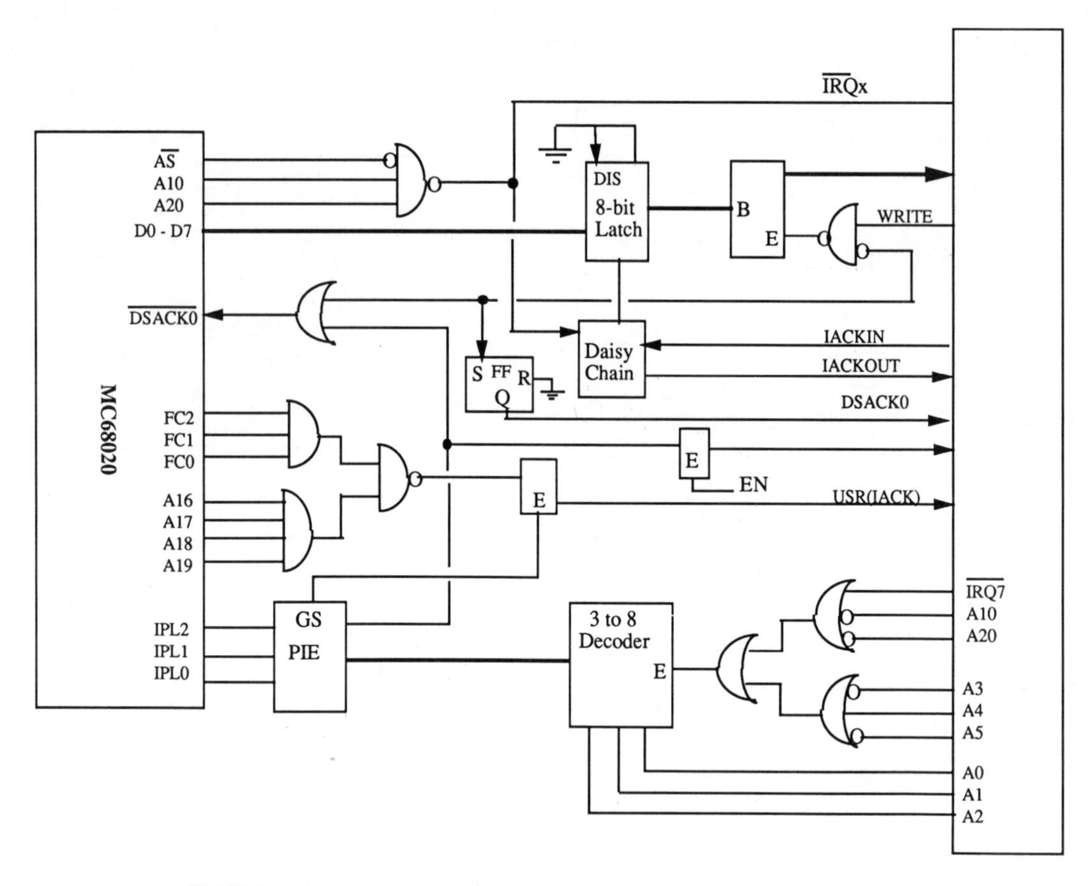

FIGURE 6.25. Interrupt circuit for the slave processor.

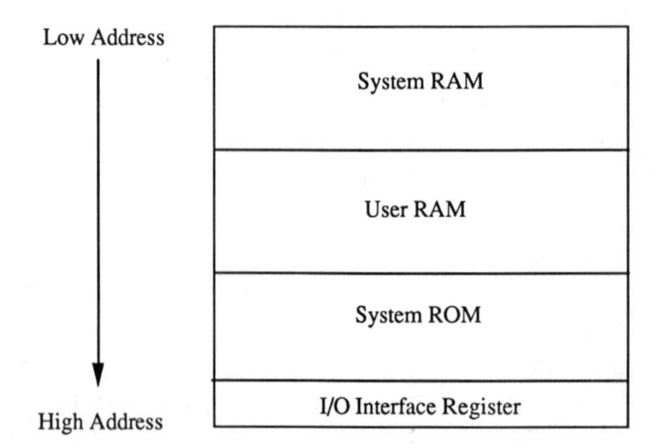

FIGURE 6.26. System memory MAP.

allowing the vector table to reside in any memory area. It also facilitates dynamic interrupt vector programming. The real vector table will reside in the first 1K of memory, and the virtual table may reside in any 1K block. During system generation, the real vector table will be loaded by the address of the exception routines.

One problem associated with placing system ROM in any segment other than the bottom of the memory is that the MC68020 looks at location $00000000 for its reset vector. This is because the Program Counter (PC) and the VBRs get cleared during a system reset. As a result, the processor always acquires the ISP value from the long-word location starting at $00000000, and the PC value from the location starting at $00000004.

This can be resolved by mapping the ROM to the lower portion of memory at reset. A simple solution is shown in Figure 6.27, where a 74F164 shift register is used to force the selection of ROM during the first eight memory cycles after RESET, to allow the processor to fetch the reset ISP and PC from ROM. When the 0H of the 74F164 shift register is low, the selection of ROM is automatic, and the selection of RAM is prohibited. Once the 0H goes high, selection proceeds in a normal fashion. The 74F164 is reset whenever $\overline{\text{RESET}}$ and $\overline{\text{HALT}}$ are both active (the system reset condition). Once $\overline{\text{RESET}}$ or $\overline{\text{HALT}}$ become inactive, a logic one is shifted into the shift register by the rising edge of the $\overline{\text{AS}}$. After eight memory cycles, the 0H goes high, and ROM returns to its normal location in the memory map.

6.6 COMMERCIALLY AVAILABLE BUSING SCHEMES

Compatibility at the microprocessor or local bus level is difficult to achieve, since each has both a very unique architecture and performance specifications. Thus,

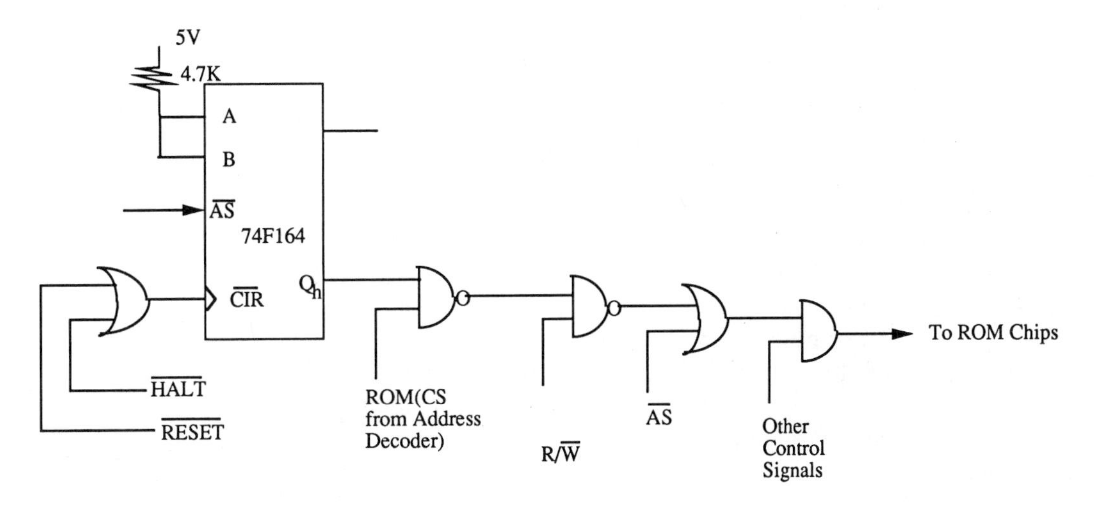

FIGURE 6.27. System ROM mapping logic during system reset.

the most viable option is standardize the bus at the back-plane level. As a result, several back-plane bus standards or de-facto standards arose for the most popular microprocessors. The designers of the de facto standards did not put much effort into optimizing the bus operation. Due to the very nature of a de facto standard, manufacturers frequently changed the specifications to meet their own requirements. Both government and nongovermental engineering and scientific bodies, namely the European International Electrotechnical Commission and American National Standards Institute (ANSI), finally realized the importance of defining legally enforceable, officially recognized standards. ANSI approached the IEEE, and suggested that the IEEE's Computer Standards Committee should create Bus Standards. Today, a number of bus standards have been adopted or are under consideration for 32-bit microprocessors. They are:

- IEEE P796, developed based on the Intel's Multibus.
- IEEE P896, called Futurebus.
- IEEE P959, based on Intel's iSBX Bus.
- IEEE P970, modeled around Motorola's Versabus.
- IEEE P1014, developed on the basis of the Motorola/Mostek/Signetics-Phillips VME bus.

Such a proliferation indicates that no one bus is suitable for all applications. Thus, system designers need to establish criteria to select the system bus for their application.

The bus structure can be described from two points of view. The *mechanical specification* and the *functional specification.* The mechanical specification describes the physical dimensions of subracks, backplanes, front panels, plug-in boards, etc. The functional specification describes how the bus works, what functional modules are involved in each transaction, and the rules that govern their behavior.

Various criteria can be established to select a bus. From the mechanical specification point of view, some of the selection criteria can be [Dexter 86]:

- The physical dimensions.
- The number of edge connectors, and their dimensions.
- The types of connectors to be used (e.g., ribbon connectors).
- The insertion life rate.
- Vulnerability to dirty environments.
- Susceptibility of the mating connectors to mechanical resonance.
- The connector mating force compared to pin count.
- The operating temperature range.
- Durability (e.g., the number of mating and unmating cycles).
- Contact protection and polarization during mating.
- Plating techniques.

From the functional structure, the criteria to be used are:

- The width of the data and address bus.
- The interrupt capabilities.
- The control structure.
- Utilities expansion.
- The bandwidth.
- The board form factor.
- The documentation.
- The board interconnect technology.
- Future system performance enhancements.
- Multiprocessor capabilities.
- The acceptable delay between the request and initiation of transmission.
- The transmission error recovery mechanism.
- The reliability of the devices controlling the bus.
- The fault tolerance (i.e., the bus continues to function correctly if one or more of the interconnected devices fail).
- Whether all devices and communications are to be treated equally (i.e., are priorities needed?).
- The maximum distance capabilities.

The nontechnical considerations are:

- The potential product volume.
- In-house facilities and capabilities.
- The availability of engineering manpower.
- The product development time frame.
- The ability to estimate real costs.

6.6.1 The Multibus

The mechanical specification of the Multibus is shown in Figure 6.28. The bus is constructed on a double-sided printed circuit board, where the signals are active low and conform to TTL logic level. The connectors on the PCBs must adhere to the following standards [Johnson 84].

1. The connectors on the bus side of the boards are called P1 and P2. P1 is the 86-pin main connector, and P2 is the 60-pin auxiliary connector.
2. Pins are numbered with odd-numbered pins on the component side, and in ascending order when going counterclockwise around the board.
3. Connectors on the non-Multibus system side of the board are called J1, J2, J3, etc. An attempt should be made to number these connectors in ascending order when going clockwise around the board as viewed from the component side.

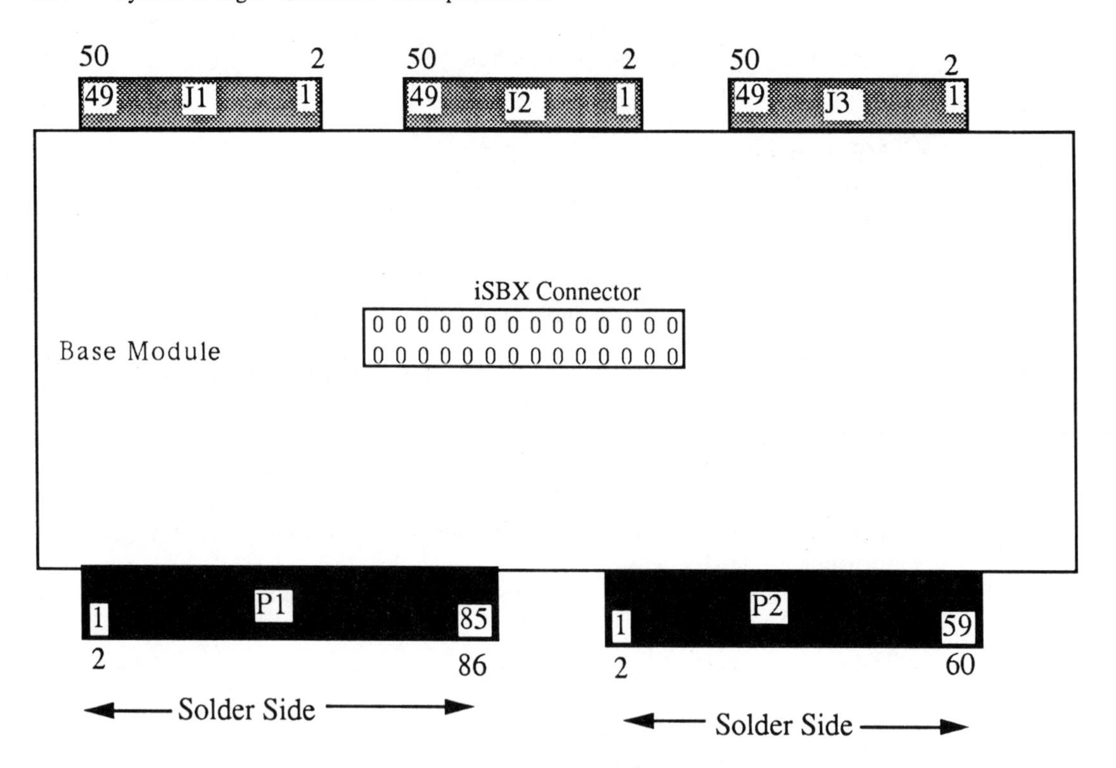

FIGURE 6.28. Mechanical specification of Multibus board.

The pin assignments for the P1 and P2 connectors are shown in Figures 6.29 and 6.30. Here, the address and data lines are driven by three-state devices, while the interrupt and various control lines are open-collector driven.

The P1 bus is composed of:

- The 16-bit bidirectional data bus, excluding the parity lines. The lines are labeled as DAT0–DAT15.
- The address lines are 20 bits, and are labeled as ADR0–ADR19, excluding the parity lines.
- The 30 control lines are grouped by function, and they are:

  six data transfer control lines: MRDC, MWTC, IORC, IOWC, XACK, BHEN
  six allocation control lines: BREQ, BUSY, BPRN, BPRO, BCLK, CBRQ
  nine device synchronization control lines: INT0–INT7, INTA
  nine utility control lines: INIT, CCLK, INH1, INH2, LOCK, and the remaining four are labeled as RESERVED.

	(COMPONENT SIDE)			(CIRCUIT SIDE)		
	PIN	MNEMONIC	DESCRIPTION	PIN	MNEMONIC	DESCRIPTION
POWER SUPPLIES	1	GND	Signal GND	2	GND	Signal GND
	3	+5V	+ 5Vdc	4	+5V	+ 5Vdc
	5	+5V	+ 5Vdc	6	+5V	+ 5Vdc
	7	+12V	+12Vdc	8	+12V	+12Vdc
	9		Reserved, bussed	10		Reserved, bussed
	11	GND	Signal GND	12	GND	Signal GND
BUS CONTROLS	13	BCLK/	Bus Clock	14	INIT/	Initialize
	15	BPRN/	Bus Pri. In	16	BPRO/	Bus Pri. Out
	17	BUSY/	Bus Busy	18	BREQ/	Bus Request
	19	MRDC/	Mem Read Cmd	20	MWTC/	Mem Write Cmd
	21	IORC/	I/O Read Cmd	22	IOWC/	I/O Write Cmd
	23	XACK/	XFER Acknowledge	24	INH1/	Inhibit 1 disable RAM
BUS CONTROLS AND ADDRESS	25	LOCK/	Lock	26	INH2/	Inhibit 2 disable PROM or ROM
	27	BHEN/	Byte High Enable	28	AD10/	
	29	CBRQ/	Common Bus Request	30	AD11/	Address
	31	CCLK/	Constant Clk	32	AD12/	Bus
	33	INTA/	Intr Acknowledge	34	AD13/	
INTERRUPTS	35	INT6/	Parallel	36	INT7/	Parallel
	37	INT4/	Interrupt	38	INT5/	Interrupt
	39	INT2/	Requests	40	INT3/	Requests
	41	INT0/		42	INT1/	
ADDRESS	43	ADRE/		44	ADRF/	
	45	ADRC/		46	ADRD/	
	47	ADRA/	Address	48	ADRB/	Address
	49	ADR8/	Bus	50	ADR9/	Bus
	51	ADR6/		52	ADR7/	
	53	ADR4/		54	ADR5/	
	55	ADR2/		56	ADR3/	
	57	ADR0/		58	ADR1/	
DATA	59	DATE/		60	DATF/	
	61	DATC/		62	DATD/	
	63	DATA/	Data	64	DATB/	Data
	65	DAT8/	Bus	66	DAT9/	Bus
	67	DAT6/		68	DAT7/	
	69	DAT4/		70	DAT5/	
	71	DAT2/		72	DAT3/	
	73	DAT0/		74	DAT1/	
POWER SUPPLIES	75	GND	Signal GND	76	GND	Signal GND
	77		Reserved, bussed	78		Reserved, bussed
	79	−12V	−12Vdc	80	−12V	−12Vdc
	81	−5V	+ 5Vdc	82	+5V	+ 5Vdc
	83	−5V	+ 5Vdc	84	+5V	+ 5Vdc
	85	GND	Signal GND	86	GND	Signal GND

FIGURE 6.29. Pin assignments of the P1 connector for the Multibus. (Courtesy IEEE).

PIN	(COMPONENT SIDE) MNEMONIC	DESCRIPTION	PIN	(CIRCUIT SIDE) MNEMONIC	DESCRIPTION
1	GND	Signal GND	2	GND	Signal GND
3	5VB	+ 5V Battery	4	GVB	+ 5V Battery
5		Reserved, not bussed	6	EEVPP	E² PROM Power
7	-5VB	- 5V Battery	8	-5VB	- 5V Battery
9		Reserved, not bussed	10		Reserved, not bussed
11	12VB	+12V Battery	12	12VB	+12V Battery
13	PFSR/	Power Fail Sense Reset	14		Reserved, not bussed
15	-12VB	-12V Battery	16	-12VB	-12V Battery
17	PFSN/	Power Fail Sense	18	ACLO	AC Low
19	PFIN/	Power Fail Interrupt	20	MPRO/	Memory Protect
21	GND	Signal GND	22	GND	Signal GND
23	+15V	+15V	24	+15V	+15V
25	-15V	-15V	26	-15V	-15V
27	PAR1/	Parity 1	28	HALT/	Bus Master HALT
29	PAR2/	Parity 2	30	WAIT/	Bus Master WAIT STATE
31	PLC	Power Line Clock	32	ALE	Bus Master ALE
33–39		Reserved, not bussed	34		Reserved, not bussed
			36	BD RESET/	Board Reset
			38	AUX RESET/	Reset switch
			40		Reserved, not bussed
41–53		Reserved, bussed	42–54		Reserved, bussed
55	ADR16/	Address	56	ADR17/	Address
57	ADR14/	Bus	58	ADR15/	Bus
59		Reserved, bussed	60		Reserved, bussed

Notes:
1. PFIN, on slave modules, if possible, should have the option of connecting to INT0/ on P1.
2. All undefined pins are reserved for future use.

FIGURE 6.30. Pin assignments of the P2 connector for the Multibus. (Courtesy IEEE).

The power and ground bus is 20 bits wide, and is composed of:

eight + 5V lines
two + 12V lines
two − 12V lines
eight signal/power ground lines

The Multibus architecture supports two address spaces, namely memory and I/O. The 20 address lines allow the CPU to directly address 1 Mbyte/word of memory. The least significant 12 address lines are used for I/O, with a maximum capacity of 1024 locations. The bus master also pulls control lines MRDC/MWTC during memory access and IORC/IOWC during I/O access. The bus architecture allows the usage of the same memory locations for RAM or ROM devices. Address conflict is resolved via control line INH1 (inhibit RAM signal), which prevents RAM memory from responding to the memory address on the system address bus. The INH1 may be used with memory I/O devices to override RAM memory, whereas control signal INH2 (inhibit ROM signal) prevents ROM memory from responding to the memory address on the system address bus. The signal allows auxiliary ROM (a bootstrap program or other utility) to override ROM devices when ROM and auxiliary ROM memory are assigned to the same memory addresses. This control line may also be used to allow memory-mapped I/O devices to override ROM memory.

The bus bandwidth at 5 MHz has a maximum transfer rate of 10 Mbytes/s as 16 bit blocks. Due to the arbitration and memory access time, a typical maximum transfer speed is on the order of 4 Mbytes/s. The Multibus supports 8- or 16-bit data transfer. The Multibus permits both 8- and 16-bit bus masters, by supporting three types of data transfer:

1. Even-byte transfers on DAT0 to DAT7.
2. Odd-byte transfers on DAT0 to DAT7.
3. Word data transfers on DAT0 to DATF.

All byte transfers use data lines DAT0 to DAT7, and lines DAT8 to DATF are not defined during byte transfers. All odd-byte transfers, which (when local to a 16-bit microprocessor) are transferred on the high-order data byte, are swapped from the local high-order data byte to the lower-order data byte while on the Multibus system bus. They are swapped back to the high-order byte once they are back on the local bus of the 16-bit microprocessor.

The control signal BHEN identifies the nature of the present transfer. In order to maintain the compatibility, the byte transfer always takes place over the least significant lines of the data bus. The following table summarizes the 8- and 16-bit data paths used for three types of Multibus transfers. Two control signals, BHEN and ADR0, control the data transfers.

Transfer Size	BHEN	ADR0	Data Bus Activity	Device Byte Transferred
Byte	H	H	DAT0–DAT7	Even
Byte	H	L	DAT0–DAT7	Odd
Word	L	H	DAT0–DAT15	Even and Odd

The Multibus system is based on a master/slave relationship. A master controls the bus via address and control lines. The slave devices cannot control the bus unless authorized by the master. Both device addresses and data are transferred under the same fully interlocked handshake control. The handshake control signals are:

Master-controlled:
Data ready: MWTC/IOWC
Data accept: XACK

Slave-controlled:
Data request: MRDC/IORC
Data ready: XACK

The address signals must be stabilized on all slaves on the bus during a read/write operation. As a result, the bus master must issue address signals at least 50 ns after the read/write command to the bus, initiating the data transfer. The bus master must maintain the address signals until at least 50 ns after the read/write command is completed.

6.6.1.1 Priority Scheme

The Multibus supports centralized independent request allocation (Figure 6.31) and decentralized serial priority schemes (Figure 6.32). The schemes are supported using the following control lines, which allow the bus master-type devices to request, be granted, and acknowledge bus mastership.

The BREQ (Bus request) signal is used in conjunction with the parallel bus priority network by a device to indicate that it intends to use the system bus for data transfers. BREQ is synchronized with BCLK.

BRPN (Bus priority-in signal) signals a bus-requesting device that no higher priority module is requesting the system bus. It is also synchronized with BCLK.

BPRO (Bus priority-out signal) is used in conjunction with serial (i.e., daisy-

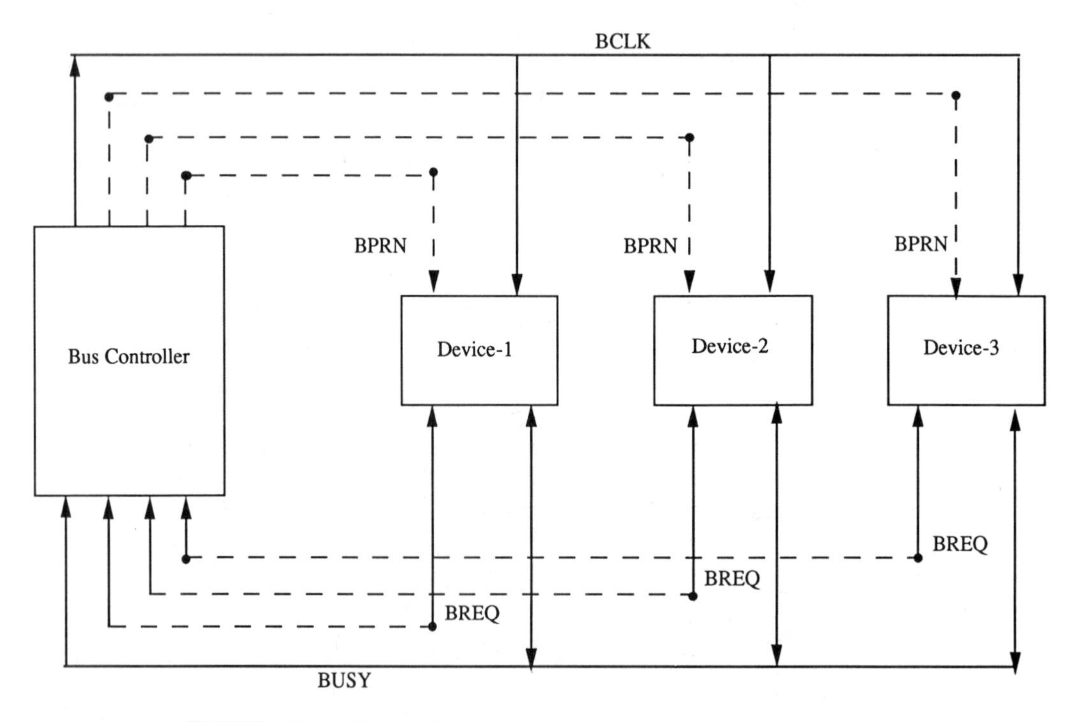

FIGURE 6.31. The Multibus centralized bus request resolution scheme.

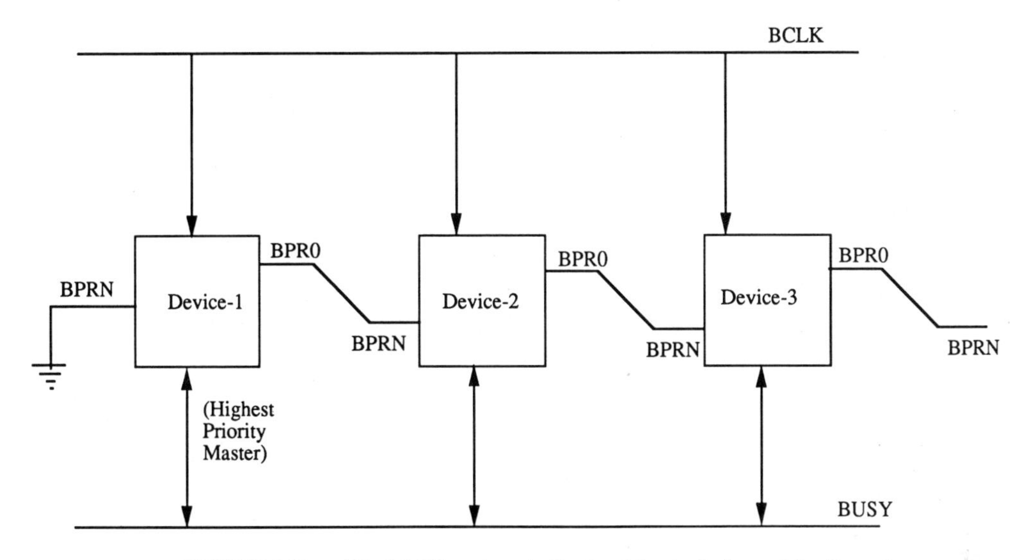

FIGURE 6.32. The Multibus decentralized serial priority bus arbitration scheme.

chain) bus priority resolution schemes. BPRO is forwarded to the BPRN input of the bus-requesting device with the next lower bus priority. Like other allocation control signals, it is also synchronized with BCLK.

The BUSY (Bus busy signal) open-collector signal line is driven by the current bus master to indicate that the bus is in use. This signal prevents all potential bus masters from gaining control of the bus. BUSY is synchronized with the BCLK.

The CBRQ (Common bus request) open-collector signal line is driven by all potential bus masters, and is used to inform the current bus master that another device wishes to use the bus. A high on CBRQ tells the bus master that no other master is requesting the bus. Therefore, the current bus master can retain the bus. This saves the bus exchange overhead required for each bus request and grant sequence for the current master.

The serial priority resolution is accomplished with a daisy-chain technique (Figure 6.32). The priority input BPRN of the highest-priority master is tied to ground. The priority output BPRO of the highest-priority master is then connected to the priority input BPRN of the next lower priority master, and so on. In this implementation, arbitration must be completed within one bus clock (BCLK), and the number of masters that can be used in a serial priority scheme is limited to three. Multibus provides the second bus request line, CBRQ, which indicates that a request is pending to those bus masters that will only release the bus on request.

Under the parallel priority scheme, the priority is resolved by a priority resolution circuit, shown in Figure 6.31. The circuit receives all BREQ signals, and supplies the encoded value of the highest-priority device to a priority decoding circuit, which then activates the proper BPRN line. The BPRO lines are not used in the parallel priority scheme. The Multibus back plane contains a trace from the BPRN signal of one card slot to the BPRO signal of the adjacent lower card slot. Therefore, for proper operation, BPRO must be disconnected from the bus on the board, or the back-plane trace must be disconnected. Due to the physical bus length limitations, a maximum of 16 bus masters can be accommodated by the parallel priority circuit.

6.6.1.2 Interrupt Operation

The Multibus interrupt lines INT0–INT7 are used by a master to receive interrupts from the bus slaves, other bus masters, or the external power fail detection logic. A master may also support various internal interrupts that do not require the service of the bus interrupt lines. Two types of interrupts are supported, namely:

Non-bus-vectored interrupts
Bus-vectored interrupts

Non-bus-vectored interrupts do not require the Multibus address lines for the transfer of the interrupt vector address. Here, the vector address is generated by the interrupt controller on the master, and transferred to the processor over the local bus. The source of the interrupt can be in the master module or in other bus modules. Modules may utilize the Multibus interrupt lines INT0–INT7 to send their interrupt requests to each other. When an interrupt request line is activated, the bus master performs its own interrupt operation and processes the interrupt.

The bus-vectored interrupts are those that utilize the Multibus address lines to channel the vector address from the slave to the bus master, using the INTA response signal for synchronization. When an interrupt from the Multibus interrupt lines INT0–INT7 occurs, the interrupt control logic on the bus master interrupts its processor. The processor then generates an INTA response signal, which freezes the state of the interrupt logic of the Multibus slaves for priority resolution. The bus master also locks the Multibus control lines, to guarantee itself the use of consecutive bus cycles. After issuing the INTA signal, the bus master's interrupt control logic puts an interrupt code on the Multibus address lines ARD8–ADRA. At this point, one of the two different vector processing sequences will commence. The differences are due to either one or two additional INTA generated by the bus master, to allow the slave devices to supply an 8-bit or 16-bit pointer value over the Multibus data lines.

If the bus master generates an additional INTA, then the second INTA causes the bus slave interrupt control logic to transmit an 8-bit pointer value over the Multibus data lines. This vector pointer is then used by the bus master to determine the memory address of the interrupt service routine.

On the other hand, if the bus master generates two additional INTAs, then the bus slave puts a two-byte interrupt vector address on the Multibus data lines (e.g., one byte for each byte value). The vector address is then utilized by the CPU to find the location of the service routine.

6.6.1.3 Utilities

The Multibus P2 connector signals provide a means for handling power failures. The logic circuits required for power failure detection and handling are optional, and the user must supply them. The other CPU-dependent control signals are transmitted over the P2 connector. The pin assignment of P2 is shown in Figure 6.28.

The iSBX provides 8- and 16-bit I/O to any SBC or board-level product. It also provides a universal I/O interface on the mainboard for I/O-mapped data between the main board and the multimodule board.

6.6.2 The VME Bus

In 1980, Motorola introduced a bus architecture to support the MC68000 microprocessor, and the future MC68020. Soon after its introduction, a number of European manufacturers proposed modifications to make it compatible with the standard Eurocard backplane. In 1981, Mostek, Motorola, and Signetics/Philips produced a modified draft standard for the backplane, and called it the VME bus. It was subsequently adopted by the IEEE as the P1014 standard.

Mechanically, the VME bus is made up of a multilayer printed circuit board with 96 pin connectors and signal paths [VME bus 85]. This is shown in Figure 6.33. Some VME bus systems have a signal PC board, called the J1 backplane, illustrated in Figure 6.34. The connector consists of three rows of pins, labeled ROWa, ROWb, and ROWc. Together, they provide the signal paths needed for basic operations.

Other VME bus systems also have an optional 96-pin second PC board, called the J2 backplane, for wider data and address transfers. The signal lines for J2 backplane are shown in Figure 6.35. Like the J1 connector it also consists of three rows of pins labeled as ROWa, ROWb, and ROWc.

All signals conforming to TTL logic levels and bus lines are driven by totem-pole, tri-state, and open-collector drivers. The bus is required to have a loaded characteristic impedence of $Z > 60$ ohms, and all lines must be terminated at both ends. The transfer speed of 12 MHz imposes a maximum data transfer rate of 48 Mbytes/s [VME bus 85].

The VME bus signal lines are:

A 16/32-line data bus: D00–D15, D16–D31

A 29/37-line address bus: A01–A23, A24–A31, and AM0–AM5 address modifier lines.

A 37/38-line control bus, consisting of the following:

- seven data transfer control lines [AS, WRITE, DS0, DS1, LWORD, DTACK, BERR]
- fourteen allocation control lines [BR0–BR3, BBYS, BG0IN–BG3IN, BG0OUT–BG3OUT, BCLR]
- ten device synchronization lines [IRQ1–IRQ7, IACK, IACKIN, IACKOUT]
- six utility control lines [SYSRESET, SYSFAIL, ACFAIL, SYSCLK, SERDAT, SERCLK]
- one reserved for future usage.

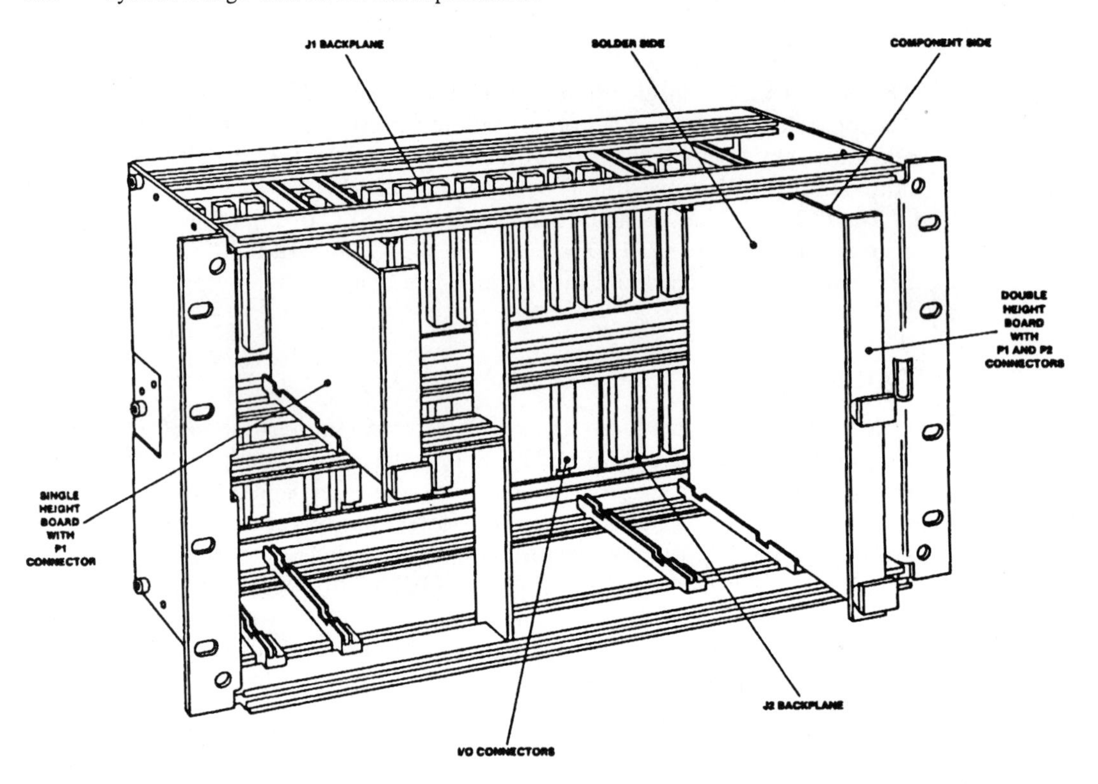

FIGURE 6.33. Mechanical specification fo the VME bus backplane. (Courtesy Motorola Inc.).

A 64-line user I/O bus [User I/O 0–63]
14/21-line power and ground lines, consisting of the following:
3/6, + 5V power lines
1, + 5V stand-by supply line
1, + 12V power line
1, − 12V power line
8/12 ground return lines

6.6.2.1 Data Transfer Protocol

In the VME bus protocol, the master initiates data transfers. The master utilizes the address bus and Address Modifier (AM) bus during addressing. The bus specification allows up to 64 different address maps for various purposes, namely:

- Program access
- Data access
- Ascending or descending address during block transfers
- I/O access
- Supervisory or privileged access
- Nonprivileged access

PIN NUMBER	ROWa SIGNAL MNEMONIC	ROWb SIGNAL MNEMONIC	ROWc SIGNAL MNEMONIC
1	D00	BBSY*	D08
2	D01	BCLR*	D09
3	D02	ACFAIL*	D10
4	D03	BG0IN*	D11
5	D04	BG0OUT*	D12
6	D05	BG1IN*	D13
7	D06	BG1OUT*	D14
8	D07	BG2IN*	D15
9	GND	BG2OUT*	GND
10	SYSCLK	G3IN*	SYSFAIL*
11	GND	BG3OUT*	BERR*
12	DS1*	BR0*	SYSRESET*
13	DS0*	BR1*	LWORD*
14	WRITE*	BR2*	AM5
15	GND	BR3*	A23
16	DTACK*	AM0	A22
17	GND	AM1	A21
18	AS*	AM2	A20
19	GND	AM3	A19
20	IACK*	GND	A18
21	IACKIN*	SERCLK(1)	A17
22	IACKOUT*	SERDAT*(1)	A16
23	AM4	GND	A15
24	A07	IRQ7*	A14
25	A06	IRQ6*	A13
26	A05	IRQ5*	A12
27	A04	IRQ4*	A11
28	A03	IRQ3*	A10
29	A02	IRQ2*	A09
30	A01	IRQ1*	A08
31	-12V	+5VSTDBY	+12V
32	+5V	+5V	+5V

FIGURE 6.34. Pin assignments of the J1/P1 connector. (Courtesy IEEE).

- User defined access
- Reserved access

Three forms of addressing modes are supported for each type of transfer, which are:

- Short-addressing AM codes, indicating that address lines A02–A15 are being used to select a range of 64 Kbytes.
- Standard-addressing AM codes, indicating that address lines A02–A23 are being used to select a range of 16 Mbytes.
- Extended-addressing AM codes, indicating that address lines A02–A31 are being used to select a range of 4 Gbytes.

The data transfer is initiated by the master. The addressed slave then acknowledges the transfer. The master terminates the operation at the end of the acknowl-

PIN NUMBER	ROWa SIGNAL MNEMONIC	ROWb SIGNAL MNEMONIC	ROWc SIGNAL MNEMONIC
1	User Defined	+5V	User Defined
2	User Defined	GND	User Defined
3	User Defined	RESERVED	User Defined
4	User Defined	A24	User Defined
5	User Defined	A25	User Defined
6	User Defined	A26	User Defined
7	User Defined	A27	User Defined
8	User Defined	A28	User Defined
9	User Defined	A29	User Defined
10	User Defined	A30	User Defined
11	User Defined	A31	User Defined
12	User Defined	GND	User Defined
13	User Defined	+5V	User Defined
14	User Defined	D16	User Defined
15	User Defined	D17	User Defined
16	User Defined	D18	User Defined
17	User Defined	D19	User Defined
18	User Defined	D20	User Defined
19	User Defined	D21	User Defined
20	User Defined	D22	User Defined
21	User Defined	D23	User Defined
22	User Defined	GND	User Defined
23	User Defined	D24	User Defined
24	User Defined	D25	User Defined
25	User Defined	D26	User Defined
26	User Defined	D27	User Defined
27	User Defined	D28	User Defined
28	User Defined	D29	User Defined
29	User Defined	D30	User Defined
30	User Defined	D31	User Defined
31	User Defined	GND	User Defined
32	User Defined	+5V	User Defined

FIGURE 6.35. Pin assignments of the J2/P2 connector. (Courtesy IEEE).

edgment. The asynchronous nature allows the slave to control the transfer time. The transfer protocols allow for 8-bit, 16-bit, and 32-bit data transfer over the data bus. The data transfer bus lines can be grouped into three categories, listed in the following table.

Addressing Lines	Data Lines	Control Lines
A01–A31	D00–D31	$\overline{\text{AS}}$
AM0–AM5		$\overline{\text{DS0}}$
$\overline{\text{DS0}}$		$\overline{\text{DS1}}$
$\overline{\text{DS1}}$		$\overline{\text{BERR}}$
$\overline{\text{LWORD}}$		$\overline{\text{DTACK}}$
		$\overline{\text{WRITE}}$

The two data strobes ($\overline{\text{DS0}}$ and $\overline{\text{DS1}}$) serve dual function, such as:

a) The levels of these two data strobe lines are used to select which byte(s) are accessed.
b) The edges of the data strobes are also used as timing signals, which coordinate the transfer of the data between the master and slave.

The master uses address lines A02–A31 to select which 4-byte group will be accessed. Four additional lines ($\overline{\text{DS1}}$, $\overline{\text{DS0}}$, A01, and $\overline{\text{LWORD}}$) are then used to select which byte location within the 4-byte group is accessed during the data transfer. Depending on the type of cycle, the master can access 1, 2, 3, or 4 byte locations simultaneously, utilizing these 4 lines. The width of the data transfer is determined by the data strobe(s) used to initiate the transfer, and by a special control signal ($\overline{\text{LWORD}}$). When $\overline{\text{LWORD}}$ is asserted, a 32-bit transfer is to take place. When $\overline{\text{LWORD}}$ is negated and both data strobes are asserted, then a 16-bit word transfer is requested. When either one of the data strobes is asserted, then one or other type of byte transfer is requested. It is important to note that the VME bus supports address pipelines by strobing the address and data with separate strobe signals. This allows a master to broadcast the address for the next cycle, while the data transfer for the previous cycle is still in progress.

6.6.2.2 Allocation Protocols

The VME bus is designed to: a) prevent simultaneous use of the bus by two masters; and b) schedule requests from multiple masters for optimum bus usage.

When several boards request the use of the DTB simultaneously, the arbitration subsystem detects these requests and grants the bus to one board at a time. The scheduling algorithm determines which board is granted the bus. The VME bus supports three algorithms:

- Prioritized (PRI)
- Round-Robin Select (RRS)
- Single-Level (SGL)

The PRI arbiter drives the following:

one bus clear line ($\overline{\text{BCLR}}$)
four bus grant lines (Slot 1 $\overline{\text{BG0IN}}$ through $\overline{\text{BG3IN}}$) if the arbiter's board also has a requester; (Slot 1 $\overline{\text{BG0OUT}}$ through $\overline{\text{BG3OUT}}$) if the arbiter's board does not have a requester.

An RRS arbiter drives the four slot 1 $\overline{\text{BGxIN}}$ or $\overline{\text{BGxOUT}}$ lines, and optionally the $\overline{\text{BCLR}}$ line.

An SGL arbiter drives only BG3IN or BG3OUT at slot 1.

Two additional lines are connected with the arbitration system during power-up and power-down sequences: $\overline{\text{SYSRESET}}$ and $\overline{\text{ACFAIL}}$.

6.6.2.3 Device Synchronization

The VME bus includes a priority interrupt mechanism to support the generation and service of interrupts. The architecture allows up to a seven-level serial or parallel priority vectored interrupt to be implemented in single or multimaster systems. The data transfer bus, arbitration bus, and priority bus are all used during the process of generating and handling interrupts. The interrupt bus is made up of seven interrupt request signal lines, one interrupt acknowledge line, and one daisy-chain interrupt acknowledge line, namely:

$\overline{\text{IRQ1}}$	Interrupt request 1
$\overline{\text{IRQ2}}$	Interrupt request 2
$\overline{\text{IRQ3}}$	Interrupt request 3
$\overline{\text{IRQ4}}$	Interrupt request 4
$\overline{\text{IRQ5}}$	Interrupt request 5
$\overline{\text{IRQ6}}$	Interrupt request 6
$\overline{\text{IRQ7}}$	Interrupt request 7
$\overline{\text{IACK}}$	Interrupt acknowledge
$\overline{\text{IACKIN}}$ and $\overline{\text{IACKOUT}}$	Interrupt acknowledge daisy-chain

The VME bus supports a seven-level independent request (Figure 6.36), a single-level daisy-chained serial request (Figure 6.37), or a multilevel parallel/ serial priority vectored interrupt scheme. The bus specification does not dictate

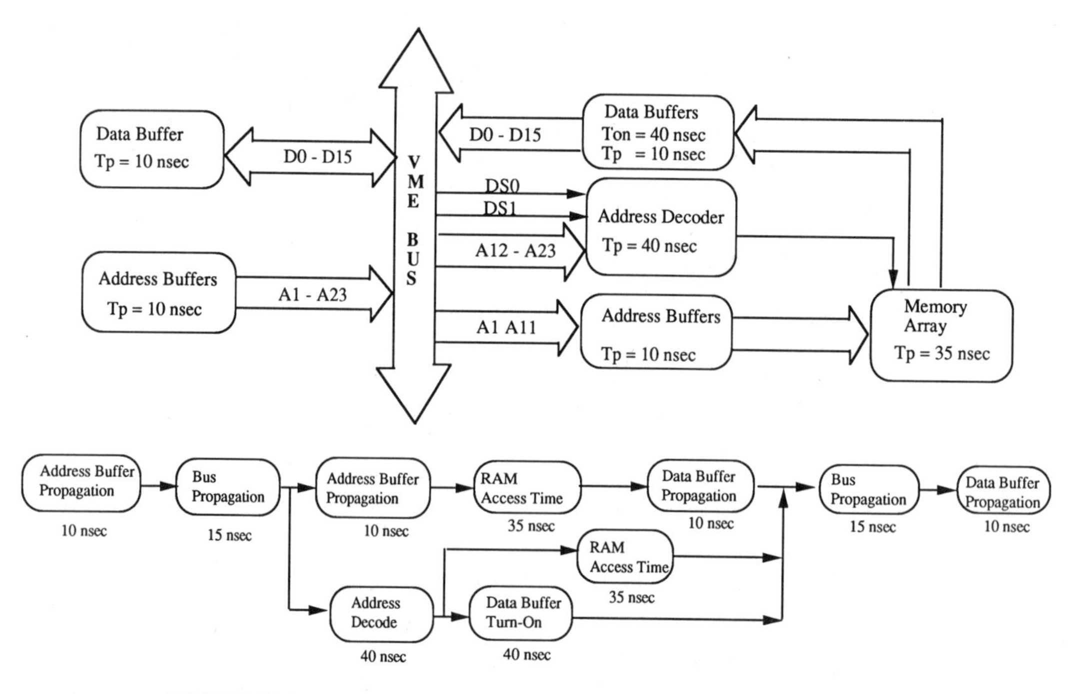

FIGURE 6.36. VME system timing.

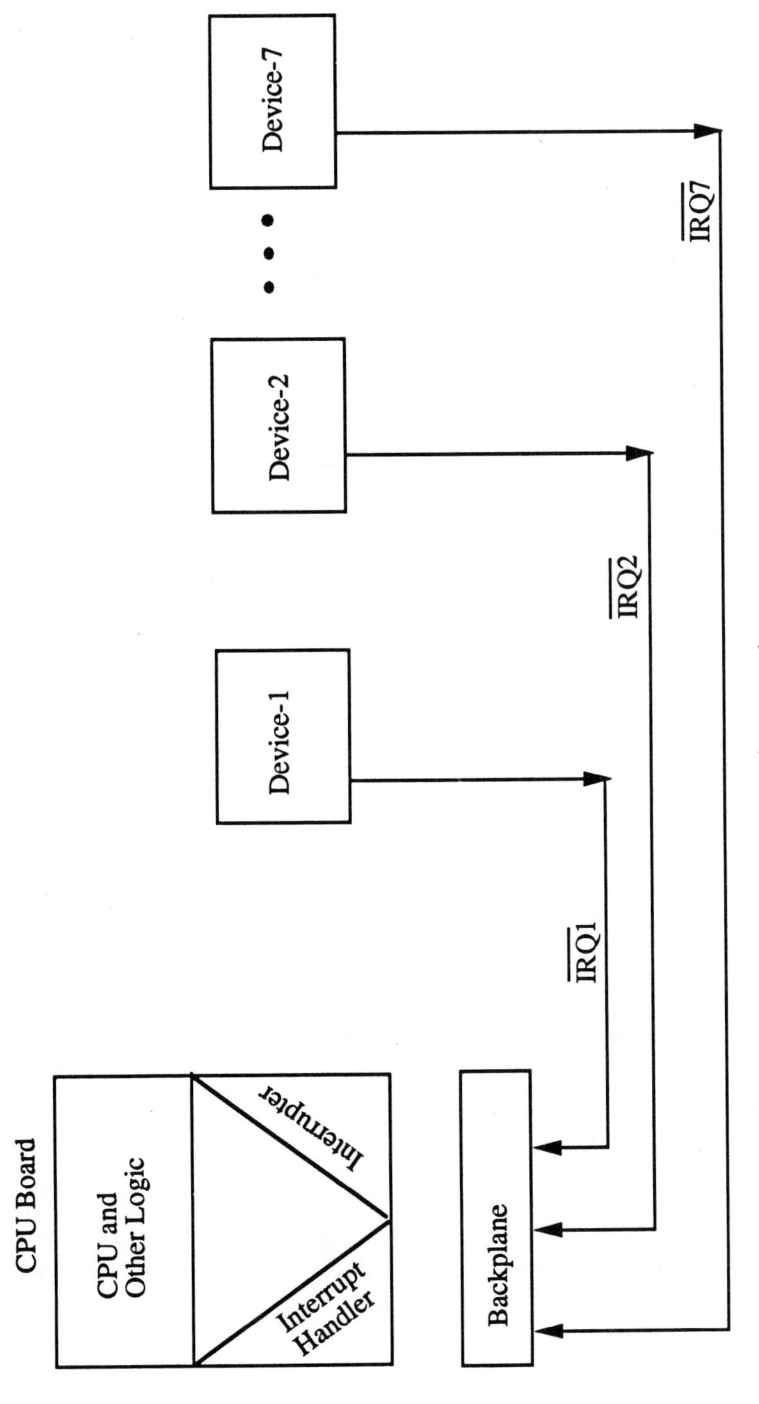

FIGURE 6.37. The VME bus seven-level independent request interrupt scheme.

what will happen during the interrupt service sequence. The servicing of the interrupt may or may not involve the use of the system bus.

Three types of functional modules are associated with the priority interrupt bus:

Interrupters
Interrupt handlers
IACK daisy-chain drivers

The interrupter executes the following algorithm:

a. It requests an interrupt from the interrupt handler, which monitors its interrupt request line.
b. IF it receives a falling edge on the interrupt acknowledgment daisy-chain input,

THEN IF
it is requesting an interrupt, and the levels on the three valid address lines correspond to the interrupt request line it is using, and the width of the requested STATUS/ID is either equal to or greater than the size it can supply,
THEN it supplies a STATUS/ID,
ELSE it passes the falling edge down the interrupt acknowledge daisy chain.

The interrupt handler is designed to accomplish the following tasks:

a. It prioritizes the incoming interrupt requests within its assigned group of interrupt request lines (highest of $\overline{\text{IRQ1}}$ through $\overline{\text{IRQ7}}$).
b. It uses its on-board *requester* to request the DTB and, when granted, the use of the DTB. It initiates an interrupt acknowledge cycle by reading STATUS/ID from the *interrupter* being acknowledged.
c. It initiates the appropriate interrupt service sequence, based on the information received in the STATUS/ID.

The IACK daisy-chain driver is another module that interacts with interrupt handlers and interrupters to coordinate the servicing of interrupts. It generates a falling edge on the interrupt acknowledge daisy-chain each time an interrupt handler initiates an interrupt acknowledge cycle.

6.6.2.4 Utility Bus

The utility bus of the VME bus supplies periodic timing, initialization, and diagnostic capabilities with the help of the bus signal lines. The members of this bus are:

SYSCLK (system clock): Is a nongated fixed frequency 16 MHz, 50% duty cycle signal. It is located in slot-1 and acts as a reference.

SERCLK (serial clock): Provides a fixed frequency, special waveform signal designed for the serial data communications task.

SERDAT (serial data): This signal line is utilized to carry serial data among the VME boards.

$\overline{\text{ACFAIL}}$ (AC fail): Indicates that AC power failure is imminent.

$\overline{\text{SYSRESET}}$ (system reset): Resets the entire system into a known state, and initiates the execution of a self-test of devices connected to the bus.

$\overline{\text{SYSFAIL}}$ (system failure): Indicates that an unexpected system failure has occurred.

The VME bus-based system will generally include a power monitor module that asserts $\overline{\text{ACFAIL}}$ when it detects a power failure. The monitor initiates a $\overline{\text{SYSRESET}}$ to reset the system on power-up. $\overline{\text{SYSFAIL}}$ is asserted by any device when it recognizes that a failure has occurred in its operation.

6.6.3 FUTUREBUS

The Futurebus, also called IEEE-896, is an advanced high-performance 32-bit backplane standardized by the IEEE Computer Society [IEEE P896]. The salient features of this bus are:

- It is processor independent, so byte orientation is not constrained.
- No central element of control exists.
- It employs a completely distributed architecture.
- It incorporates drivers and receivers to solve the classic bus driving problem.
- Its architecture is robust enough to include new features into the specification to grow in functionality and speed.

When comparing the bandwidth capabilities against various other 32-bit buses, the Futurebus offers superior performance. A unique feature of this bus is support of live insertion and removal of boards. It is also not confined to a message-passing architecture, but utilizes a fully compelled handshake protocol [Balakrishnan 84].

The arbitration and parallel protocols on the Futurebus are fully distributed. It includes support for fault-tolerant systems, and provides direct support for cache memory systems.

The bus architecture can be divided into three major segments (Figure 6.38):

- P896.1 hardware model: This defines the hardware features of the bus.
- P896.2 firmware model: This defines the firmware, including provisions for caching protocols.
- P896.x software: This defines the software interface to the operating system.

6.6.3.1 Hardware Model

The hardware model consists of the mechanical and electrical specifications. It defines the module size, module pitch, maximum component height, margin

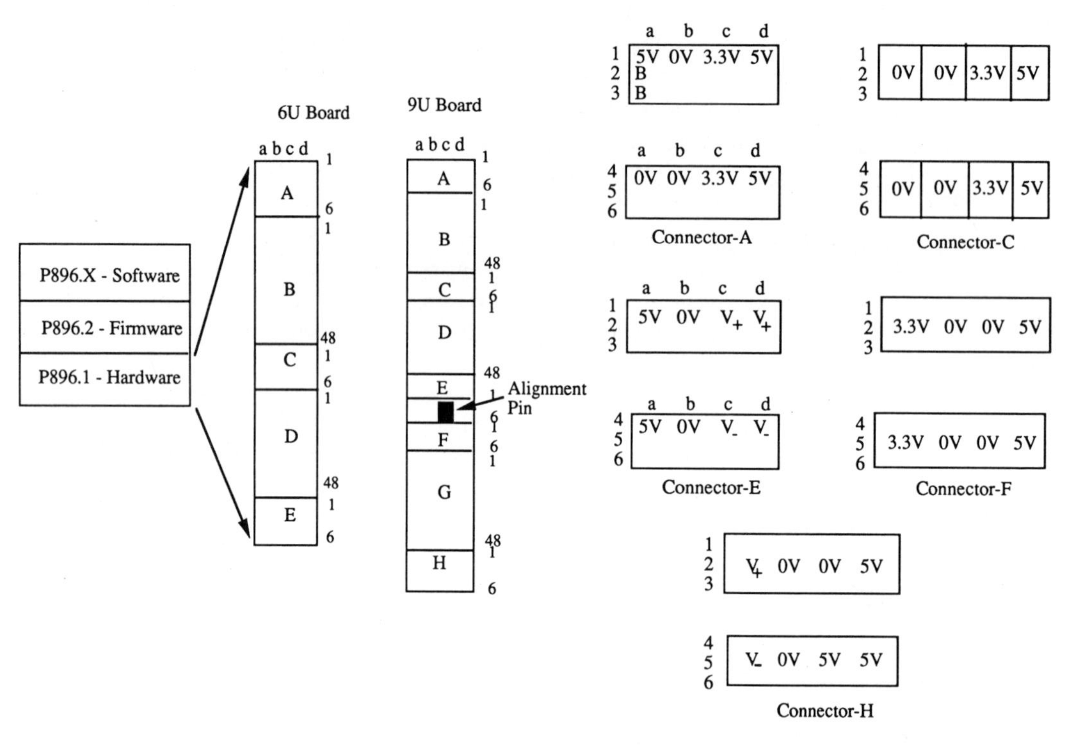

FIGURE 6.38. Architectural model and connector diagram of the P896.

from the edge of the board, connector, I/O, front panel, backplane, electrical, signal, and power requirements.

Futurebus provides a 64-bit architecture with a compatible 32-bit subset, and data extension to 128 and 256 bits. Figure 6.38 depicts the connector assignments. Futurebus modules use two types of boards. A 6U board may be configured with two 192-pin signal connectors, labeled B and D, while the 9U board supports three 192-pin connectors, B, D, and G. The specification calls for the B connector to have a 32/64-bit bus. An extension to a 128-bit bus is optionally available on the first 108 pins of connector D. If a 128-bit Futurebus is not used, then connector D will be user definable. The pin-outs for connector D are shown in Figure 6.39 and for connector B in Figure 6.40.

6.6.3.2 Arbitration Protocol

The arbitration on the Futurebus is asynchronous and fully distributed. It requires no central elements for control. Arbitration occurs concurrently with the data transfer transactions, and contributes no overhead as long as the data connection and transfer outlast the arbitration time.

The Futurebus arbitration mechanism operates in parallel with data transfers on the bus. Each module in the system has a unique arbitration number that is used in a parallel contention algorithm during the competition to become the bus

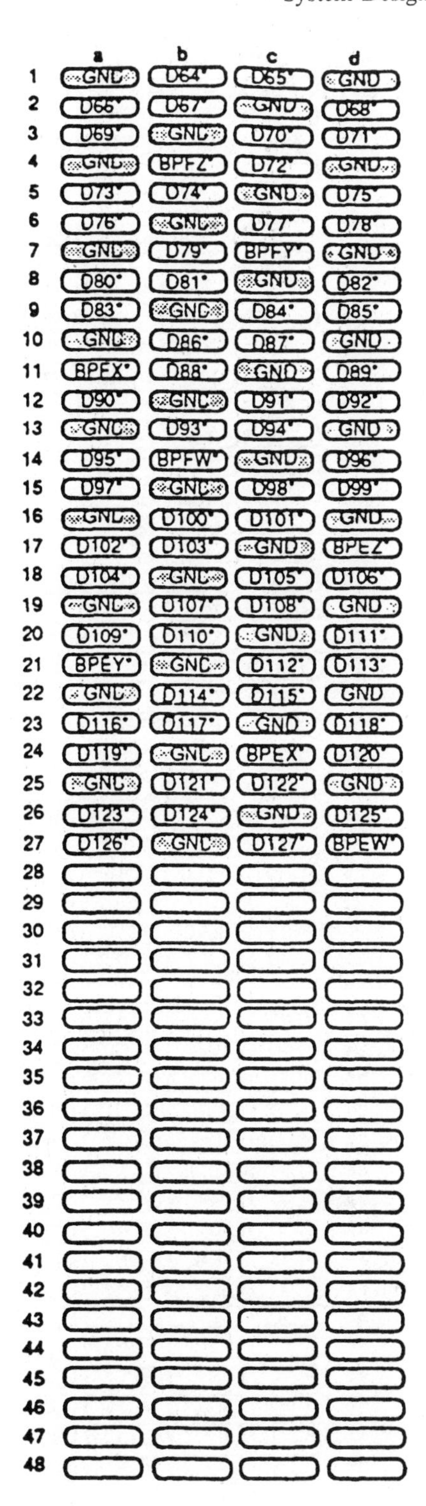

FIGURE 6.39. Futurebus D-connector pinout for 128-bit datapath extension. (Courtesy IEEE).

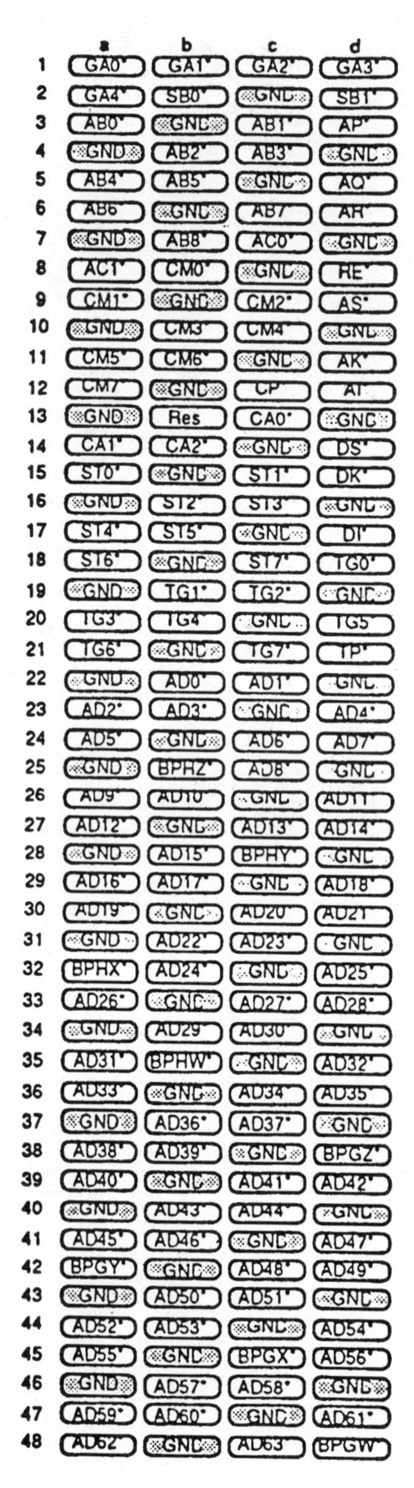

	a	b	c	d
1	GA0*	GA1*	GA2*	GA3*
2	GA4*	SB0*	GND	SB1*
3	AB0*	GND	AB1*	AP*
4	GND	AB2*	AB3*	GND
5	AB4*	AB5*	GND	AQ*
6	AB6*	GND	AB7*	AR*
7	GND	AB8*	AC0*	GND
8	AC1*	CM0*	GND	RE*
9	CM1*	GND	CM2*	AS*
10	GND	CM3*	CM4*	GND
11	CM5*	CM6*	GND	AK*
12	CM7*	GND	CP*	AI*
13	GND	Res	CA0*	GND
14	CA1*	CA2*	GND	DS*
15	ST0*	GND	ST1*	DK*
16	GND	ST2*	ST3*	GND
17	ST4*	ST5*	GND	DI*
18	ST6*	GND	ST7*	TG0*
19	GND	TG1*	TG2*	GND
20	TG3*	TG4*	GND	TG5*
21	TG6*	GND	TG7*	TP*
22	GND	AD0*	AD1*	GND
23	AD2*	AD3*	GND	AD4*
24	AD5*	GND	AD6*	AD7*
25	GND	BPHZ*	AD8*	GND
26	AD9*	AD10*	GND	AD11*
27	AD12*	GND	AD13*	AD14*
28	GND	AD15*	BPHY*	GND
29	AD16*	AD17*	GND	AD18*
30	AD19*	GND	AD20*	AD21*
31	GND	AD22*	AD23*	GND
32	BPHX*	AD24*	GND	AD25*
33	AD26*	GND	AD27*	AD28*
34	GND	AD29*	AD30*	GND
35	AD31*	BPHW*	GND	AD32*
36	AD33*	GND	AD34*	AD35*
37	GND	AD36*	AD37*	GND
38	AD38*	AD39*	GND	BPGZ*
39	AD40*	GND	AD41*	AD42*
40	GND	AD43*	AD44*	GND
41	AD45*	AD46*	GND	AD47*
42	BPGY*	GND	AD48*	AD49*
43	GND	AD50*	AD51*	GND
44	AD52*	AD53*	GND	AD54*
45	AD55*	GND	BPGX*	AD56*
46	GND	AD57*	AD58*	GND
47	AD59*	AD60*	GND	AD61*
48	AD62*	GND	AD63*	BPGW*

FIGURE 6.40. Connector B pinout—64/32-bit address and data paths. (Courtesy IEEE).

master. When two or more modules compete for the bus, the module with the largest arbitration number wins. Two arbitration mechanisms are supported:

Unrestricted mode: which has 14-bit arbitration numbers and requires two arbitration cycles. This is because the arbitration mechanism is based on an 8-bit competition number and 8 priority bits.

Restricted mode: which has 8-bit arbitration numbers and requires one cycle, and is limited to 2 priority bits.

6.6.3.3 Arbitration Logic

Lines AB6–AB0 represent the arbitration bus, and carry the arbitration vector. The Arbitration Condition (AC) line, indicates things such as a bus request during various stages of arbitration. Three arbitration synchronization lines (AP, AQ, and AR) form a three-wire handshake, where any module can assert a line while releasing another in a circular handshake arrangement. The reset line indicates that arbitration process can begin.

The arbitration logic is shown in Figure 6.41, where each logic element performs a function on an individual bit [Taub 84]. On the left-hand side is the arbitration number, provided by the module. On the right-hand side is the bus interface, which provides the wired-OR combination. If a module asserts a bit, the output is active low. The comparator detects the difference between the module bit and the bus. The difference causes the lower logic to be disabled. Only one master will match the bus signals, and will be declared the winner of that arbitration cycle.

6.6.3.4 Acquisition and Control

The acquisition and control mechanism is based on preemption. Let us assume that a potential master has won arbitration and become the 'master-elect.' While this module is waiting for the current master to complete its bus tenure, another potential master with a high priority can displace or preempt the master-elect from gaining access to the bus. The bus also has a parking mode, which means that a master will remain as the bus master after completing a bus cycle, as long as no other potential master wishes to use the bus. A 'fairness gate' opens, allowing all requesters to enter. Each is serviced in order based on an assigned identity. When the final requester has acquired the bus, the 'fairness gate' reopens so that requesting modules can begin arbitration again.

6.6.3.5 Protocol Architecture

Figure 6.42 illustrates the components of the protocol architecture. The parallel protocol lines consist of a 32-bit bus, which is segmented into four byte lanes, W, X, Y, and Z, each with its own parity bit. There is also a tag bit ($\overline{TG}$) associated with the bus, which can be defined by the system designer. The command lines CM0–CM5 provide information from the master to the slave during the connection phase. There exists an accompanying parity ($\overline{CP}$) bit. The status lines ST0–ST3, from the slave to the master, inform the master of the slave's interpretation of the previous command [IEEE P896].

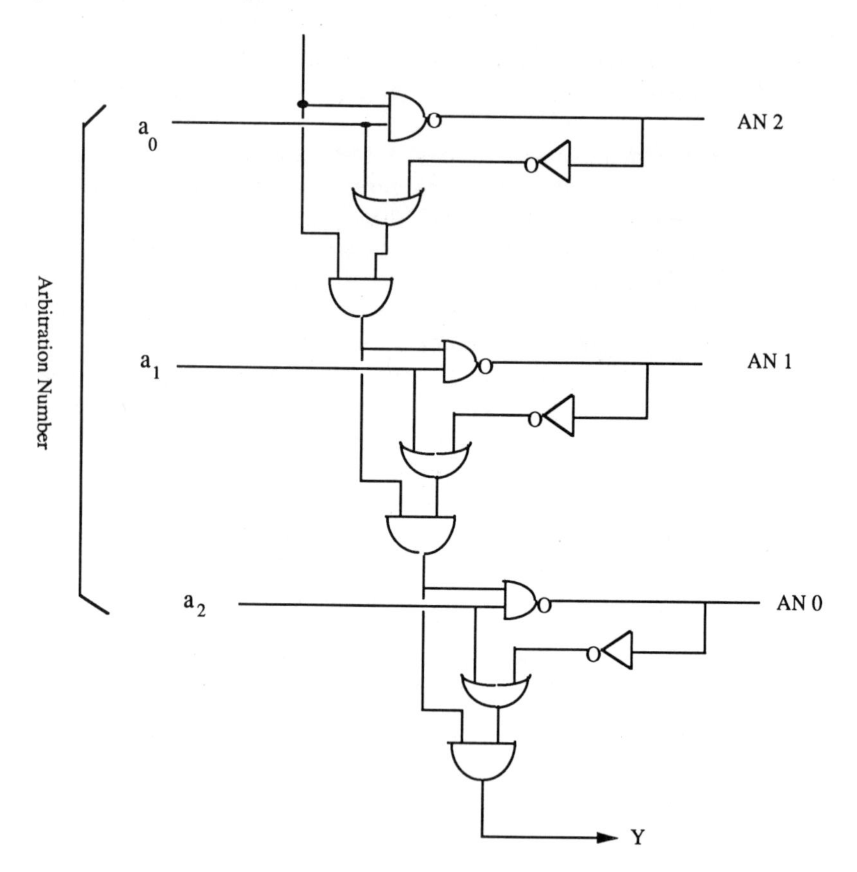

FIGURE 6.41. Arbitration logic.

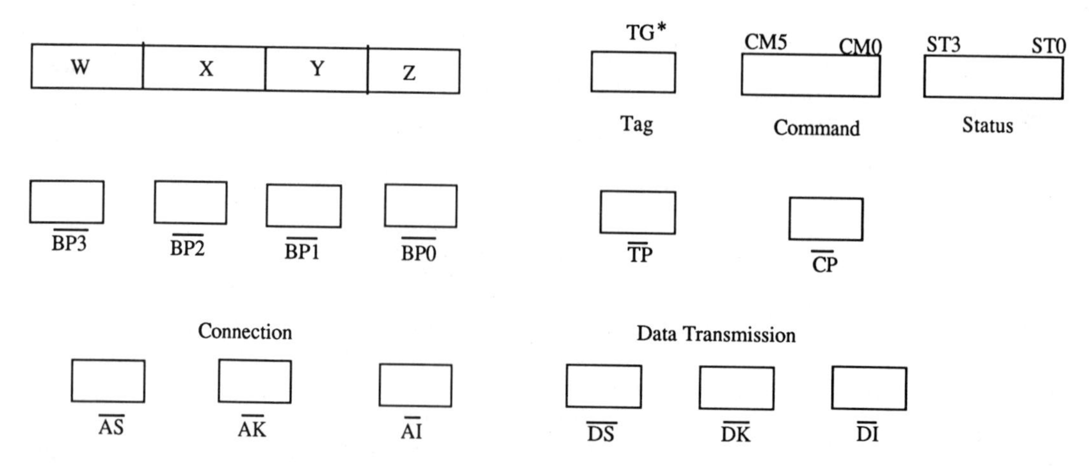

FIGURE 6.42. Parallel protocol synchronization lines.

The address transfer begins with the master placing an address on the AD lines and a command on the CM lines. After a skew delay, the master asserts the Address Strobe ($\overline{AS}$). The addressed slave responds by placing its status on the ST lines and asserts an Address Acknowledgment ($\overline{AK}$). If the master has chosen to use the broadcast mode—by asserting BC during the address transfer—the addressed slave also releases the Address Acknowledgment Inverse ($\overline{AI}$). The data transfer phase begins after the establishment of the connection.

The signal $\overline{AS}$ synchronizes the address generated by the master. All of the slaves respond with $\overline{AK}$ and $\overline{AI}$. The slowest module will finally allow the $\overline{AI}$ signal line to go high, since it is an open-collector line. The protocol then proceeds to the data transfer phase. At the end of the transaction, $\overline{AS}$ is released. All slaves then respond by asserting $\overline{AI}$ and releasing $\overline{AK}$.

The transaction on the Futurebus is controlled by the connection command codes. Six bits are transmitted from the master to the slave. The command codes carry different information on them during the connection phase than they do following the transfer, where there will be a write operation. The meanings of the codes are:

LK–Indicates to the slave to lock the transaction to prevent access through other ports.
BT–Allows a block transfer handshake to be used.
BC–Allows a broadcast handshake to be used.
EC–Allows the modules to move to an entirely new protocol.

During the data transfer sequence, CC remains a cache function bit. WR is a true read/write signal. LW, LX, LY, and LZ are lane deselect signals, which allow a module to write to any combination of the byte lanes.

The two-edge protocol requires two transitions of $\overline{DS}$ and $\overline{DK}$ for block transfers. The four-edge protocol passes a signal piece of information, using four transitions. The four-edge protocol is required to transfer intermixed read and write operations.

6.6.3.6 Slave Status Indication

The slave communicates its status to the master during each address and each data transfer. The slave utilize the slave status lines (ST2–ST0). The statuses are listed in the following table.

ST2	ST1	ST0	Description
0	0	0	Illegal code reserved
0	0	1	Valid action
0	1	0	Busy
0	1	1	Access error; The slave cannot perform the operation, or the request is a violation of the rules.
1	0	0	End of data; The slave is unable to send or receive more data.
1	0	1	Illegal code/reserved
1	1	0	Parity error
1	1	1	Error code; An unknown type of error has occurred. This generally occurs during broadcast.

6.6.3.7 Data Transfer

Once the connection has been established, the master may proceed with the transfer of data to or from the slave. The master can create two data transfer sequences: A block transfer or a single transfer [Borrill 84].

The block transfer consists of multiple data transfers sending a number of data words to or from a slave without sending additional address information across the bus. The slave must keep track of the address internally by incrementing it after each data transfer. The protocol does not require the master to provide the length information at the beginning of the sequence. Thus, a block transfer may span the boundaries of two or more memory boards. If a master detects an end-of-data, it simply drops the AS/AK lock, puts the next address on the AD lines, and reasserts AS to establish a connection with the next slave.

In block transfer mode, the DS/DK handshake pair are immediately ready to synchronize the passage of the next piece of data. The slave provides the data upon the release of DK, and the master acknowledges it on the release of DS. A piece of data is passed with each pair of transactions of DS and DK, as shown in Figure 6.43.

A single transfer protocol is used for normal random access of a piece of information of from one to four bytes in parallel. The transfer begins on the

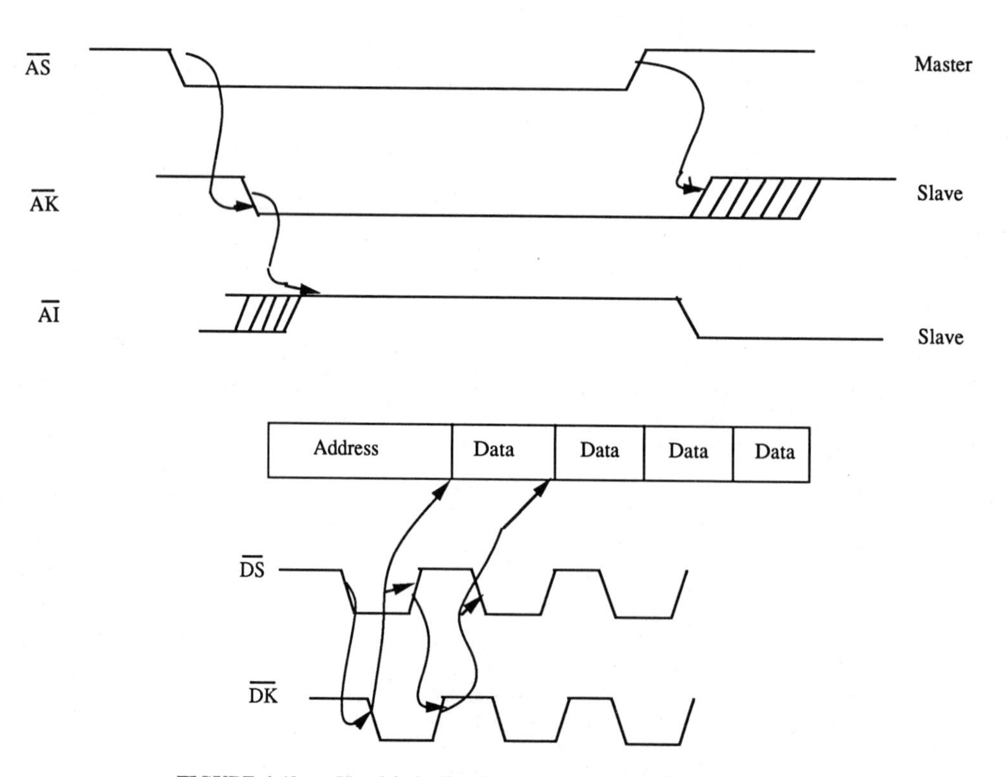

FIGURE 6.43. Handshake for the transfer of an information block.

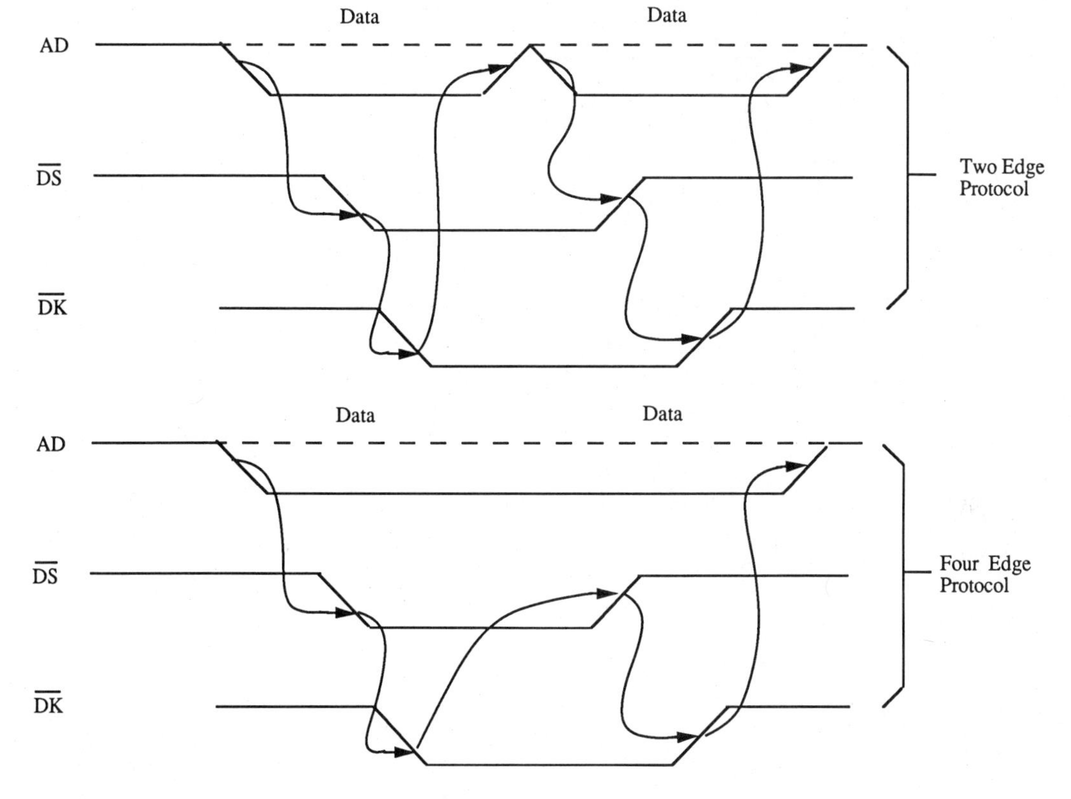

FIGURE 6.44. Single data transfer.

assertion of DS or DK, or the release of DI. For a read, the slave responds by putting the data from the requested location on the AD lines, and asserts DK. The master acknowledges receipt by releasing DS. The slave then cleans up by removing the data and releasing DK. The DS/DK handshake pair is now in their original state and are ready to synchronize the passage of the next piece of data as shown in Figure 6.44. The diagram illustrates the conventional four-edge type and the advanced two-edge variety. The four-edge handshake uses both the active and inactive edges of the strobe and acknowledgment signals, whereas the two-edge type saves time by using only one edge from each signal.

6.6.4 Comparisons

A formal comparison of computer buses are difficult, since each was designed for different applications, and a meaningful benchmark test is extremely difficult, if not impossible, to devise. However, a number of comments can be made to highlight the differences existing among them. The comparison is illustrated in the following table.

Criteria	VME bus	Multibus	Futurebus
Standard Number	P1014	P695	P896.1
Address/Data Path	Non-MUX	MUX	MUX
Byte Orientation	Big Endlan	Little Endlan	Not Constrained
Bus Timing	Async	Sync	Async
Centralized Service	Bus Timer Interrupt Handler Arbiter Power Monitor System Clock Bus Time-out	Clock Power Monitor	Distributed Architecture
Bandwidth (Mbytes/s)	20–57	40	120
Cache: Write-Through	Limited	Limited	Fully Supported
Cache: Write-Back	Not Supported	Not Supported	Fully Supported
Live Insertion	Not Available	Not Available	Fully Supported

Chapter 7

Exception Processing

Exception processing is termed as the deviation from normal processing associated with the execution of the instructions. A deviation from a normal processing sequence may occur due to an internally generated error condition, or an external request or error condition. The exception processing prepares the MC68020 to manage the exception condition, but does not include the execution of the specific service routine.

This section describes the processing states and privilege states of the processor, and then discusses how it processes interrupts, traps, and the actions taken by the processor in response to exception conditions.

7.1 STATES OF THE MC68020

The MC68020 always remains in one of the three processing modes shown in Figure 7.1:

- User mode
- Supervisory mode
- Halted mode

The programs executing under supervisory mode have a higher priority than the programs operating under user mode. The processor is capable of executing all instructions from the supervisory mode, which it cannot do from the user mode. These distinctions are supported by the underlying architecture. Now, the user programs may only access their own code and data areas, and can be re-

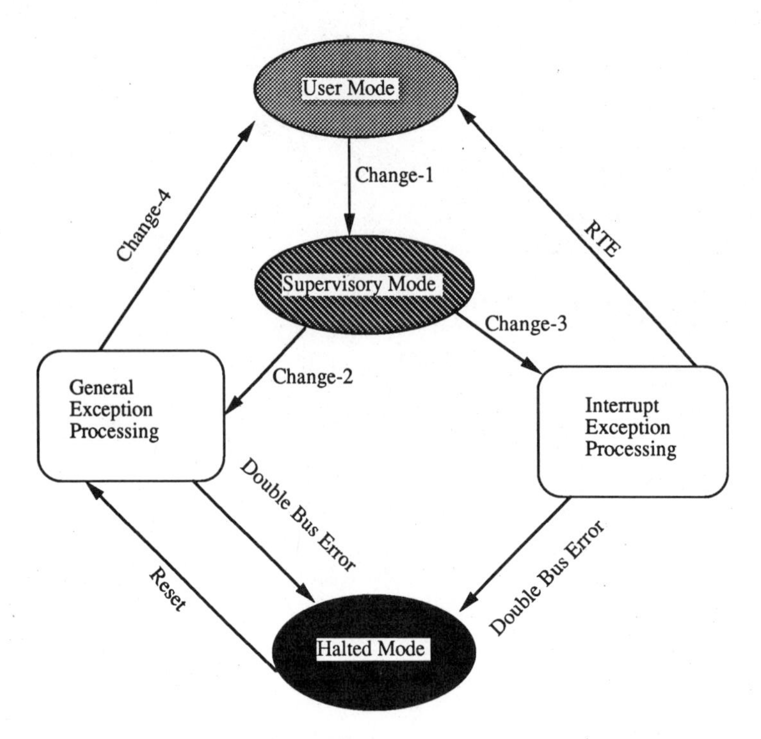

FIGURE 7.1. Processing states of the MC68020.

stricted from accessing other information. User program behavior can be controlled, and errors can not affect the supervisor's operation.

The only way that the processor can migrate from the user to the supervisor privilege level is through exception processing. A transition from the supervisor to user state can be caused by one of the following instructions: RTE, Move to SR, ANDI to SR, EORI to SR, and ORI to SR.

A special case of the normal processing is the *halted* mode. It is an indication of catastrophic hardware failure. As a result, only an external $\overline{\text{RESET}}$ can restart a halted processor.

Exception processing may be initiated internally by the processor itself, or can also be initiated by externally generated events. Examples of internally generated exceptions are:

- Unusual conditions during instruction execution.
- Trap instructions.
- The initiation of tracing.
- Other exception conditions.

Exceptions may also be triggered by external events, such as:

- Interrupts
- Bus errors

- Resets
- Coprocessor primitive commands

In the MC68020, exception processing is designed to provide efficient context switching, so that the processor can quickly and gracefully manage unusual conditions.

7.2 THE EXCEPTION PROCESSING PARADIGM

Figure 7.2 gives an overview of the components involved in exception processing. On the left side are the steps involved for general exception processing. The middle of the diagram contains the vector table. It consists of the locations in memory, called vectors, that the MC68020 accesses to determine the address of the service routines. Entries in the table are the pointers to the exception service rou-

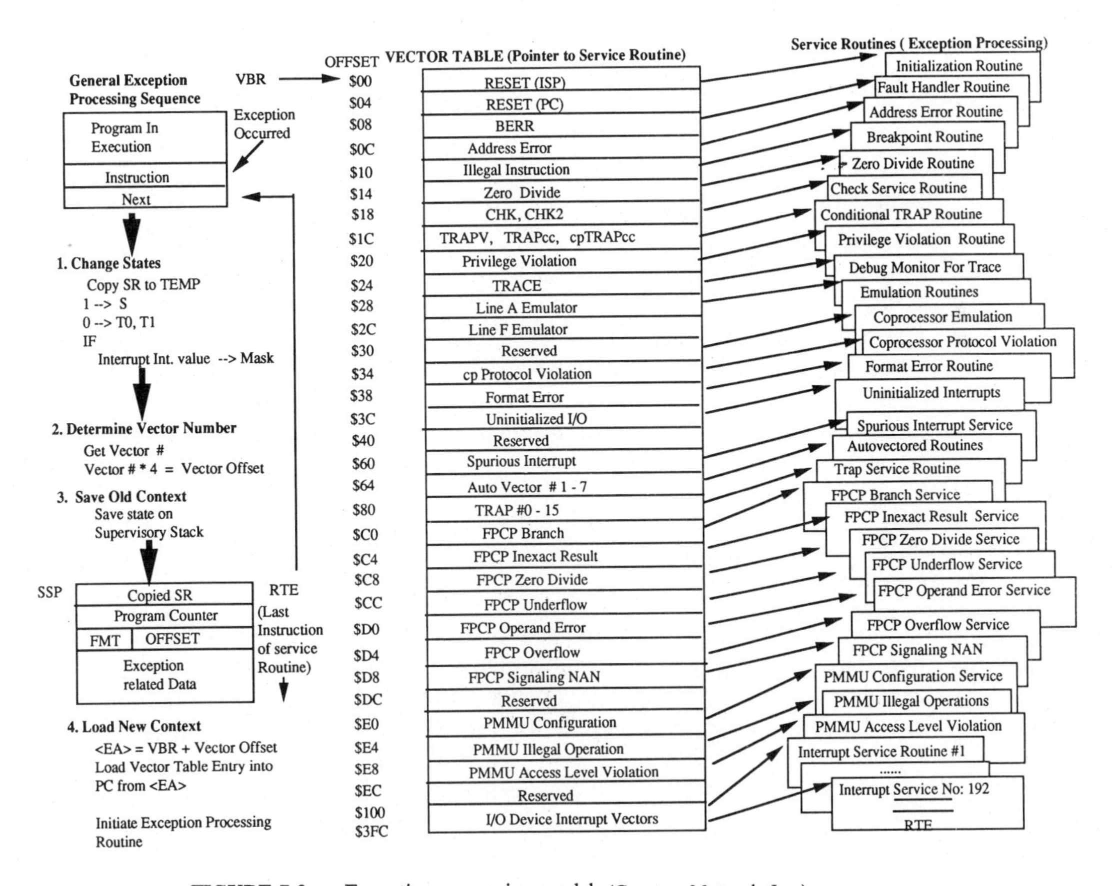

FIGURE 7.2. Exception processing model. (Courtesy Motorola Inc.)

tines. The exception processing service routines are shown on the right-hand side of the diagram.

7.2.1 Exception Vectors

For each exception function performed by the processor, there exists a vector number assigned to it. The vector numbers are stored as a block of memory known as the *exception vector table.* The table contains up to 256 vectors, labeled as vector numbers 0 through 255. The Vector Base Register (VBR) points to the base of the vector table, and the address of the exception vector entry is calculated by adding the contents of the VBR to the vector offset.

Figure 7.2 shows the assignments of exception vectors in the vector table. The first 64 vectors are defined by the processor, and the remaining 192 vectors are user definable. Since there is no protection on the first 64 vectors, external devices may use them in any way that the system designer deems appropriate. The features of the exception vectors are:

- All vectors must reside in the supervisor data memory area and are 2-word in length, with the exception of the reset vector.
- Vector 0, 4-word in length, is assigned to the hardware reset function, and must be stored in the supervisor program memory.

The vector table is divided into five logical groups, which are:

Group 1: Vector numbers 0 through 15 are reserved for system functions.
Group 2: Vector numbers 16 through 23 are reserved by Motorola for future use.
Group 3: Vector numbers 24 through 47 are reserved for interrupt and trap vectors.
Group 4: Vector numbers 48 through 63 are reserved for coprocessor-related exception processing. However, vector numbers 59 through 63 are yet to be defined.
Group 5: Vector numbers 64 through 255 are kept reserved for user-definable interrupt vectors.

7.2.2 Calculation of the Exception Vector Address

We have already seen that the vector table has 256 entries, and that the VBR points to the base of the vector table. The address of the exception vector entry is calculated by adding the contents of the VBR to the vector offset. The processor calculates the vector offset by multiplying the vector number by four. It is the responsibility of the system designer to store the address of the service routine corresponding to an exception condition to the proper location of the vector table. The exact algorithm of the service routine should be decided by the system designer, implemented by the application programmer, and loaded into the memory as a part of system reset routine.

For example, if a device generates a vector number $0F, then the processor will calculate the vector address as:

$0F*4 = $3C

and if the VBR contains an address $2000000, then the processor will find the address of the 'uninitialized interrupt' service routine at the location $200003C. The MC68020 will then expect the address of the uninitialized interrupt service routine at that location.

7.2.3 General Exception Processing Sequence

The exception processing sequence is composed of four primary steps, which are invoked automatically by the microcodes in the MC68020. The steps are [Motorola 85]:

1. Change state
2. Determine the vector number
3. Save the old context of the program
4. Restore the context of the old program at the completion of the exception

During the first step, a temporary copy of the status register is made, the status bit is set (i.e., S = 1), and trace bits are cleared (i.e., T0 = 0 and T1 = 0). All subsequent activities will be suspended, putting the CPU in the supervisory mode and inhibiting the tracing of the exception handler. If this exception is caused by an interrupt, then the interrupt mask bits are updated to the level of the interrupt to be serviced.

At the second step, the MC68020 determines the vector number and vector offset. If the source of the exception was an interrupt, then the CPU initiates an interrupt acknowledgment bus cycle to get the vector number. However, if the exception is initiated by a coprocessor, then a response primitive provided by the coprocessor contains the vector number. For any other exception, the processor generates the vector number internally, as supplied by the microcode. The vector offset is determined by multiplying the vector number by four.

During Step 3 of exception processing, the processor preserves the old context. The internal temporary copy of the SR, the PC, format code, vector offset, and various relevant information are saved on the active stack. The supervisory stack is used because the S bit is set during Step 1.

The CPU marks the type of stack frame created with a *format code.* This code is used during the return from an exception, to determine which states are to be restored. The microcode of the MC68020 may generate one of six different stack frames. The return from exception processing restores the pre-exception context, and returns control to the next instruction in the execution sequence. All exceptions except RESET exception processing follow Step 3.

At the fourth step, the new context is loaded from the supervisory memory. The MC68020 determines the address of the exception vector by adding the vector offset to the value in the vector base register. The PC is loaded with the exception vector from this address location. The new program counter value is used to prefetch two long-words to fill the instruction pipe. Normal instruction decoding and exception processing then resumes from the exception service routine.

7.2.4 Exception Priority Sequence

A question that often arises is what happens if two or more exceptions occur simultaneously? For example, if an interrupt occurs while a bus error exception processing is in progress? The MC68020's architects solved these problems by grouping the exceptions and assigned priorities to them. Table 7.1 list the groups, and the members of each group, including their priorities.

This priority scheme is very important for determining the order in which exception handlers are invoked during multiple exception situations. A general rule is, the lower the priority of an exception, the more quickly it will be serviced. For example, if trap, trace, and interrupt become pending simultaneously, then trap will be serviced first, followed by the trace, and interrupt will be serviced at the end. When the processor resumes normal instruction execution, it will be an interrupt handler, which returns to the trace handler, which returns to the trap

TABLE 7.1. Exception Priorities.

Group Number	Exception and Relative Priority	Exception Processing Characteristics
0	0.0—Reset	Abort all processing and do not save the context.
1	1.0—Address Error 1.1—Bus Error	Suspend processing and save the internal context.
2	2.0—BKPT #n, CALLM, CHK, CHK2, cp Midstream, cp Protocol violation, cp TRAPcc, DIV-by-zero, RTE, RTM, TRAP #n, TRAPV	Exception processing is a part of instruction execution.
3	3.0—Illegal Instruction • Line-A • Unimplemented Instruction • cp Preinstruction • Privilege violation	Exception processing begins before the instruction is executed.
4	4.0—cp Postinstruction 4.1—Trace exceptions 4.2—Interrupt exception	Exception processing begins when the current instruction or previous exception processing is completed.

exception handler. This rule does not apply to the reset exception. Its handler is executed first, because of its highest priority, and above all, reset clears all other exceptions.

The recognition time of the exceptions are shown in Table 7.2, along with their class.

The Class 1 type exceptions will be initiated at the end of the present clock cycle. The Class 2 type will be recognized and acted upon at the end of the current bus cycle. The Class 3 type will be recognized at the end of the current instruction cycle. Lastly, the Class 4 type exceptions will be recognized within an instruction cycle. However, the recognition of an exception does not guarantee immediate processing. The processing sequence will be dictated by the priority level, as listed in Table 7.1.

7.3 EXCEPTION PROCESSING BY TYPE

Exceptions originate from various sources, and each exception has characteristics that are unique to it. This section analyzes the sources of exceptions, their unique features, and the processing sequence.

7.3.1 Reset Exception

The assertion of the $\overline{\text{RESET}}$ signal initiates this highest level of exception for the MC68020. It is designed for system initiation, and for recovery from a catastrophic failure. Any processing in progress at the time of the reset is aborted and nothing can be recovered. This exception is recognized within the clock. The

TABLE 7.2. Recognition Time.

Recognition Time	Class	Exceptions
At the end of the clock cycle	1	*Reset *Address Error *Bus Error
At the end of the bus cycle	2	*Halt *Bus Arbitration *Illegal Instruction *Privilege Instructions *Protocol Violation
At the end of the current instruction cycle	3	*Trace Exception *Interrupt Exception
With an instruction cycle	4	*TRAP, TRAPV, TRAPcc, *cpTRAPcc, CHK, CHK2 *Zero Divide, Format Error

flowchart in Figure 7.3 describes the sequence of steps executed by the processor during this exception processing:

1. Put the processor into the supervisory state by setting S = 1.
2. Clear both trace bits in the status register, which disables tracing.
3. Place the processor into the supervisory interrupt state by clearing the M bit in the SR register.
4. Set the interrupt priority mask into Level 7, the highest level (nonmaskable interrupt).
5. Clear the contents of the VBR.
6. Disable the trace bits, which will prevent further tracing (T0 = 0 and T1 = 0).
7. Clear the cache control register, invalidating all of the cache entries.
8. Generate a vector number to reference the reset exception vector, which is two long-words at offset zero from the supervisory program address space.

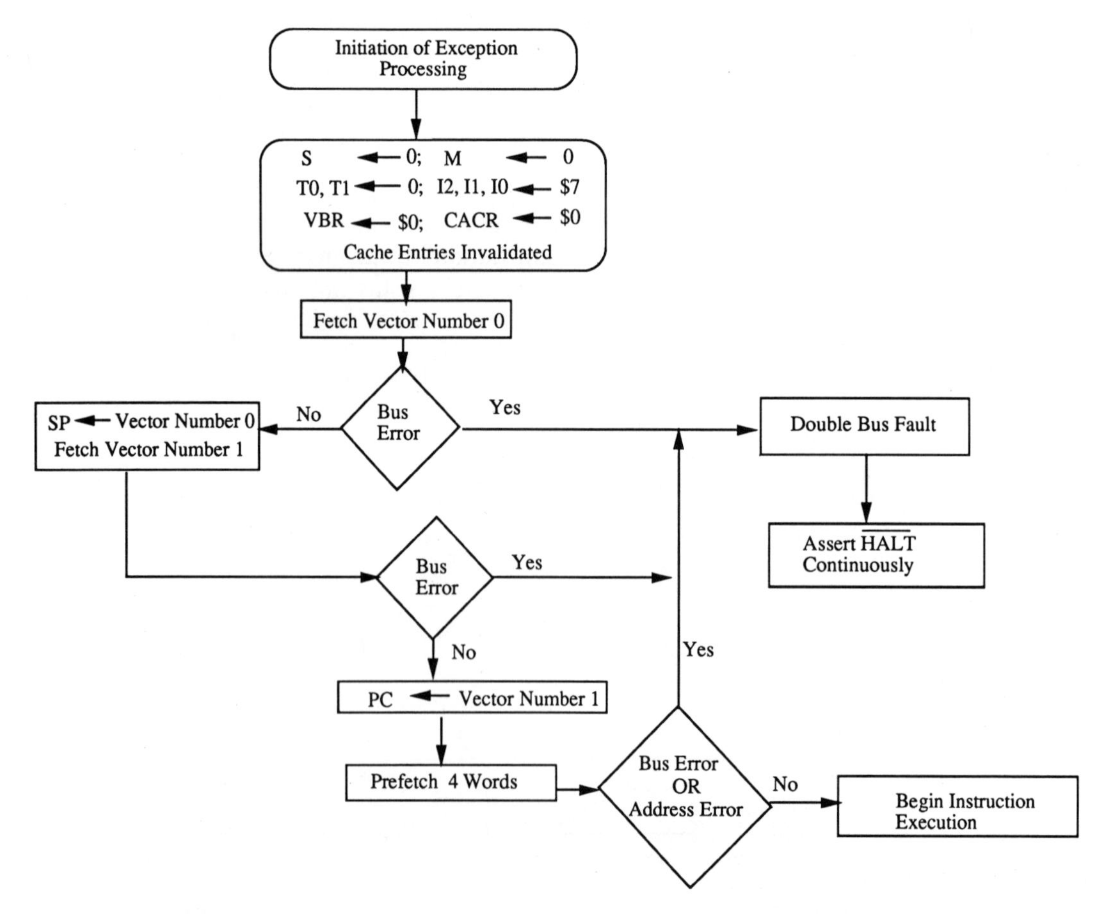

FIGURE 7.3. Reset exception processing.

9. Load the first long-word of the reset exception vector into the interrupt stack pointer.
10. Fetch the second long-word of the reset exception vector into the program counter.
11. The processor prefetches four words from the location referenced by the program counter and the program execution begins.

If the processor detects a bus error during Step 9, Step 10, or Step 11, then it realizes that a catastrophic error has taken place in the system and the processor enters into halt state.

However, the execution of the RESET instruction does not cause a loading of the reset vector. It does, however, put a reset pulse on the reset line to reset external devices. This allows the software to reset the system to a known state, and then continue execution of the next instruction from the instruction sequences.

7.3.2 Interrupt Exceptions

Seven levels of interrupt priorities are supported by the MC68000 family of processors, with the exception of the MC68008. The MC68008 supports three interrupt levels, namely Levels Two, Five, and Seven. Level Seven has the highest priority and Level Zero has the lowest.

Interrupt requests made to the processor do not initiate an immediate exception processing, but are simply made pending. Pending interrupts are recognized between instruction executions. The interrupt signals must maintain their request level until the MC68020 acknowledges the interrupt, in order to guarantee their recognition. If the priority of the pending interrupt is lower than or equal to the current priority, then the CPU ignores the interrupt request and continues with the current execution sequence. However, if the priority of the pending interrupt is greater than the current priority level, then the exception processing sequence is initiated. Figure 7.4 shows the flow of the interrupt exception processing, which involves:

1. Copying the status register into a temporary internal register.
2. Entering into the supervisor state ($S = 1$).
3. Reset the trace bit to 0.
4. Setting the interrupt mask level to the level of the interrupt being acknowledged.
5. Saving the old context into the interrupt stack.
6. Generating the vector address from the vector number.

If the external logic requests automatic vectoring by asserting $\overline{\text{VPA}}$, then the processor generates a vector number internally, whose number corresponds to the interrupt level number. Otherwise, the MC68020 assumes that the interrupt requesting device would provide the vector number.

Once the vector number is acquired, the processor saves the exception vector

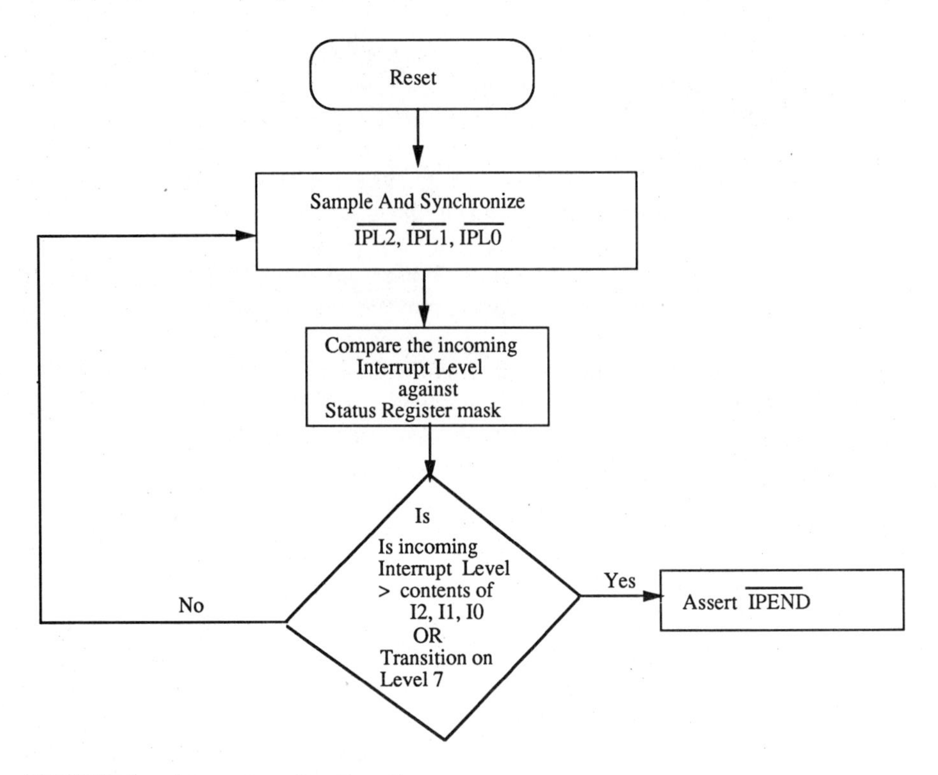

FIGURE 7.4. Interrupt-pending flow diagram.

offset, the PC, and the internal copy of the SR on the active supervisory stack. The saved value of the PC is the logical address of the instruction that would have been executed had the interrupt not occurred. If the interrupt was acknowledged during the execution of a coprocessor instruction, additional internal information is saved on the stack, so that the MC68020 can continue executing the coprocessor instruction when the interrupt handler completes execution. The mechanism of 'context save and restore' during interrupt service processing will be described in detail in the next section.

7.3.3 Spurious Interrupt

During the interrupt acknowledgment cycle, if no device responds by asserting $\overline{\text{DTACKx}}$ or $\overline{\text{VPA}}$, then the watchdog timer circuitry would normally assert $\overline{\text{BERR}}$ to terminate the vector acquisition process. The processor separates the processing of this event from a normal bus error by forming a short-format exception stack, and saves it onto the supervisory stack. The concept of the stack frame will be described in Section 7.4. The processor then fetches the spurious interrupt vector (vector number 24) and initiates the execution of the service routine written for this purpose.

7.3.4 Uninitialized Interrupt

Most of the MC68000 family of peripherals provide for programmable interrupt vector numbers to be used during the interrupt request/acknowledge mechanism of the system. If their vector numbers are not initialized after reset, then the MC68000 compatible peripheral will automatically supply vector number 15, the uninitialized interrupt vector. Upon recognition of this vector number, the processor will invoke the appropriate routine to handle this type of events. This provides a uniform mechanism for recovering from a programming error. Peripheral designers can take advantage of this feature to implement various mechanisms to support their needs.

7.3.5 Address Error

We have already seen that the MC68020 requires that instructions reside at the even byte boundary. An address error will occur during the prefetch of an instruction from an odd address. This address error-related exception is analogous to an internally initiated bus error exception. Here, a bus cycle is not initiated. Rather, exception processing begins immediately. The exception processing sequence is basically the same as that for the bus error exception to be described in Section 7.5. The processor generates the vector number 3, and the vector offset in the stack frame refers to the address error vector. The MC68020 enters into a *halted* state if the address error occurs during exception processing of a bus error, address error, or reset. We will see in Section 7.4 that the processor will generate either a short or a long stack frame.

7.3.6 Illegal/Unimplemented Instruction

An illegal instruction is a 16-bit binary pattern that does not represent one of the valid Op-words in the MC68020's instruction set. The MOVEC instruction with an undefined register specification field in the first extension word may also cause an illegal/unimplemented instruction. During instruction execution, if the above conditions are detected, then an illegal instruction exception is initiated by the processor. Examples of illegal instructions are:

```
MOVE.W  D0, $2001    ; trying to move into the odd byte boundary
MOVE.W  D0, $6(A0)   ; where A0 has an odd number
```

The exception processing for illegal or unimplemented instructions are similar to trap exception processing. At the end of the instruction fetch, the processor initiates instruction decoding. When the processor decides that the execution of an illegal instruction is being attempted, it copies the SR into temporary register, enters into a supervisory state, and disables the instruction tracing. The CPU generates vector number 4, 10, or 11, depending on the exception type. The processor saves the illegal/unimplemented instruction vector offset, current PC, and the copy of SR, before entering to execute the appropriate service routine. The saved value of the PC holds the address of the illegal or unimplemented instruction. This allows the execution of the instruction at the address contained in the exception vector. Thus, it is the responsibility of the service routine software developer to adjust the stacked PC, if the instruction is emulated in software or is to be skipped upon returning from the handler.

Vector number 4 is generated by the processor when it attempts to execute an illegal instruction.

When the processor attempts to execute an unimplemented instruction with an A-line Op-code (i.e., with instruction word bits [15:12]=$A), an exception vector 10 is generated, permitting efficient emulation of unimplemented instructions.

During the execution of an instruction, if the processor detects that the bits [15:12] of the first word are equal to $F, then the undefined patterns in the following words are treated by the processor as unimplemented instructions with F-line Op-code. If the same instruction execution is initiated during user mode, then a privilege violation is initiated. The exception vector number 11 is generated for such an unimplemented instruction with an F-line Op-code.

If the instruction bits [15:12] are equal to $F and bits [11:9] are not equal to $0, then the Op-word is labeled as a coprocessor instruction. If the MC68020 detects a coprocessor instruction, it runs a bus-cycle referencing CPU space type $2, and addresses one of the seven coprocessors. If the addressed coprocessor is not present in the system, then the cycle terminates with the assertion of a bus error signal. In such an event, the processor initiates an unimplemented instruction (F-line Op-code) exception, which allows the system designers to implement a software emulator for the coprocessor, with an F-line handler.

7.3.7 Trap Instructions

Trap exceptions are caused by instructions. The general format for a trap instruction is:

TRAP #n

where n = the vector number

This instruction always forces exception processing, and is an extremely valuable tool for implementing system calls to be used by user programs. TRAPcc, TRAPV, and cpTRAPcc instructions initiate exceptions if the program detects an arithmetic overflow, an out-of-bound value, etc. during their execution.

A trap causes the processor to create an exception stack frame, place it onto the supervisory stack, and fetch the address of the service routine pointed to by the vector number. The vector number is an integer with a range of 0 through 15, and is used to calculate the hexadecimal vector address, using the formula:

Vector address = <vector base register> + 4* vector number

The trap instruction can be utilized in various ways. The 16 possible traps allow a user program to call the operating system for service, which must be executed at the supervisor level. For example, TRAP #2 might be used for the printer, which is controlled by the operating system, and TRAP #1 used for the hard disc controller, etc.

7.3.8 Other Trap Instructions

Certain types of arithmetic errors can be detected and trapped by the processor. For example, the signed divide (DIVS) and unsigned divide (DIVU) instructions will force an exception if a division operation is attempted with a divisor of zero. The CPU will initiate exception processing after fetching the address of the service routine from vector number 6. The trap on overflow (TRAPV) instruction tests the overflow ($V = 1$) flag in the condition code register and jumps to a specific vector location if V is found to be set. If V is clear, execution continues with the next sequential instruction.

7.3.9 Privilege Violation

System security was always in the minds of the designers of the MC68000 family of processors. This motivated them to create two classes of instructions, one known as privilege instructions and the other as nonprivilege instructions. An attempt to execute one of the privilege instructions while in the user mode causes a privilege violation exception. A privilege violation can also occur if a coprocessor requests a privilege check, and the processor happens to be at the user level. The privilege instructions are:

ANDI to SR	MOVEC
EORI to SR	MOVES
cpRESTORE	ORI to SR
cpSAVE	RESET
MOVE from SR	RTE
MOVE to SR	STOP
MOVE USP	

The privilege violation exception processing sequences are similar to the illegal instruction's exception processing. The processor initiates an exception processing before the execution of the privileged instruction. The processor generates the privilege violation exception vector, which is vector 8, and saves the privilege violation vector offset, the current PC, and a copy of the SR prior to the initiation of the exception processing in the supervisory stack. The saved value of the PC is the address of the first word of the instruction responsible for privilege violation. Instruction execution resumes from the location pointed to by the privilege violation exception vector.

7.3.10 Tracing

The MC68020 supports instruction-by-instruction tracing to support the program development process. The processor doesn't suspend the normal execution, but rather executes the trace service routine after the processing of each instruction. This allows a debugger program to monitor the execution of the target program under test.

The trace bits T1 and T0 in the status register control tracing. Table 7.3 lists the trace mode selection. The trace mode remains disabled when both T bits are cleared, and instruction execution proceeds normally. If $T1=0$ and $T0=1$, then any change in the program flow will initiate a trace exception processing. Instructions traced in this mode include all branch instructions, jumps, instruction traps, returns, and coprocessor instructions that modify the program execution flow. The status register modifications are also included, because the MC68020 must prefetch instruction words to fill up the pipe any time that an instruction that can modify the status register is executed.

If $T1=1$ and $T0=0$ are set at the beginning of the execution of an instruction, then a trace exception will be generated at the completion of that instruction.

The trace function for $T1=1$ and $T0=1$ is not defined, and they are reserved by Motorola for future use.

In general, a trace exception extends the boundary of any traced instruction. That is, the execution of a traced instruction is not complete until the trace exception processing is completed. If the instruction is not executed, either due to an interrupt or an illegal or privilege instruction violation, then the trace exception will not be initiated. The trace exception will be deferred until after the execution of the suspended instruction resumes. Again, trace exception does not occur if the instruction is aborted by a reset, bus error, or address error exception.

The exception processing for a traced instruction starts after the successful completion of the instruction, and before the execution of the next instruction in the original program. The exception processing steps are [Motorola 85]:

1. Copy the SR into an internal temporary register.
2. Set $S=1$ and enter into the supervisory state.
3. Clear the T0 and T1 flags and disable further tracing.

TABLE 7.3. Tracing Control.

Trace	Bits	Remarks
T1	T0	Tracing function
0	0	Tracing is disabled
0	1	Trace on change of flow (e.g., BRA, JMP, etc.)
1	0	Trace on instruction execution (any instruction)
1	1	Undefined

4. Internally generate vector number 9 for trace exception.
5. Save the trace exception vector offset.
6. Save the PC value.
7. Stack the information saved in Steps 1 through 6 in the supervisory stack.

The stacked PC value is the logical address of the next instruction to be executed. This allows the normal program execution to resume after the necessary prefetch from the address in the trace exception vector.

7.4 STACK FRAMES

During the exception processing sequence, the MC68020 generally saves information about the current context on the supervisory stack, called the *stack frame.* The stack frame refers to the template of the information pushed onto the stack. The CPU marks each stack frame with a format code during its inception. The RTE instruction retrieves the format code to determine which internal locations are to be restored and how much information is to be removed from the stack. Figure 7.5 lists the format codes of six different stack frames generated by the MC68020 for this purpose.

A minimum of four words of information are saved, which are the: SR, PC, format code, and vector offset. Various exceptions may save additional information. However, these four words always remain in the same location relative to the stack pointer, regardless of the exception types. Figure 7.6 illustrates stack frame templates, exception types supported by the stack frames, and the reference point for the stack frame counter. The left side of this diagram shows the stack frames. The middle column of the table lists the exception types that are generated by each of the stack frames. The right-hand column shows the information represented by the PC. For exceptions caused by the interrupt or trap

Format ID	Format Type
0000	Short frame - 4 Word Length
0001	Throwaway frame - 4 Words Length
0010	Instruction Exception frame - 6 Word Length
0011 to 0111	Reserverd by Motorola for Next Generation Processor
1000	Bus Fault frame applicable for MC68010 - 29 Words Length
1001	Coprocessor mid-instruction frame - 10 Words Length
1010	Short Bus Fault frame for MC68020 - 16 Words Length
1011	Long Bus Fault frame for MC68020 - 46 Words Length
1100 to 1111	Reserverd by Motorola for Next Generation Processor

FIGURE 7.5. Definition for the MC68020 Stack Frame Format.

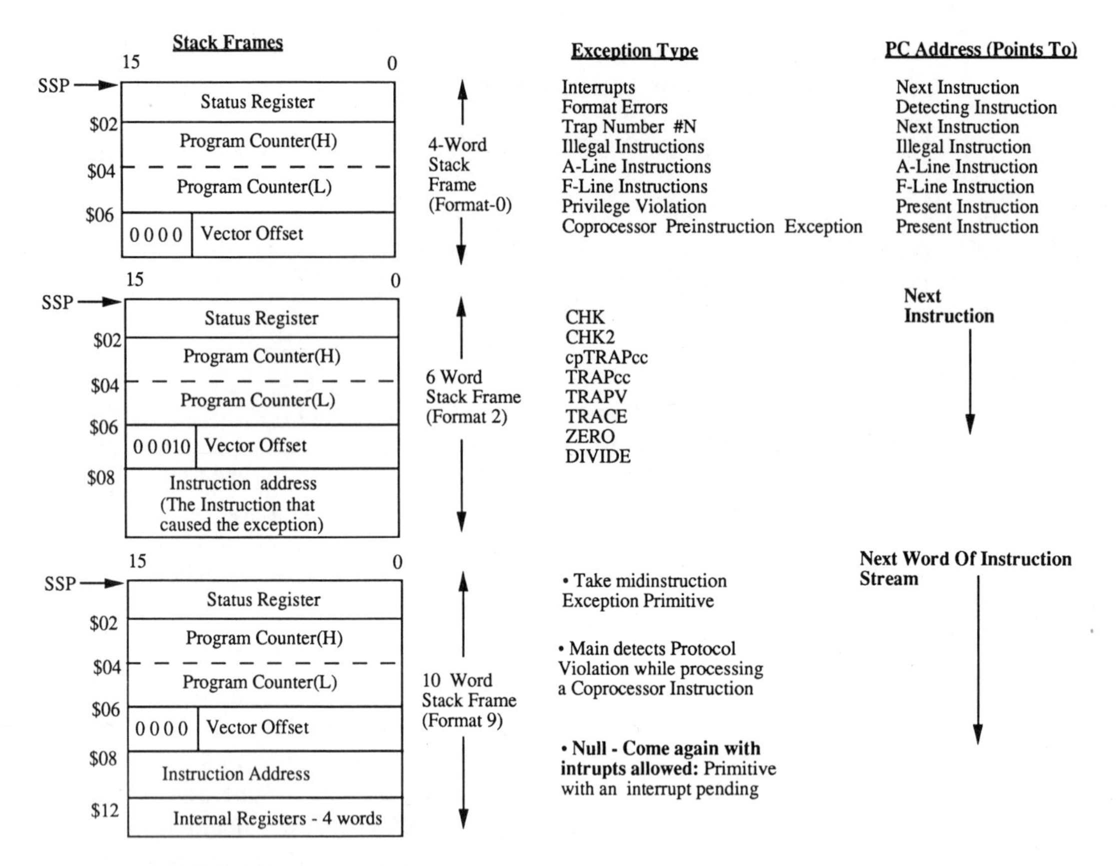

FIGURE 7.6. Stack frame templates. (Courtesy Motorola Inc.)

instructions, the PC represents the return address of the next instruction to be executed after the service has been performed. In the case of format errors, illegal instructions, line A, line F, breakpoint Op-code, and privilege violations, the program counter points to the address of the instruction that caused the exceptions.

A number of exception events generate a six-word stack frame, in addition to the PC. The address of the instruction that caused the exception is also saved. This is necessary because, for example, the conditional trap instruction can be one, two, or three words in length, where the stacked PC points to the next instruction to be executed when the service is completed. On the other hand, the instruction address points back to the location of the instruction that caused the exception. Again, exceptions taken during communication between the main processor and the coprocessor use ten-word stack frames. The stack frame saves various undocumented internal registers of the MC68020. They are saved to permit the MC68020 to continue from the middle of the coprocessor dialog once the exception has been serviced.

The vector offset is also saved on the stack in support of the implementation of the generic interrupt service routine. A generic service routine is one that provides service to a common set of I/O devices, for which slightly different operations are required, and where each device provides a unique vector number. When the CPU acknowledges the interrupt request, all vectors point to the same routine. This generic routine can be designed to identify the originator of the service request by looking at vector offset from the stack, instead of executing a routine to pull the devices.

7.4.1 Normal Four-Word Stack Frame

We have seen that the MC68020 always saves a minimum of four words of information, namely: SR, PC, format code, and vector offset, on the stack during exception processing. These four words remain in the same locations relative to the stack pointer, regardless of the exception type. The four-word stack frames whose format is '0000' are created by interrupts, trap number n, illegal instructions, A/F line emulator traps, privilege violations, and the coprocessor preinstruction exceptions.

7.4.2 Six-Word Stack Frame

The format of a six-word frame is '0010', and is created during the exception processing of:

- A coprocessor postinstruction exception.
- CHK and CHK2 exceptions.
- cpTRAPcc, TRAPcc, and TRAPV exceptions.
- Trace and divide by zero exceptions.

Here, the instruction address points to the address of the instruction that triggered the exception. The PC value acts as a pointer to the next instruction to be executed, once the RTE completes the context restoration.

7.4.3 Coprocessor Midinstruction Exception Stack Frame

This stack frame, with a format code of '1001', is created by three exceptions related to coprocessor operations. The first is called 'take midinstruction exceptions'. This happens during the dialog with the coprocessor, while the MC68020 is executing a coprocessor instruction.

The second exception occurs when the MC68020 detects a protocol violation during the processing of a coprocessor instruction.

The last exception occurs when a 'NULL, come again with interrupts allowed' response is received, and the main processor detects a pending interrupt. For this stack frame, the PC is the address of the next word to be fetched from the instruction stream. The instruction address value is the address of the first word of the instruction that was executing when the exception occurred.

7.4.4 Format Error Exception

The microcode of the MC68020 always performs a validation check on instruction and data values for a control operation, including the coprocessor state frame format word and stack frame format. The stack format is verified as a part of the RTE instruction execution. The RTE instruction compares the internal version of the processor against that contained in the frame residing in the memory for bus-cycle fault format frames. This check ensures that the processor can correctly interpret the internal state information from the stack frame.

The CALLM and RTM both check values in the option and type fields in the module descriptor and module stack frame. If these fields contain incorrect values, or an illegal access attempt is detected by the external memory management unit, then an illegal call or return is being requested. In such cases, the processor will initiate format error exception processing.

The cpRESTORE instruction passes the format word of the coprocessor state frame to the coprocessor for validation. Being a slave to the main processor, if the coprocessor cannot recognize the format value, it asks the MC68020 to initiate format error exception processing.

If any of the checks described earlier determine that the format of the stacked data is invalid, then the processor will initiate a format error exception. This exception saves the context of the processor in a short-format stack frame and generates exception vector number 14. It should be noted that the instruction that detected the format error is stored in the short frame. The processor will continue to execute from the location directed by the format exception service routine.

7.4.5 Multiple Supervisory Stack

When the MC68020 is in the supervisory state, the M bit in the SR identifies the currently active supervisory stack. If the system is dedicated to one task, a single stack is sufficient. However, in a multitasking system, having a second supervisory stack is extremely convenient. In such an environment, the operating system maintains a system table, to keep track of the status of each task. It maintains all of the information needed for task management in a block of RAM called the Task Control Block (TCB). When the operating system switches from Task A to Task B, it must preserve the status of Task A in A's TCB, and switch to Task B by pulling information from B's TCB. All information related to Task A should be preserved in A's TCB. Assume that an exception occurs when Task A is executing. Now, the stack frame created by the MC68020 during exception processing will contain information directly related to Task A.

One approach is to have the supervisory stack pointer pointing to the TCB of the currently executing task. This approach keeps the information related to the task in the corresponding TCB. It is possible that additional exceptions may occur before the first one has been completely serviced. For example, a task may execute a trap instruction. While the trap service routine is executing, an interrupt occurs, perhaps even multiple interrupts. In such a case, multiple stack frames will be created. Except for the first stack frame, subsequent frames do not contain task-related information, as the PC was pointing to the service routine, not to the task. Furthermore, a significant amount of memory is required for this stacking. Thus, TCB has to be very large to accommodate a worst-case stack size.

The MC68020 offers a second supervisory stack pointer called the Interrupt Stack Pointer (ISP) to overcome the above problems. The Master Stack Pointer (MSP) could be used to point to the TCB of the currently active task (Figure 7.7). The second stack pointer becomes active only if an interrupt occurs. If more exceptions occur before the interrupt service routine is completed, then the stack frames are preserved in the system memory, not in the TCB. As a result, only one worst-case stack frame area must be allocated in the system memory. The MSP points to the task-related information, and the ISP references system-related information. The M bit allows the separation of task-related and asynchronous I/O-related information.

The MSP now points to the active task's TCB. When a task switch is required,

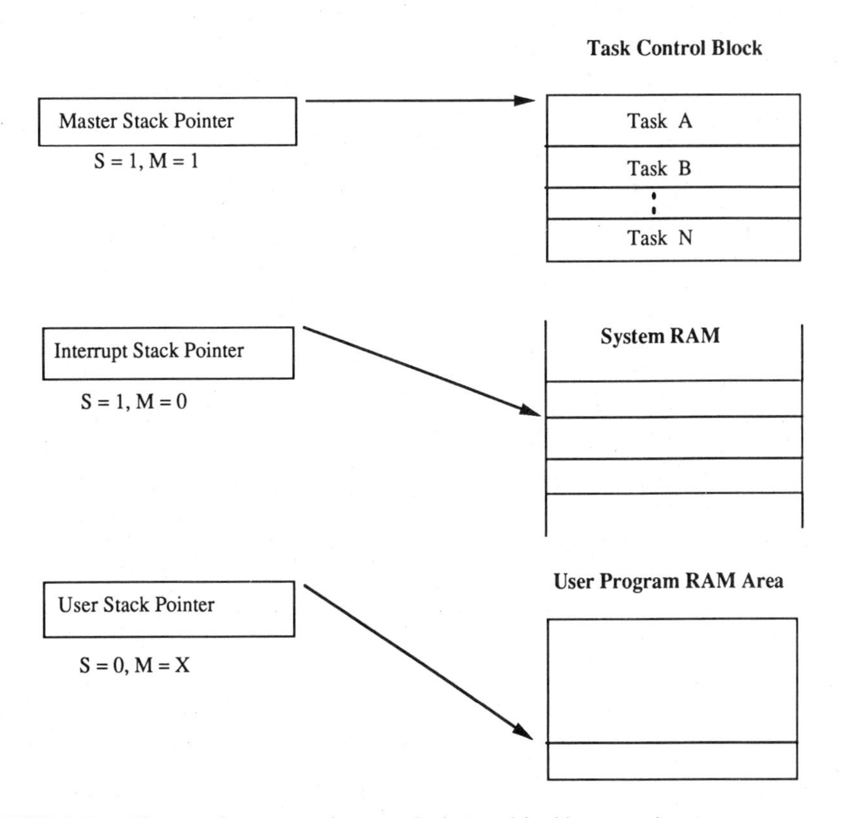

FIGURE 7.7. The use of two supervisory stacks in a multitasking operating system.

the MSP is redirected to point to the new task's context (i.e., a new TCB). The ISP maintains a valid independent stack space for interrupt-related information, and any temporary storage that the interrupt service routines may require. When an interrupt is processed by the MC68020, the stack space associated with the active task is not disturbed by the interrupt stack operations. This simplifies the task scheduler's job, and eliminates the need for a large TCB to handle the worst-case interrupt information.

7.4.6 Throwaway Stack

In order to clearly understand this two-stack concept, let us examine the exception processing when an interrupt occurs. Let us also assume that our operating system utilizes the two system stack pointers. When a user task is dispatched, the OS sets M = 1 and utilizes the MSP to point to the task's TCB. If an interrupt is recognized while the CPU is in the middle of this dispatch, it first copies the SR to an internal temporary location. The processor sets S = 1, clears the T1 and T0 bits, and acquires the vector number from the interrupting device. The previously stacked SR had S = 0 and M = 1. The PC is stacked, along with a format code of '0000', as shown in Figure 7.8. This stack frame is created with the task's TCB pointed to by the MSP. Since an interrupt exception is being processed and M = 1, the current SR, with S = 1 and M = 1, is copied into an internal temporary register. The M bit then gets cleared, which makes ISP the active stack pointer. The MC68020 proceeds to stack the information again, but this time on the interrupt

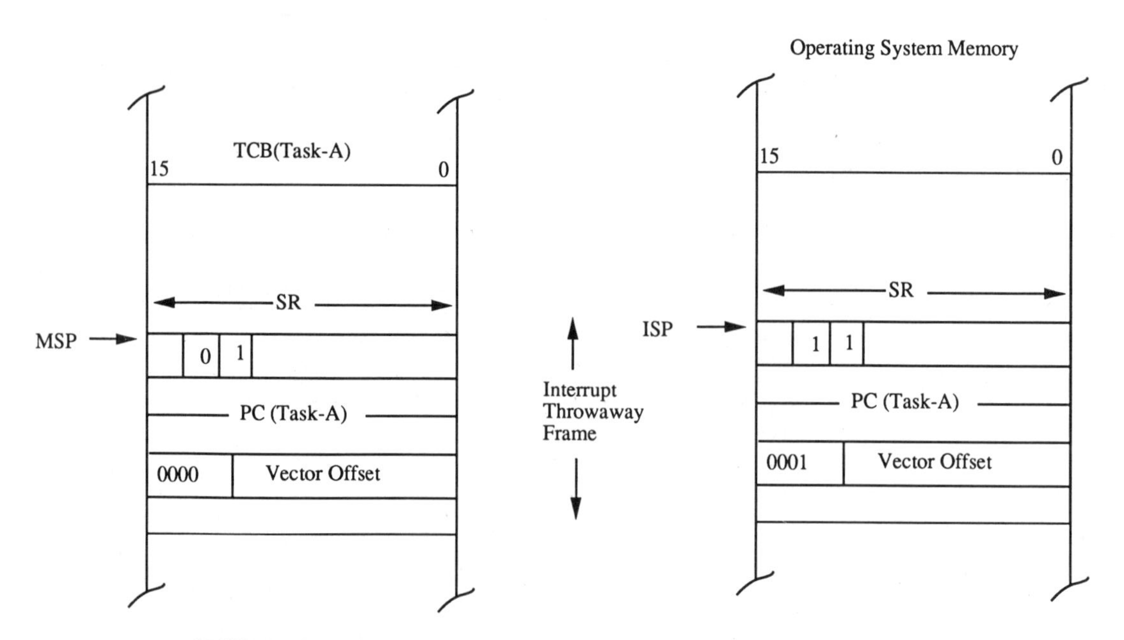

FIGURE 7.8. Throwaway stack frame.

stack. This frame will have a different status register value (S = 1), along with the same PC value and the same vector offset, but with a new format code of '0001'.

This is called a throwaway stack frame. We will see how this stack frame is thrown away when we examine the operation of RTE instruction. Notice the differences between the two stack frames. The differences lie in the format code and the status of the S bit. Any additional space required by the interrupt service routine for the local variable or nested variable is stacked on the interrupt stack.

7.5 RETURN FROM EXCEPTION PROCESSING

Once the MC68020 completes the processing of all pending exceptions, it resumes normal instruction execution from the address contained in the vector referenced by the last exception. The processor must gracefully restore the previous context prior to the initiation of exception processing. The RTE instruction restores that context.

When the processor executes an RTE instruction, it retrieves the stack frame, using the active stack pointer to determine whether it contains a valid frame, and the nature of the context to be restored for this frame. Since the CPU builds one of six different stack frames when an exception occurs, the RTE instruction must be able to restore the context from any one of them. The RTE is often the last instruction of an exception service routine. It begins by reading the stacked values of the SR register from the memory and copying them into a temporary internal location. Then it reads the stacked format code at an offset of +6 from the stack pointer. This format code is first validated. This valid format determines subsequent actions to be taken to restore the context from the stack.

If the format code is '0001', then it is a throwaway frame and it is removed from the stack. Once SR is loaded from that context, the frame is thrown away by adding +8 to the stack pointer (i.e., by deallocating the memory). The SR value is read from the throwaway frame and is placed into the temporary internal register. If M = 1, on the throwaway stack frame (which would typically be the case), then the MSP would become active at this time, and the CPU again reads the stacked SR. If the MSP is the active pointer, and it is pointing into the previously interrupted user task's TCB, then the MC68020 starts to restore the user's task context. Next, the format code is read from the location at stack pointer +6. It would typically be '0000', and so the PC is restored. A value of +8 is added to the SP, and the value of the temporary internal register is routed into the SR. Instruction execution finally resumes in the user task. In such an environment, in order to use two supervisory stack pointers, the CPU naturally generates a throwaway stack frame when M = 1 and an interrupt occurs.

The throwaway frame created during an interrupt provides the mechanism to automatically switch from the ISP to the MSP. A single 'return from exception' instruction manages all stack frames, including the throwaway frame. This mechanism greatly simplifies the life of the programmer. We have focused on the behavior of the RTE instruction when it encounters a format code of '0000' or

'0001'. These frames are created when an interrupt occurs. Other format codes have unique sequences of their own, and are utilized to automatically restore additional information from the stack. The algorithm resides in the microcode of the processor.

7.6 BUS ERROR

The bus cycle is aborted by the processor when the bus error signal is asserted during an operand access. The CPU may take two approaches to manage the bus error condition:

1. It can give up the effort to restart the instruction, and can come back to complete it at a later time.
2. It can keep a footprint of the place where it had to cease the execution of the instruction, and restart the execution from the place where it left off at a later time.

The first approach is favored by many of today's microprocessors, because of its simplicity. The second approach is more complex, but is more desirable for real-time applications. The designers of the MC68000 family of processors chose to implement the second approach.

In the case of a bus error, the MC68020 initiates exception processing. If the CPU is at an instruction boundary, the internal state of the machine is relatively simple to capture, because only a small amount of information is required to be saved. If the CPU is in the middle of an instruction execution, an extensive amount of internal information must be preserved, so that the instruction can be continued to completion at a later time. For example, let us consider the case:

```
MOVE.W   (A1), (A2)
```

If the bus error occurs during the instruction prefetch, the processor will create a short-stack frame (Figure 7.9a), because the MOVE instruction will be brought into the CPU as a part of the previous instruction. However, if the bus error occurs during the read bus cycle of the source operand, then the processor creates a long-stack frame (Figure 7.9b) (this is because the processor has processed a large amount of information, and its internal pipeline units are full). Because the sequencer is no longer at the instruction boundary, more internal state information must be saved. However, if the bus error is in response to the write bus cycle for the destination operand, then either a long- or short-stack frame will be created.

The bus error exception processing follows the general exception processing sequence. The status register is first copied into a temporary register, the processor enters into supervisory mode, and tracing is disabled. A vector number is automatically generated by the processor, which points to the bus error vector location in the exception vector table. In addition to this, the processor keeps an internal copy of the first word of the instruction being processed, and the address

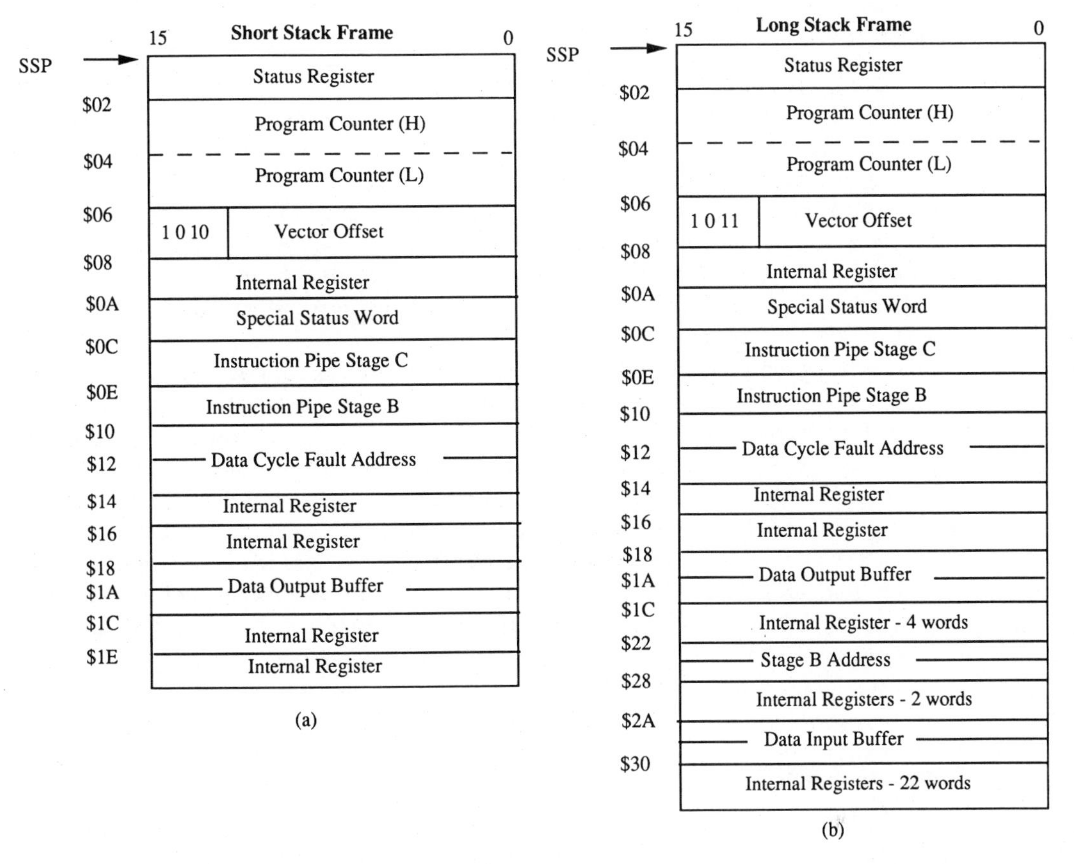

FIGURE 7.9. Bus error fault stack frame: a) Short stack frame; b) Long stack frame.

(Courtesy Motorola Inc.)

that was being generated by the aborted bus cycle. The processor also keeps a snapshot of the special status word on the stack frame at an offset of $A.

If a bus error occurs during the last step of exception processing, while the processor was either reading the exception vector or fetching an instruction, then the program counter will have the address of the exception vector. Although this information is not sufficient for a complete recovery from bus error, it does facilitate software diagnosis. Finally, the processor resumes instruction execution at the address contained in the vector. It is the responsibility of the "error handler routine" to clean up the stack and determine from where to resume the program execution.

However, if a bus error occurs during the exception processing of a bus error, address error, and read operation, then the processor is halted and all further processing ceases. This simplifies the detection of a catastrophic system failure, since the processor removes itself from the system, rather than destroying the memory contents. This situation is called a double bus fault. Only an externally generated reset signal can bring the processor back to an operating mode.

7.6.1 Short Bus Error

Because the MC68020 is capable of overlapping instruction execution with the bus controller, and thus, if the write bus cycle involves many 'WAIT STATES', then the sequencer may have been able to proceed further at the time that the bus error is recognized. A long-stack frame would be created, in such a case. It is also possible that the sequencer was at an instruction boundary when a bus error occurred, resulting in a short-stack frame. The SR, PC, format code, and vector offset are always stacked to the same position relative to the SP, for all stack frames. The format code of the short bus error stack frame is '1010'. A total of 32 bytes is saved in the stack frame.

The saved program counter represents the address of the next instruction of the instruction stream. When the faulted bus cycle was detected, the information in Pipe B and Pipe C are stacked at SP + 14 and SP + 12, respectively (the data into Stages C and D of the pipe are from memory locations at PC + 2 and PC + 4, respectively). The address on the bus is saved at SP + 16 when the faulted bus cycle occurred. If the fault occurs during a write bus cycle, then the data that the CPU was driving on the bus is saved at SP + 24. The contents of various internal registers are also saved in the stack frames. This information is not documented. The stack frame is also created during an address error, caused by the prefetch from an odd address. The vector offset would indicate an address error. An illegal address behaves much like a internally generated bus error.

7.6.2 Long Bus Error

The long bus error is created when the MC68020 is not at an instruction boundary and detects a bus cycle fault. The format code is '1010' and 92 bytes are saved on the stack. The PC represents the address of the instruction in execution when the fault occurred, which may not be the instruction that caused the fault. The long bus error stack frame contains the same data items as the short bus error stack frame, from the SP up to the SP + 32. The long frame holds additional information. The address where the information in Stage B came from is stored at SP + 36. The information in Stage C came from Stage-B's address minus 2. The stack frame location at SP + 44 holds a value representing the data input buffer.

When the RTE instruction detects a format code of '1010' in the stack frame, then the processor (in addition to checking the validity of the format code) also checks bits 12 through 15 of the word from location SP + 54. These four bits contain the version number of the CPU that created the stack frame. This number must match the version number of the MC68020 restoring the stack frame. This validity check is conducted to ensure that, in a multiple processor system, the information is properly interpreted by the RTE instruction.

If the version number does not match, a format error exception processing takes place. If it is inaccessible, bus error exception processing occurs, and an additional long-stack frame is created. The programmer of the service routine has two choices for completing faulted bus cycles. The RTE instruction can rerun the faulted bus cycle. The faulted bus cycle can also be emulated in software, and the faulted bus cycle skipped. The default case is that the RTE instruction rerun the faulted bus cycle.

7.6.3 Special Status Word

The format of the *special status word* is shown in Figure 7.10. It indicates whether the aborted operation took place during a read or a write operation, whether the processor was processing an instruction or not, and the nature of the instruction, including the status of the function code pin-outputs during bus error.

The special status word gets stored at location SP + 10 for both the short and long bus error stack frame, and provides detailed information useful for completing the faulted bus cycle entirely or not at all. The information in this special status word indicates the cause of the exception. That is, whether it is an instruction access or data access that caused the fault, or both. Twelve bits are defined in the special status word. The function code values are saved in bits 0 through 2. Bit 3 identifies the nature of the instruction. If bit 3 is set to 1, then the processor was processing a group 0 or group 1 exception. Bit 4 indicates the type of the bus cycle (read or write) and, if it is set, then the processor was performing a read operation, otherwise it was engaged in a write operation.

The fault Stage C (bit 15), and fault Stage B (bit 14), indicate that the CPU attempted to use the information in Stage C or Stage B of the pipe, and found a fault bit set. An address error exception will always create a stack frame, but

15	14	13	12	11	10	9	8	7	6	5	4	3	2	1	0
FC	FB	RC	RB	0	0	0	DF	RM	$R/\overline{W}$	SIZ2	SIZ0	NAT	FC2	FC1	FC0

FC2 - FC0 = Function Space

SIZ2, SIZ1 = Size Code for Data Cycle

$R/\overline{W}$ = Read/Write Data Cycle

RM = Read/Modify/Write on Data Cycle

DF = Fault/Rerun Flag for Data Cycle

RB = Return Flag for Stage B of the Instruction Cycle

RC = Return Flag for Stage C of the Instruction Cycle

FB = Fault on Stage B of the Instruction Pipe

FC = Fault on Stage C of the Instruction Pipe

NAT = Nature of Instruction

FIGURE 7.10. Format of the special status word.

with the fault Stage B and fault Stage C bits always clear. The rerun Stage C (bit 13) and rerun Stage B (bit 12) indicate whether the word in Stage C or Stage B is valid. If the rerun Stage C or Stage B is set, either the information is invalid or there is a prefetch pending to that stage. The RTE instruction checks the rerun Stage C and rerun Stage B bit to indicate whether or not to fetch the information for the stages of the pipe from the instruction stream, or to use what it finds in the stack frame. If, during the return from exception, the processor finds the rerun bit clear, it assumes that there is no prefetch pending for the stage, or that software has repaired or filled the image of the stage. The data fault, the F-bit, indicates whether a bus error occurred during a data access.

If the read/write bit is set, indicating a faulted read bus cycle, then the programmer may choose to place the correct data in the data input buffer and clear the data fault bit, or fix the problem in memory and leave the data fault bit set, letting the processor rerun the data access again during the RTE instruction. The first approach might be useful for systems providing virtual I/O capability [MTTA2 87]. For a faulted write bus cycle, the read/write bit would be cleared, and the data output buffer holds data to be written. The location is indicated by the fault address, and the address space is defined by the function code bits saved in the special status word.

7.6.4 Return from Exceptions During a Bus Fault

The data from the bus fault stack frame can be used to control the action taken by the processor during the execution of the RTE instruction. In the case of either an instruction fault or a data fault, if the programmer decides to let the RTE instruction rerun the faulted bus cycle, then the rerun bit is left set in the special-status word as it was. However, the problem that caused the bus fault must be fixed, so that the fault does not reappear when the processor reruns the faulted bus cycle. In a virtual memory environment, this may mean that the service routine must copy another page from the virtual memory, program the memory management unit to map the new page in memory, and execute the RTE instruction to resume the program execution. However, if no space was available for the new page, then the operating system would first have to determine which page to overwrite in memory. It would also have to determine whether the existing page in memory had been modified and, if so, update the old copy on the disk. It is not necessary for the programmer to alter the information on the stack. If the rerun bit was left as set by the processor, then the appropriate bus cycle will be rerun by the RTE instruction.

The only alternative is for the programmer to emulate the faulted bus cycle by executing explicit instructions to repair the image of the information on the stack and clear the rerun bit, so that the CPU does not rerun the faulted bus cycle during the execution of the RTE instruction.

The read-modify-write cycle requires special consideration. If the choice is made to let the RTE instruction rerun the faulted bus cycle, then the CPU begins from the first operand read access and the entire read-modify operation is rerun. If the choices are made to perform the access in software, then the entire operation must be completed, including repairing the condition code register and any data registers involved. The routine must read the Op-code of the instruction, determine whether it is test and set, compare and swap, or CMP2 and SWAP2 operands, and which registers are being used during compare and update values.

7.7 COPROCESSOR EXCEPTION PROCESSING

The F-line codes were not completely defined for the MC68020. For the F-line codes that are not defined, exception processing will occur and control will be transferred to the F-line routine for the purposes of emulation. The defined F-line instruction causes the MC68020 to initiate communication with the coprocessor whose identity is defined in the F-line Op-code.

If the coprocessor responds with $\overline{\text{DSACKx}}$, a dialogue is established. If the MC68020 receives a bus error when it initiates the coprocessor communication, an F-line exception occurs. This enables the system to be fault-tolerant. If the coprocessor is not in existence or is only partially implemented, a transition is made to a service routine to emulate the coprocessor actions.

The coprocessor interface uses vector number 13 for a coprocessor protocol violation. The coprocessor violation service routine is entered when the coprocessor informs the main processor that interface registers have been accessed in the wrong order. The protocol violation service routine is also entered when the MC68020 receives an undefined response primitive. The two Motorola-designed coprocessors use various vector table entries. The floating-point arithmetic coprocessor has been defined for the MC68020, and use vector numbers 48 through 54. These coprocessors are designed to return one of these vector numbers in the response register when an exception related to operations occurs. The Page Memory Management Unit (PMMU) uses vector numbers 56, 57, and 58, for reporting an exception-related operation.

Coprocessor exception processing will be described in detail in Chapters 10 and 11.

Chapter 8

Memory System for the MC68020

The bottlenecks in most von Neumann architecture have traditionally occurred because a processor could only read a single unit of information from memory during each access cycle. Therefore, to be able to match the processor's cycle time, a system would have to use very high-speed devices that, in most cases, could not be justified in terms of cost. A solution has been to configure a hierarchical or multilevel structure containing several types of memory devices with various cost and performance characteristics.

Performance can be affected by such interrelated factors as program behavior with respect to memory references, access times and sizes of each information transfer, memory management, and the memory hierarchy. We will discuss many of these concepts in this chapter, and examine how they are supported in the MC68020.

8.1 MEMORY HIERARCHY

The central issues in designing the memory system are [Stone 87]:

a) Bringing the information from the outside world into memory.
b) Buffering the information into the memory until required by the CPU.
c) Computing the output information and buffering it in memory until it can be forwarded to the outside world.
d) Transferring the outgoing information from memory to the outside world.

These problems led designers to define a hierarchical memory structure, as illustrated in Figure 8.1. Four layers are defined, with the top-most layer having

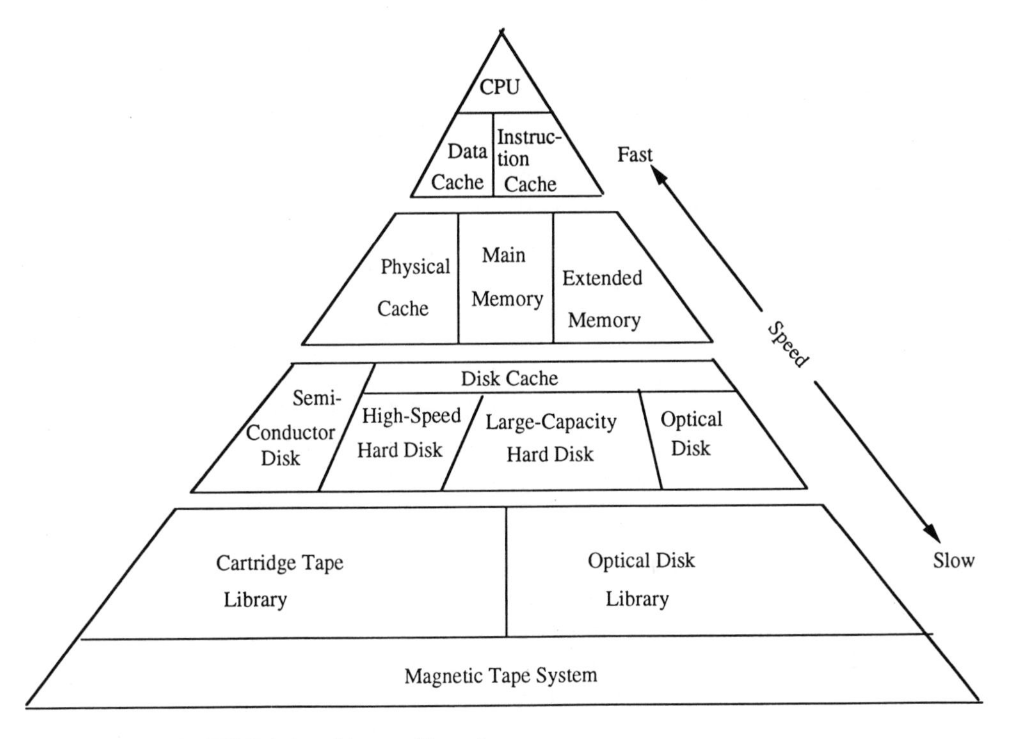

FIGURE 8.1. Memory hierarchy.

the fastest speed, and the lowest level having the slowest speed. The CPU also contains ROM written once when the chip is fabricated. It holds the micropro-gram that gets executed during instruction processing. This part of the memory may or may not exist in RISC processors. The CPU also contains registers, so that it can temporarily store results without going to the memory. The CPU can access these registers within one clock period.

The next level is that of on-chip cache memory, which may or may not exist in the CPU. Recently-used data and their neighbors are held here with the antici-pation that they will be used in the near future. Today's high-performance micro-processor-based systems always contain some form of cache memory. Multiple levels of cache memory are used to bridge the speed gap that exists between the CPU and the main memory. The main memory is the primary storage area for active data and programs. It may be organized as one linear memory block or may be distributed over several physical blocks.

The secondary memory is slower than the main memory. It is often used as an immediate backup to the main memory, and to implement the virtual memory technique. We have seen recent technological breakthroughs in the implementa-tion of the secondary memory. Disk technology has pushed the capacity near to thousands of gigabytes, and the optical disk will soon extend that capacity to nearly a trillion bytes.

The last layer of this architecture is designed to hold archive data. The 8mm cartridge tape library and optical disk library are often used to hold large amounts of data. High-capacity magnetic tape systems are also becoming commonplace in high-performance computer systems.

It is useful to understand the concepts behind the hierarchical memory system before we discuss the memory management architectural of the MC68020 microprocessor.

8.1.1 Properties of Program Locality

The computer system designer has observed the existence of program locality. It indicates that memory accesses are clustered within a small region of memory during any short period of time. Program locality has two aspects: *temporal* and *spatial.* The first, the locality of time, means that data that will be in use in the near future is likely to already be in use. This behavior can be expected from program loops, where both the data and the instructions are reused. The second property, the locality of space, indicates that portions of the address space that are in use generally consist of a fairly small number of individual contiguous segments of that address space. That is, the program's loci of reference in the near future are likely to be near the current loci of reference.

The characteristics of temporal locality have shown a strong tendency for program references to be grouped in time, and are responsible for the invention of virtual memory and the subsequent design of high-speed caches. Both of these exploit the properties of locality by storing a copy of the program in a special segment of memory. Virtual memory increases the size of the system by segmenting the program into pages, which are individually loaded from secondary memory into the main memory. A cache optimizes the CPU throughput by also storing a segment of program in a high-speed buffer that matches the speed of the processor.

8.1.2 Balancing the Memory Bandwidth

Once program behavior is understood, the main memory can be structured to optimize the performance of the processor. The goal in the design of an n-level memory hierarchy is to achieve a performance close to that of the fastest memory (Mc), at a cost close to that of the cheapest memory (Mp). The performance of the memory system is indicated by the effective hierarchy access time per memory reference, which depends on program behavior, the access time, the memory size of each level, the granularity of information transfer, and their management policies.

Four parameters are used to model the performance of each memory module within the hierarchy, which are:

W = Width of a memory unit (bits).
N = Number of words (bits/memory unit).

M = Total number of memory units.
T = Access time of each unit.

Thus:

Memory size = W*N*M (bits)
The total memory bandwidth = (W*M)/T (bit/sec)

Therefore, the cost of the memory is proportional to W*N*M / T.

The system designers want the size and bandwidth to be sufficiently large so that the system can manage programs and data in the most efficient manner. That is, the size must be big enough to hold sufficient programs and data to keep the CPU busy most of the time, without wasting too much time accessing the memory. Again, the bandwidth must be high enough to maximize the access of program, data, and I/O resources. Thus, W, N, and M are to be as large as possible, and T to be as small as possible, for good system performance. This is what the designers want to achieve during the implementation of memory hierarchies. The hierarchy allows them to store large quantities of information by spreading it out over slower devices, so that the cost per bit of storage becomes less and less.

The parameters N and T are technology dependent. Given that the system designers and users want large, fast memories, the parameters N and T are constraints imposed by physical properties and current manufacturing techniques. Again, parameters W and M are somewhat dependent on the microprocessor to be selected in the design and its overall architecture.

For example, the MC68020 has a cycle time of 50 nsec and can access 32 bits of data from memory during each bus access. Therefore: the maximum bandwidth of this microprocessor is 32/50 = 640 Mbit/sec.

In reality, the utilizable bandwidth will be lower, because the access time of the primary memory may be slower. The bandwidth will decrease, due to the sequential nature of the hard disk, floppy disk, or magnetic tape. Delays associated with the bulk memory are due to the latency time dictated by the mechanical limitations.

8.1.3 Optimizing the Hierarchy

Once the software behavior is understood, then the main memory can be organized to optimize processor performance. Thus, the effective access time is the sum of the average access time in each of the levels of the hierarchy, and is given by:

$$T_e = t_1 + t_2 + \ldots + t_k + \ldots + t_i$$

where T_e is the effective access time from the processor to the ith level of the hierarchy, and t_k is the individual average access time at each level. In general, t_k includes not only the wait time caused by memory conflicts at level k, but also the delay in the interconnection network between the levels k and (k−1). The probability of this conflict occurring depends on the number of processors present in the system, the number of memory modules at each level, and the interconnection network connecting the processors and memory modules.

To model the performance of a hierarchy, we assume that the probability of finding the information in the memory of a given level is characterized by a success function S. Thus, S depends on the granularity of the information transfer, the capacity of memory at that level, the management policy, etc. However, for some class of management strategy, S is most sensitive to memory size. Because copies of information at the highest hierarchical level are assumed to exist in the levels below that level, the probability of finding the data at the highest levels is F = 1 − S, where F is the miss ratio. Therefore, in a two-level system, effective access time would be equal to:

$$T_e = S^* t_{k1} + (1-S)t_{k2}$$

$$\text{where } t_{k1} = t_k \text{ at Level 1}$$
$$t_{k2} = t_k \text{ at Level 2.}$$

If the hierarchy consists of one level of infinite size, then the probability of accessing this data at Level 1 is 100%. However, memory size greatly impacts the probability of finding data at a given level. Thus, the probability at each level is expressed in terms of success and miss ratios.

For example, to find the effective access time for a two-level hierarchy, if the success ratio at Level 1 is 0.99, then the probability of finding the data at Level 2 is 1−0.99 = 0.01. Thus, the effective access time is:

$$T_e = 0.99^* t_{k1} + 0.01^* t_{k2}$$

Again, a 10% decrease in the success ratio (0.99 to 0.98) almost doubles the effective cycle time, and divides the net performance in half when the cycle time ratio is 10. If the cycle time ratio is 20, then that same 10% decrease increases the effective cycle time by almost a factor of 4.

A large memory structure with only one level is very expensive for many systems. A multilevel organization is more cost effective. Therefore, the goal is to organize the hierarchy so that the highest performance can be achieved at the lowest cost. Since the success ratio is a function of the memory size at each level, the implication is that the larger the memory at a given level, the higher the success rate at that level [NEC 91].

8.2 FUNDAMENTALS OF MEMORY SYSTEMS

In order to understand the memory system and how the processor accesses it, we need to understand a few other concepts first. One of them is the physical location, and another is the logical address. The memory is organized as a set of storage locations numbered sequentially, starting from location zero, up to the maximum available size. The number associated with one of these physical locations is known as the *physical address,* and the set of all physical addresses is commonly called the *physical address space.*

The *logical address* on the other hand, is an address in an instruction, an address viewed by a program. Thus, the *logical address space* is the collection of all logical addresses that can be referenced by a program. The organization of the logical address space is defined as the *memory architecture.*

The organization of the physical address space is determined by the memory technology of the time, and the designer's notion of future technology, and the cost, performance, and various other attributes. However, the organization of the logical memory is determined by the structure of the programs that will run in memory. In the early microprocessors and, in fact, for all early computers, the logical address space was identical to the physical address space.

8.2.1 Linear Memory

The most common architecture for memory is a linear contiguous address space, where the address starts at location zero and proceeds in linear fashion, with no gaps or breaks, to the upper limit imposed by the total number of bits in the address bus of the processor. This model is shown in Figure 8.2, where a program often consists of several procedures/subroutines and all of the data located within this single address space. Here, the logical address space of a linear memory has the same basic organization as the physical memory. The simplest implementation of linear address space can be found in the IBM PC (Intel 8088 CPU-based). Here, the processor can generate 65,536 distinct addresses. The addresses are directly utilized by the memory hardware to locate and fetch data from memory devices.

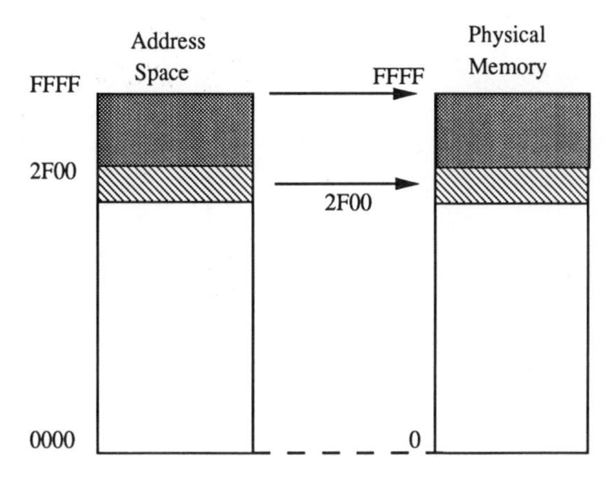

FIGURE 8.2. Linear memory.

The limitations of linear address space become apparent when several programs need to share the same machine in order to maximize the utilization of the system's resources. First, nothing prevents one program from accessing or even overwriting data in another program. This is possible because all programs share the same address space. Any program can potentially access any location in the memory, and no program is protected from unauthorized access by another program. Second, all programs must fit inside the relatively small address space.

Figure 8.3 illustrates the problems that pertain to uncontrolled access to this simple memory organization. In this example, nothing blocks the Program A from overwriting Program B, or another part of its own code or data. A trivial programming mistake can destroy any program. Under such a memory organization, even complex software partitioning schemes fail to guarantee the security of the execution of multiple programs. To overcome these liabilities, logical-to-physical address translation mechanisms (i.e., memory mapping) for memory management was invented.

8.2.2 Mapping of Linear Memory

Mapping is the process of translating logical addresses into physical addresses. In the previous example, the logical addresses are simply equated into physical

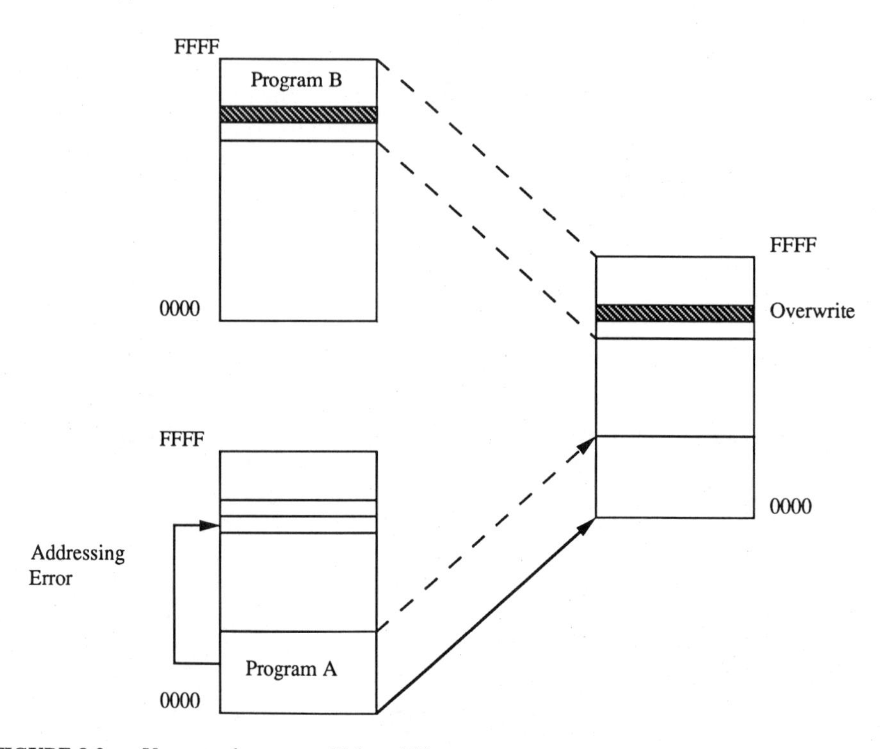

FIGURE 8.3. Unmapped memory: Vulnerability.

addresses. However, mapping can be implemented in such a way that the logical address can be assigned to any arbitrary physical address. Thus, mapping is a mechanism for relocating the logical address space within the physical address space. For example, the entire logical address space of a program with locations 0000H–FFFFH can be mapped into physical locations 20000H–2FFFFH. Now, a program referencing a logical location 2F00H actually fetches data from physical location 22F00H (i.e., 20000H + 2F00H). This is illustrated in Figure 8.4.

Mapping is an extremely powerful tool for the implementation of multiprogramming systems. It was first developed for mainframe computers, then found its way to minicomputer, and today it is an integral part of any high-performance microprocessor. Under mapping strategy, a program's logical address space is completely independent of any other program's logical address space. As a result, many programs can share a physical memory without the risk of interfering with each other. Mapping logic transforms all logical address spaces into physical address spaces, and provides the required protections. However, the transformation process is completely transparent to the programs or to the programmer. Figure 8.5 shows a multiprogramming mapped environment.

8.2.3 Page-Based Mapping

It is obvious from Figure 8.5 that mapping the entire logical address space into physical space is an inefficient technique. This mapping technique generally leaves holes or unused spaces in the physical memory, because programs cannot be fitted into them [Deitel 83]. The unused memory spaces cannot easily be com-

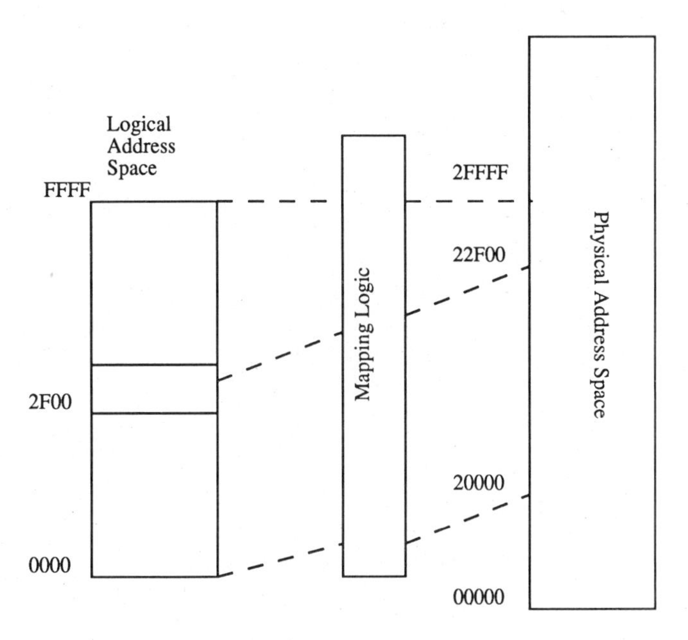

FIGURE 8.4. Mapping technique.

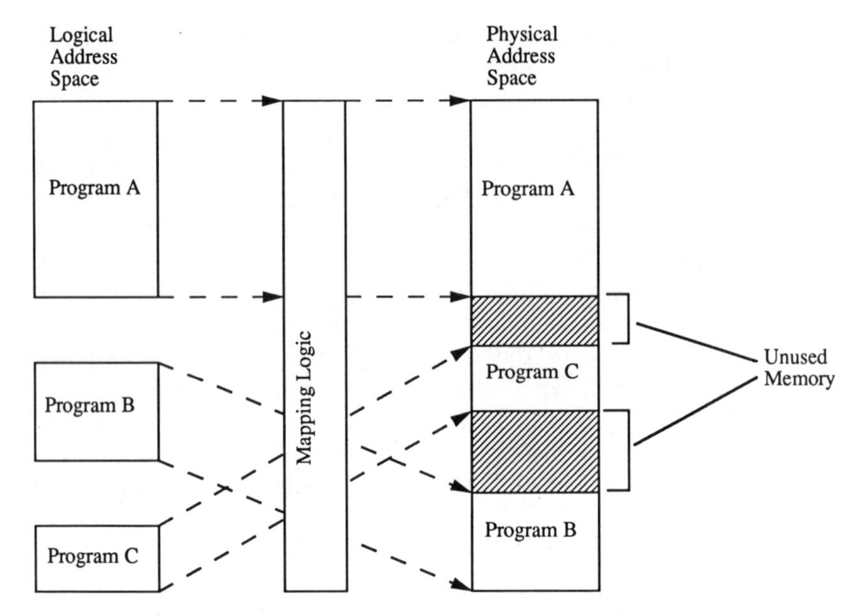

FIGURE 8.5. Mapping technique in multiprogramming.

bined to create a sufficiently larger space so that another program can be fit into that space. This is solved by *page-based mapping,* where the logical spaces are divided into fixed-size blocks called *pages.* The physical address space is also divided into pages of equal size. A large program need not be relocated into one contiguous chunk of physical memory, which might be difficult to find in a multiprogramming environment. However, it is easier to find several blank pages of memory in different areas of memory. Thus, it is more feasible to find ten blocks of 1 K pages than one block of 10 K physical memory.

8.2.4 Access Rights

The page mechanism also includes features for memory protection within a logical address space. Attributes can be assigned to each page, which indicate how the pages are to be utilized. The attributes can allow read-only, read and write, or they can prevent any access at all. They may also carry timing information utilized by the memory management algorithm.

8.2.5 Segmented Memory

The main motivation behind segmented memory is that programs are not written as one linear sequence of instructions and data, but rather as parcels of instructions and parcels of data. For example, an application can consist of a main code section and various separate procedures. Data could be organized into arrays or tables, linked lists, or any number of complex data structures. Again, these blocks of codes or data could come in different sizes.

The segmented memory architectures were developed to support this program structure. The logical address space is broken down into many linear address spaces, each with a specified size and length. These linear address spaces are called *segments.* An item within a segment is accessible via a two-level mapping process. The first component, *the segment selector,* identifies the segment itself. The second component, *the displacement,* specifies the offset from the base of the segment to the item being selected. Now each segment can be used to accommodate a program or data module. Thus, a program can have the main procedure in a segment, each additional procedure in its own segment, and each major data structure in its own segment. Here, the logical address space reflects the logical organization of the program, as shown in Figure 8.6.

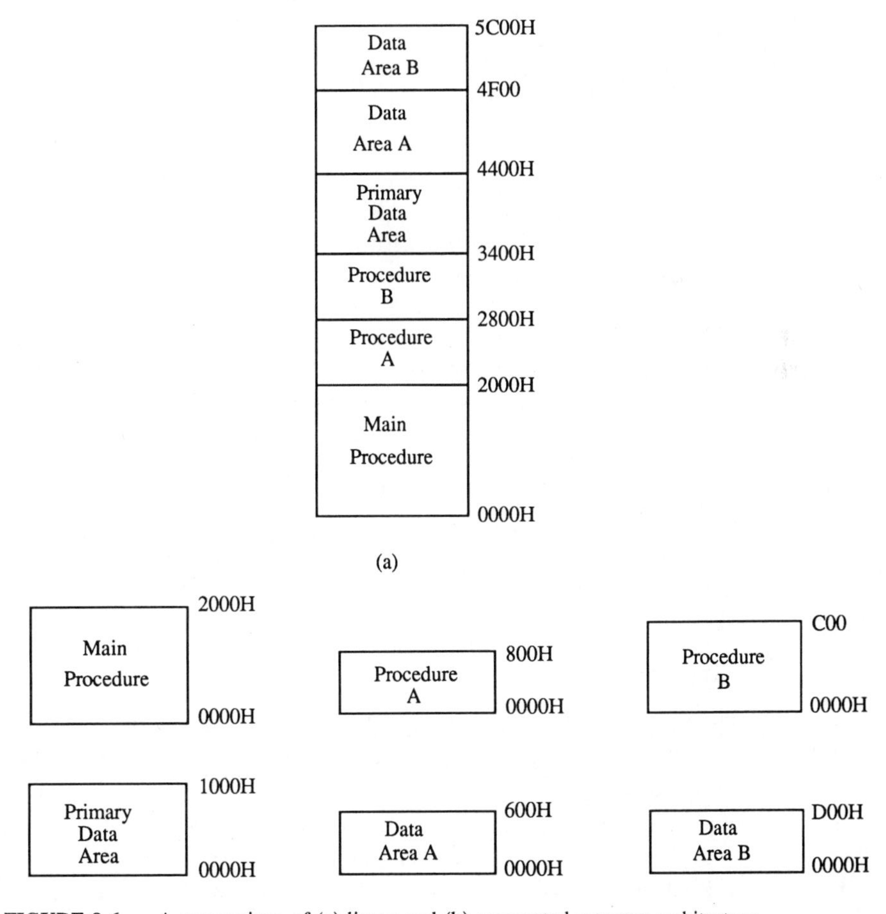

FIGURE 8.6. A comparison of (a) linear and (b) segmented memory architecture.

In contrast, the linear address space is, by definition, logically structureless. The protection mechanisms are usually based on fixed-length pages, whose size is determined by the underlying hardware, and which have no necessary relationship to the logical structure of the programs.

8.3 VIRTUAL MEMORY SYSTEM

Due to the cost and physical size constraints of memory devices, designers generally cannot create a physical memory equal to the size of the logical memory. They utilize a *virtual memory* mechanism to get around the limits of the physical memory size. In a virtual memory scheme, it appears to users as if the entire logical address space is available for storage. But, in fact, only a small portion of the logical address space is mapped onto physical space for any given instance. The rest of the portions are not present in the main memory at all. Instead, they are stored in secondary storage, such as a disk, whose cost per bit is more economical.

8.3.1 Implementation of Virtual Memory

The key to the virtual memory concept is in disassociating the addresses referenced by a running program (task) from the address available in the physical memory. This concept is shown in Figure 8.7. It appears to the program that it has all of the required memory for its execution. The main memory contains data and instructions to be referenced by the program. The secondary memory holds the material that does not fit into the main memory. The virtual address generated by the CPU gets mapped by the hardware/software technique to locations, if they are present in the main memory. Otherwise, a memory fault is generated and program execution is suspended until the desired data is brought into the main memory from secondary memory.

The mapping is performed via software, or by special-purpose hardware called a Memory Management Unit (MMU). Today, most high-performance systems support the virtual memory technique using an MMU that is either a part of the processor itself, or is placed outside the CPU and the processor utilizes a special protocol to communicate with it. Table 8.1 illustrates the memory organization supported by various 16/32-bit microprocessors.

8.3.2 Block Mapping

The basic intent of memory mapping is to transform the larger address space of the CPU into the smaller memory area normally available in the system. If the mapping is implemented on a word-by-word or byte-by-byte basis, then the mapping information may require as much or more memory than the executing programs themselves. Therefore, information is grouped into *blocks,* and the system keeps track of the locations of various blocks. The larger the block size, the smaller the fraction of real storage to be assigned for storing the mapping infor-

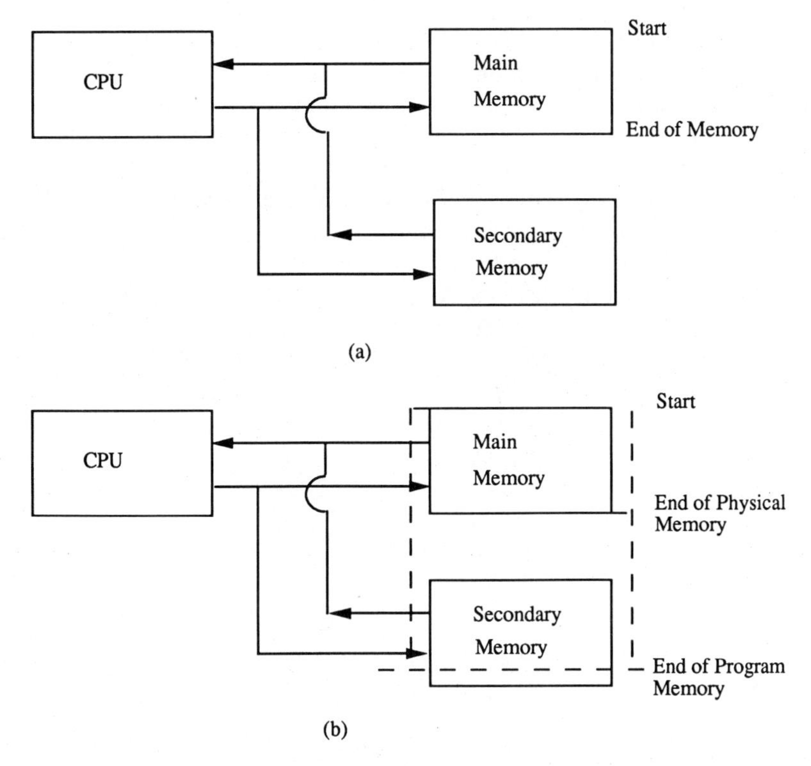

FIGURE 8.7. Virtual memory concept. A system (a) without and (b) with virtual memory.

TABLE 8.1. Memory Organizations Supported by Various Popular Microprocessors.

Processor	Linear	Segmented	Paged
Intel:			
80286, 432	yes	yes	no
80386, 80486	yes	yes	yes
Motorola:			
68000, 68010	yes	no	—
68020	yes	no	+MMU
68030, 68040	yes	no	yes
National:			
16032, 32032	yes	—	+MMU
Zilog:			
Z8000	yes	yes	—
Z80,000	yes	yes	—
AT&T:			
32100, 32200	yes	no	+MMU
NCR:			
NCR/32	yes	—	—

mation. However, larger blocks take longer time to be swapped between the main and secondary memory.

The block mapping or direct mapping technique is shown in Figure 8.8. Here, a program specifies the block where the item resides, including the displacement of the item from the beginning of the block. The algorithm works as follows. The operating system or specialized software maintains a Block Map Table (BMT) for each executing program in memory. This table contains one entry for each block of the program, and entries are kept in sequential order. A specific register (the block table origin register)) within the CPU is loaded with the physical address of the BMT. The block number, B, is added to the base address, *a,* to find an entry to the BMT. This entry contains a real address, *b,* for the specified block in the memory. The dispalcement, D, is then added to the block start address, *b',* to determine the real address ($r = b' + D$) of the data in the memory. For efficient execution of the program, the mapping must be performed dynamically.

Under paging implementation, the block size becomes equal to the size of the page. The BLT is called as the Page Description Table (PDT) and additional information, such as the page-resident bit (r), and time bit (t), etc. are added for

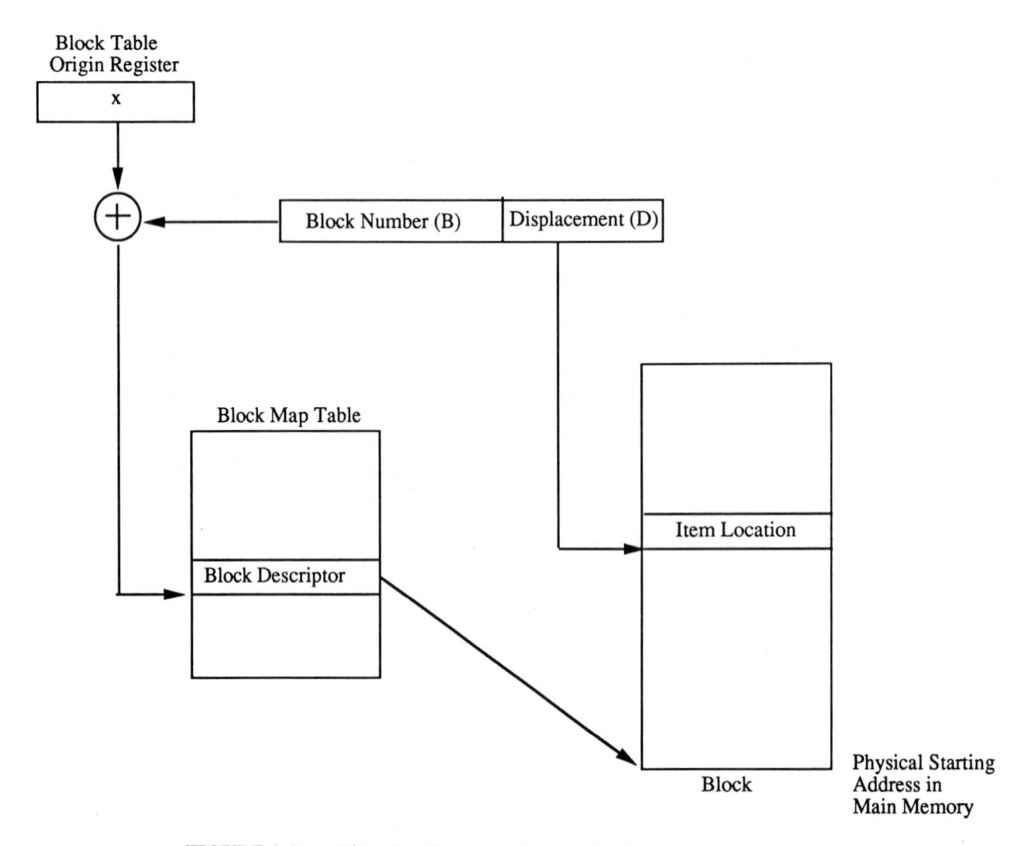

FIGURE 8.8. Virtual address translation with block mapping.

efficient management of the system's resources. The *r* bit indicates whether a particular page is in the memory, and *t* indicates when and how long the page has been in the memory.

8.3.3 Direct Mapping for Paging

The address translation mechanism can be accelerated by placing the entire PDT into primary memory, which has a faster access time. Here, the entries of the PDT are searched to determine the physical address. The technique is shown in Figure 8.9.

An executing program generates a virtual address (p, d). The entries within the PDT are searched for page p. If a match is found, then it returns p' as the frame number, corresponding to virtual page p, and the p' value is added with the displacement, d, to form the real address r, which corresponds to the virtual address generated by the program. In the event of a no match, a conventional direct mapping technique is invoked. The address, a, from the page origin register is added to p, to identify the appropriate entry for page p in the direct mapping page map table in the main memory. The table indicates that p' is the page frame corresponding to virtual page p, and p' is concatenated with displacement d, to form the real address, r, corresponding to the virtual address generated by the running program. To reduce the access time, the PDT is generally placed in cache memory.

Due to the cost, designers generally put only a small portion of the complete

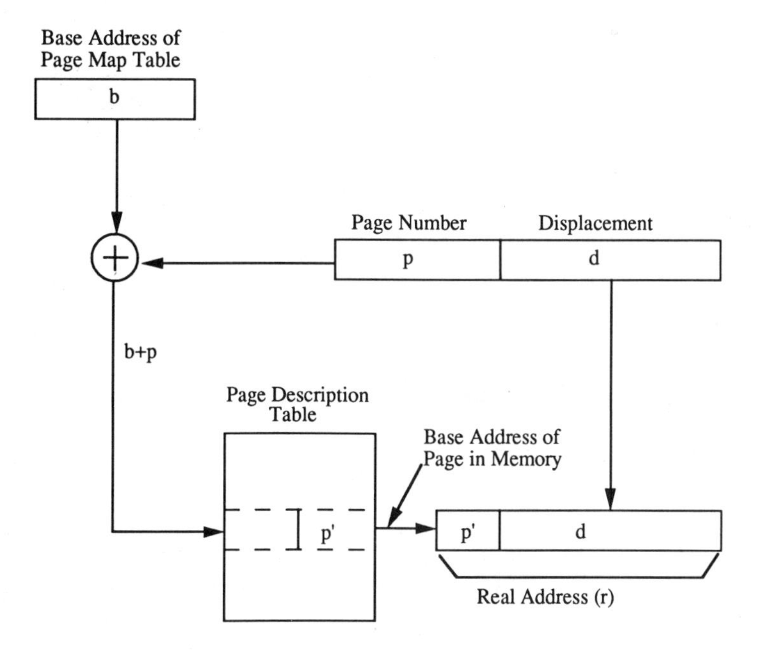

FIGURE 8.9. Address translation by direct mapping for paging system.

PDT into cache memory. The most recently referenced page entries are kept in the cache, which is based on the premise that a page referenced during the recent past is likely to be referenced again in the near future.

8.3.4 Dynamic Address Translation Using Multilevel Mapping

The new generation of 32-bit microprocessors have a large addressing capability. The virtual memory technique allows any program to become as large as 2^{32}. If the system designer decides to create a page size of 2K, then the PDT can potentially have 4,194,304 entries. To overcome this problem, designers resort to multilevel mapping. The idea is to break up the large field into smaller fields. The technique is shown in Figure 8.10.

The virtual address is divided into three logical groups. A common technique is to designate the higher-order field as a segment number (*s*), the next field as a page number (*p*), and the least significant field as the displacement (*d*). For example, in case of 32-bit micros, the least significant 10 bits are used as the dis-

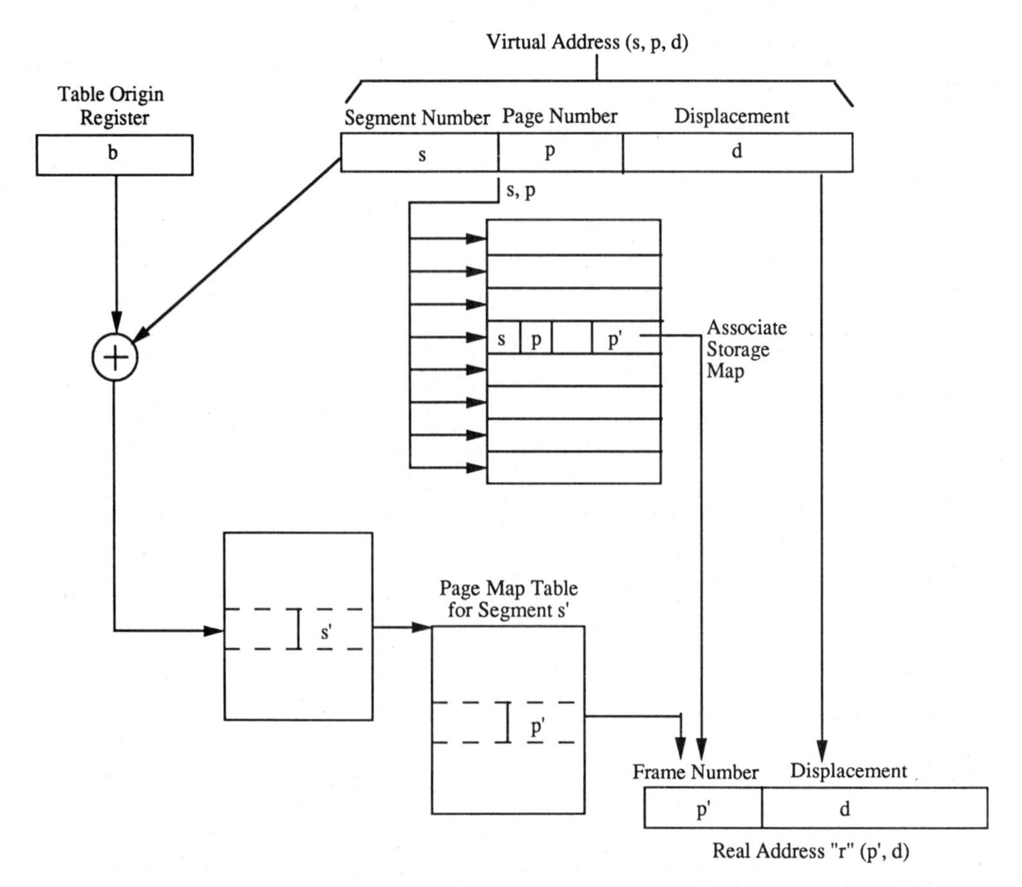

FIGURE 8.10. Virtual address translation with a multiple-level hierarchy.

placement. The remaining 22 bits are broken into an 11-bit segment number and an 11-bit page number. Thus, a very large page table has been broken into many pieces, each not larger than 2,048 entries. The smaller tables need not reside in the memory. Hence, the memory only contains the active portions of the page table.

The virtual address generated by the running program is given by three fields (s,p,d). As before, the most recently referenced pages will have entries in the memory called associative memory. An associative search is launched to locate (s,p). In the event of a match, the page frame, p', is obtained from the PDT. The p' is then added to the displacement field, d, to determine real address r, corresponding to virtual address $v = (s,p,d)$.

The disadvantages of two-level mapping is the second level of lookup. Moreover, both levels can have page faults during lookup. The translation process may fail for various reasons. For example, a segment table search for a match may fail to indicate that segment s is not in the main memory. This will cause a memory fault, and control will be handed over to the operating system to locate the required segment in the secondary memory. The operating system may have to replace an existing page that belongs to another program, in order to make room for the incoming segment required for the execution of the current program. The same situation may arise during the reference to the PDT. This would cause a page fault, and the operating system would gain control, locate the page in secondary memory, and load the page, after swapping another page belonging to another program.

8.3.5 Instruction Restartability During the Context Switch

Since an instruction might go through several steps of its execution before encountering a page fault, finding a mechanism to restore the machine state and resume execution becomes very complicated. Two techniques are generally utilized to support the 'state save' and 'state restore' process. They are a part of instruction restart and instruction continuation mechanism.

Under the instruction restart strategy, the original values of any internal registers or memory locations must be restored to their previous condition, so that execution of the instruction can be resumed from the start, after the cause of the page fault has been corrected. There are some costs associated with the restart method. Certain addressing modes are extremely difficult to correctly preserve and restart (e.g., memory indirect addressing). Another problem is related to I/O processing. For example, if a long-word memory move is taking place that crosses a logical page boundary, a fault will occur and a new page needs to be swapped. Unfortunately, if the destination long-word of memory overlapped the original source, then some of the original good data will be destroyed during the overwrite. Thus, the completion of the memory transfer will leave a partially duplicated original long-word in the destination location. This problem is commonly known as the operand overlap problem.

In *instruction continuation,* the execution of an instruction will be suspended, and the internal states are saved, before the CPU initiates an exception process (i.e., tracking the problem). The initial state is restored at the completion of the repair. The access is initiated for the second time and the execution resumes. The main drawback is that the page fault handler must be able to resume a suspended instruction in midstream; so that parts of it will not appear to have been executed twice, as previously alluded to with the operand overlap problem.

In the case of the MC68000 family, the breaks in the microroutines allow for the definition of boundary conditions, which are required to check the validity of the machine before acknowledging a fault and initiating exception processing. In reality, this behaves like an instruction restart at the microinstruction level. The MC68010 and the next generation of processors follow an algorithm during the implementation of the fault detection and correction mechanism. The steps of this algorithm are:

- Identify a fault
- Enter into exception processing
- Preserve the internal states
- Correct the fault
- Restore the internal states
- Resume processing

8.4 CACHE MEMORY

The idea of cache memories is similar to that of virtual memory, where some active portion of the main memory is stored in duplicate in a high-speed cache memory. When the CPU generates a memory request as a result of executing a program, the request is presented to the cache. If the cache fails to respond to the request, then this event is called a *cache miss.* The request is then forwarded to the main memory. If the main memory fails to respond, then a page fault will occur and the request will be forwarded to the secondary memory. Thus, the difference between the cache and the virtual memory is a matter of implementation. However, cache implementations are totally different from virtual memory implementations, because of the speed requirements of a cache.

The primary motivation behind cache memory is to improve the system throughput by eliminating the speed differences that exist between different levels of the memory hierarchy. The more global objectives are to optimize the system throughput within certain cost, size, physical dimension, and power limitations. Thus, the main goals are:

1. To minimize the probability of not finding a target memory reference in the cache (the miss rate).
2. To minimize the time to access instruction/data that resides within the cache.
3. To minimize the delay time, in the event of a miss.
4. To minimize the overhead of information in the main memory and maintain instruction/data coherency.

Designers take various approaches to achieve these goals. They are constantly making trade-offs with parameters like:

- The size of the cache memory
- The placement policy
- The memory update policy (write-through vs copy-back).
- The positioning technique (real and/or virtual cache)
- The coherency policy
- The single vs split implementation
- The replacement scheme

8.4.1 Operation of a Cache

The CPU can have access to data in two ways (Figure 8.11).

In the first, the CPU checks the cache for the desired data. In the case of a *hit,* the data is accessed without any additional delay. Here, the effective access time, Tacc, and the access time of the cache, Tcache, become equal. That is:

$$\text{Tacc} = \text{Tcache}$$

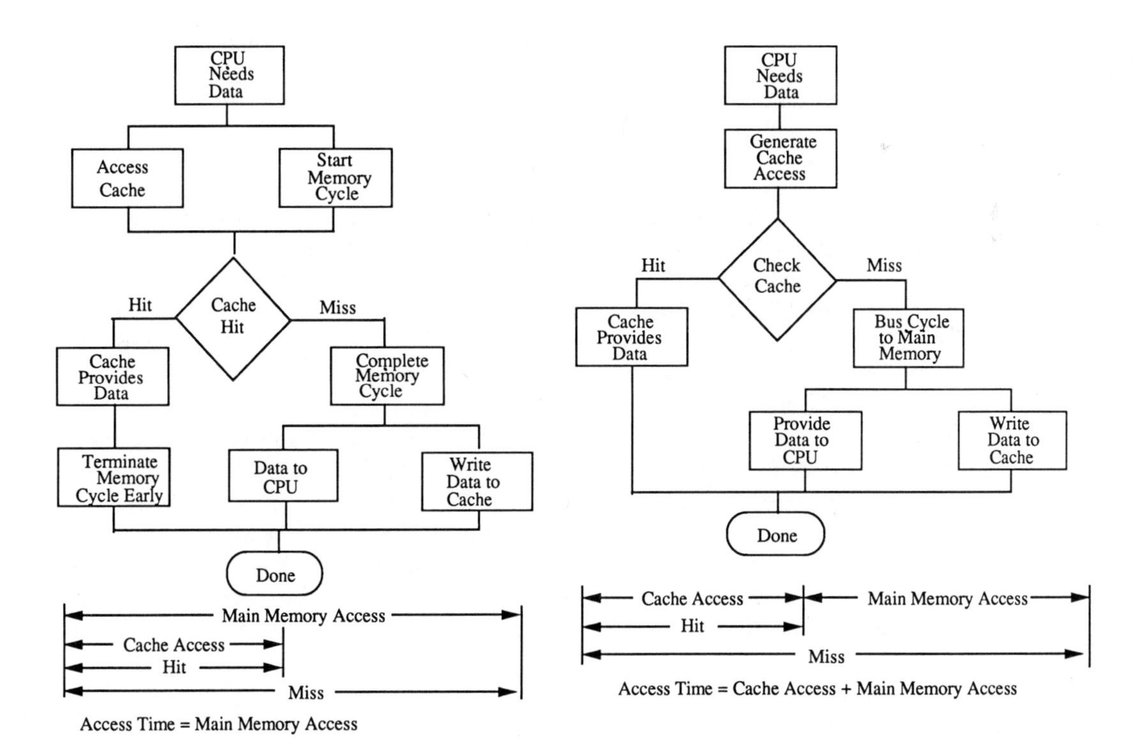

FIGURE 8.11. Cache access strategy. (Courtesy Motorola Inc.)

In the event of a *miss,* the main memory has to be accessed, and the effective access time becomes:

$$Tacc = Tcache + Tmain$$

where Tmain is the access time of the main memory.

Using the second method, the CPU presents the address simultaneously to both the cache and the main memory. If there is a hit, data is forwarded to the CPU within the access time of the cache (Tcache), and the main memory cycle is terminated. On a miss, the data access is determined by the main memory access time (Tmain). That is:

$$Tacc = Tmain$$

The *hit rate* is defined as the percentage of memory references that are matched within the cache. The hit rates are affected by the size, the physical organization of the cache, the cache updating algorithm, and the reference pattern of the program being run. Thus, the hit rate is a measure of the success of the cache system. The effective access time is expressed as:

$$Tacc = h * Tcache + (1 - h) * Tmain$$

where h = the hit ratio.

If the main memory is 10 times slower than the cache, then a decrease in the hit ratio from 0.99 to 0.98 results in an increase in Tacc of roughly 10%. Thus, a small change in the hit ratio is amplified by the ratio of main-memory cycle time to the cache-memory cycle time [Stone 87].

8.4.2 Cache Bandwidth

The cache bandwidth is the rate at which information can be transferred into and out of the cache. The rate can be enhanced by increasing the data path width, interleaving the cache for concurrency, and decreasing the access time. However, the wider the bus, the faster the data transfer rate. The number of fetches required to load a block of a given size depends on the bus width. Interleaving the cache can keep the bus width low, while maintaining the bandwidth.

Again, the bandwidth can be increased by reducing the amount of data requested to execute a cache miss cycle. This can be accomplished by the *burst transfer mode.* It requires a single address for each 16 bytes of data, rather than separate addresses for each 4 bytes of data. Burst data transfers are implemented in high-performance microprocessors such as Intel's 80486, the MC68040, NEC's V80, etc. Burst mode allows a 16-byte cache block to be transferred during a cache miss, minimizing the cache miss data transfer time and increasing the system bus bandwidth.

8.4.3 Cache Partitioning

A cache may be partitioned into several independent caches in order to segregate various types of references. Splitting the cache generally improves its bandwidth and access time. In a pipelined system, the processor is usually partitioned into two units, namely the I-unit and the E-unit.

The I-unit performs instruction fetch and decode. It then forwards the decoded instruction into the E-unit, which executes the instruction. By splitting the cache into two units, the Data (D) and Instruction (I) caches, designers will be able to physically place the instruction cache next to the I-unit, and the data cache next to the D-unit. This permits simultaneous access, and reduces access time. It is interesting to note that the MC68020 utilizes only the D-unit, whereas the MC68030 and MC68040 architecture supports both the D-unit and the I-unit.

8.5 CACHE ORGANIZATION

The cache memory is organized into two parts:

- The Cache Director (CD)
- The Data Storage Memory (DSM)

The DSM is partitioned into a number of equal-sized blocks, called 'cache blocks.' The directory records are composed of block tags and access control bits. The address tags contain the block addresses of the blocks that are currently in the cache. This is shown in Figure 8.12, with an example where the reference

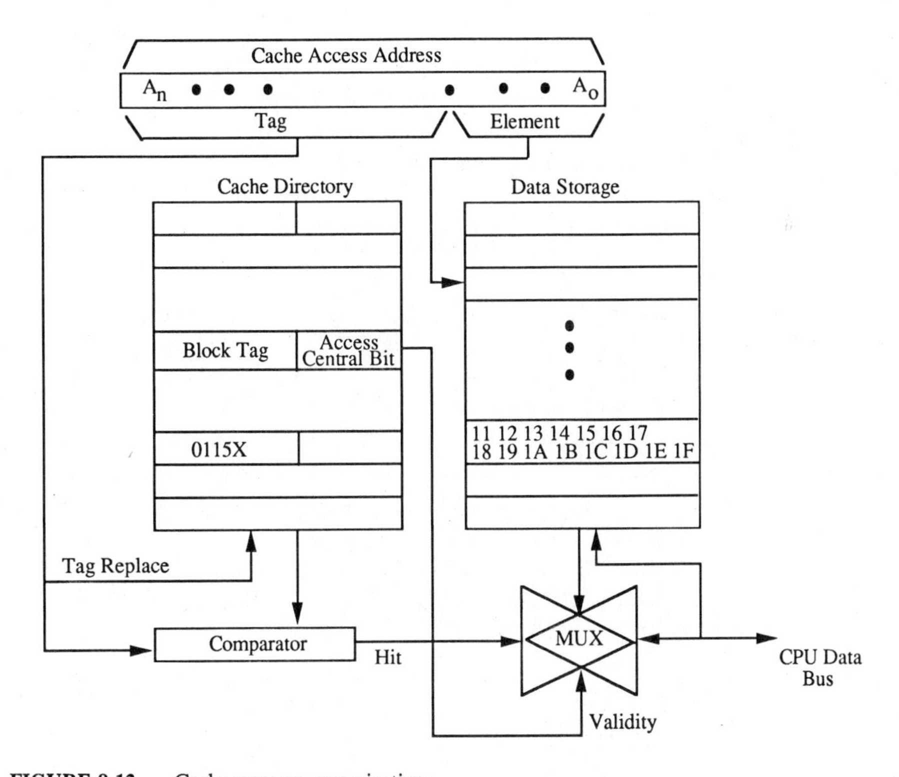

FIGURE 8.12. Cache memory organization.

to address 0115A matches the tag 0115X, where the X represents any hex digit. Since there is a match, the desired element resides in the cache. The data associated with tag 0115X have an address range of 01150–0115F. Therefore, the access must be made to the tenth element, whose address is 0115A. This element, which has a value of '1A', is copied to the data register of the cache and compared against the validation bits, before being forwarded to the CPU for processing.

In this example, an address is mapped into a cache entry by a technique referred to as associativity. Cache associativity is divided into two classes:

- Single set associativity or direct mapping
- Full associativity

8.5.1 Single-Set Associative Cache (Direct Mapping)

If a given address can be mapped into only one location within the cache, then the cache is considered to be direct mapped (single-set associative). This is illustrated in Figure 8.13. Here, a memory address is divided into three fields. The *Element field* indicates the word within a block, the *Block field* indicates the cache block that may contain the program, and the *Tag field* is used for comparison against the tag of the block in the cache to determine if there is a match (hit). Since the block field of the access address points to a unique cache location, a direct-mapped cache needs only one comparator to determine whether the tagged location contains valid information or not.

Single-set associative mapping is the simplest and least expensive cache organization. However, if two or more program blocks, used alternatively, happen to map onto the same cache block, then the cache hit ratio will drop drastically. However, the effect of this problem can be minimized through innovative design techniques.

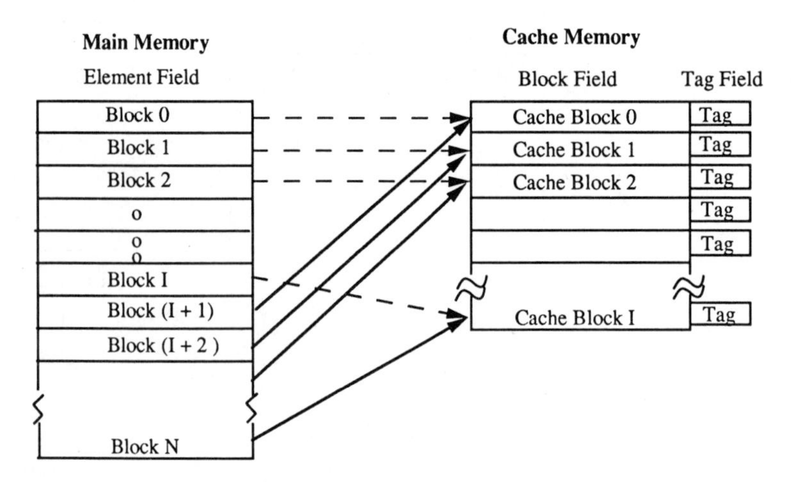

FIGURE 8.13. Single-set associative cache organization.

8.5.2 Fully Associative Cache

The counterpart of the single-set associative cache is the fully associative or content-addressable cache. In this organization, any program block can potentially be mapped into any block [Hwang 84]. A memory address is divided into two fields, as illustrated in Figure 8.14. The *Word field* indicates the instruction word within a block, and the Tag field is compared against the tags of all blocks in the cache to determine if there is a hit. A content-addressable memory is utilized for this purpose. It requires a comparator for each cache entry, in order to perform compare operations in parallel. The mapping flexibility permits the development of a wide variety of replacement algorithms. Although the fully associative cache eliminates the high block contention, it encourages longer access time because of the associative search. From the hardware design perspective, it is very expensive to implement this technique.

8.5.3 Set Associative Cache

The set associative organization represents a compromise between direct and fully associative mapping. In this scheme, shown in Figure 8.15, the cache memory blocks are divided into S sets, with N block frames per set (N = M/S, where M = the total number of block frames in the cache). A block 'I' in memory can be in any block frame belonging to the set 'I modulo S'. Various schemes are devised for mapping a physical address into a set number. The simplest and the most common is the 'bit-selection' algorithm. Here, the address is logically divided into three fields. The Word field, indicates, as before, the word within a block. The Set field identifies the set of cache blocks that may contain a referenced program block. The Tag field is compared against the tags of all blocks within the set, to determine if there is a hit or miss. The set size dictates the

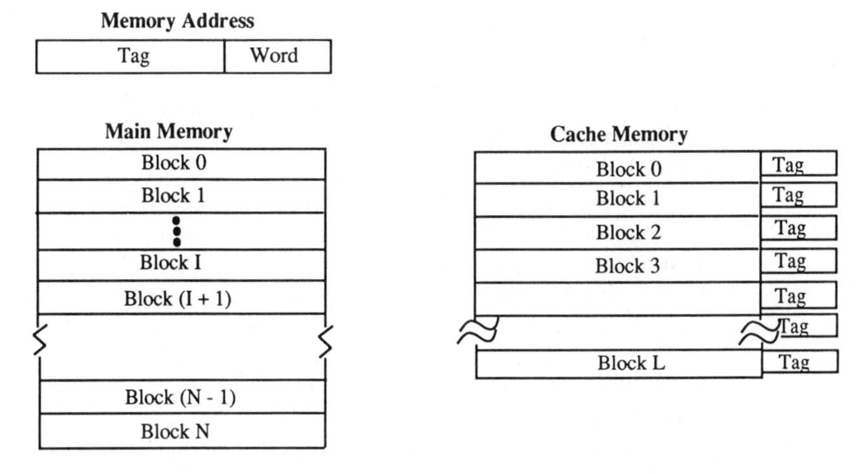

FIGURE 8.14. Fully associative cache organization.

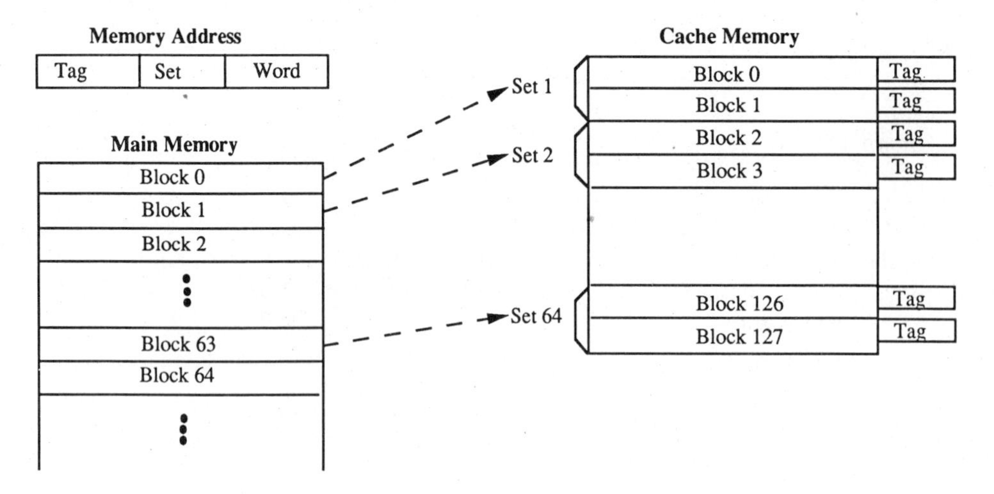

FIGURE 8.15. Set-associative cache organization.

number of simultaneous compare operations to be performed. A comparator is utilized for each set, and comparisons are performed on the set in a sequential manner.

This technique reduces the number of required comparison circuits. It also helps to minimize excessive replacement of cache entries (known as trashing) caused by programs that tend to access different operands frequently, operands whose addresses correspond to the same location in a direct-mapped cache.

8.5.4 Cache Coherency

The cache contains a replica of a portion of the main memory. If one copy is altered and the other is not, then two different sets of data become associated with the same address. The system must include some mechanism to ensure that this does not happen. In order to maintain data integrity, it is important that the main memory be modified as well. Two basic approaches are available to achieve this goal, which are: *write-through,* and *copy-back.*

In the write-through strategy, all memory write operations are performed in cache and in the main memory simultaneously. This simplifies the updating policy, because the main memory always contains the recent copy of the data. Each write operation is treated as a cache miss, and the CPU must wait for the main memory to be updated. This problem can be solved by a technique known as *buffered write-through.* In this technique, the write access to the main memory is buffered, so that the CPU can begin a new cycle before the write cycle to the main memory is completed. When a write access is followed by a read access that is a cache hit, the read access can be performed while the main memory is being updated. However, if a single write access is buffered, two consecutive writes to the main memory will force the CPU to wait. Again, a write followed by a read miss will also require the processor to wait.

Under the copy-back technique, the cache is continuously updated, and the main memory gets updated at a later time. The main memory gets updated when data leaves the cache (i.e., when a miss requires that a cache location be updated or when the CPU performs context switching). Here, the Tag field of each block in the cache includes a 'bit-called altered bit'. This bit is set if the block has been written with new data and therefore contains data that is more recent than the corresponding data in the main memory. Before overwriting any block in the cache, the cache controller checks the altered bit. If it is set, the controller writes the block to the main memory before loading new data into the cache.

The copy-back is faster than write-through, because the number of times an altered block must be copied into the main memory is usually less than the number of write accesses. However, the copy-back technique has some limitations. First, the cache controller logic for copy-back is more complex than for write-through. During the memory update, the controller must reconstruct the write address from the tag, and perform the write-back cycle as well as the requested access. Secondly, all altered blocks must be written to the main memory before another device can access these blocks in the main memory. Thirdly, in the event of a power failure, the data in the cache is lost. Therefore, there is no way to tell which locations of the main memory contain stale data. The main memory, as well as the cache, must then be considered volatile, and provisions must be made to save the data in the cache in the case of power failure.

8.6 DESIGN ISSUES

The primary factors that influence cache performance are the cache hit rate and the cache management policies. The hit rate is influenced by the program characteristics and design parameters like:

- Cache memory size
- Transfer block size
- Cache replacement policy

The hit rate primarily depends on the locality of reference for code and data (i.e., the extent to which code and data entries are near to each other). In general, the compiler determines the placement of object code and data. The hit ratio also depends on multitasking and the degree to which context switching occurs. The design parameter issues are summarized in the following material.

8.6.1 Cache Memory Size

The cache size, like the block size, strongly affects the miss ratio. However, the cache size is subject to various constraints. Clearly, the larger the cache, the lower the miss, and the better the performance. A large-size cache requires a high fan-in and fan-out for the gates. This results in a longer rise time, thus large caches tend to be slightly slower than their smaller counterparts. Cache size is also limited by the available real-state inside the chip and the circuit board area. The larger-size cache costs more and requires additional power and cooling.

It is important to estimate how a change in the cache size will affect the miss ratio, which is very sensitive to the work-load in a time-sharing system, and workstations typically serving one user at a time will result in a different miss ratio. The pragmatic way of deciding the cache size is to look at the application domain, select important parameters, and ultimately decide on the size based on simulation.

8.6.2 Block Size

Block size is one of the most important parameters in the design of a cache memory system. If the block size is too small, the look-ahead and look-behind are reduced. Therefore, the hit rate is reduced, particularly for programs that do not contain many loops. However, large block sizes suffer a number of pitfalls.

Larger blocks decrease the number of blocks that fit into a cache. Since each block fetch overwrites older cache contents, a cache memory containing a smaller number of blocks will result in a situation where data may be overwritten immediately after it is fetched.

As the blocks become larger, each additional word gets further from the requested word, and is therefore less likely to be accessed by the processor. Again, large blocks require a wider bus between the cache and the main memory, as well as between static and dynamic memory, resulting in increased cost.

8.6.3 Cache Replacement Policy

When a miss occurs and a new block has to be brought in, a decision must be made as to which old blocks need to be overwirtten, in the event of a cache overflow. The main objective of the replacement algorithm is to retain the lines likely to be referenced in the near future, and discard lines that are no longer required, or whose next access is in the more distant future. In a direct mapped cache, the data pointed to by the index always get replaced. On the other hand, in the set-associative case, a replacement decision has to be made. As in the virtual memory management technique, the common replacement schemes are Random, FIFO, LRU, etc.

8.6.3.1 Least Recently Used (LRU) Policy

Under this policy, on a cache miss the least recently referenced block of the resident set get replaced. For smaller set sizes, the LRU policy may be implemented efficiently in hardware so that it can operate at cache speed.

8.6.3.2 First In First Out (FIFO) Policy

At a miss, this policy replaces the longest resident block. The reasoning behind this is that this block has had its chance, and it is the time to give another block a chance. Unfortunately, FIFO is likely to replace a heavily used block, because the reason that a block may have been in cache for a long time may well be that it is the most often referenced block.

8.6.3.3 Random (RAND) Policy

This policy replaces a randomly chosen block from a set in the cache.

It is generally considered that, on the average, LRU performs better. However, various theoretical studies reveal that RAND performs better than LRU or FIFO, in the case of an instruction cache.

8.7 CACHE LOCATION

The physical location of a cache within the computer system is another design criterion. This involves the implementation of cache memory in the data and address paths, relative to the CPU and the Memory Management Unit (MMU). If the cache is placed between the CPU and MMU, then it deals with a virtual address and this is called a logical cache. On the other hand, if the cache memory is placed between the MMU and the main memory, then it is known as a physical cache.

8.7.1 Logical Cache

The most important benefit of locating a cache memory on the logical bus is the speed of access. Here, the cache is accessed directly, using the virtual address. The entry corresponding to the logical address generated by the CPU gets loaded into the tag array. During the subsequent bus cycles, cache tag-compare logic compares the logical address with the tag array contents. In parallel, the MMU initiates a logical-to-physical address translation, which reduces the access time in the event of a miss. If a cache hit occurs, the MMU operation is terminated, and the cache forwards the data to the CPU. A cache controller allows the CPU to operate with no wait states, in the event of a hit. With the advent of the appropriate technology, more CPUs are incorporating the cache and MMU on the chip (e.g., MC60030), and implementing this strategy.

The complexity of the logical cache design centers around the stale data problem. Since the logical cache doesn't monitor physical accesses, it is possible that another bus master could update the data value in the main memory and cause the cache entry to become stale. To overcome the stale data problem in a logical cache, a simple strategy can be followed concerning a task's relationship with I/O operations initiated for that task. When DMA activities are initiated into a task's address map, the cache should be flushed prior to allowing that task to access the new information. Flushing the cache ensures that entries loaded into the cache contain the most recent data values. Therefore, the task-related information in the cache should be flushed by the task dispatch handler, and it should become an integral part of normal task management strategies.

In a virtual cache, the possibility may still exist for several copies of the same physical block to be present in the cache under different names. This is known as the *synonym problem.* The solution is to avoid multiple copies of the same block in the same cache by detecting the synonym when it occurs and enforcing consistency. This can be achieved by the Inverse Translation Buffer (ITB), which works in a manner opposite that of the TLB technique. The ITB is accessed on a physical address and indicates all of the virtual addresses in the cache associated with that physical address. The content of ITB is then written back into the memory.

8.7.2 Physical Cache

The physical cache benefits primarily from the use of the physical (real) addresses to map information in the cache. This technique allows the cache controller to monitor all activities that require a main memory update. Since an unlimited number of logical addresses could be generated, all of which reference the same main memory location, the updating of the physical cache means copying the contents of the physical location.

The primary disadvantage of a physical cache is that all CPU access must suffer the Memory Management Unit's (MMU) translation time. For example, the MC68020 requires assertion of the data transfer acknowledge signal ($\overline{DTACK}$) 40 nsec after the address strobe ($\overline{AS}$) is asserted, in order to perform a no-wait-state bus cycle. Based on the 120 nsec worst-case translation time for the MMU, it is clearly not feasible for a physical cache to compare the translated address and determine (in time to signal the completion of a no-wait-state bus cycle) whether a hit occurred or not. Therefore, the processor will have to wait for an extra clock cycle.

8.8 ON-CHIP CACHE ORGANIZATION OF THE MC68020

The MC68020 microprocessor includes a 256-byte on-chip instruction cache that is referenced by logical (virtual) addresses. By physically including the cache within the CPU, two important performance goals are accomplished:

1. The external bus activity is reduced. In the MC68000/68010-based system, the CPU utilizes approximately 80 to 90 percent of the available bus bandwidth. This leaves little bandwidth for another bus master (e.g., another CPU in a multiprocessing system, or DMA devices), which results in performance degradation.
2. The effective access time is improved by placing a high-speed cache between the CPU and the main memory system.

Again, by including an instruction-only cache, the designers avoided the stale data problem. This allows the designers the capability to flush out the entire cache instantaneously. Thus, the on-chip cache of the MC68020 provides a substantial increase in CPU performance, and allows much slower and less expensive memories to be used for the same qualitative system performance.

8.8.1 Cache Organization and Operations in the MC68020

The MC68020 is the first microprocessor where the CPU and the cache memory share the same piece of silicon. This is because the designers wanted to keep the cache organization as simple as possible. It utilizes a direct memory mapped scheme, organized as 64 lines, where a given address corresponds to only one location within the cache.

The cache consists of two parts, a tag array and a data array, shown in Figure 8.16. The parts reside in different locations of the MC68020. The tag array stores the full address. Each value in the tag array has a corresponding value in the data array. The tag array is composed of three parts:

- The tag field
- The index field
- The block size

The tag field is made up of:

- A valid field.
- The function code bits.
- The 12 most significant address bits.

The valid bits indicate whether the information in the data array is actually associated with the cache access address, or is a leftover from some previous program.

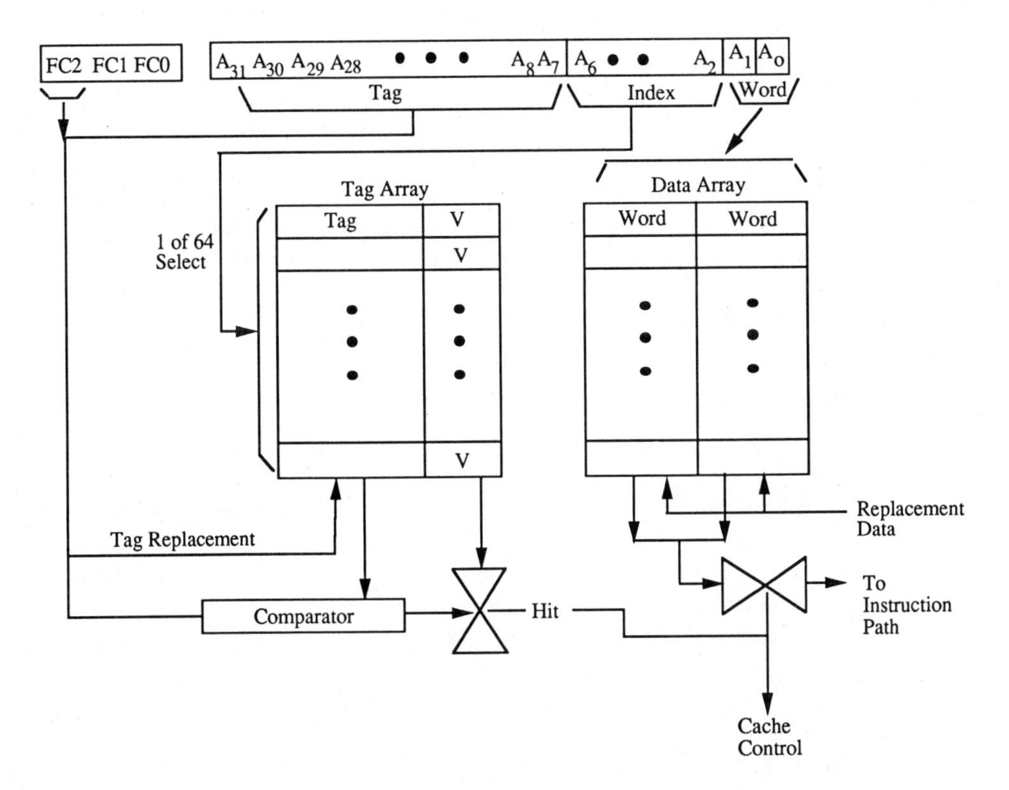

FIGURE 8.16. MC68020 cache organization.

The function code indicates the CPU's mode of operation. That is, whether it is in user or supervisory mode, and accessing data or instructions.

The index field is derived from the least significant address bits, A7 through A2. The address bits select one of 64 lines and its associated tag.

The block size specifies the number of words in each line of the data array. It contains 32 bits (or two words) of instruction data, where they are independently replaceable.

As the MC68020 initiates a bus cycle, the comparator compares the cache access address against the data in the tag array. A *hit* is said to have occurred in the event of a match. The cache directory is scanned first, using the A7 through A2 address bits as the index field. This selects one of the 64 entries in the cache. The cache control logic then uses the address bits A8–A31, including the FC2 bit, to compare against the tag field of the selected entry. A cache hit occurs in the event of a tag match. In the event of a match, the logic sets the valid bit in the cache directory, and the word selected by the address bit A1 is supplied to the instruction pipe.

When the address and the function code bits do not match against the tags, or the requested entry is not valid, a *miss* is said to occur. The bus controller initiates a long-word prefetch operation for the required instruction word from the memory. The new instruction words are then automatically written into the cache entry, and the valid bit gets set by the cache control logic, provided the entry is cachable. Since the MC68020 always prefetches 32 bits of instruction from the main memory, both entries of the cache will be updated as a result, regardless of which word was responsible for a miss.

8.8.2 Cache Control

The cache control circuitry of the MC68020 can have direct access to the cache directory and entries. The supervisory program can set bits in the CACR to exercise control over cache operations. This register contains the address for a cache entry to be cleared.

8.8.3 Cache Control Register (CACR)

The CACR register is shown in Figure 8.17. It is a 32-bit register, which can be written or read by the MOVEC instruction. It is also modified by a reset. The least significant eight bits are defined for the MC68020. The unused portions (i.e., bits 8 through 31) are always read as zero, and ignored during the write operation. In order for future compatibility, programmers should not set these bits during write operations. Only four of the eight least significant bits are used to control the operation of the instruction cache. These bits are shown in Figure 8.17.

8.8.3.1 Cache Clear (C)

Bit 3 of the CACR is set to clear all entries in the cache. The operating system and equivalent software set this bit to clear instructions from the cache prior to

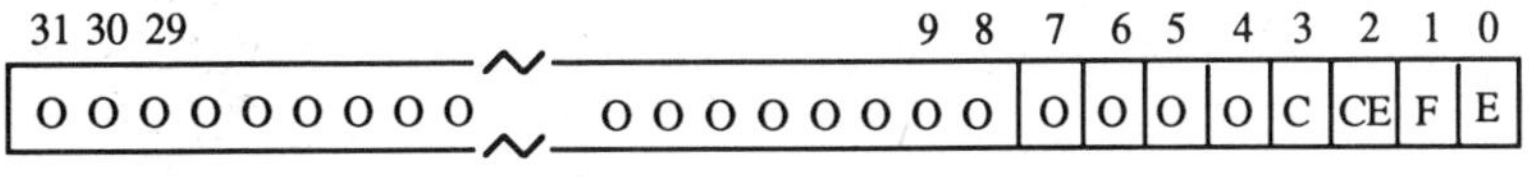

31 30 29 ...	... 9 8	7	6	5	4	3	2	1	0
0 0 0 0 0 0 0 0 0 0	0 0 0 0 0 0 0 0 0	0	0	0	0	C	CE	F	E

C - Clear Cache
CE - Clear Entry
F - Freeze Cache
E - Enable Cache

FIGURE 8.17. Organization of the cache control register.

a context switch. The CPU clears all valid bits in the cache directory at the time that a MOVEC instruction loads a 1 into the C bit of the CACR, regardless of the status of other bits in the register. This bit is always read as zero.

8.8.3.2 Clear Entry (CE)

Bit 2, CE, is set to clear an entry in the cache. The index field of the cache address register specifies the entry to be cleared. The CPU invalidates the specified long-word by clearing the valid bit for the entry at the time that a MOVEC instruction loads a 1 into the CE bit of the CACR, regardless of the status of the E and F bits. The CE bit is also read as zero.

8.8.3.3 Freeze Cache (F)

Bit 1, the F bit, is set to freeze the operation of the cache. If this bit is set and a miss occurs in the cache, then the entry is not updated. If Z is cleared, then a miss in the cache causes the entry to be filled. The reset operation clears this bit. This bit can also be utilized by emulators to freeze the cache during emulation function execution.

8.8.3.4 Enable Cache (E)

Bit 0, the E bit, is set to enable the cache. The enable function is necessary during system debug and emulation. The cache becomes disabled when it is cleared. Again, a reset operation will clear this bit. In general, the operating system (or equivalent software) enables the cache to indicate a normal operation, but can disable it during debugging. Clearing this bit will force the CPU to access external memory, but does not clear the previous entries in the cache. The previously valid entries remain valid, and can be used again by enabling the cache.

8.8.4 Cache Address Register (CAAR)

This 32-bit register is utilized to select an entry within the cache. The organization of this register is shown in Figure 8.18. The MC68020 uses this register for the CE function, which is accessible by the MOVEC instruction. The index field (bits 7–2) contains the address for the "clear cache entry" operations. These fields identify the index and the long-word of a cache line. It is important to note that all 32 bits of the CAAR are reserved by Motorola for future use.

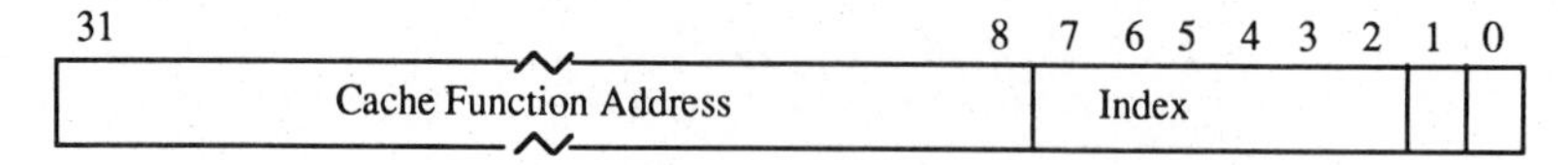

FIGURE 8.18. Organization of the cache address register.

8.8.5 Cache Initialization

During a hardware reset, the CPU clears all of the valid bits in the cache directory. It also clears the cache enable bit (E) and cache freeze bit (F). This effectively disables the cache.

8.9 INSTRUCTION EXECUTION TIMING

The cache memory efficiency is directly linked to the length of a program loop. If the loop fits with the 256 contiguous bytes address range, then it will achieve a high hit rate. On the other hand, programs that frequently jump over 256 bytes worth of instructions may not hit the cache often. A cache is also of some use to program loops between 256 and 512 bytes long, since such loops will only use some index values once.

It is a relatively straightforward task to determine the instruction and operation of the MC68000/68010/68012 in terms of the external clock cycle. The same thing cannot be said about the MC68020/MC68030/MC68040. One can only provide execution and operation guidelines, but not exact times for all possible circumstances. This approach is used because the execution time depends on factors such as: the preceding and following instructions, the instruction stream alignment, the redundancy of operands and instruction words in the caches, operand alignment, instruction prefetch, and instruction execution overlap.

The calculation of the instruction execution time is described in subsection 4.3.3, and the timing list is given in Appendix D.

Chapter 9

Coprocessor Interfacing

Webster defines a coprocessor as something "together or with" a main processor. Microprocessor manufacturers define them as, 'any circuit that meets its interfacing requirements and implements some functions that are not supported within the main processor.' However, we will define a coprocessor as 'an intelligent device that enhances the capability and performance of the processor.' The role of a coprocessor is to relieve the main processor from the burden of performing some specialized task or function. This concept extends the capabilities and performance of a general-purpose processor to include a particular application without unduly encumbering the main processor architecture.

We will discuss the concept of coprocessor and various protocols that are created to support them. The MC68000 family coprocessor protocol will be studied through examples.

9.1 COPROCESSOR CONCEPTS

In today's microprocessor architecture, the main processor provides general-purpose integer arithmetic and data movement functions required by the majority of application systems. For example, the MC68020 provides general-purpose integer arithmetic and data movement functions, and coprocessors are utilized to support floating-point arithmetic, graphic operations, memory management, etc. Such a separation of functionality provides benefits to both the chip manufacturer and system designer.

From the point of view of the system designer, coprocessors provide more flexibility to design a unique system, and provide assurance of future system enhancements by adding a coprocessor as they become available. From the point of view of the chip manufacturer, separating specialized functions onto individual pieces of silicon is more desirable, because:

1. Devices with more general application features can be made for the major portion of the market. For example, a general-purpose microprocessor is more easily produced and marketed than one with special functions, such as floating-point arithmetic and graphics built in, since a specialized processor will cost more to manufacture and will have a more limited market appeal.
2. By separating functions onto individual devices, functions can be implemented in a superior way.
3. A coprocessor scheme allows easy expansion of the product line without expensive and time-consuming redesign of existing devices. An added advantage is that, if the coprocessor interface is properly designed, a new main processor can be designed, with a higher performance, that will still utilize the older coprocessor. Again, existing processors can be enhanced by the addition of a newly designed coprocessor.

9.1.1 Coprocessor Architecture and Protocols

Due to the rapid growth in the number and complexity of applications, the concept of the coprocessor is becoming increasingly important. Some of the earlier coprocessors were involved in the field of signal processing, where fast fourier transforms in real time were required. Since then, coprocessors are designed to perform complex tasks like:

- Floating-point arithmetic
- Matrix manipulation
- Fast fourier transform
- Graphics data processor
- Memory management unit
- Communication controllers

The interactions between the main processor and the coprocessor needs to be transparent to the programmer (i.e., no knowledge of the intricate communication protocol between the main processor and coprocessor is required). Thus, the goals of coprocessor designers are to provide capabilities to the user so that they do not appear as hardware external to the main CPU.

The designers of coprocessors follow three distinct architectures, which are based on:

- Intelligent monitoring
- Special signals and instructions
- Special instructions

Various implementations exist for the three architectures. Three main microprocessor vendors, National Semiconductor, Intel, and Motorola, are each proponents of their own approaches, because they are closely tied to their microprocessor architectures.

9.1.2 Intelligent Monitoring

This architecture is shown in Figure 9.1, where each coprocessor monitors the data bus. Here, the coprocessor examines every instruction from the system data bus, to determine whether it is destined for it. In the event of a match, it starts the execution of the instruction and uses the system bus. The main processor may enter into a wait mode, or can continue its operation, based on the circumstances. This architecture requires that all coprocessors must have a smart interface capable of capturing and decoding instructions. Here, the coprocessors are responsible for error handling, context switching, and recovery, in addition to their processing functions. This architecture forces the entire system to operate at the speed of the slowest coprocessor.

The advantage of this architecture is that all processors operate at the same clock speed and, as a result, the communication protocol followed by the interface becomes simple. Again, no additional bus cycle is required to pass the instruction word to coprocessor, which enhances system performance.

National Semiconductor's coprocessors are based on this architecture.

9.1.3 Special Signals and Instructions

In this architecture (Figure 9.2), the coprocessor acts as a slave to the main processor. The main processor communicates with the coprocessors via special signal lines. The main processor fetches the Op-code from the memory, decodes it, and then determines which coprocessor should execute it. It then transfers that instruction to the appropriate coprocessor over special signal lines. This provides

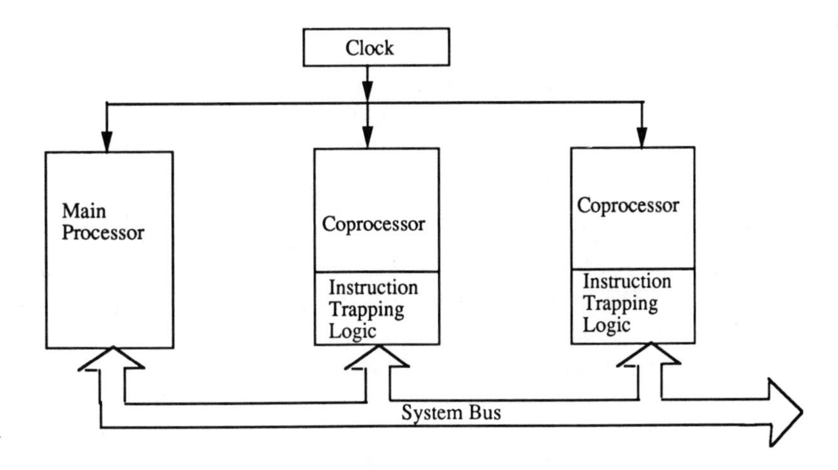

FIGURE 9.1. Coprocessor with an intelligent interface.

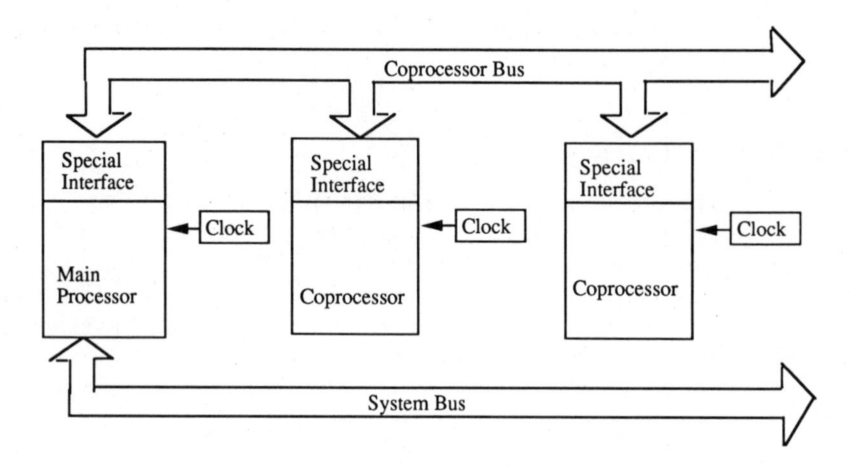

FIGURE 9.2. Coprocessor protocol via special signals and instructions.

coprocessor designers the freedom to create any type of coprocessor interface. The only restriction is to conform to the special signaling protocol and pin specifications.

The main advantage of this architecture is that each coprocessor can operate at its best speed and with its own clock, thus enhancing reliability. As a result, the coprocessor interface becomes simple.

The main disadvantage is that the main processor must have a prior knowledge of all of the coprocessors available in the system.

Intel's coprocessors fall in this class.

9.1.4 Special Instructions

In this architecture, special instructions are created in support of the coprocessors. These instructions have a different format from that of the main processor. During instruction execution, if the main processor detects a coprocessor instruction, it activates a special protocol to forward the instruction to the appropriate coprocessor. The CPU asks the coprocessor to execute the instruction. It performs all of the effective address calculations on behalf of the coprocessor. In the event of any trouble, the coprocessor requests the main processor to solve them. This architecture is illustrated in Figure 9.3. The processor communicates with the coprocessors through a well-defined set of registers, which are generally memory mapped to the processor's address space.

The advantage of this architecture is that the main processor talks to the coprocessors via a well-defined protocol, and the coprocessors will have their own local clock and can operate at their own speed.

Motorola's coprocessors follow this strategy.

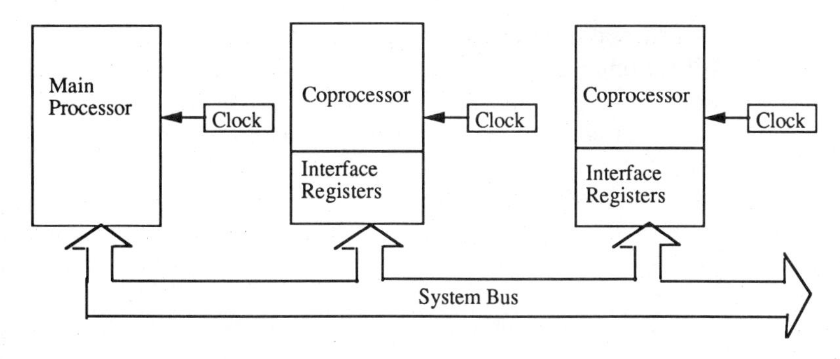

FIGURE 9.3. Coprocessor protocol using special instructions.

9.2 COPROCESSOR PROTOCOL FOR THE MC68000 FAMILY

The programming model of the MC68000 family consists of well-defined instructions, a register set and the address space available for software development. The coprocessor extends this basic capability by including additional instructions, registers, and data types that are not directly supported by the main CPU. In the MC68000 architecture, the programmer does not need to know the intricacy between the CPU and the coprocessor, because the protocol is implemented in hardware. This makes the coprocessor resources transparent to the user.

In the MC68000 architecture, standard peripheral hardware is accessed through interface registers mapped within the memory space of the processor, and the user utilizes standard instruction sets to communicate with the peripherals. In general, one needs to know the operational aspects of the device or may use predeveloped device drivers that hide those details (e.g., operational details). That is, peripherals could conceivably provide capabilities equivalent to the coprocessor for many applications. In such cases, the system designer must design hardware that supports the standard asynchronous protocol, protection mechanisms, and the device drivers.

9.2.1 Interface Features

In general, coprocessor interface consists of two parts:

- The hardware interface between the main processor and the coprocessors in the system.
- The communication protocol, which is utilized to transfer commands and data between the processor and the coprocessors, in order to complete the desired function.

The hardware interface between the main processor and all coprocessors in a system is an extension of the MC68000 family bus. Coprocessors are memory mapped, and can communicate with the main processor using synchronous or asynchronous protocols. Since information between the CPU and the coprocessor is exchanged via standard bus cycles, the coprocessor can be implemented in any technology, and can be interfaced as 8-, 16-, or 32-bit devices. Thus, a coprocessor can be implemented as a VLSI chip (residing on the mother board or on a separate board on the system), as a separate system board, or even as a complete system.

Since the coprocessor can communicate with the processor using the asynchronous protocol they can, as a result, execute instructions at different speeds. The system designer can select the speed of the CPU and of the coprocessor to optimize the performance of a given system. On the other hand, if the system designer decides to interface the coprocessor as a synchronous device, then all coprocessor signals and data exchange must be synchronized with the CPU.

9.2.2 Coprocessor Interfacing with the MC68020

An example system, consisting of a CPU, memory, I/O, coprocessor, and the bus expansion block, is shown in Figure 9.4. Here, all devices are memory mapped, and their selection is based solely on the address presented to the devices including the activation of the control lines during each bus cycle. During instruction execution, the CPU supplies a 3-bit function code value that identifies the

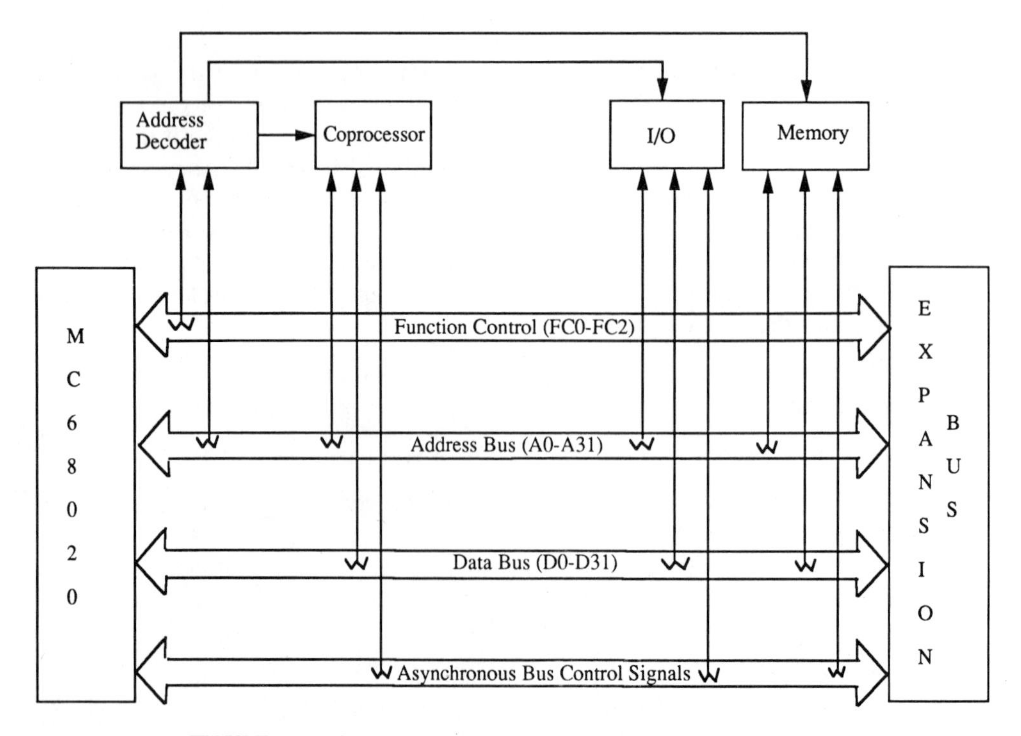

FIGURE 9.4. MC68020 system configuration with a coprocessor.

address space of the current bus cycle. Four out of the eight available address spaces are assigned for normal program execution and I/O functions. The fifth address space is reserved for processor-related functions such as 'interrupt acknowledgment' or 'communication with coprocessor'. By creating a separate address space for the coprocessor, the MC68000 architects have provided a complete 4 Mbyte space to access coprocessor-related memory and I/O, and have provided tools to the hardware designer to implement special hardware protection mechanisms.

The CPU and coprocessor communicate with each other via a *mailbox*. The MC68020 allows a total of eight mailboxes. This specifies the maximum number of coprocessors that can be directly attached to the CPU. The mailbox is composed of 11 memory-mapped register locations for passing commands and data between the communicating parties. Figure 9.5 illustrates the memory map of this register set. It is important to realize that these interface registers are used purely for communication purposes, and bear no relationship to the programming model of a coprocessor.

9.2.3 Interface Classification

The MC68000 processor family supports two types of coprocessor interfaces, non-DMA and DMA. The non-DMA coprocessors operate as bus slaves. The

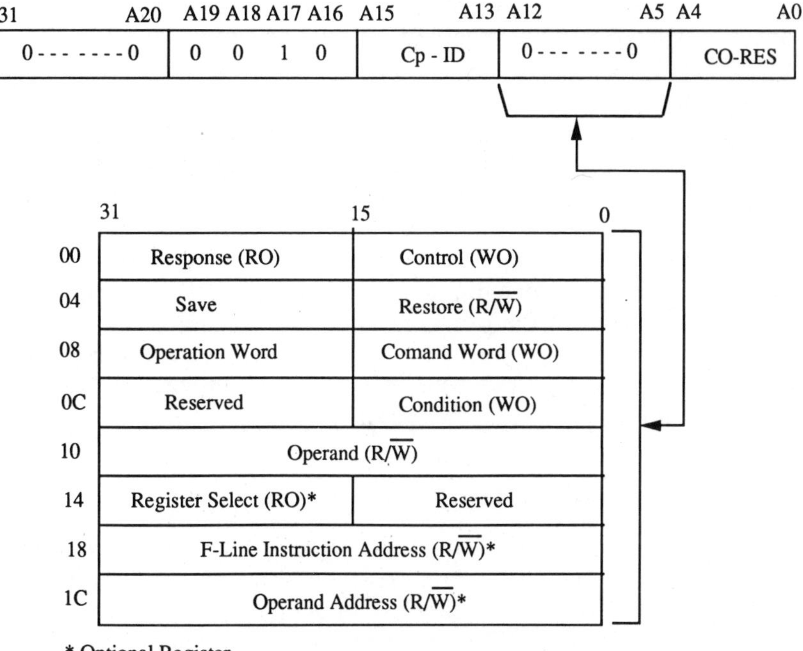

FIGURE 9.5. Coprocessor interface register map.

DMA coprocessors also operate as bus slaves, but are capable of operating as bus masters by directly controlling the system bus.

If a coprocessor does not require a large bus bandwidth, or requires special services not directly provided by the main processor, then it can be efficiently implemented as a non-DMA device. In such a case, all external bus-related functions required by the coprocessor are performed by the CPU. During program execution, if the CPU detects a coprocessor instruction, it will supply that to the coprocessor. It will also fetch the operand from the specified effective address, and then write the operand value into a designated register of the specific coprocessor. Then, the CPU will read the response of the selected coprocessor from another register, to determine what action is required of the CPU in order to assist the coprocessor to complete the desired instruction.

The DMA class of coprocessor can operate as bus master. The coprocessor generates all control, address, and data signals necessary to request and obtain the bus, and then performs DMA transfers over the system bus. DMA coprocessors, however, still have to act as bus slaves when they require information or services from the CPU, using the MC68000 coprocessor interface protocol.

9.2.4 Concurrent Operation Support

The programming model of the MC68000 processor family is based on the assumption that instructions are executed sequentially from the memory in a nonconcurrent way. The coprocessor should also maintain the model of sequential, nonconcurrent instruction execution for consistency. Consequently, the user can assume that the images of registers and memory affected by an instruction can only be updated due to the execution of the next sequential instruction accessing these registers or memory locations [MTTA287].

The coprocessor interface also allows concurrency, and it is the responsibility of the coprocessor designer to implement that concurrently, while maintaining the fundamental programming model based on noncurrent instruction execution. For example, if the coprocessor determines that instruction "X" does not use the resources required by instruction "Y", then instructions "X" and "Y" can be executed concurrently. Thus, the required instruction interdependences and sequences of the program are always maintained. For example, the MC68882 coprocessor supports concurrent instruction execution, while the previous generation, the MC68881, does not. However, the MC68030 can operate concurrently with the coprocessor MC68881 executing instruction.

9.2.5 Communication Protocol

The designers of coprocessors are given the freedom to design their own instructions. When the MC68020 encounters a coprocessor instruction, it initiates communication with the coprocessor and executes the necessary protocol to complete that instruction. This permits the user/programmer to know only the instruction set and the register set defined for the coprocessor, not the actual implementation details.

The MC68000 family instruction set was created in mid-1970, and the designers made a provision for future expansion. They set aside an eighth of the Op-code map. They are the $A and $F values of the most significant 4 bits. The coprocessor interface was the first architectural implementation that utilized the $F opcode for coprocessor instructions. Figure 9.6 shows the format of the first word of the coprocessor instruction.

The F-line Op-code consists of 1's from bits 12 through 15. Bits 9 through 11 of the F-line encode the coprocessor identification code (Cp-ID). The MC68020 utilizes the Cp-ID field to indicate which coprocessor the instruction refers to. The Cp-ID codes of 000 through 101 are reserved by Motorola for its current and future coprocessors. The Cp-ID codes 110 and 111 are reserved for user-defined coprocessors. The currently available Motorola coprocessor codes are given in the following table.

Cp-ID	Coprocessor
000	MC68851 page memory management unit
001	MC68881/MC68882 floating arithmetic coprocessor

It is important to realize that by default, the Motorola assembler will use Cp-ID "001" to generate instruction operation codes for MC68881 or MC68882 instructions, and Cp-ID = 000 is reserved for the MMU. The designers of coprocessors should not utilize these Cp-ID to design their system.

The bits 0 through 8 of the coprocessor instruction Op-word are dependent on the particular instruction being implemented. These bits are logically divided

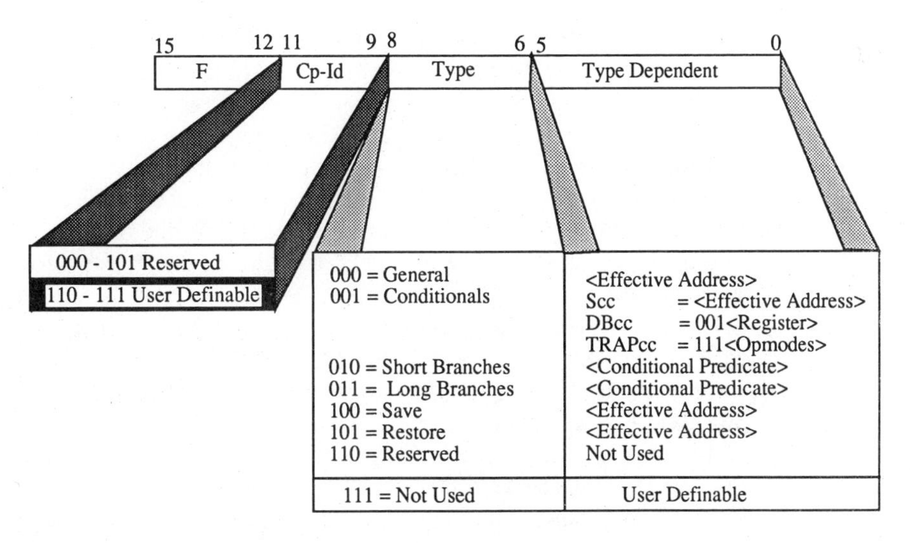

FIGURE 9.6. Coprocessor F-line Op-code. (Courtesy Motorola Inc.)

into two classes, namely *type* and *type-dependent*. The definition of these bits is illustrated in Figure 9.6.

Whenever the MC68020 encounters an F-line word as the first word of an instruction, it initiates a coprocessor communication over the coprocessor interface. Based on the instruction type, the processor will initiate the communication by writing/reading one of the CIR's (Coprocessor Interface Registers). If a protocol error is detected during the first bus cycle of the coprocessor communication sequence, then the MC68020 assumes that the specified coprocessor is not present in the system, and the processor will initiate F-line exception processing. This allows a software package to emulate the features of the coprocessor in a manner transparent to the program.

However, if the coprocessor responds with an acknowledgment, then the dialog continues, with the MC68020 reading a response from the coprocessor. This response may ask the processor to perform further actions on behalf of the coprocessor. The actions may involve performing condition tests, getting operands, or taking an exception. Once a dialog is initiated, all additional interaction is initiated by the processor, but in response to commands given by the coprocessor. Thus, once a coprocessor instruction is detected in the instruction stream, and communication with the proper coprocessor is established, the MC68020 becomes a slave to the coprocessor until the instruction is completed [Beims 85].

9.2.6 Interface Logic

Figure 9.7 illustrates the design of an interface logic that interfaces coprocessors at asynchronous non-DMA devices. The design of synchronous DMA interfaces is very much similar. The Cp-ID signal lines, A15–A13, and address lines, A16–A19 and the FC0–FC2, are decoded to select the coprocessor. The system designer can also use several coprocessors of the same type, and assign unique Cp-IDs to them.

In such an implementation, the MC68020 will use the standard asynchronous bus cycle to access the coprocessor interface register sets. This requires that the coprocessors satisfy the address, data, and control signal requirements of the MC68020. During coprocessor instruction execution, the processor drives the function control pins FC0–FC2 to "111" to indicate that it is performing the CPU space bus cycle. The MC68020 places the Cp-ID over the address pins A15–A13. Once a coprocessor is selected, other address lines are then used to select the coprocessor's interface register set.

The address lines A19–A16 indicates that the MC68020 has initiated a CPU bus cycle for the coprocessor. The MC68020 asserts zeros on the address lines A31–A20 and A12–A5 during a coprocessor access. The currently defined CPU space cycles for the MC68020 are the: interrupt acknowledge, breakpoint acknowledge, module support operations, and coprocessor access cycles. The MC68010 and MC68012 can emulate the coprocessor access cycles in CPU space using the MOVES instruction.

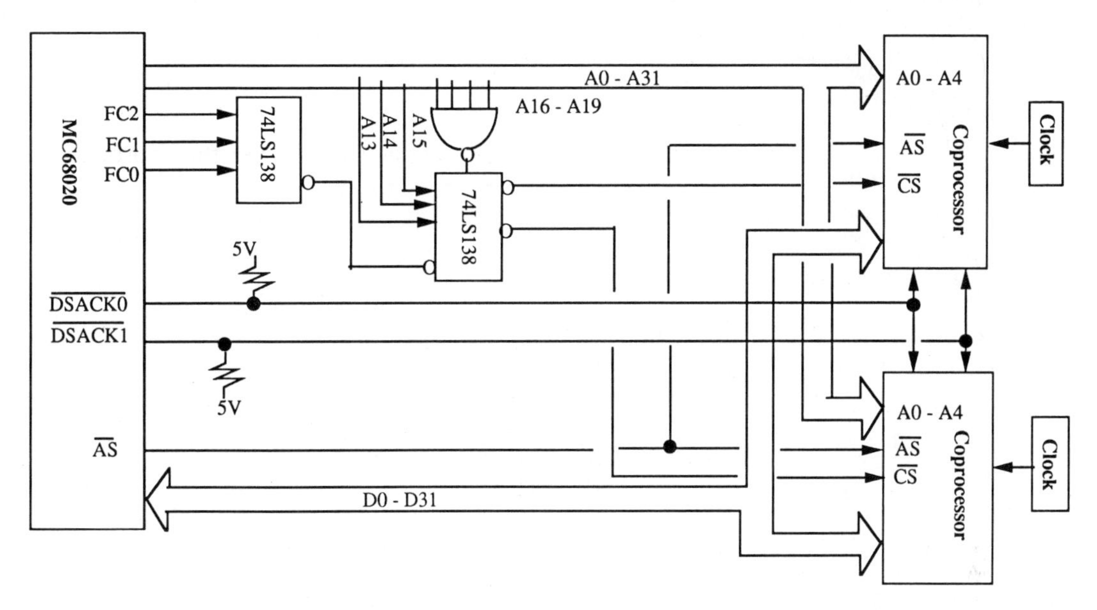

FIGURE 9.7. Coprocessor interface logic design.

9.2.7 The Mapping of Coprocessor Interface Registers

After the initiation of the coprocessor protocol and selection of a specific coprocessor, the MC68020 is required to access the registers defined in the Coprocessor Interface Register (CIR) set to communicate with the coprocessor. The map of the CIRs are shown in Figure 9.8. This CIR set is also known as the coprocessor mailbox, or simply as the mailbox. A mailbox consists of 16 sequentially located registers. The MC68020 uses the address lines A4–A0 to specify a register within the mailbox. The interface registers within the mailbox are logically divided into two classes, namely: *user registers* and *supervisory registers.* The supervisory registers can only be accessed by the operating system, operating under supervisory mode. The user registers can be accessed by the user task or operating system. Since the interface is designed for sequential operation, only one register within the mailbox can be accessed at one time. However, multiple coprocessors can be accessed at the same time.

9.3 COPROCESSOR INSTRUCTION FORMATS

The instruction types of the MC68000 family coprocessor interface can be separated into three groups:

- General (coprocessor defined)
- Conditional
- System control (context save and restore)

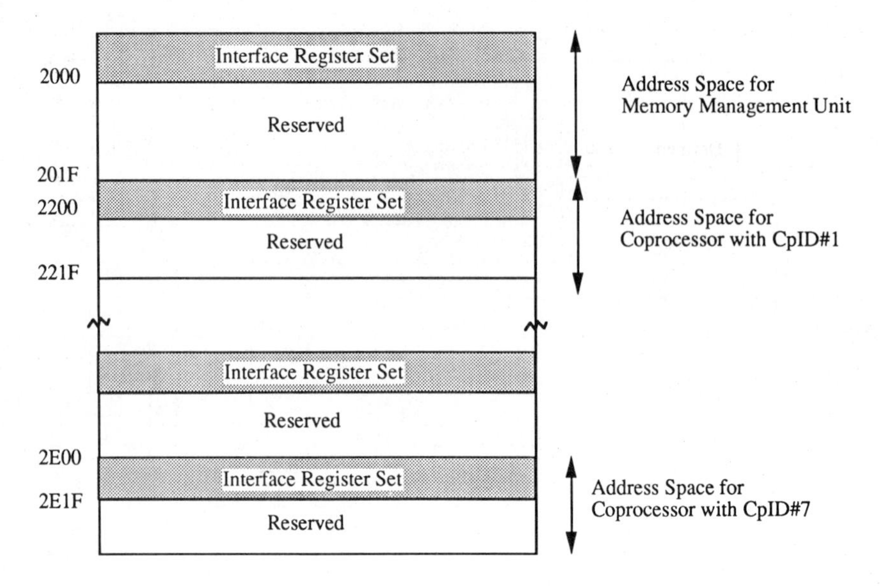

FIGURE 9.8. Coprocessor address map in the MC68020 CPU space.

The coprocessor instruction types, their operand size, and their operations are summarized in Figure 9.9. The group name identifies the type of operations provided by the instructions within the group. The instruction group identifies the coprocessor interface register accessed by the processor, to initiate instructions and communication protocols between the MC68020 and the coprocessor required for instruction execution. During the execution of the general and conditional groups, the coprocessor uses the coprocessor response primitive codes to initiate instruction and communication protocols between the main processor and the coprocessor, which are necessary for the completion of the instruction. The coprocessor then performs the conditional test and makes decisions. Again, during context save and restore, the coprocessor utilizes the coprocessor format codes to signal its status to the main processor.

9.3.1 General Instruction Format

The format for a general instruction is illustrated in Figure 9.10. The general instructions are undefined to the main processor, and are specific to the coprocessors. When the MC68020 detects a general instruction, it writes the coprocessor command word to the command CIR of the coprocessor. It then reads the response from the response CIR. By utilizing a set of primitive commands passed through the response CIR, the coprocessor proceeds to execute the instruction with the help of the processor. The syntax cpGEN is used by the Motorola assembler for all general instructions. However, the actual syntax, and the syntax used,

Instruction	Operand Size	Operation	Notation
cpGEN	User Defined	Pass comand word to coprocessor and respond to coprocessor primitives	cpGEN ⟨parameters defined by coprocessor⟩
cpBcc	16, 32	If cpcc true Then PC + d → PC	cpBcc <label>
cpDBcc	16	If cpcc false Then (Dn - 1 → Dn; If Dn ≠ 1 Then PC + d → PC)	cpDBcc Dn, <label>
cpScc	8	If cpcc true Then 1's → Destination Else 0's → Destination	cpScc <ea>
cpTRAPcc	None 16, 32	If cpcc true Then TRAP	cpTRAPcc cpTRAPcc #<data>
*cpSAVE	None	Save internal state of coprocessor	cpSAVE <ea>
*cpRESTORE	None	Restore internal state of coprocessor	cpRESTORE <ea>

*Privileged instructions, for O.S. context switching support

- MAIN and CO each do what they know how to do best

MAIN: Tracks instruction stream
Takes exceptions
Takes branches, etc.

CO: Does graphics manipulations,
Calculates transcendentals, floating point,
Does matrix manipulations, etc.

- When MAIN obtains an F-line opword, it cooperates with a coprocessor to complete execution of the instruction

FIGURE 9.9. Coprocessor instructions. (Courtesy Motorola Inc.)

may vary for the assembler or compiler, but all will generate the same object code for the coprocessor instruction.

The minimum size of the coprocessor type of instruction is two words. The bits [15:12] of the first word are an F-line operation code with a value of 1111. The Cp-ID field, bits [11:9], is used to select the desired coprocessor to execute the instruction. Bits [8:6] = 000 indicate that the instruction is a member of the general instruction category. Bits [5:0] encode the effective address mode, if any

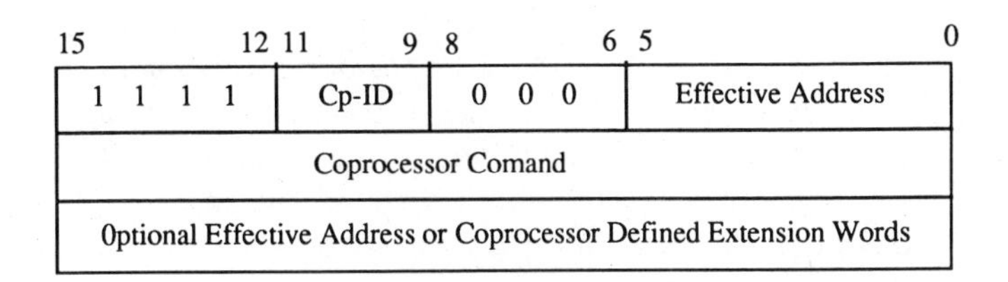

FIGURE 9.10. Coprocessor general instruction format (cpGEN).

is present. During the execution of the Cp-GEN instruction, the coprocessor may utilize the RESPONSE primitive to request that the MC68020 calculate an effective address, required for the execution of the instruction. If the MC68020 receives a RESPONSE primitive, it will then utilize bits [5:0] of the F-line operation code, to determine the effective addressing mode being requested.

The second word of the instruction format is the coprocessor command word. This word is written to the designated coprocessor's Command CIR, to initiate the execution of the instruction by the coprocessor.

The Cp-GEN may also include extension words that follow the coprocessor command word in the instruction format. These words provide additional information required for the coprocessor instruction. For example, if a coprocessor wants the MC68020 to calculate an effective address during coprocessor instruction execution, information required for the calculation must be included in the instruction format as effective address extension words.

The protocol utilized during the execution of the general instruction is shown in Figure 9.11. After detecting a coprocessor instruction, the main processor initiates communication with the coprocessor by writing the instruction word into the appropriate CIR. The coprocessor retrieves the CIR contents and initiates the execution of the instruction. The MC68020 does not decode the instruction for the coprocessor, rather the coprocessor design determines its interpretation.

During the execution of a cp-GEN instruction, if the coprocessor requires the service of the main processor, it transmits that request by placing coprocessor response primitive codes in the response CIR. The main processor reads the contents of the coprocessor's response CIR register and acts accordingly. However, if the coprocessor completes the instruction, or no longer requires the service of the main processor to execute the instruction, then it releases the processor as a part of the RESPONSE primitive. Now, the main processor can continue to execute the next instruction from the instruction stream. However, in the event of a pending 'trace instruction', the MC68020 does not terminate communication with the coprocessor until the coprocessor indicates that it has completed all processing associated with the Cp-GEN instruction. This protocol allows a wide range of operations to be implemented in the general instruction category.

9.3.2 Conditional Instruction Format

During the execution of a conditional instruction, the coprocessor evaluates a condition and returns a decision to the main processor in the form of true/false. The conditions are related to the operation of the coprocessor, and therefore are evaluated by the coprocessor. Upon receiving a decision, the main processor completes the execution of the instruction, which may involve changing the execution sequence, setting a byte, or performing a TRAP operation.

The main processor initiates the execution of these instructions by writing a condition to the conditional CIR into the coprocessor's mailbox. After evaluating the condition, the coprocessor returns a true/false indicator to the main processor by placing a null primitive in the response CIR. The main processor com-

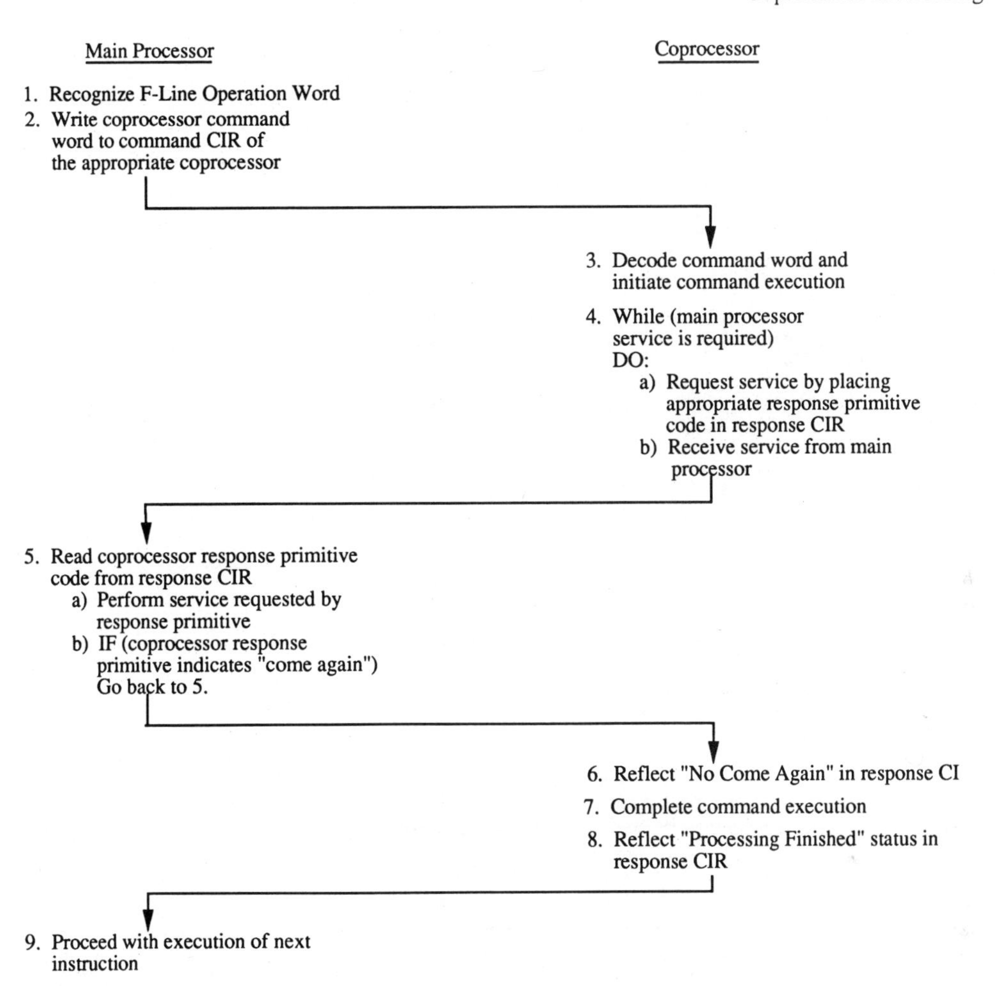
Main Processor
Coprocessor
1. Recognize F-Line Operation Word
2. Write coprocessor command word to command CIR of the appropriate coprocessor
3. Decode command word and initiate command execution
4. While (main processor service is required) DO:
a) Request service by placing appropriate response primitive code in response CIR
b) Receive service from main processor
5. Read coprocessor response primitive code from response CIR
a) Perform service requested by response primitive
b) IF (coprocessor response primitive indicates "come again") Go back to 5.
6. Reflect "No Come Again" in response CI
7. Complete command execution
8. Reflect "Processing Finished" status in response CIR
9. Proceed with execution of next instruction

FIGURE 9.11. Coprocessor interface protocol for cpGEN instruction.

pletes the coprocessor instruction execution, in accordance with the coprocessor's request.

The format for the coprocessor conditional instructions is shown in Figure 9.12. For example, the coprocessor branch on condition CpBcc.W instruction is two words in length, whereas the CpBcc.L is three words.

The first word of the CpBcc instruction F-line word is shown in Figure 9.12(a), which has the following meaning:

Bits [15:12] = 111, indicates an F-line instruction.
Bits [11:9] = The coprocessor ID.
Bits [8:6] = Identifies either word (010) or long-word (011) displacement.
Bits [5:0] = The coprocessor condition selector field.

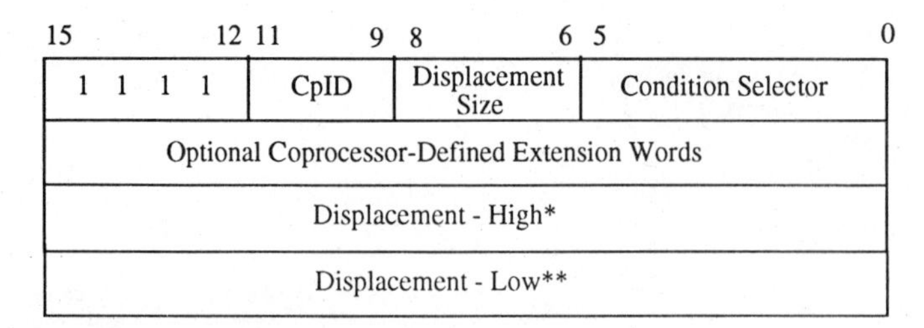

* Displacement - High: One Word for CpBcc.W
** Displacement - Low: Also Exist for CpBcc.W

a) Branch on Coprocessor Condition Instruction

15 — 12	11 — 9	8 — 6	5 — 0
1 1 1 1	CpID	0 0 1	Effective Address
			Condition Selector
Optional Coprocessor-Defined Extension Words			
Optional Effective Address Extension Words (0 - 5 Words)			

b) Set On Coprocessor Condition (CpScc)

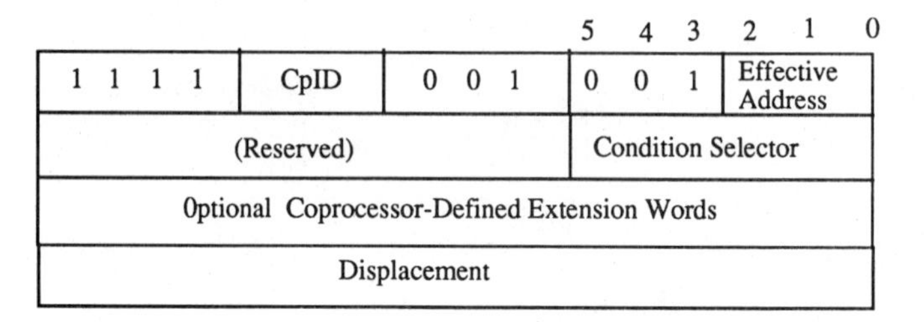

c) Test Coprocessor Condition, Decrement and Branch Instruction Format (cpDBcc)

1 1 1 1	CpID	0 0 1	1 1 1	OP-Mode
Reserved			Condition Selector	
Optional Coprocessor-Defined Extension Words				
Optional Word				
Or Long Word Operand				

d) Trap On Coprocessor Condition (cpTRAPcc)

FIGURE 9.12. Coprocessor conditional instruction format: (a) Branch on coprocessor condition instruction; (b) Set on coprocessor condition (cpScc); (c) Test coprocessor condition, decrement, and branch instruction format (cpDBcc); (d) Trap on coprocessor condition (cpTRAPcc).

The MC68020 writes the entire operation word to the condition CIR to initiate the execution of a branch instruction by the coprocessor. The coprocessor utilizes the bits [5:0] to decide the type of the condition to be evaluated. If the coprocessor requires additional information to evaluate the condition, then the branch instruction format may include that as extension words. However, the number of possible extension words is determined by the designers of the coprocessor. The remaining word(s) of the CpBcc instruction format contains the displacement used by the main processor to calculate the destination address when the branch is taken.

The set on coprocessor condition, cpScc, instruction format is shown in Figure 9.12(b). The main processor sets or resets a flag, according to a condition evaluated by the coprocessor, whose operation is similar to the operation of an Scc instruction.

The F-line operation word of cpScc has the following meaning:

Bit [15:12] = 111; Identifies the F-line instruction.
Bit [11:9] = The Cp-ID.
Bit [8:6] = 001; Identifies the CpScc instruction.
Bit [5:0] = Identifies the encoded representation of the effective addressing mode.

The second word written into the CIR to initiate the cpScc contains the following meaning:

Bit [15:6] = Reserved by Motorola for future use, and should be made zero in order to be compatible with the future product.
Bit [5:0] = The coprocessor condition selector.

The third word contains additional information to facilitate the evaluation of the condition by the coprocessor.

The remaining zero to five word(s) of the cpScc instruction contain various additional information required to calculate the effective address specified by bits [5:0] in the F-line operation word.

The test coprocessor condition decrement and branch, cpDBcc, instruction is also similar to the DBcc instruction of the MC68000. Its format is shown in Figure 9.12(c). It uses a coprocessor-evaluated condition and a LOOP counter in the main processor. It is efficient for implementing DO-UNTIL constructs used in high-level languages.

The F-line operation word of the cpDBcc has the following meaning:

Bit [15:12] = 111; Identifies the F-line instruction.
Bit [11:9] = The Cp-ID.
Bit [8:3] = 001001; Identifies the CpDBcc instruction.
Bit [2:0] = Identifies the main processor data register used as a loop counter during instruction execution.

The second word written into the CIR to initiate the cpDBcc conveys the following meaning:

Bit [15:6] = Reserved by Motorola for future use, and should be made zero in order to be compatible with the future enhancement.
Bit [5:0] = The coprocessor condition selector.

The third word will contain necessary information required for the evaluation of the condition by the coprocessor.

The last word of the cpDBcc instruction contains the two's complement displacement of the 16-bit value. It is sign-extended to long-word size during the calculation of the destination address.

The trap on coprocessor condition, cpTRAPcc, instruction, is also similar to the TRAPcc instruction of the MC68020. Its format, shown in Figure 9.12(d), uses a coprocessor evaluated TRAP condition by the main processor.

The first word of the cpTRAPcc operation word has the following meaning:

Bit [15:12] = 1111; Identifies the F-line instruction.
Bit [11:9] = The Cp-ID.
Bit [8:3] = 001111; Identifies the CpTRAPcc instruction.
Bit [2:0] = Op-mode (Identifies the number of optional operand words in the instruction format).

Where

Op-mode = 010 means one optional word is required.
Op-mode = 011 means two optional words are required.
Op-mode = 100 means no optional word is required.

The second word written into the CIR to initiate the cpDBcc conveys the following meaning:

Bit [15:6] = Reserved by Motorola for future use, and should be made zero in order to be compatible with future enhancement.
Bit [5:0] = The coprocessor condition selector.

The last word(s) will contain additional information required to evaluate a condition. The size on the extension word(s) is encoded in the Op-mode field.

9.3.3 System Control Format

In a multitasking environment, the coprocessor may have to change its context asynchronously, and the coprocessor may be interrupted at any point during the execution of general and conditional instructions. The coprocessor context save (cpSAVE) and restore (cpRESTORE) instructions are designed to manage that situation. To save the previous context, the microcode of the MC68020 automatically generates a "coprocessor state frame," The format of the frame is shown in Figure 9.13. The coprocessor may communicate its status information to the processor via coprocessor format codes. The main processor then saves the context.

9.3.3.1 Context Save

The main processor initiates the context saving operations after recognizing the F-line operation. The processor initiates the cpSAVE instruction after the coprocessor communicates status information associated with the context save operation to the main processor, by placing the coprocessor format code (discussed in Section 9.4) into the save CIR. However, if the coprocessor is unable to immediately suspend its current operation when the processor reads the save CIR, then the coprocessor will return the 'not ready' format code. In such an event, the processor will service any pending interrupts, return to check the save CIR, and reinitiate the cpSAVE instruction. After sending the 'not ready' format code to the main processor, the coprocessor should resume either completing the currently executing instruction, or suspending its execution. The coprocessor will then place a format code in the save CIR and supply its internal states. The lower byte of the coprocessor format word specifies the amount of state information to be transferred from the coprocessor.

During the execution of the cpSAVE instruction, the MC68020 calculates the state frame's effective address from the information in the operation word of the instruction, and stores a format word at that address. The MC68020 then writes long-words from the coprocessor state frame into descending memory addresses, beginning with the address specified by the sum of the effective address and the format word length field multiplied by four.

The content of the state frame describes the operational context of the in-

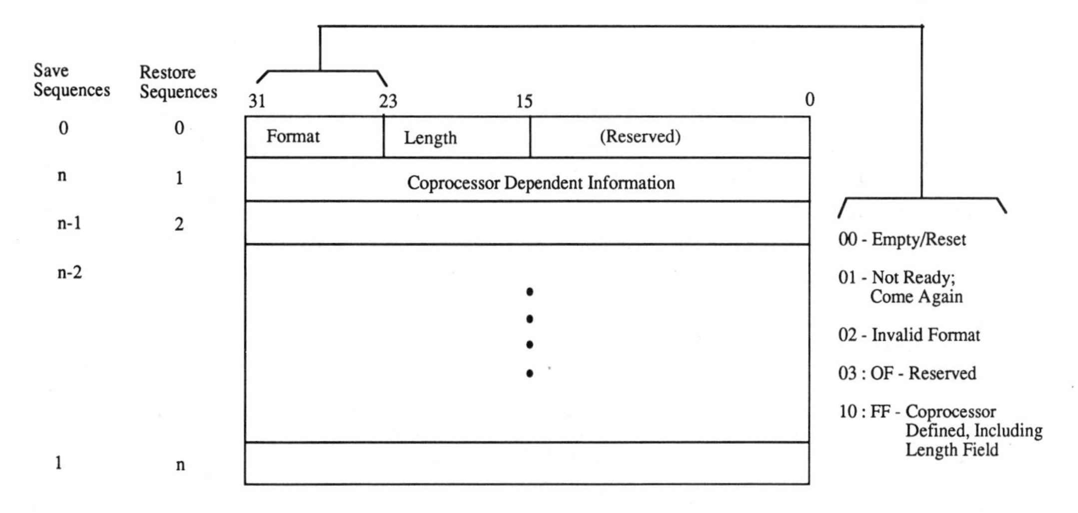

FIGURE 9.13. Coprocessor state frame format in memory.

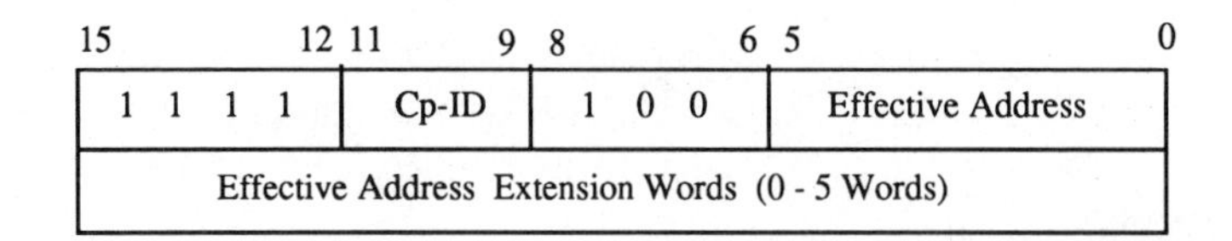

FIGURE 9.14. Coprocessor context save instruction format.

volved coprocessor. The context includes the relevant internal registers and/or status information necessary for the coprocessor to continue its operation at the point of suspension. The state frame may also contain all register values that appear in the coprocessor programming model, and that can be directly accessed using the coprocessor instruction set. For example, if CpGEN instructions are utilized to save the program visible state of the coprocessor, then the cpSAVE and cpRESTORE instructions should only transfer the program invisible state information, to minimize interrupt latency during a save or restore operation [Motorola 85].

The format of the cpSAVE instruction is illustrated in Figure 9.14. The first word of the instruction in the F-line operation code contains:

Bit [15:12] = 1111; Identifies the F-line instruction.
Bit [11:9] = The Cp-ID.
Bit [8:6] = 100; Identifies the CpSAVE instruction.
Bit [5:0] = Identifies the effective address of the memory location where the state frame will be saved.

Up to five effective address extension words may follow the cpSAVE instruction operation word. Thus, if the main processor requires additional information for the calculation of the effective address specified in bit [5:0] of the Op-word, then that information will be included in the effective address extension words.

9.3.3.2 Context Restore

The coprocessor context restore instruction, cpRESTORE, forces a coprocessor to terminate any current operation and restore a state associated with a different context or execution. During the execution of the cpRESTORE, the coprocessor communicates its status information to the main processor by placing appropriate format codes in the restore CIR. The format of cpRESTORE is shown in Figure 9.15.

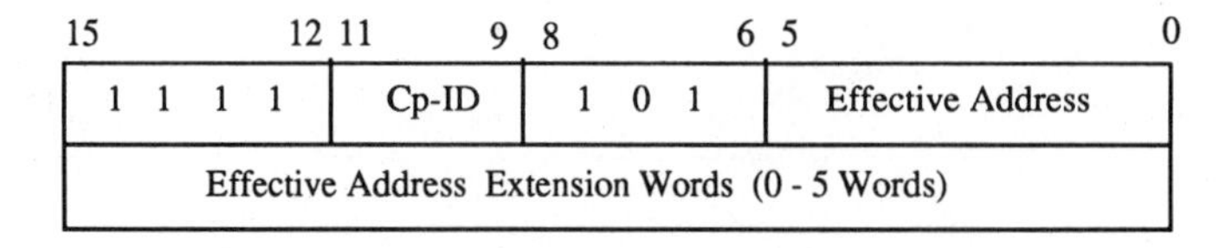

FIGURE 9.15. Coprocessor context restore instruction format.

The format of the first word of the instruction is listed below:

Bit [15:12] = 1111; Identifies the F-line instruction.
Bit [11:9] = The Cp-ID.
Bit [8:6] = 101; Identifies the CpRESTORE instruction.
Bit [5:0] = Identifies the effective address of the memory location from where the state frame will be restored.

Up to five effective address extension words may be retrieved following the cpRESTORE instruction operation word. If the main processor requires additional information for the calculation of the effective address specified by bit [5:0] of the Op-word, then this information is included in the effective address extension words.

The cpSAVE and cpRESTORE are privileged instructions. During the execution of these instructions, the MC68020 checks the status of the supervisory bit in the status register. If the processor finds itself in the user mode during the execution of these instructions, it then initiates privilege violation exception processing, without accessing any of the coprocessor interface registers.

9.4 COPROCESSOR RESPONSE PRIMITIVES

The protocol between the MC68020 and the coprocessor during normal instruction processing is shown in Figure 9.16 [MTTA2 87]. The MC68020 initiates the activities by writing an instruction command word to the command CIR or conditions selector in the condition CIR. In response, the coprocessor communicates status information and requests service from the main processor. The response primitive set can be considered to be a special class of executable instructions issued by the coprocessor. They are read by the main processor from the coprocessor response CIR, whose format is shown in Figure 9.17. Table 9.1 gives a summary of the primitive command set. These primitives allow the coprocessor to utilize various resources resident on the MC68020.

These special instructions mean various things, ranging from "I am busy, please try again" to "Evaluate the effective address for me" or "Get the data for me and place it into my operand register." The fields in the response register have special meanings, and vary from primitives to primitives. The constant fields for all primitives are Come Again (CA) and Pass Program Counter (PC) bit. The CA bit informs the MC68020 that it should perform the services indicated by the remainder of the primitives, then reread the response register value. If the CA bit is cleared, no services are required of the main processor after the completion of service requested in this primitive, and the MC68020 is free to continue with normal instruction execution.

Bit 14, the PC bit of the response register, indicates that the main processor is to pass its program counter value to the coprocessor via the instruction register, prior to the initiation of any requested service. Bits [13:8] identify the requested

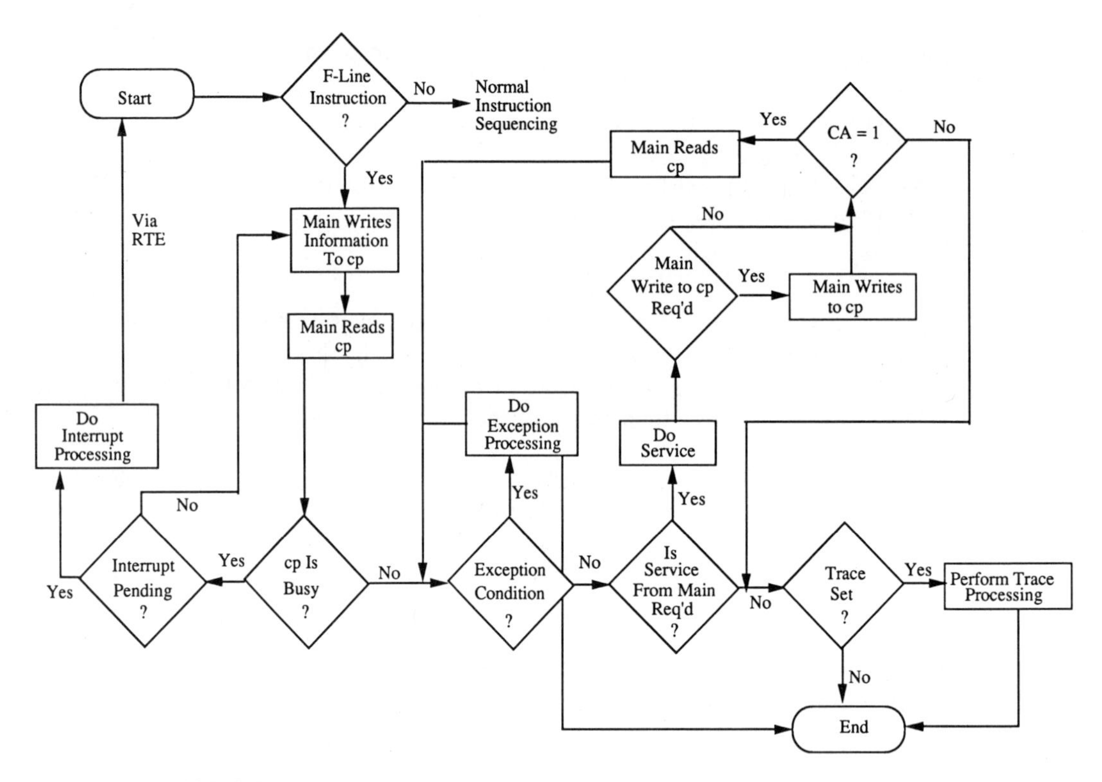

FIGURE 9.16. Coprocessor instruction flow. (Courtesy Motorola Inc.)

function to be performed. Bits [7:0] indicate the specific parameters to the requested function.

Many of the primitives allow for operations to be performed in or out of the coprocessor, indicated by Bit [13], the Direction Bit (DR). If DR = 0, the direction of the transfer is from the main processor to the coprocessor. If DR = 1, the direction of the transfer is from the coprocessor to the main processor. If the operation indicated by a given response primitive does not involve an explicit operand transfer, the value of this bit depends on the particular primitive encoding. The coprocessor response primitives are illustrated in Figure 9.18.

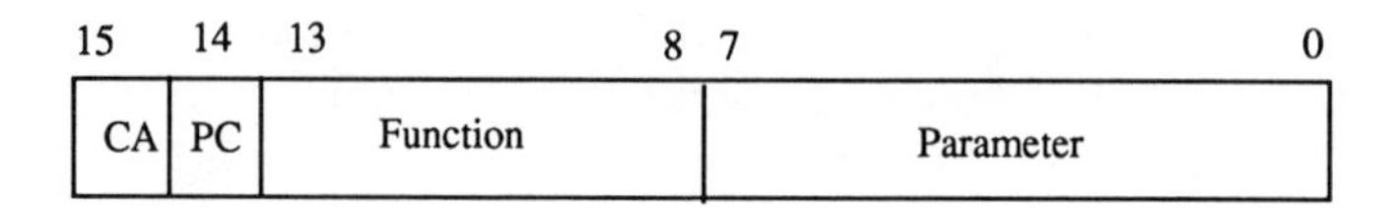

FIGURE 9.17. Coprocessor response register.

TABLE 9.1. Classification for Coprocessor Responses.

Processor Synchronization	*Busy Executing Previous Instruction *Busy Executing Current Instruction *Proceed with the Next Instruction, If No Trace Pending *Proceed with the Next Instruction, If Trace Pending *Proceed with the Execution, Condition True or False
General Operand Transfer	*Evaluate and Pass <ea> *Evaluate <ea> and Transfer Data *Write to Previously Evaluated <ea> *Take Address and Transfer Data *Transfer to/from Top of Stack
Register Transfer	*Transfer Single CPU Register *Transfer Single CPU Control Register *Transfer Multiple CPU Registers *Transfer Multiple Coprocessor Registers *Transfer CPU SR and/or PC Value
Instruction Manipulation	*Transfer Operation Word (OP-Word) *Transfer Words from Instruction Stream
Exception Processing	*Take Privilege Violation If S = 1 *Take Preinstruction Exception *Take Midinstruction Exception *Take Postinstruction Exception

9.4.1 BUSY Primitive

Thirteen generic types of response primitives are available to the coprocessor. The BUSY response primitive causes the main processor to reinitiate a coprocessor instruction. It is designed for coprocessors capable of operating concurrently, but do not have buffering capabilities to write to their command register. In response to this primitive, the main processor will check for and process any pending interrupts, and will then restart the general or conditional coprocessor instruction that it had attempted to initiate earlier. Thus, the first write to the command register is totally ignored by both the main processor and coprocessor.

The MC68020, MC68030, and MC68040 respond to the BUSY primitives as a special case that can occur during a breakpoint operation. This occurs when:

- A breakpoint acknowledge cycle initiates a coprocessor F-line instruction.
- The coprocessor returns the BUSY primitive in response to the instruction initiation.
- An interrupt is pending.

Whenever the above conditions are met, the processor reexecutes the breakpoint acknowledge cycle at the completion of the interrupt exception processing. In a system that uses a breakpoint to monitor the number of passes through a loop by incrementing or decrementing a counter, this may not work correctly

Primitive	15	14	13	12–0
BUSY	1	PC	1	
Transfer Multiple Co-processor Registers	CA	PC	dr	length (bytes)
Transfer Status and/or ScanPC	CA	PC	dr	SP
Supervisor Check	CA	PC	0	
Take Address and Transfer Data	CA	PC	dr	length (bytes)
Transfer Multiple Main Processor Registers	CA	PC	dr	
Transfer Operation Word	CA	PC	0	
NULL	CA	PC	0	IA, PF, T/F
Evaluate and Transfer Effective Address	CA	PC	0	
Transfer to/from Main Processor Register	CA	PC	dr	D/A, register
Transfer to/from Main Processor Control Register	CA	PC	dr	
Transfer to/from Top-of-Stack	CA	PC	dr	length (bytes)
Transfer from Instruction Stream	CA	PC	0	length (bytes)
Evaluate Effective Address and Transfer Data	CA	PC	dr	valid <EA>, length (bytes)
Take Pre-Instruction Exception	0	PC	0	vector number
Take Mid-Instruction Exception	0	PC	0	vector number
Take Post-Instruction Exception	0	PC	0	vector number
Invalid (Reserved)	CA	PC	0	
Write to Previously Evaluated Effective Address	CA	PC	1	length (bytes)
Invalid (Reserved)	CA	PC	1	

NOTES: dr: 0 – Into Co-processor
1 – Out of Co-processor
D/A: 0 – Data Register
1 – Address Register

Valid <EA>: 000 – Control Alterable
001 – Data Alterable
010 – Memory Alterable
011 – Alterable
111 – Any Mode Allowed (No Restrictions)
100 – Control
101 – Data
110 – Memory

FIGURE 9.18. Coprocessor response primitives. (Courtesy Motorola Inc.)

under these conditions. This special case may cause several breakpoint acknowledge cycles to be executed during a single pass through a loop.

9.4.2 NULL Primitive

The NULL primitive is used to maintain synchronization between the main processor and the coprocessor. The format of a NULL primitive is:

Bit [15]—Come Again
Bit [14]—Pass Program Counter (PC)
Bit [8]—Interrupt Allowed (IA)
Bit [1]—Processing Finished (PF)
Bit [0]—True/False (TF) Condition

A NULL primitive with the CA bit set is referred to as a NULL Come Again primitive. The only action taken by the main processor is to check the status of the IA bit and either handle or not handle an interrupt based on this bit state prior to rereading the response register.

If the CA and PF bits are cleared, then this is a NULL Release primitive. It indicates that no more main processor services are needed by the coprocessor to complete the current instruction. As a result, the main processor is free to continue with its normal execution sequence.

If the PF bit is set, then it indicates that the coprocessor has completed all processing associated with the current instruction. This is a NULL-Done primitive. The T/F bit is only valid for a NULL primitive issued in response to conditional coprocessor instructions.

9.4.3 Transfer Operation Word Primitive

This primitive allows the coprocessor to get a copy of the F-line instruction that caused the main processor to open communication with the coprocessor, into its Op-word interface register. It is used with general and conditional types of instructions. As a response to this primitive (after reading from the response CIR), the main processor transfers the operation word of the currently executing coprocessor instruction to the operation word CIR of the coprocessor.

This primitive uses the CA and PC bits. If this primitive is used with CA = 0 during a conditional category instruction, then the MC68020 initiates protocol violation exception processing.

9.4.4 Supervisor Check Primitive

The supervisor check primitive allows the coprocessor to execute privileged instructions. When the processor reads the supervisory check primitive from the response CIR, it tests the S bit within the SR register. If the S bit is set, the processor takes no action. However, if the bit is cleared, the processor writes $0001 to the control CIR, which causes the coprocessor to abort processing, and the main processor initiates protocol violation exception processing.

The Supervisor Check primitive allows the mapping of privileged instructions within the coprocessor general and conditional instruction categories. This primitive should be the first one issued by the coprocessor during the dialog, for an instruction that is classified as privileged.

9.4.5 Coprocessor Exception Processing

An important feature of the coprocessor interface is the manner in which exceptional conditions are managed. Whenever any exception occurs during a coprocessor operation, the coprocessor utilizes one of the three primitives to request the main processor to begin exception processing. This exception is processed in the same way as the processor would if the exception had been detected during the execution of a main processor instruction. This eliminates the need for the coprocessor to use interrupts to report exceptions to the main processor, and allows the coprocessor to be included on the same chip (found in the MC68030 and MC68040), with no modifications to the exception handling mechanism.

Three primitives are supported by the MC68000 coprocessor interface, which will cause the main processor to take exception processing on behalf of the coprocessor. The primitives differ in the way that stack frames are saved by the MC68020 during exception processing. During the execution of an RTE instruction from the exception handler, the MC68020 will either restart the instruction during which the primitive was received, and continue the instruction, or proceed with the execution of the next instruction in the instruction sequence. A single vector number is used by the main processor during the exception processing of all of the primitives. The primitive set consists of:

- Take Preinstruction Exception
- Take Midinstruction Exception
- Take Postinstruction Exception

9.4.5.1 Take Preinstruction Exception

This exception primitive initiates exception processing using a coprocessor-supplied vector number and the preinstruction exception stack frame format. It applies to general and conditional category instructions. The format of the response register and the stack frame are shown in Figure 9.19. During coprocessor instruction execution, if an illegal command word is received in the command CIR, or an illegal condition is received in the condition CIR, then the coprocessor will issue a preinstruction exception using the F-line trap vector. The MC68020 will first acknowledge the coprocessor exception request by writing $0002 into the control CIR. The processor will then proceed with exception processing by collecting the excepting vector number from bits [7:0]. The main processor uses a four-word stack frame to save the context of the processor.

The preinstruction exception processing primitive is an important vehicle for handling exception conditions during cp-GEN instructions. The coprocessor is

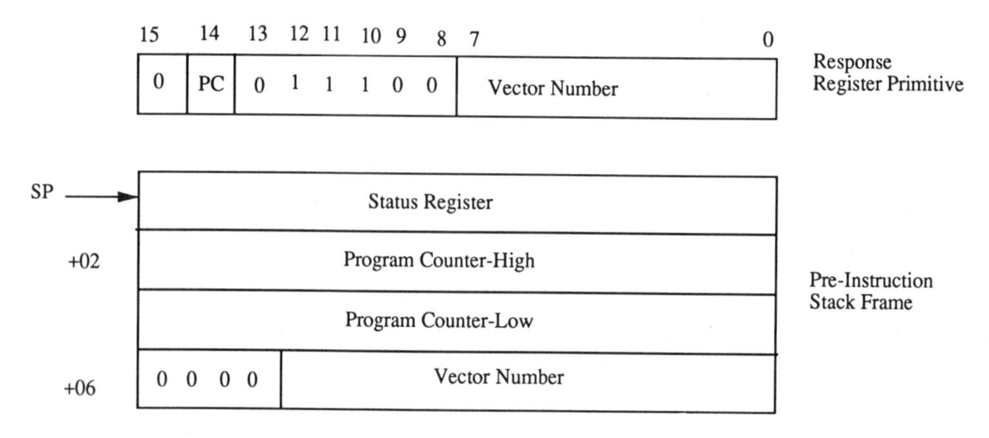

FIGURE 9.19. Preinstruction exception.

allowed to operate concurrently with the processor, and it is then very possible that the coprocessor will detect an exception condition due to a computational error or trap condition after it has released the main processor. In this case, the coprocessor will report the exception as a preinstruction exception, and will wait until the main processor attempts to initiate the next general or conditional instruction before the exception can be processed, after the main processor acknowledges the receipt of a request to perform exception processing by writing an acknowledge code to the coprocessor's control CIR. Now, the coprocessor can rerun the take preinstruction exception primitive. This will allow the main processor to proceed with exception processing related to the previous concurrently executing coprocessor instruction, and then return to reinitiate the coprocessor instruction during which the exception was signaled.

9.4.5.2 Take Midinstruction Exception

The take midinstruction exception primitive initiates exception processing using a coprocessor-supplied exception vector number. This primitive also applies to the general and conditional class of instructions. Upon recognition, the main processor acknowledges the coprocessor exception request by writing an acknowledge mask ($0002) to the control CIR. The vector number of the exception is copied from bits [7:0] of the primitive, and the main processor uses the ten-word stack frame format to save the context. The format of the response register primitive and the stack frame format are shown in Figure 9.20.

The MC68020 saves the PC value (operation word address of the coprocessor instruction during which the primitive was received), the SR, and the vector number in this stack frame. The ScanPC field contains the value of the MC68020's ScanPC when the primitive was received. It is important to note that, if no primitive caused the evaluation of the effective address in the coprocessor instruction operation word prior to the exception request primitive, the value of the effective address field in the stack frame remains undefined.

The coprocessor uses this primitive to request exception processing when it completes or aborts an instruction while the main processor is awaiting a normal response. The response is a release for a general category instruction. For a conditional category instruction, it is an evaluated false/true condition indicator. Thus, this exception processing is compatible with the MC68000 family exception processing.

9.4.5.3 Take Postinstruction Exception Primitive

This primitive applies to general and conditional category instructions. Here, the main processor initiates exception processing using a coprocessor-specified exception vector and the postinstruction exception stack frame. The format for the postinstruction exception primitive, and the accompanying six-word stack frame generated by the main processor, is shown in Figure 9.21.

When the MC68020 receives this primitive, it sends an acknowledgment to the coprocessor by writing an exception mask ($0002) to the control CIR of the coprocessor. The MC68020 then performs exception processing, as described in Chapter 7.

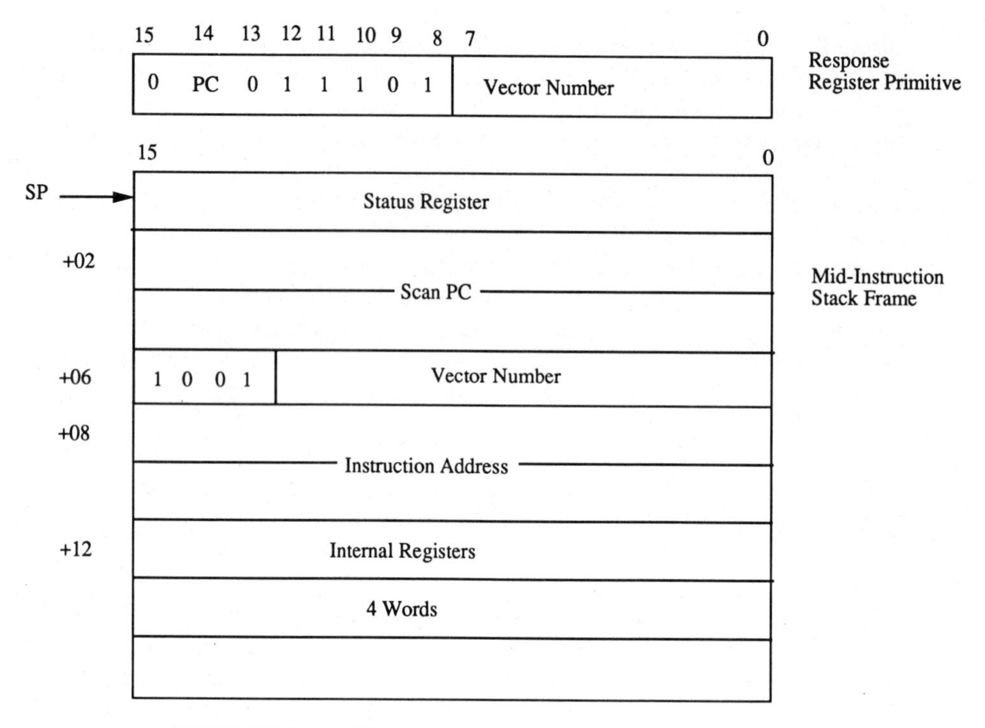

FIGURE 9.20. Midinstruction exception.

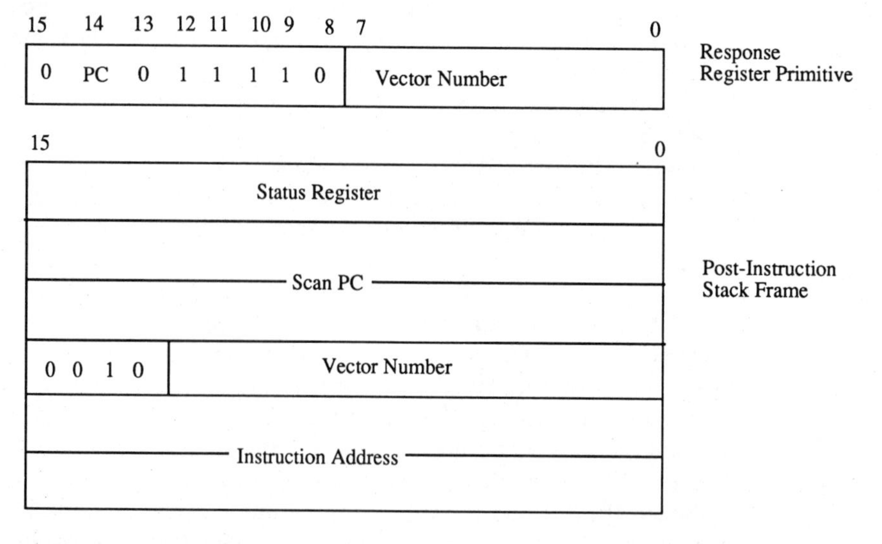

FIGURE 9.21. Postinstruction exception.

The value in the main processor ScanPC during the receipt of this primitive is stored in stack frame location 2. The value of the PC saved is the F-line Op-word address of the coprocessor instruction during which the primitive is received.

The coprocessor uses this primitive to request exception processing as it completes or aborts an instruction while the main processor is awaiting a normal response. The expected normal response for a general category instruction is "Release," and for conditional category instruction is "Evaluate True/False Condition." As a result, the operation of the MC68020 in response to this primitive is compatible with the standard MC68000 family instruction-related exception processing.

9.5 COPROCESSOR EXCEPTION PROCESSING

Problems may arise during the execution of coprocessor instructions. These problems may be detected by the main processor or coprocessor. Problems may be caused by: a protocol violation, an illegal command or condition, a failure to execute instructions, a trace, a bus error, an address error, etc. Exceptions detected by the coprocessor or by the main processor on behalf of the coprocessor are classified as coprocessor-detected exceptions. These exceptions can occur during MC68020 and coprocessor interface operations, internal operations, or other system-related operations of the coprocessor. Coprocessor-related exceptions are classified into five groups, namely:

- Coprocessor-detected protocol violations.
- Coprocessor-detected illegal command condition words.
- Coprocessor data processing exceptions.
- Coprocessor system-related exceptions.
- Format errors.

9.5.1 Coprocessor-Detected Protocol Violations

The coprocessor-detected protocol violation exceptions are caused by a communication failure between the main processor and coprocessor across the coprocessor interface. The coprocessor raises a flag as soon as it detects an out-of-sequence access by the main processor to its interface register set.

The specification of the MC68000 coprocessor protocol specifies that the main processor will always access the Op-word, operand, register select, instruction address, or operand address CIRs in a specific sequence, with respect to the operations of the coprocessor. In addition, the specification requires the coprocessor to detect a protocol violation if the main processor accesses any of these five registers when it is expecting an access to either the command or condition CIR. The coprocessor is incapable, however, of signaling a protocol violation to the main processor during the execution of CpSAVE and CpRESTORE instructions. However, the coprocessor needs to signal the protocol violation to the main processor during the next coprocessor instruction execution sequences.

The concept behind the coprocessor-detected protocol violation is that the coprocessor should always acknowledge an access to one of its interface registers. If the coprocessor determines that the access is not a valid one, it should assert $\overline{\text{DSACKx}}$ and signal a protocol violation whenever the main processor reads the response CIR during the next sequence. If the coprocessor fails to assert $\overline{\text{DSACKx}}$, the main processor waits for the assertion of that signal indefinitely. In such situations, some time-out mechanism should be in place to take the main processor out of this mode.

Coprocessor-detected protocol violations can be signaled to the main processor with the take midinstruction exception primitive. In addition, it should be encoded with the coprocessor protocol violation exception vector number 13, to maintain consistency with the main processor-detected protocol violations. However, if no modifications are made to the stack frame within the exception handler, then (after an RTE instruction) the MC68020 will read the response CIR.

9.5.2 Coprocessor Data Processing Exceptions

Exceptions related to the internal operations of the coprocessors fall into the category of data processing-related exceptions. An example of this is the divide by zero exception, defined by the MC68000 processor, and should be signaled to the main processor using one of the three take exception primitives containing an appropriate exception vector number. The restarting point identifies which of the three exception primitives will be used to signal an exception.

9.5.3 Coprocessor-Detected Illegal Command or Condition Words

If the coprocessor fails to recognize the values written into the command CIR or condition CIR, then this situation is known as a coprocessor-detected illegal command or condition. The coprocessor should initiate a take preinstruction exception primitive in the response CIR. Upon receiving this primitive, the main processor will initiate a take preinstruction exception sequence. At the end of exception processing, the MC68020 will try to reinitiate the instruction that caused the exception. Thus, the designer of the coprocessor should ensure that the state of the coprocessor is not unrecoverably altered by an illegal command or condition exception, if emulation of the unrecognized command or condition word is to be supported.

All of the MC68000 family of coprocessors from Motorola signal illegal command and condition words by returning the take preinstruction exception primitive, with the F-line emulator exception vector number 11 [Motorola 85].

9.5.4 Coprocessor System-Related Exceptions

The system-related exceptions are initiated by a coprocessor when it detects bus-related problems, interrupts, or other exceptions external to it. The actions taken by the coprocessor and the main processor depend on the nature of the exception.

When an address error is detected by the DMA-type coprocessor, the coprocessor needs to store all relevant information for the main processor exception handler routine into system accessible registers. The coprocessor needs to place one of the three take exception primitives, encoded with an appropriate exception vector number, in the response CIR. The selected primitive depends on the point at which the coprocessor instruction exception was detected, and the point at which the main processor should restart after it returns from exception processing.

9.5.5 Format Errors

Format errors are the only coprocessor-detected exceptions not forwarded to the main processor with the response primitives. During the execution of a cpRESTORE instruction, the main processor writes a format word to the restore CIR. The coprocessor first decodes this word to identify its validity. If the format word is not valid, the coprocessor places the invalid format code in the restore CIR. When the main processor reads the invalid format code, it aborts the coprocessor instruction by writing an abort mask ($0001) into the control CIR. The main processor then performs exception processing, using a four-word preinstruction stack frame and the format error exception vector number 14. Upon returning from the exception routine the MC68020 will restart the cpRESTORE instruction. However, if the coprocessor returns the invalid format code when the main processor reads the save CIR to initiate a cpSAVE instruction, the main processor performs format error exception processing.

9.6 COPROCESSOR INSTRUCTION EXAMPLE

An example of a floating-point coprocessor instruction is shown in Figure 9.22 [MTTA2 87]. It performs a floating-point move of a 96-bit packed BCD data from memory, pointed to by address register A0, to a floating-point coprocessor register FP0. Let us examine the execution of this instruction, bus cycle by bus cycle basis.

The MC68020 detects an F-line instruction. The processor determines that it is a general type of coprocessor instruction. Therefore, it writes the word following the F-word into the coprocessor command register. This bus cycle occurs at CPU space 2, with a coprocessor ID of #1, and with an offset of $A. This selects the command register within the coprocessor's mailbox. The coprocessor accepts the command word $4C01 and returns $\overline{\text{DSACKx}}$. The coprocessor determines the nature of the service that it will require from the MC68020, and concludes that it must ask the MC68020 to calculate the effective address and get 12 bytes of data from the memory location pointed by the <EA>. As a result, the coprocessor places $960C into its response register. The MC68020 reads the contents of the coprocessor response register and determines that it is required to transfer 12

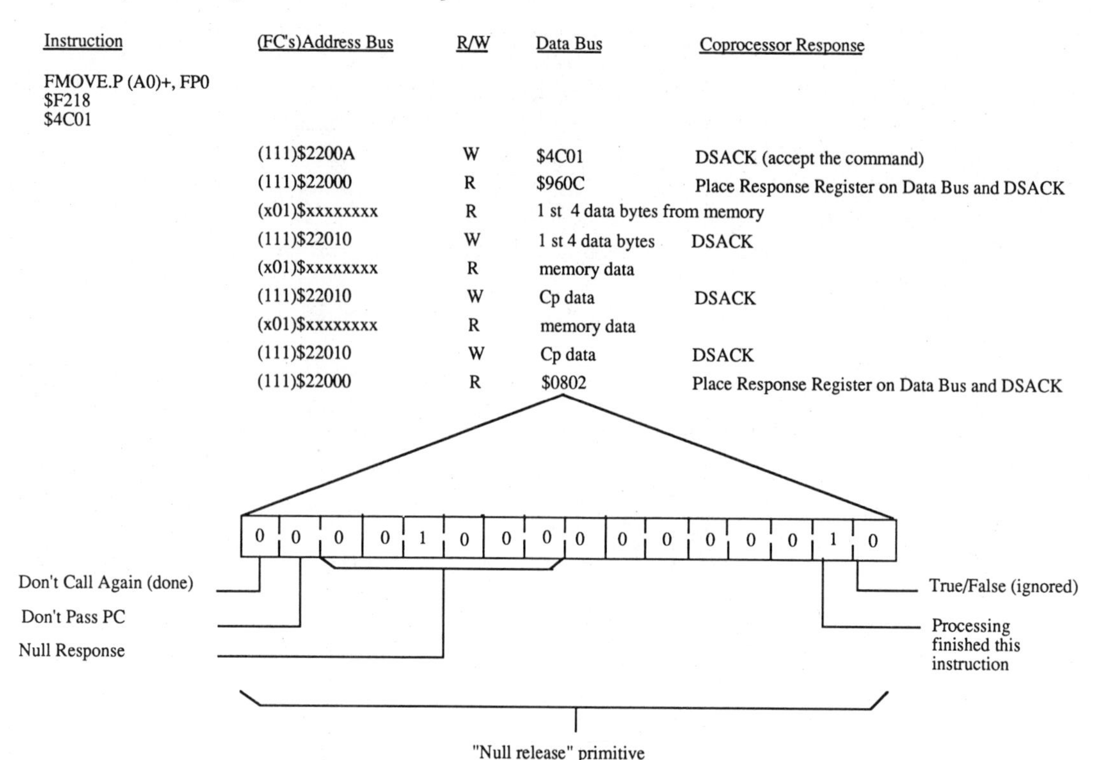

Instruction
(FC's)Address Bus
R/W
Data Bus
Coprocessor Response
FMOVE.P (A0)+, FP0
$F218
$4C01
(111)$2200A W $4C01 DSACK (accept the command)
(111)$22000 R $960C Place Response Register on Data Bus and DSACK
(x01)$xxxxxxxx R 1 st 4 data bytes from memory
(111)$22010 W 1 st 4 data bytes DSACK
(x01)$xxxxxxxx R memory data
(111)$22010 W Cp data DSACK
(x01)$xxxxxxxx R memory data
(111)$22010 W Cp data DSACK
(111)$22000 R $0802 Place Response Register on Data Bus and DSACK
0 0 0 0 1 0 0 0 0 0 0 0 0 0 1 0
Don't Call Again (done)
Don't Pass PC
Null Response
True/False (ignored)
Processing finished this instruction
"Null release" primitive

FIGURE 9.22. FPCP instruction example

(Courtesy Motorola Inc.)

bytes of data from the memory pointed to by the address register A0 to the operand interface register of the coprocessor.

Thus, we will observe a read bus cycle instruction taking place from the data space, followed by a write bus cycle instruction to the coprocessor in the CPU space. The read and write sequences will be repeated three times. A total of 12 bytes of data are transferred, and the address of A0 is incremented by 12 at the conclusion of this instruction execution.

Finally, since the CA bit was set during the initial read of the response register of the coprocessor, the MC68020 reads the content of the response register to examine if any additional service is required. The coprocessor issues the null release primitive, signaling that no more services are required. Now, the MC68020 determines that it is free to continue with the execution of the next instruction from its instruction stream.

Chapter 10

Memory Management Concepts in the MC68020

In a simple microprocessor-based system, a processor is directly connected to the memory. No memory mapping or protection functions are provided. The address generated by the processor identifies the physical locations to be accessed. Any location in the address space that does not contain a memory device cannot be used by the processor. This simple mechanism has proved to be inadequate for the execution of multiple concurrent tasks, since there is no mechanism to protect the memory of one task from getting corrupted by another task. It is also inadequate for hosting virtual systems that allow the uniform use of address space that is larger than the physical address space. These issues are addressed in great detail in Chapter 8.

Thus, memory management is a necessary prerequisite for most high-performance systems. As hardware costs decrease, and the software community demands more memory, chip designers are coming forward with intelligent devices to manage the execution of complex tasks like address translation, page mapping, housekeeping, protection, etc. at a high speed, releasing the main processor to do other important tasks.

In this chapter, we will focus our attention to the architecture of the Paged Memory Management Unit (PMMU) MC68851, its interface with the MC68020, and its operation. The MC68851 is a good example of the implementation of one of today's memory management devices, designed to support the memory management functions required by a large number of microprocessor-based systems.

10.1 PAGED MEMORY MANAGEMENT UNIT MC68851

The MC68851 is fabricated using HCMOS technology. It is designed to perform fast logical-to-physical address translation, and to provide access control, protection mechanisms, and extensive support for paged virtual systems. The device is designed to work with the MC68020 as a coprocessor and implements a paged memory management environment, where pages are swapped in and out of memory based on need.

When interfaced with the MC68020 as a coprocessor, it supports a logical extension to the program control and processing abilities of the main processor. A set of translation, protection, and breakpoint registers are included to control the operations of the memory management function. To users, these registers are a logical extension to the processor's internal registers.

The main features of the MC68851 are [Motorola Inc. 1986]:

- Fast logical-to-physical address translation.
- Support for up to eight levels of hierarchical protection.
- Full 32-bit logical and physical addresses, with 4-bit function code.
- Programmable page size, ranging from 256 to 32K bytes.
- Fully associative 64-entry on-chip address translation cache.
- Automatically updates the on-chip translation cache from an external translation table.
- Multitasking and task-switching capabilities.
- Instruction breakpoint capability for software debugging and program control.
- Support for logical and/or physical data caches.
- Support for multiple logical and/or physical bus masters.

10.1.1 System Configuration for an MC68020-Based Virtual System

A virtual memory system consisting of the MC68020 and MC68851 is shown in Figure 10.1. Here, the address generated by the main processor is commonly known as the *logical address* and the address generated by the memory management unit is called the *physical address.* The logical address is generated by the processor, and is monitored by the MC68851 on its logical address inputs. The MC68851 performs an address translation and privilege checking on the logical address and, if valid, asserts the translated physical address values over the physical address bus. This physical address is then used to access the memory or peripheral devices. Thus, all accesses to physical devices are controlled by the MC68851. As a result, tasks can be prevented from accessing the resources used by other tasks. Under the direction of the operating system, the MC68851 performs the logical-to-physical address mapping, and allows tasks to utilize the entire address space of the processor without the knowledge of the physical attributes of the system.

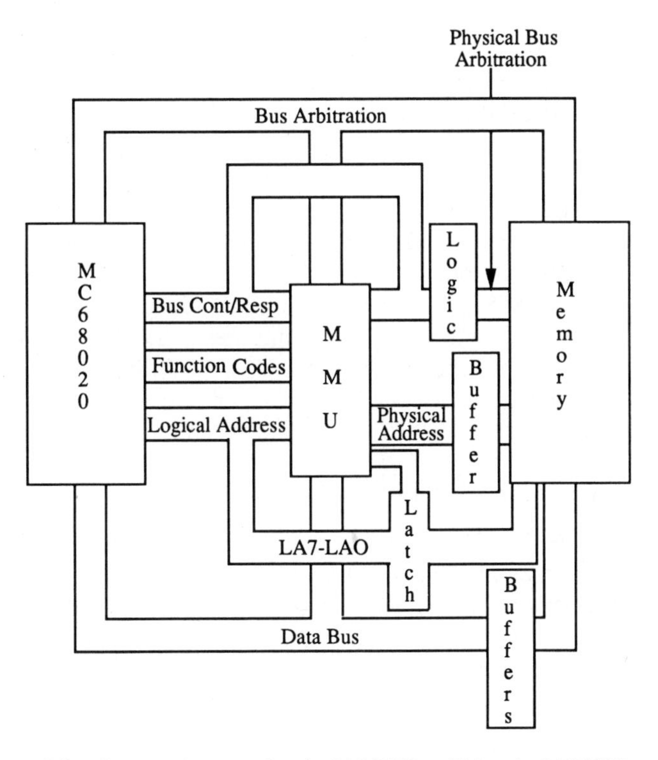

FIGURE 10.1. Virtual memory system for the MC68020, utilizing the MC68851.

The features of the MC68851 are implemented by eight major components, which are:

- The Bus Interface Unit (BIU)
- The Address Translation Cache (ATC)
- The Root Pointer Table (RPT)
- The Execution Unit (EU)
- The control store
- The control logic
- The Address Translation Sense Circuit (ATSC)
- The register decode logic

The BIU controls the interface logic to both the logical and physical buses, including the bus arbitration state machine for both buses.

The ATC maintains the most recently used 64 translation descriptors, including the required control circuitry to monitor the access rights and create new ATC entries.

The RPT contains a cache to retain the eight most recently used values of the CPU root pointer, and a task alias associated with each of the stored values.

The EU performs address calculations for accessing the translation tables, contains the MC68851 register sets, and controls table search activities, including instruction execution.

Two-level microcodes are used to implement the control store section of the MC68851. Address generation, the table search algorithm, and the instruction sets are all implemented in microcodes.

The control logic section implements the remaining logic for the MC68851. This includes the decode for the control store, the register decode outputs, and the control points in the EU.

The ATSC monitors the logical address bus for any changes in status. A change in the logical address initiates activity in the ATC, which first checks the access rights and generates the corresponding physical address.

The register decode logic selects and manages the operations of various internal registers under the direction of the control logic.

10.1.2 Programming Model of the MC68851

The MC68851 appears to the user as a logical extension of the MC68020's programming model when it is interfaced with the processor as a coprocessor. Thus, the MC68020/MC68851 pair appears as one entity, which has registers for data storage, address pointers, general control, translation and protection control, breakpoint functions, etc.

The programming model of the MC68851 is shown in Figure 10.2 and contains:

- Three 64-bit root pointer registers, CRP, SRP, and DRP. They point to the root of user programs (CRP), supervisory programs (SRP), and DMA (DRP) translation table.
- A 32-bit Translation Control (TC) register, which contains configuration information for the MC68851.
- A 16-bit Cache Status Register (PCSR), to provide information concerning the MC68851's internal translation cache.
- A 16-bit Status Register (PSR), which contains status and access right information for a given logical address.
- Three 8-bit protection control registers (CAL, VAL, and SCC) are used during privilege checking.
- A 16-bit Access Control (AC) register, which contains configuration information for the privilege mechanism.
- Eight 16-bit Breakpoint Acknowledge Data (BAD0–BAD7) registers, to provide replacement Op-code during the MC68020 breakpoint acknowledge cycles.
- Eight 16-bit Breakpoint Acknowledge Control (BAC0–BAC7) registers, which enable and count functions for the instruction breakpoint capabilities of the MC68020 and MC68851.

10.1.3 Root Pointer Formats

The three root pointer registers, CRP, SRP, and DRP share the same format. The format is illustrated in Figure 10.3.

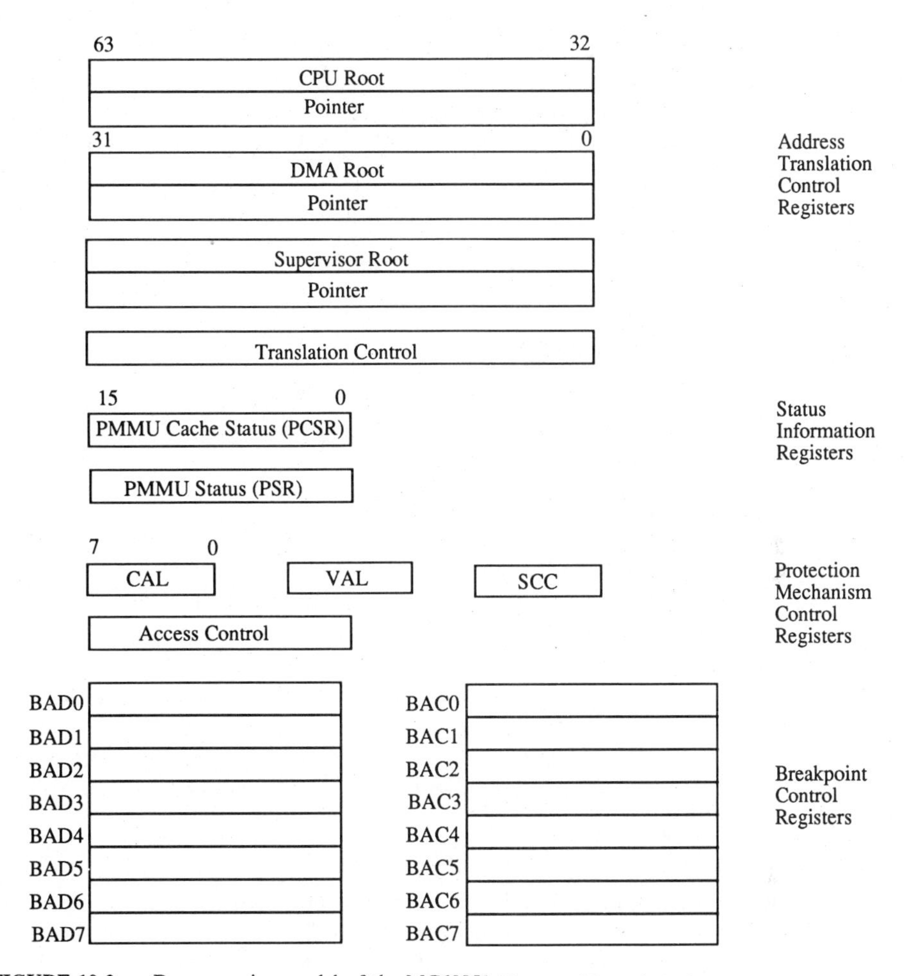

FIGURE 10.2. Programming model of the MC68851 (Courtesy Motorola Inc.)

The CPU Root Pointer (CRP) points to the root of the translation tree for the currently executing task in the main processor. The operating system reloads the CRP to point to the root of the translation tree before invoking a new task for execution. The CRP works in conjunction with the Root Pointer Table (RPT). Thus, updating the CRP may cause ATC entries to be invalidated and the PCSR register to be updated.

The Supervisory Root Pointer (SRP) points to the operating system's translation table. The SRE bit of the Translation Control Register (TCR) controls the operation of the SRP. If the SRE bit is cleared, then the SRP is not used. Instead, the CRP is used to translate supervisory access. However, if the SRE is set, then the SRP points to the root of the translation table to be utilized during translation.

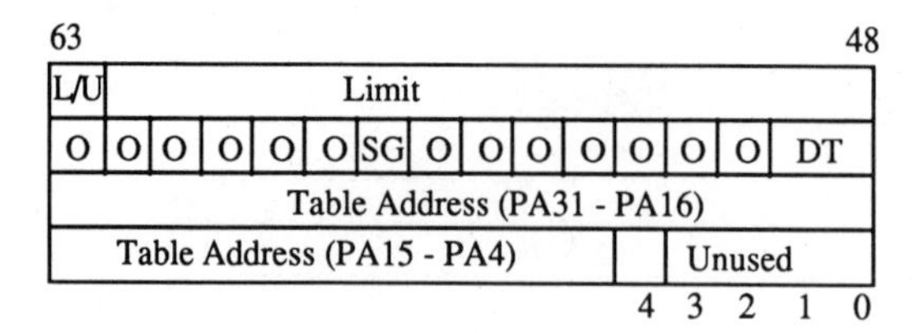

Note: L/U — Lower or Upper Page Range

SG — Shared Globally

DT — Descriptor Type

Limit — Limit on the Table Index for this Table Address

Table Address — Address of the Table at the Next Level or Page Offset If DT = 1

FIGURE 10.3. Format of the Address Translation Control (ATC) registers.

The DMA Root Pointer (DRP) manages the space of an alternate bus master, which could be a DMA controller. This address space is separate from that of the main processor. When FC3 = 1, then the DRP points to the root of the translation table for this separate address space.

The root pointer to be used during the generation of the logical address depends on the function code status and the SRE bit of the TC register. Table 10.1 lists the combinations used in selection of the root pointer.

The various fields of the registers are explained in the following subsubsections.

10.1.3.1 Lower/Upper (L/U) Field

Bit [63] the L/U field specifies the upper or lower limits to be used as indices for the next level of the translation table. If L/U = 0, then the limit field contains the unsigned upper limit of indices. L/U = 1 specifies the unsigned lower limit of the indices.

10.1.3.2 Limit Field

The limit field, bits [62:48], specifies a maximum or minimum value for the index to be used at the next level of the table search, which in turn determines the size of the next level of the transition tree.

TABLE 10.1. Section Mechanism for Root Pointers.

Function codes			
FC3	FC2	SRE	Root Pointer Used
0	0	0	CRP
0	0	1	CRP
0	1	0	CRP
0	1	1	SRP
1	x	x	DRP

10.1.3.3 Shared Global Field

Bit [41], the SG field, tells the MC68851 that the logical-to-physical address mapping specified by this root pointer is the same for all tasks, when set, and a single descriptor for the translation is to be maintained in the ATC. However, if this field is set to 0, then it tells the MC68851 that the logical-to-physical mapping identified by the root pointer is unique for this particular task.

10.1.3.4 Descriptor Type Field

The DT field bit [33:32], specify the type of descriptor contained in the root pointer or in the first level of the translation table pointed by that root pointer. Table 10.2 lists the meanings associated with the DT fields.

10.1.3.5 Table Address Field

The table address field bits [31:4] specify the physical base address of the translation table for a root pointer. If the DT=1 (i.e., it is a page descriptor), then the value in the table address field identifies a constant offset to the logical address during the creation of a page descriptor by the MC68851. The constant offset can have a zero value.

10.1.3.6 Unused Bits

Bits [3:0] of the root pointer are not used by the MC68851. These bits may be used for operating system activities.

The rest of the unused bits of the address translation control registers format are set to zero.

10.1.4 Adddress Translation Control Register Format

The format of the Address Translation Control (ATC) register is shown in Figure 10.4. It is used to configure the address translation mechanism of the MC68851. The meanings of the various control fields are described in the following subsubsections.

TABLE 10.2. Root Pointer Descriptor.

DT[33]	DT[32]	Results
0	0	Invalid; Table address does not point to a valid translation table.
0	1	Page descriptor; No transition table exists and an ATC entry is created using this root pointer.
1	0	Valid 4-byte; The root of the translation tree contains a short format descriptor.
1	1	Valid 8-byte; The root of the translation tree contains a long format descriptor.

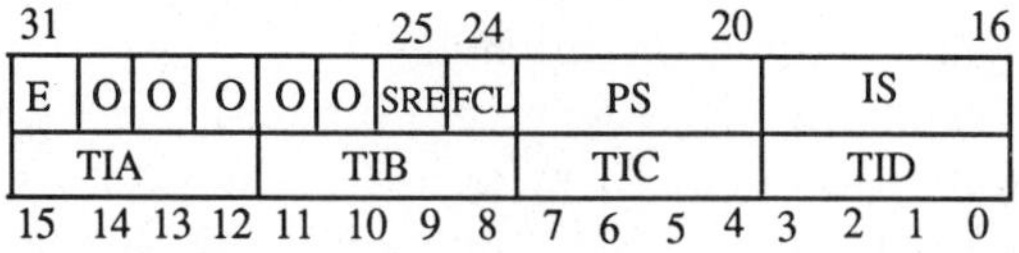

31						25	24	20	16
E	O	O	O	O	O	O	SRE	FCL	PS	IS

TIA	TIB	TIC	TID
15 14 13 12	11 10 9 8	7 6 5 4	3 2 1 0

E — Enable

SRE — Supervisor Root Pointer Enable

FCL — Function Code Look-up Enable

PS — Page Size

IS — Initial Shift

TIA,TIB,TIC, TID — Table Indices

FIGURE 10.4. Translation Control Register (TCR) format.

10.1.4.1 Enable (E) Field

Bit [31], when set, invokes the address translation mechanism of the MC68851, and allows execution of the PLOAD, PTEST, and CALLM instructions. When cleared, the MMU disables its translation operation. It also blocks the execution of all PTEST, PLOAD, and CALLM/RTM instructions, and requires the CPU to initiate exception processing.

When E = 0, the contents of the logical address bus gets routed to the physical address bus, the MC68851 asserts $\overline{PAS}$ (Physical Address Strobe) for all non-CPU space cycles, and $\overline{CLI}$ is asserted by the MMU for all CPU space cycles that do not access the MMU device. An externally generated reset, or instruction-generated reset, clears the E bit.

10.1.4.2 Supervisory Root Pointer Enable (SRE) Field

The SRE, Bit [25], when set, directs the MC68851 to perform address translation using the translation tree pointed to by the SRP. When cleared, the MC68851 utilizes the CRP for supervisory space translation.

10.1.4.3 Function Code Look-Up (FCL) Field

The status of the FCL field determines whether or not the top-level table in the address translation tree should be indexed by the function code during the use of the CRP or SRP. When FCL = 0, function code look-up is not used in the search process. Instead, the first look-up index is created, using the portion of the logical address space specified by the IS and TIA fields. However, a function code look-up is always performed when the MC68851 executes a table search using the DMA root pointer.

10.1.4.4 Page Size (PS) Field

The PS field specifies the page size to be utilized by the MC68851. This field must contain a valid number, or else an MMU configuration exception will be generated by the coprocessor. The valid numbers are:

Number	Page Size
$8	256 bytes
$9	512 bytes
$A	1 Kbyte
$B	2 Kbytes
$C	4 Kbytes
$D	8 Kbytes
$E	16 Kbytes
$F	32 Kbytes

10.1.4.5 Initial Shift (IS) Field

The IS field determines how many most significant bits in the logical address should be ignored by the MC68851 during an address table search operation. This field may take integer values from 0 through 15, and specifies the number of bits to be discarded from the logical address, starting at the most significant bit, bit 31.

Although the MC68851 will ignore higher-order logical address bits during the table search, as required by the value in the IS field, it is a good design practice to tie the unused address lines to Vcc via pull-up registers.

10.1.4.6 Table Index (TIA, TIB, TIC, and TID) Fields

The table index fields specify the bit size of the logical address to be used as indices into the translation tables at each level during a table search operation. This field takes an unsigned integer value from 0 through 15.

10.1.5 PMMU Cache Status Register (PCSR) Format

Figure 10.5 illustrates the format of this 16-bit read-only register. The value of the PCSR is updated whenever the CRP (CPU Root Pointer) register is updated by the PMOVE or PRESTORE instruction. The operating system reads the contents of this register to acquire the ATC information. Three fields of this register are of any significance, and are listed in the following subsubsections.

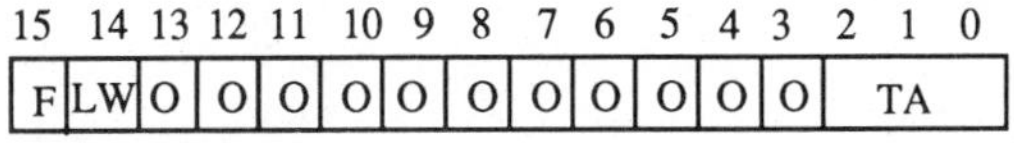

15	14	13	12	11	10	9	8	7	6	5	4	3	2	1	0
F	LW	O	O	O	O	O	O	O	O	O	O	O	TA		

F — Flush

LW — Lock Warning

TA —Task Alias

FIGURE 10.5. Cache Status Register (TCR) format.

10.1.5.1 Task Alias (TA) Field
Bits [2:0] contain the current internal task alias maintained by the MC68851.

10.1.5.2. Flush (F) Field
Bit [15], when set, indicates that entries in the task alias shown in the TA field have been flushed. Otherwise this bit is cleared.

10.1.5.3 Lock Warning (LW) Field
Bit [14] is set when all entries in the ATC but one have been locked.

10.1.6 Access Control Register

This 16-bit register configures the various access control mechanisms supported by the MC68851. The format of this register is shown in Figure 10.6. It enables and disables the access level during the MC68020's module call/return instruction.

The ALC (Access Level Control) field determines the number of upper logical address bits used as access level information, including their invocation. The encoded meaning of this field is:

$0—Access level checking is disabled.
$1—Two access levels are used.
$2—Four access levels are used.
$3—Eight access levels are used.

The MDS (Module Descriptor Size) field identifies the boundaries within which a module descriptor is permitted to fall. The meaning of this field is:

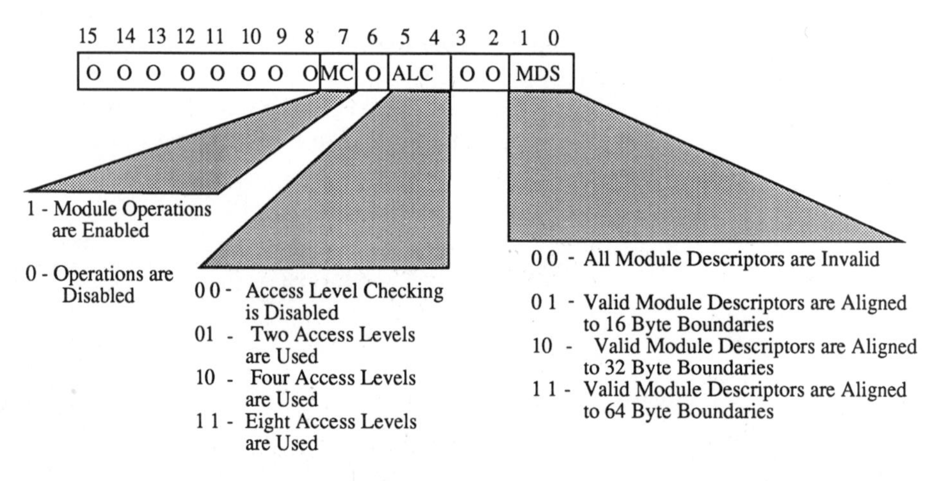

FIGURE 10.6. Access Control Register (ACR) format.

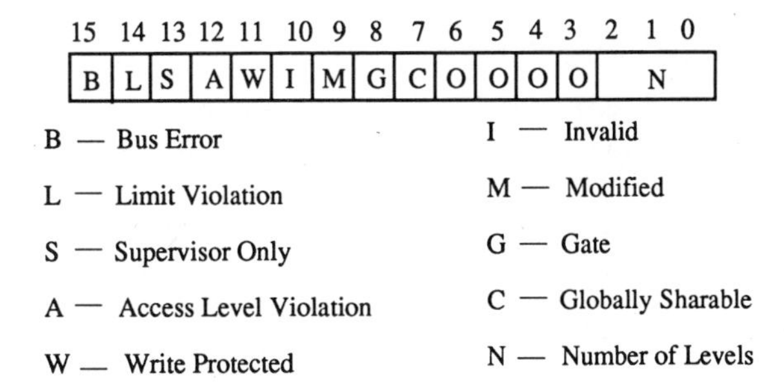

FIGURE 10.7. PMMU status register format.

$0—All module descriptors are invalid.
$1—Valid and aligned to 16-byte boundaries.
$2—Valid and aligned to 32-byte boundaries.
$3—Valid and aligned in 64-byte boundaries.

10.1.7 PMMU Status Register (PSR) Format

The format of the 16-bit PSR is shown in Figure 10.7. It is used by the operating system to identify the source of a fault. The contents of the PSR can be changed using the PTEST instruction.

10.1.8 Breakpoint Acknowledgment Control (BAC7–BAC0) Registers

The MC68851 is equipped with eight 16-bit breakpoint acknowledgment registers. They contain the enable and count functions for the instruction breakpoint acknowledgment mechanism. The format of this register is shown in Figure 10.8.

The BPE (Break Point Enable) field bit [15], when set, means that the breakpoint instruction corresponding to this register is enabled. The BPE bit is cleared at reset.

Bits [7:0] of the BACx registers contain unsigned integers that specify the number of times a specific breakpoint acknowledgment data register should be returned to the CPU. When the value in this field becomes negative, the MC68851 terminates the breakpoint acknowledgment cycle by asserting a bus error.

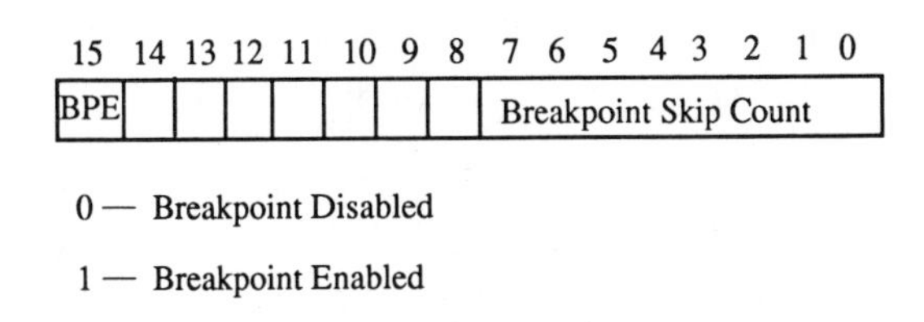

FIGURE 10.8. Breakpoint acknowledgment control register.

10.2 INSTRUCTION SETS

The MC68851 was designed to work as a coprocessor for the MC68000 family of processors. Thus, its instruction set appears to the programmer as an extension to the MC68020's instructions. Instructions are used by programmers to access registers and functions supported by the MC68851.

The instructions of the MC68851 fall into three groups, namely:

- Loading and storage of PMMU registers.
- Testing of access rights and conditional branches based on the results.
- PMMU control functions.

All instructions are privileged except PVALID. It is also important to remember that the MC68851 participates in the execution of the general CALLM and RTM instructions. It also provides a breakpoint acknowledge function in support of the MC68020's breakpoint instructions.

10.2.1 Data Movement (PMOVE)

The PMOVE moves data to and from the MC68851's registers. The operand data size can be byte, word, long-word, and double-word lengths, based on the PMMU's registers.

10.2.2 Parameter Validation (PVALID)

The PVALID compares the access right of a logical address against the current access level, and generates a TRAP if the address requires a higher privilege than its current privilege level. This instruction can be used by a routine to verify that an address passed to it by a calling routine is a valid address. For example, if a routine has passed parameters on the stack, the following example [Motorola 85] can be utilized to verify that the calling routine has the necessary privilege to use the parameters:

```
PVALID ([A7,OFFREF])       ; Validate the address.
MOVE ([A7,OFFREF]), D1     ; Use the address.
```

However, if the programmer frequently uses this data, the operation can be accelerated by loading it into a register. This is shown below:

```
LEA ([A7,OFFREF]), A1    ; Calculate the address and pass it to the register.
PVALID (A1)              ; Validate the address.
MOVE (A1), D1            ; Now it is ready for use.
```

10.2.3 Address Attributes Testing (PTEST)

This instruction is generally used during bus error handling. The PTEST instruction searches the translation table and loads the status and access rights information of a logical address used during the write access into the MC68851's status register. The results of the search are available in the PSR, or (optionally) the physical address of the last descriptor fetched may be returned. This instruction allows the operating system to determine the reason for faults generated by a write cycle to a specific logical address.

An example of PTEST, in the case of a bus error, is given by:

PTESTW #1, ([A7,OREF]), #7, A1

This instruction requires the MC68851 to search the translation tables for an address in the user data space (#1), and to examine the protection information as if a write cycle is taking place. The specific address is stored at OREF from the current stack pointer ([A7,OREF]).

10.2.4 Cache Pre-loading (PLOAD)

The PLOAD instruction searches the translation table and loads the ATC with an entry to translate the address, using the address and the function code as pointers.

The Preload can be executed for read or write attributes. If a write attribute is selected, then the assembler syntax to be used is PLOADW, and for the read attribute, it is PLOADR.

PLOADW requires the MC68851 to perform the table search and update all history information of the translation tables, as if a write operation to that address took place. Similarly, for PLOADR, the history information in the translation table is updated as if a read operation had occurred.

10.2.5 Cache Flushing

Four instructions are available for cache management. They are: PFLUSH, PFLUSHA, PFLUSHR, and PFLUSHS.

PFLUSH invalidates the translation cache entries, using the logical address, function codes, and effective address. This instruction allows the operating system to easily update entries from the ATC after making modifications to the external translation tables.

PFLUSHA clears all entries from the translation cache.

The PFLUSHR instruction invalidates the root pointer table and translation cache entries pointed to by the root pointer.

PFLUSHS invalidates entries from the ATC by the logical address and/or the function code, including globally shared entries.

10.2.6 State Save and Restore

The MC68851 provides various instructions to allow the user to save the current context of the machine and restore it at a later time. The instructions are PSAVE and PRESTORE.

The PSAVE instruction saves CRP, SRP, CAL, SCC, breakpoint registers, and the internal states of the MC68851, in order to support fast context switching and the MC68020 virtual memory/virtual machine capabilities.

PRESTORE restores the internal state of the MC68851 that was saved using the PSAVE instruction.

10.2.7 Conditionals

The conditional instructions provide the operating system with a vehicle by which the program flow can be controlled utilizing MC68851 conditions. These instructions are utilized to test for status bits (B, L, S, A, W, I, G, and C) in the PSR register and perform their complements. The instructions are:

BPcc (Branch Conditionally) branches conditionally on an MC68851 condition.
PDBcc (Decrement and Branch) tests the MC68851 condition, decrements the main processor register, and branches.
PScc (Set Conditionally) sets the operand according to the MC68851's condition.
PTRAPcc (Trap Conditionally) tests a condition, and causes an exception if the condition is true.

10.3 ADDRESS TRANSLATION

The MC68851 can perform logical-to-physical address translation of up to 4 gigabytes of logical address space without any support from the MC68020. The logical-to-physical address translation occurs within one clock period for most of the cases. The PMMU achieves this performance by implementing the entire function in hardware. The translation timing diagram is illustrated in Figure 10.9. It indicates that the physical address strobe is asserted one clock after the assertion of the logical address strobe by the MC68020. The MC68851 takes one clock period to complete the access right check and perform address translation, before generating a valid physical address.

10.3.1 Address Translation Cache

The MC68851 achieve this throughput by storing recently used logical-to-physical translated addresses in a high-speed memory. This memory is called the Address Translation Cache (ATC). It is a 64-entry, fully-associative array, which contains logical addresses, including their corresponding physical translations. It is equipped with 64 separate comparators, so that comparisons and translations can be performed in tandem. The cache replacement algorithm is implemented in hardware. The functional implementation of the address translation mechanism is illustrated in Figure 10.10. This diagram shows the logical address generated by the MC68020, the ATC with the task alias, RPT, and the cache.

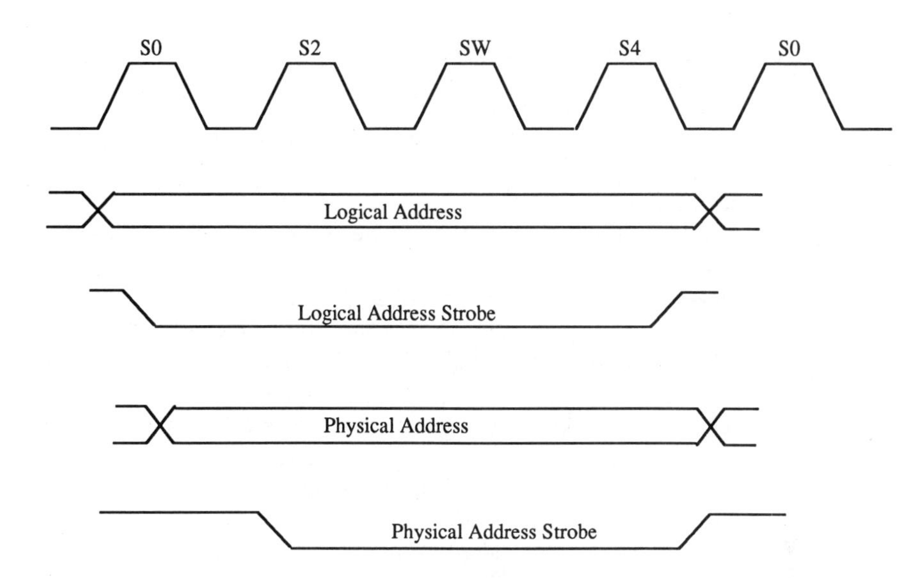

FIGURE 10.9. MC68851 address translation timing diagram.

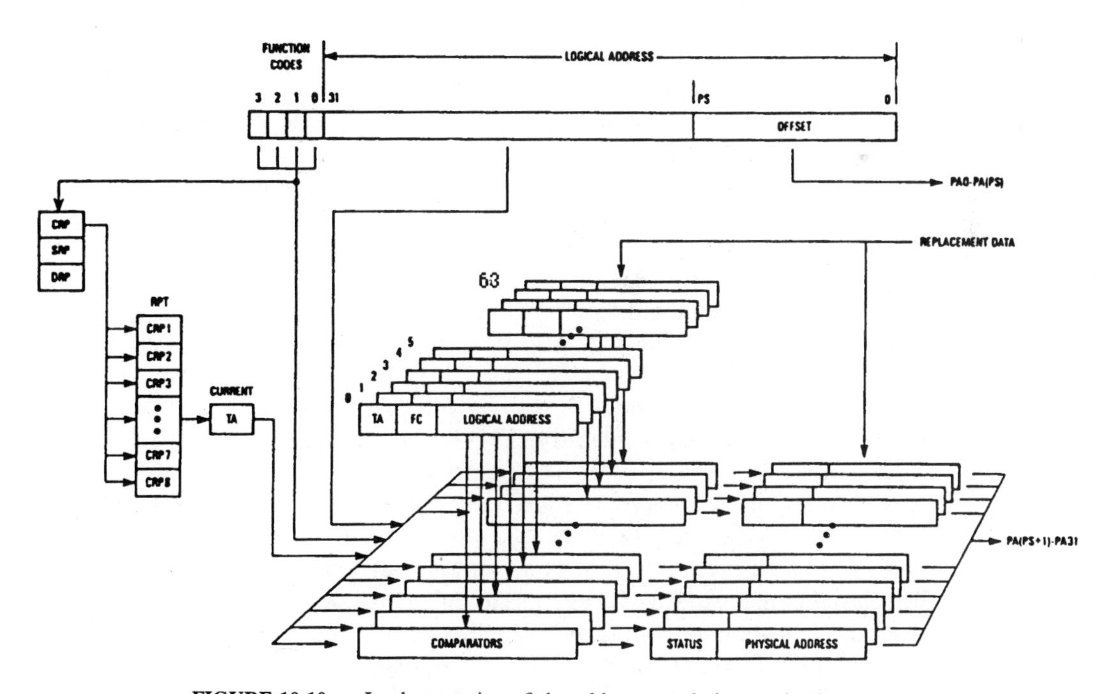

FIGURE 10.10. Implementation of the address translation mechanism. (Courtesy Motorola Inc.)

10.3.2 ATC Access Sequence

When a bus cycle is initiated by a logical bus master, the logical address and function code are routed to the ATC, where they are simultaneously compared against all current entries. If one of the ATC entries matches (i.e., a hit has occurred), then the ATC drives the corresponding physical address out onto the physical address bus that interfaces with the memory devices. If the MC68851 encounters no exceptional conditions (i.e., no write violations), then it asserts the physical address strobes ($\overline{\text{PAS}}$), and declares the address a valid one.

To enhance performance in a multitasking environment, the MC68851 internally maintains CRP values of the eight most recently executed tasks in the root pointer table. The MC68851 assigns each of these tasks a task alias for tagging ATC entries. The mapping of task-to-task alias values, and the assignment of task alias values, are performed by the MC68851's internal logic.

Whenever an entry is made in the ATC, the RPT indexes a 3-bit value, corresponding to the currently saved CRP value in the TA field of the ATC entry. The TA field is then considered as a part of the logical address, to determine whether a hit has occurred in the ATC.

The task alias mechanism utilizes the root pointer and the root pointer caching function of the MC68851. Whenever the MC68020 initiates a new task, the operating system loads the PMMU's CPU root pointer register with the pointer needed for the address translations for this new task. The CRP of MC68851 then contains the address of the root of the translation table for the currently executing task in the physical memory.

To switch (start a new task or restart a suspended task) a task, the operating system loads the task's translation table pointer into the CRP register of the MC68851. It then compares the CRP register value against the entries of the RPT (Root Pointer Table). If no match is found in the RPT, then a new entry is created into the RPT, and the task alias associated with that entry is assigned to the current task. If the MC68851 discovers no empty location into the RPT, then the new entry overrides the oldest entry from the RPT, utilizing a replacement algorithm (see Subsection 10.3.4). It also flushes the entries in the ATC that are currently identified with the task alias, since they are associated with the old task.

If a match is found during the search of the RPT, then the MC68851 already has a task alias needed to identify the ATC entries belonging to the would-be current task. In such cases, no ATC entries are flushed, because they can be reused during the address translation for the current task. Hence, when a task is restarted, it is very likely that the ATC may contain all required entries for the task to begin its execution. However, without task aliasing, the new task would be delayed until a working set of translation descriptors are loaded into the ATC from the transition table in memory.

In addition to the CPU root pointer, the MC68851 maintains exclusive root pointers for the supervisory (SRP) and DMA-class devices, including coprocessors (DRP) that require logical-to-physical address translations. Again, since the CRP and RPT reside in an on-chip cache, the search process does not involve any external bus references.

10.3.3 Flushing the RPT

Entries are automatically discarded from the RPT by the replacement algorithm, without an explicit command from the operating system. When a task is erased, system software (operating system software) should ensure that all ATC entries for it have been invalidated by executing the PFLUSHR instruction, giving the CRP value of the destroyed task as the operand. In addition, it should invalidate the corresponding RPT entry, thus improving utilization of the RPT.

10.3.4 TAG Replacement Algorithm

The MC68851 contains circuitry that automatically determines which tag/data pair are to be used for a new ATC entry. This is implemented via an algorithm known as a pseudo-LRU replacement algorithm. The steps are:

1. Locate an invalid entry and use it.
2. If no invalid entry is found, select an entry whose L bit is not set.
3. Replace the entry.

During an ATC replacement operation, the algorithm attempts to locate the entry that was used the longest time ago. This enables the ATC to maintain a close approximation to a proper working set of page descriptors.

10.4 THE ADDRESS TRANSLATION TABLE

We have already seen (in Chapter 8) that, in a paged virtual memory system, the logical address is divided into a number of fixed-size pages. Corresponding to each of these logical pages, mapping exists into the physical page in the memory. In order to allow each logical page to have a unique mapping into the physical memory, it is necessary to provide a translation descriptor corresponding to each page in the memory. The translator descriptors are put into a table called the *descriptor table.* A *linear table* is the simplest organization, where an MMU would use the logical page address as an index in this table for an entry. This mechanism is not suitable for an MC68020-based system which is capable of supporting logical address spaces with several million pages.

The alternative is to use a *tree structure,* which holds the mapping information, where a portion of the logical address space is mapped at each level of the translation tree. The MC68851 utilizes such a tree structure to organize the address translation table for each task.

10.4.1 Translation Table Organization

The translation table for the MC68851 can be organized to a maximum of five levels, and requires five separate index fields to locate a logical-to-physical mapping. The levels of the translation tree are determined by the contents of the

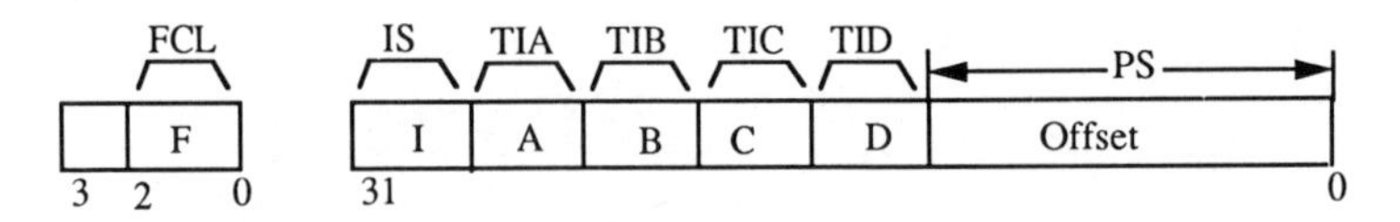

FIGURE 10.11. Derivation of the table index fields.

Translation Control (TC) register. The *IS field* of this register is used to set the size of the logical address space. The Page Size (PS) field determines the page size to be used. Using the IS and PS fields, a system designer will be able to specify the number of pages in the system. The number of pages is given by $2^{(32\text{-IS-PS})}$. However, if the logical address space is mapped by the function code (FC3:FC0) signals, then eight separate logical address spaces of size $[2^{(32\text{-IS})}]$ would be created giving a total logical address space size of $2^{(32\text{-IS-}+3)}$, with a $2^{(32\text{-IS-PS}+3)}$ number of pages.

A logical page address is determined by the FCL bit and the TIA, TIB, TIC, and TID fields of the TC register (Figure 10.4). The mapping of these fields to the logical address generated by the MC68020 is shown in Figure 10.11. The table index fields (TIA, TIB, TIC, and TID) specify the number of bits of the logical address to be used at each level of the translation tree. Here, the A,B,C, and D fields of the logical address, specified by the IS, TIA,TIB, TIC, and TID of the TC register, must follow the constrained listed in Table 10.3, where IS + TIA + TIB + TIC + TID = 32.

Figure 10.12 illustrates a simple address translation table tree, including a logical address translated using this tree, where the 32-bit logical address is divided into three fields; A (12 bits), B (10 bits), and PS (10 bits). Here, the function code look-up is suppressed. As a result, indexing by function is not used. This division is generally set during system initialization time by writing a value, for example: $80A0CA00 to the TC register, where E = 1, SRE = 0, FCL = 0, PS = 1010, IS = 0000, TIA = 1100, TIB = 1010, TIC = 0000, and TID = 0000. This diagram also displays the sequence of descriptors utilized to translate the logical address $00C023xx.

10.4.2 Translation Table Format Description

The logical-to-physical mappings for a system can be described to the MC68851 using two different translation descriptor formats. The descriptors are based on

TABLE 10.3. Logical Address Generation Scheme.

Field	Starting Bit Position	Width Restriction
A	31-IS	1–15 (TIA must be non-zero)
B	31-IS-TIA	0–15 if TIB = 0, then TIC = TID = 0 is required
C	31-IS-TIA-TIB	0–15 if TIC = 0, then TID = 0 is required
D	31-IS-TIA-TIB-TIC	0–15

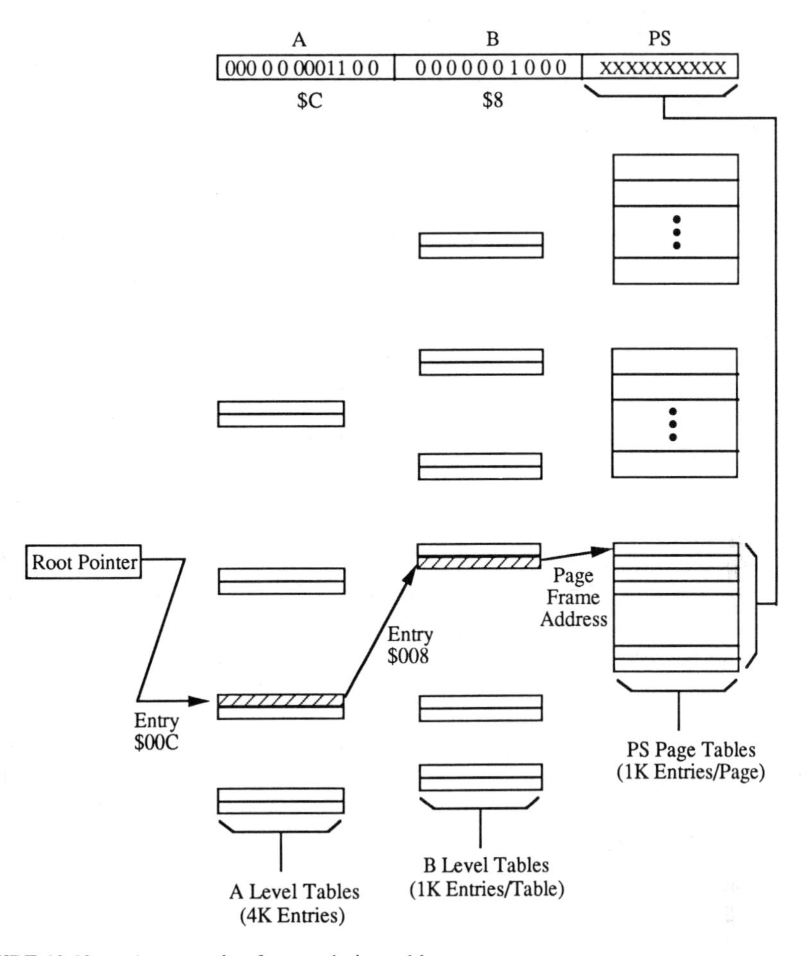

FIGURE 10.12. An example of a translation table tree.

eight bytes, called the *long format,* or four bytes, called the *short format.* They are illustrated in Figure 10.13, and can be mixed to create various sized tables of the translation tree.

10.4.3 Table Search Algorithm

Whenever the main processor (e.g., the MC68020) generates a logical address that does not have a corresponding translation resident in the ATC, the MC68851 performs bus operations to load the mapping from the translation tables in the memory pointed to by the relevant root pointer. To perform this search operation, the MC68851 simultaneously aborts the logical bus cycle, signals the main processor to retry the operation, and requests the mastership of the logical bus. After detecting a free bus, the MC68851 completes the bus arbitration sequence, gains the bus mastership, and initiates a search of the translation tables pointed

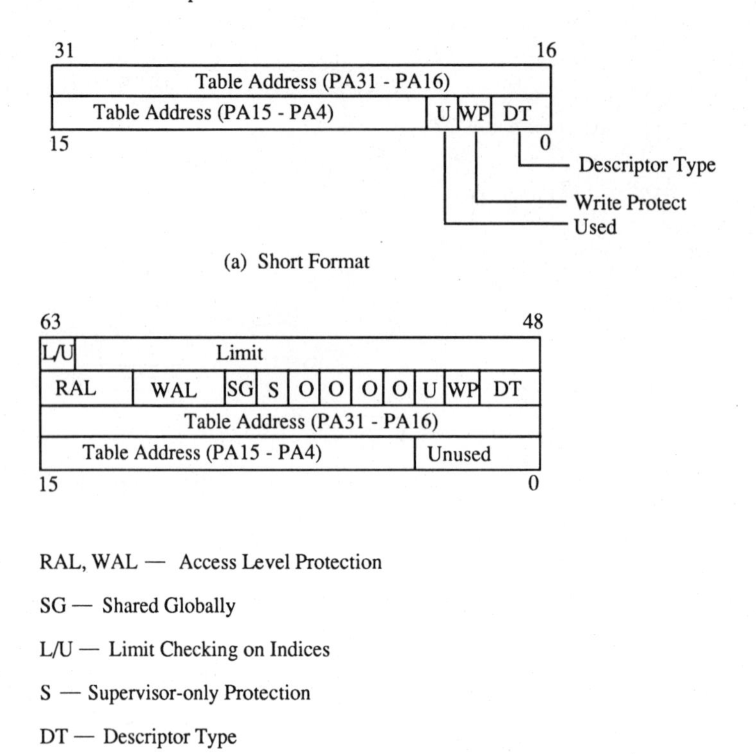

FIGURE 10.13. Table descriptor format: (a) Short format; (b) Long format.

to by the current root pointer. The MC68851 locates the translation descriptor, updates the ATC, and returns to the logical bus master of the main processor to retry the previously aborted bus cycle. The MC68851 again receives the logical address, verifies the access right, and produces the correct physical address.

The MC68851 automatically searches the translation tables when a translation misses in the ATC, and does so completely in hardware, without any software support. Such hardware helps the MC68851 to minimize the table search overhead when compared to a software-directed memory management implementation.

10.5 PROTECTION MECHANISM

The MC68851 supports various protection mechanisms to facilitate the implementation of a fully protected system. Two protection techniques are available, which can be used either independently or collectively. The first protection mechanism uses the *function code outputs* of the logical bus master and the type of operand that is being accessed. The other mechanism separates the user logical address spaces into discrete regions of definite privileges to the memory with a hierarchical structure.

10.5.1 Protection Using Address Space Encoding

Under the MC68000 architecture, bus-master class of peripherals declare their context of operation on a cycle-by-cycle basis. The function code identifies the current privilege level (user or supervisory level) and the nature of the operand (program and data) that it is fetching. Thus, the function code signals become the basis for the first level of protection.

Since all mapping and protection information used by the MC68851 is contained in the translation tables, which reside in the physical memory, the first method uses the function code as the logical index in the first level of the address translation table. This strategy creates four separate address spaces, namely:

- Supervisory program space
- Supervisory data space
- User program space
- User data space

We have seen that the processor will not allow the user mode program to access either a supervisory program and/or data space. Thus, supervisory programs/data can be protected from any user mode accesses by not placing any valid logical-to-physical mappings in the user branch of the translation table that references supervisor-only information.

However, it is possible to separate the supervisor and user address spaces without separating the program and data. The operating system can achieve this by enabling the SRE bit (this enables the root pointer) and clearing the FCL bit in the TC register, which will suppress the function code look-up. Now, the MC68851 will translate the supervisory mode references using the address translation table pointed to by the SRP register, and user references using the CRP register. Here, the MC68851 will allow the creation of several classes of protection, utilizing the SRP and CRP registers. The protections are:

- No access.
- Only the supervisor has complete read and write privileges.
- Only the supervisor has read-only privileges.
- The supervisor and user both have complete read and write privileges.
- The supervisor and user both have read-only privileges.
- The supervisor and user both have data-only privileges.
- The supervisor and user both have program-only privileges.

10.5.2 Protection Using the Access Level Protection Mechanism

An MC68020-based system can use the access level protection mechanism of the MC68851 to implement eight additional levels of protection. Here, routines operating at different access levels can have different privileges to memory, and a facility is provided to closely control access level changes. The MC68020 module call and return instructions coordinate with the MC68851 to allow a task to alter its access level under the control of the operating system.

The value of the ALC bits within the AC register selects the access levels of the MC68851 (Figure 10.6). Access levels are associated with the logical addresses, pages in the logical address spaces, and tasks. In a system using eight access levels, Level 0 represents the most privileged or protected level, and Level 7 represents the least protected. The protection levels are hierarchical, which means that if access to an address is permitted for a given level, it is also permitted for all levels having greater privileges. Again, if an access to an address is denied for a given level, it is also denied for all levels with less privileges.

The access level of a logical address (contained in the most significant one, two, or three bits of the logical address, determined by the ALC field of the AC register) is interpreted as the level of privilege requested during an access using the address.

However, a page within the logical address space (corresponding to a page in the physical memory) has two access levels associated with it. They are the read and write levels. All access levels of information for a page reside in the address translation tables. The long format page and table descriptors contain the Read Access Level (RAL) and Write Access Level (WAL) fields. During any search of the address translation table, the access level bits of the logical address are compared against the RAL and WAL fields for all long descriptors. The effective read access level of the page is the most privileged of all encountered RAL fields, and the effective write access level is the most privileged of all WAL and RAL fields.

Before the operating system invokes a task for execution, it loads the address (physical address) of the translation table's root into the MC68851's CRP for that task. Since it is possible for a task to generate a logical address with any access level, the PMMU compares the access level against the access level of the task stored in the CAL register. The MC68851 aborts the access, and initiates a bus error, if it detects that the task has attempted to use an access level more privileged than it is permitted.

10.5.3 Transfers Between Access Levels

In order to access a higher privileged level code and/or data than its current level, a task needs to request them, utilizing the MC68020's CALLM and RTM instructions. These instructions allow a routine to transfer execution control to a module operating at the same or higher level of privilege, and to return from that module after the completion of the module function. Whenever the MC68020 executes a CALLM instruction that requests an increase in the access level, it automatically communicates with the MC68851 access-level mechanism, via the access-level control CPU space cycles, to determine the validity of the requested change. The PMMU checks the request against the module descriptor for that operation, and indicates the validity of that request to the MC68020. The

RTM instruction operates similarly, except that control is always passed from a task to a task having the same or lower levels of privilege.

10.6 HARDWARE INTERFACING

The MC68851 is designed to interface with the MC68000 family of processors as a coprocessor. We have learned that the coprocessor interface has two components:

1. The hardware connection between the main processor and the coprocessor.
2. The communication protocol to be used to transfer commands and data between the main processor and coprocessor, in order to complete a desired function.

The following subsections describe the implementation of these parts of the MC68851 architecture in detail.

10.6.1 The Hardware Connection

The MC68851's signal lines provide the basic mechanism for electrically connecting it to the MC68020. The signal lines can be grouped into twelve functional areas, shown in Figure 10.14. They are the:

- Logical address bus
- Physical address bus
- Shared address bus
- Function code
- Data bus
- Transfer size
- Bus control
- Bus exception control
- Physical bus arbitration
- Logical bus arbitration
- Excitation control
- Miscellaneous control

Hardware and system designers need to understand the purpose of the functional groups in order to interconnect this device with the MC68020 or an equivalent device, to construct a complete workable system.

10.6.1.1 Logical Address Bus

The MC68851 uses the signal lines [LA31:LA8] to accept a logical address for translation or for internal operations. For an MC68020-based system, these logical signal lines are connected to the corresponding address line signal of the CPU (i.e., LA31 is connected to A31, LA30 is connected to A30, etc.). The logical address bus should also be connected to the address lines of all logical bus masters.

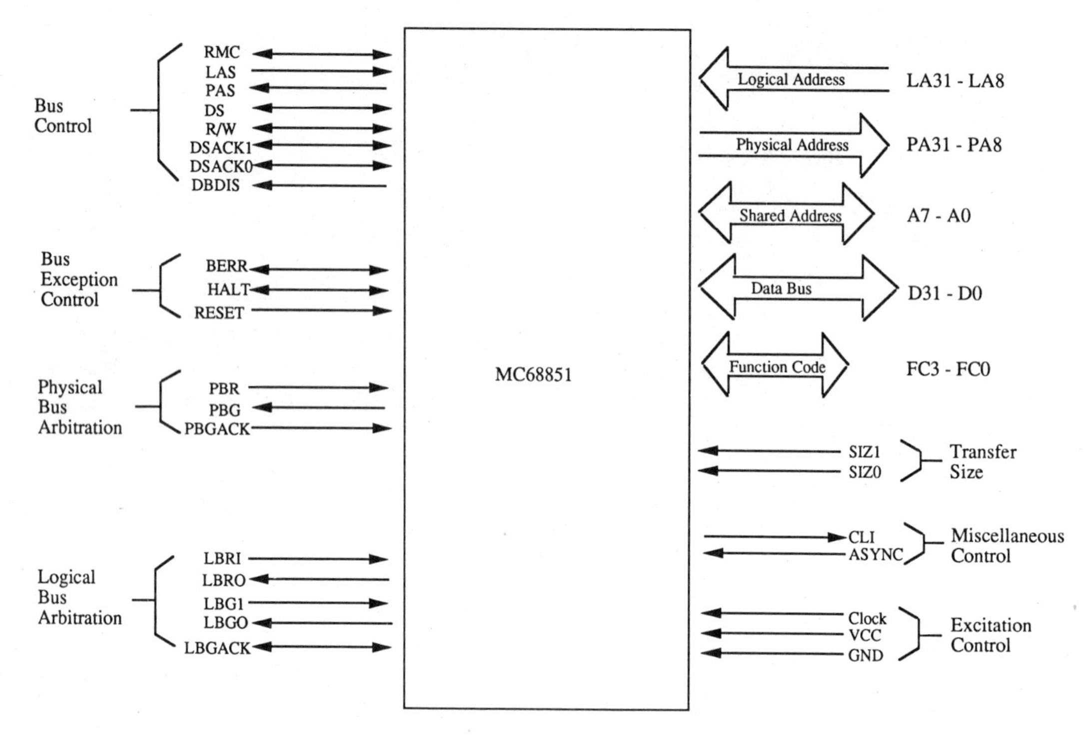

FIGURE 10.14. Functional signal groups for the MC68851.

For a system where the logical address is less than 32 bits, the unused signal lines should be tied to a +5V supply, using pull-up registers.

10.6.1.2 Physical Address Bus

These three-state output signal lines, PA31–PA8, provide the physical addresses for both address translations and MC68851-initiated bus operations.

10.6.1.3 Shared Address Bus

These eight three-state bidirectional signal lines, A7–A0, form the shared address bus between the functions of the logical and physical buses. The MC68851 monitors the status of these lines (A7–A0) during logical-to-physical address translation, because the main processor uses them to identify the internal registers of the MC68851 utilizing the CPU space. The MC68851 drives the system address bus when it gains the bus mastership.

10.6.1.4 Function Code

The function code signal lines, FC3–FC0, for the MC68851 are three-state bidirectional lines. During the address translation sequence, they reflect the address space being accessed by the current bus master. The MC68851 uses the function code associated with a bus cycle as an extension to the logical address during the creation of entries in the address translation cache. The function codes are also used to implement first-level protection.

The address spaces generated by the function codes are listed in Table 10.4. Function code FC3 indicates the status of the DMA access in progress. The remaining three function code lines, FC2–FC0, convey the same information as conveyed by the function code lines of the MC68020 processor. When the MC68851 becomes the bus master, it asserts a value of $5 over these lines, to indicate the supervisory data space.

10.6.1.5 Data Bus

These 32 three-state bidirectional signal lines, D31–D0, carry data between the MC68851 and other devices. This bus may be dynamically sized through the use of $\overline{\text{DSACKx}}$ signals, transferring 8, 16, 24, or 32 bits of information during a bus cycle.

During the externally generated hardware reset operation, the MC68851 receives the configuration information over the data bus lines D7–D0. This information allows the MC68851 to set the bus size to be utilized during the coprocessor protocol and the *decision time* window for determining whether or not an ATC hit has occurred. The timing for $\overline{\text{PAS}}$ assertion and the decision as to whether the $\overline{\text{CLI}}$ signal should be asserted for all MC68851-initiated bus operations are also extracted from the configuration information.

10.6.1.6 Transfer Size

These two three-state signal lines are used to implement the dynamic bus sizing capabilities of the MC68851.

10.6.1.7 Bus Control

Eight signal lines implement the necessary bus control functionalities. They are:

Read-Modify-Write ($\overline{\text{RMC}}$): This three-state bidirectional signal is used to indicate whether the bus cycle in progress is an indivisible read-modify-write cycle or not.

When the MC68851 is the bus master, $\overline{\text{RMC}}$ may be asserted to indicate that

Table 10.4. Address Space Generation Utilizing Function Tables.

Function Code Value	Cycle Type
$0	Undefined—Reserved for future use
$1	User data space
$2	User program space
$3	Reserved for user definition
$4	Undefined—Reserved for future use
$5	Supervisory data space
$6	Supervisory program space
$7	CPU space
> = $8	Alternate bus master—User definable

the operation in progress should not be interrupted by any other bus access. The MC68851 will suspend all arbitration for the physical bus as long as this signal remains asserted.

Logical Address Strobe ($\overline{LAS}$): This input signal line indicates that the contents of the logical address bus, function code, and R/$\overline{W}$ lines are valid. It is also used to specify that SIZEx signals are valid when the MC68851 is accessed as a slave device.

Physical Address Strobe ($\overline{PAS}$): This is a three-state output signal for the MC68851. When asserted, it validates the contents of the physical address bus. The $\overline{PAS}$ line also indicates that the function codes, R/$\overline{W}$, and SIZEx are being validated by the MC68851.

Data Strobe ($\overline{DS}$): This bidirectional three-state signal is used to control the flow of information on the data bus. $\overline{DS}$ becomes input to the MC68851 when it is selected by the processor. However, whenever the MC68851 becomes the bus master, $\overline{DS}$ indicates that the slave device should drive the data bus during the read cycle, and the PMMU has placed valid data on the bus in the case of a write cycle.

Read/Write (R/$\overline{W}$): This bidirectional three-state line is used to signal the direction of information transfer over the data bus.

Data Transfer and Size Acknowledgment ($\overline{DSACK1}$ and $\overline{DSACK0}$): These bidirectional three-state signal lines are used to normally terminate a bus cycle and to declare the port size of the responding device during that bus cycle.

When the MC68851 arbitrates for the logical bus, it monitors these signal lines. After receiving a bus grant from the CPU, the MC68851 waits until the $\overline{LBGACK}$, $\overline{LAS}$, and both $\overline{DSACKx}$ signals are negated before asserting logical bus grant acknowledgment. This is necessary to ensure that the previous slave device has been disconnected from the bus.

Data Buffer Disable (DBDIS): This active-high output provides an enable signal to external data buffers connected to the MC68851 data bus. It is also used to disconnect buffers that exist between the MC68851/MC68020, including the buffers driving the physical memory during the MC68851's bus mastership.

10.6.1.8 Bus Exception Control

Three signals, $\overline{RESET}$, $\overline{HALT}$, and $\overline{BERR}$, together implement the bus exception control function.

Reset ($\overline{RESET}$). This is an input signal to the MC68851, and (when asserted) places the device into a known state. When $\overline{RESET}$ is applied, the PMMU disables the address translation mechanism, clears all breakpoints, sets the internal state to idle, and accepts configuration information over the data bus.

Halt ($\overline{HALT}$). This is a bidirectional three-state signal. $\overline{HALT}$ becomes an input signal as the MC68851 becomes the bus master. The MC68851 will stop bus activity at the completion of the current bus cycle, when $\overline{HALT}$ is asserted. It will also place all control signals into a high impedance state, and the physical address bus will retain its previous values. However, the bus arbitration lines function normally, even though the MC68851 is halted. Only an externally generated $\overline{RESET}$ can take the MC68851 out of this state.

When the MC68851 is translating addresses, HALT is used as an output, in conjunction with the BERR and/or LBRO, to signal the current logical bus master to *relinquish* or *relinquish and retry* the operation.

Bus Error (BERR). This bidirectional three-state signal line is used to specify that a bus cycle should be discontinued, due to an abnormal condition.

The BERR is an input to the MC68851 when it becomes the bus master. When asserted by an external device, it means that there has been some problem with the current bus cycle. The source of the problem may be related to the malfunction of a device or to some application-related error.

When the MC68851 operates as a coprocessor and is performing an address translation, the BERR is used as an output to the logical bus master. The MC68851 can assert the BERR because of any of the following conditions:

1. The BERR bit is set in the matched ATC entry.
2. A write or read-modify-write cycle is initiated into a write-protected page.
3. An instruction breakpoint is detected and the associated count register is found to have a zero, or is disabled.
4. As a part of relinquish and retry operation, if:
   a) The required address mapping is not resident in the ATC.
   b) A write operation occurs to a previously unmodified page.
   c) A read from the responding CIR causes a suspended PLOAD or PTEST instruction to be restarted.
   d) A module call operation references a descriptor, where the appropriate status is not resident in the ATC.
5. An RMC cycle is attempted, and a corresponding descriptor with the appropriate status is not resident in the ATC.
6. The access-level protection mechanism detects an access violation.

The bus error signal is coupled with the HALT signal to determine whether the current bus cycle should be re-executed or aborted.

10.6.1.9 Physical Bus Arbitration

The three signal pins PBR, PBG, and PBGACK are used by the MC68851 to determine the system's current bus master. The bus arbitration protocol of the MC68851 is very similar to that of the MC68000 family of processors.

Physical Bus Request (PBR). The bus request signals from all potential physical bus masters are Wire-OR to this input pin. In such a case, the MC68851 is considered as the default bus master of the physical bus. Any device requiring an access to the bus must arbitrate for bus mastership.

Physical Bus Grant (PBG). This output tells the potential bus master that the MC68851 will release the ownership of the physical bus at the completion of the current bus cycle.

Physical Bus Grant Acknowledgment (PBGACK). This input signal for the MC68851 indicates that a requesting device has gained the mastership of the physical bus. This signal should not be asserted until the following conditions are met:

a) A physical bus grant signal has been received through the arbitration sequence.
b) The $\overline{PAS}$ signal is negated. This reflects that no external device is using the physical bus.
c) The $\overline{DSACKx}$ signals are negated, indicating that no external device is still driving the data bus.
d) $\overline{PBGACK}$ is negated and no other device is still claiming bus mastership.

The physical bus request and grant process is very similar to the system bus request and transfer protocol. Like the bus arbitration protocol for the MC68020, $\overline{PBGACK}$ must remain asserted as long as a device other than the MC68851 is the bus master.

10.6.1.10 Logical Bus Arbitration

The logical bus arbitration protocol is slightly more complex than the physical bus arbitration protocol. Five signal lines are utilized to identify which device will gain the ownership of the logical bus. The signals are:

Logical Bus Request In ($\overline{LBRI}$). This input signal line tells the MC68851 that a device with a higher priority than the current logical bus master, or the PMMU itself, is requesting the bus.

Logical Bus Request Out ($\overline{LBRO}$). This output signal is asserted by the MC68851 to request the bus. This signal is used as a part of the relinquish operation, including the relinquish and retry operation.

The request input to the logical bus arbiter (usually the main processor) should be generated from Wire-OR requests, input to $\overline{LBRI}$, and logically ORed with the $\overline{LBRO}$ output of the MC68851.

Logical Bus Grant In ($\overline{LBGI}$). This input signal to the MC68851 is generated by the MC68020 to inform that it will release the bus ownership after the completion of the current bus cycle. Otherwise, if an alternate master is currently the owner of the bus, then the MC68020 will not claim the bus after its release.

Logical Bus Grant Out ($\overline{LBGO}$). This output pin indicates that the MC68851 has recognized and synchronized the assertion of $\overline{LBGI}$ by the MC68020. The MC68851 is now passing the bus grant to an alternate logical bus master or to arbitration priority circuitry.

Logical Bus Grant Acknowledge ($\overline{LBGACK}$). This bidirectional three-state signal indicates that a logical bus master, other than the CPU, has taken control of the logical bus.

This signal line is asserted by the MC68851 to declare that it is the current logical bus master. The $\overline{LBGACK}$ is also monitored by the MC68851 as an input to decide when it can become the bus master.

10.6.1.11 Excitation Control

Three input signal lines fall into this category. They keep the MC68851 alive. They are:

- Power
- Ground
- Clock

Power (Vcc). It utilizes a TTL-compatible 5V supply.

Ground (GND). This line should be tied to the system ground plane in order to eliminate noise. One should use a sufficient number of bypass capacitors between the Power and GND lines whose typical value is 0.1 pf.

Clock (CLK). The MC68851 clock input is a TTL-compatible signal that is internally buffered to generate internal clocks for the memory management logic. It must be of a constant frequency, and conform to the minimum and maximum period and the pulse width. The specification of the clock is listed in Table 10.5.

10.6.1.12 Miscellaneous Control

Two signal lines, $\overline{\text{ASYNC}}$ and $\overline{\text{CLI}}$, are the member of this group.

Asynchronous Control ($\overline{\text{ASYNC}}$). When a logical bus master does not present logical bus control signals with exact timing specifications to the MC68020, then this input must be driven with the proper setup and hold time. This informs the MC68851 that input synchronization needs to take place.

When the MC68851 operates in a synchronous mode, it utilizes a known signal relationship in order to perform faster translations. However, if the logical bus master fails to conform to this specification (i.e., has a different control strobe timing and/or different operating frequency), it must assert $\overline{\text{ASYNC}}$ prior to initiating bus activity.

Cache Load Inhibit ($\overline{\text{CLI}}$). During address translation, this three-state output signal is asserted by the MC68851 if the matched address translation cache entry has its CI (Cache Inhibit) bit set. Assertion of this output informs the external caches that the data associated with the current bus cycle is noncacheable.

To maintain the distinction between CPU space and other address spaces (e.g., supervisor programs, supervisor data, etc.), the MC68851 does not assert $\overline{\text{PAS}}$ during CPU space cycles. Since $\overline{\text{CLI}}$ is asserted on the falling edge of the clock, all external qualification of $\overline{\text{CLI}}$ must be synchronized with $\overline{\text{LAS}}$ and CPU space indicators to provide a CPU space address strobe. The CPU space cycles are utilized to access the MC68851's registers and are decoded internally. As a result, they do not generate any physical bus activity. Note that, if the MC68851 is not

TABLE 10.5. Clock Specifications for the MC68851.

	MC68851RC12		MC68851RC16		
Characteristics	Min.	Max.	Min.	Max.	Unit
Frequency of Operation	8.0	12.5	8.0	16.67	MHz
Cycle Time	80	125	60	125	nsec
Clock Pulse Width	32	93	24	95	nsec

the current physical bus master, then $\overline{CLI}$ is not asserted until ownership of the physical bus is returned to the MC68851.

When the MC68851 performs a table search operation, it continuously asserts $\overline{CLI}$ in order to prevent the caching of translation table information. However, this table search is disabled during a reset configuration.

10.6.2 Communication Protocol

When the MC68851 is interfaced with the MC68020 as a coprocessor, it follows the MC68000 family's coprocessor protocol. No external decoding logic is required to interface the MC68851 as a coprocessor. The selection of the MC68851 is generated internally by the device. It gets activated whenever the proper function codes, logical addresses, and related control signals appear on the system bus.

This allows the MC68020 to communicate with the MC68851 via a set of coprocessor interface registers (CIR) internal to the MC68851. A programmer does not notice the presence of these CIR registers. However, when the MC68020 is not used, then these CIRs need to be explicitly referenced by the processor that emulates the behavior of the MC68020.

10.6.3 Coprocessor Interface Registers

Figure 10.15 illustrates the bit pattern to be placed on the address bus to access the MC68851 as a coprocessor within the CPU space. Therefore, the MC68851's internal logic decodes the function code signals FC3–FC0 (0111), the CPU space type fields A19–A16 (0010), and the Cp-ID field A15–A13 (000). During the access of CIRs, the MC68020 asserts 0 to the field position 'x', which is ignored by the device. The MC68851 decodes the address bits A4–A0 to determine the appropriate coprocessor interface register that the CPU wishes to access. The address of the CIRs, and their attributes, are shown in Figure 10.16.

The MC68020 and MC68851 follow the coprocessor protocol described in Chapter 9. The MC68020 initiates communication with the MC68851. The main processor writes data into one of the CIR registers, and waits for a response from the MC68851. In response, the MC68851 writes data into its response CIR, and asserts $\overline{DSACKx}$ to terminate the protocol.

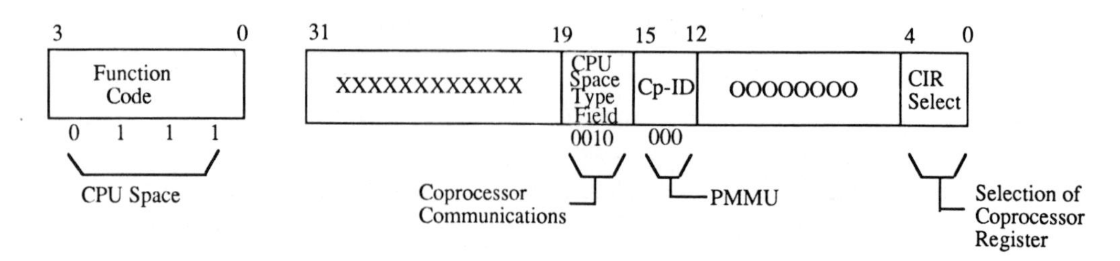

FIGURE 10.15. Encoding of the coprocessor interface address bus.

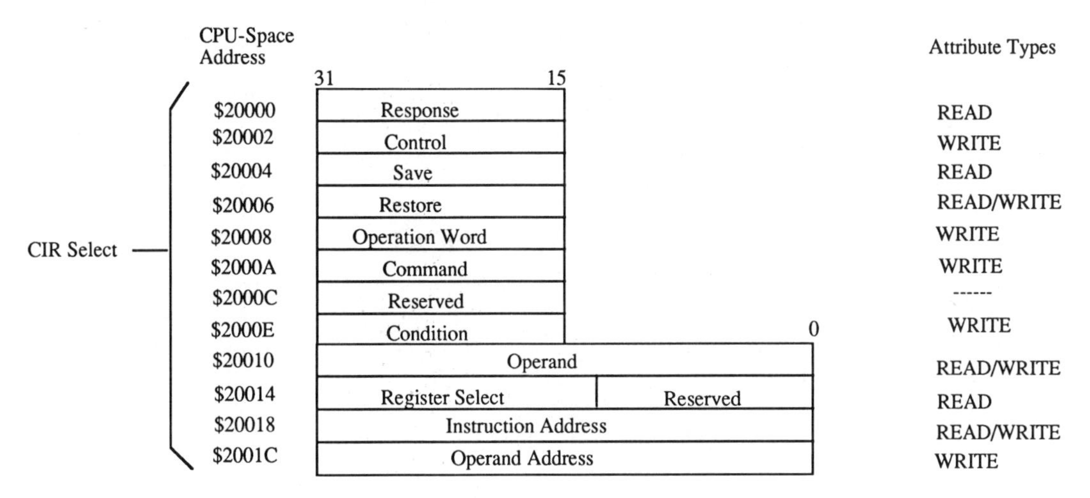

FIGURE 10.16. Coprocessor interface register map.

10.6.4 Access Level Control

The MC68851 contains a set of Access Level Control Registers (ALCRs), by which MC68020 and MC68851 communicate during CALLM and RTM operations. These registers are not part of the MC68851's programming model. Rather, they are used as communication ports with specific functions associated with them. In general, programmers don't access them, since they are implemented as internal hardware and controlled by the microcodes of the MC68020 and MC68851.

The ALCRs are shown in Figure 10.17. The internal decoder logic of the MC68851 decodes the function code signals FC3–FC0 (0111), the CPU space type fields A19–A16 (0010), and address bits A6–A0 to select the appropriate access-level control interface register. The MC68020 initiates a dialogue using these registers. In response, the MC68851 supplies an appropriate response, and asserts $\overline{\text{DSACKx}}$ to terminate the bus cycle.

10.6.4.1 Current Level (CL) ALCR ($10000)

This 8-bit read-only register has an offset of $00 with respect to ACLRs. This is read by the MC68020 during the execution of a CALLM instruction. The CPU needs this to save the access information of the module call stack frame into the CAL (Current Access Level) (bit [7:5]) and VAL (Valid Access Level) (bit [3:1]) fields of the calling module.

10.6.4.2 Access Status (AS) ALCR ($10004)

This is also an 8-bit read-only register, containing the status of the access-level change, requested by the MC68020. During the execution of the CALLM instruction, the MC68851 reads the information from the Current Access Level (CAL) register, the Increase Access Level (IAL) register, and the Stack Change Control

	31	23 ... 0
$00	CAL	(Unused, Reserved)
$04	STATUS	(Unused, Reserved)
$08	IAL	(Unused, Reserved)
$0C	DAL	(Unused, Reserved)

	31 ... 0
$40	Function Code 0 Descriptor Address
$44	Function Code 1 Descriptor Address (User Data)
$48	Function Code 2 Descriptor Address (User Program)
$4C	Function Code 3 Descriptor Address
$50	Function Code 4 Descriptor Address (Supervisor Data)
$54	Function Code 5 Descriptor Address (Supervisor Program)
$58	Function Code 6 Descriptor Address
$5C	Function Code 7 Descriptor Address (CPU Space)

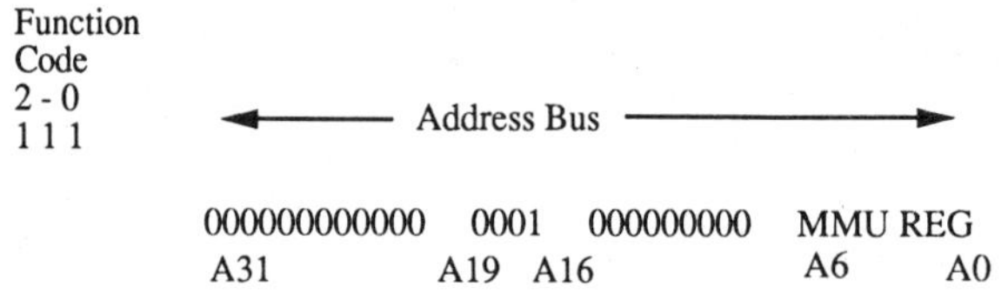

Function Code 2 - 0: 1 1 1

← Address Bus →

000000000000 0001 000000000 MMU REG
A31 A19 A16 A6 A0

FIGURE 10.17. MC68851 access-level control interface register map. (Courtesy Motorola Inc.)

(SCC) register, to determine whether or not the requested module call is valid or not. A stack change generally takes place before the program control is passed to the called module.

During the execution of the RTM instruction, the MC68851 uses the information contained in the CAL register and decreases the access level in the DAL register. It then decides the validity of the requested module return operation.

10.6.4.3 Increase Access Level (IAL) ALCR ($10008)

The MC68020 use this 8-bit write-only register to request an increase of access right during a module call operation. The MC68851 compares the access level contained in the highest-order bits of the IAL against the corresponding bit of the CAL register. If the result of the comparison is less than or equal, then the requested change is considered valid. Otherwise, the MC68851 will ask the MC68020 to initiate format exception processing.

10.6.4.4 Decrease Access Level (DAL) ALCR ($1000C)

The MC68020 requests an access right decrease during the module return,utilizing this 8-bit write-only register. During the execution of an RTM instruction, the MC68020 writes the saved access level field value from the module call stack frame to the DAL register, to reverse the operation performed by reading the CL register during the CALLM instruction.

10.6.4.5 Descriptor Address ALCRs ($10040–$1005C)

These eight 32-bit registers are used by the MC68020 to pass the module descriptor address to the MC68851 during a type $01 module call. A descriptor address ALCR exists for each of the eight MC68020 address spaces.

When the MC68851 locates a translation descriptor for the page containing the module descriptor (either in the ATC or from a search of the external tables), it first verifies that the G bit of the page descriptor is set, indicating that the page is allowed to contain module descriptors. In the event of a success, it examines the low-order bits of the module descriptor address, to ensure that the descriptor begins on an appropriate byte boundary, as determined by the MDS field (bits [1:0]) of the AC register. For module descriptor of sizes of 16, 32, and 64 bytes, the lowest-order four, five, or six bits of a module descriptor address must be zero, in order for that address to be valid. The MC68020 will initiate format exception processing if these requirements are not satisfied.

If the MC68851 failed to locate a translation for the module descriptor (the table search terminated due to encountering an invalid descriptor or a bus error during the table search), then it updates the access status ALCR to indicate that a format exception should be taken by the MC68020.

10.7 APPLICATION EXAMPLES

Figure 10.18 represents the block diagram of a simple MC68020-based system that includes the MC68851 and MC68881 (described in Chapter 11). Both devices are connected to the processor as coprocessors. The interconnection of the MC68020 and the MC68881 is discussed in Chapter 11.

10.7.1 Hardware Design

The schematic illustrates the important interconnections among the MC68020, MC68851, memory, and other peripheral devices [Motorola 86]. In this system, the MC68020 is the logical bus master, and the MC68851 is the physical bus master. The system utilizes the previously designed reset, halt, interrupt acknowledgment, timeout, and clock circuitry explained in various chapters.

The address strobe for the CPU space is generated by decoding the function codes (FC3–FC0), $\overline{AS}$, $\overline{BGACK}$, and $\overline{CLI}$ signals. The latch 74F373 is required for holding the data byte appearing on signal lines PA7–PA0, because (during the address translation sequences) the MC68851 drives the PA31–PA8 and guarantees the hold time for these signals with respect to $\overline{PAS}$ (the physical address strobe). However, it does not drive the lower byte of the address bus, and leaves it to the logical master (the MC68020) to drive these lines. Thus, when $\overline{LAS}$ is negated, the lower address values are latched, and the output of the 74F244 will retain the same values until the assertion of $\overline{LAS}$.

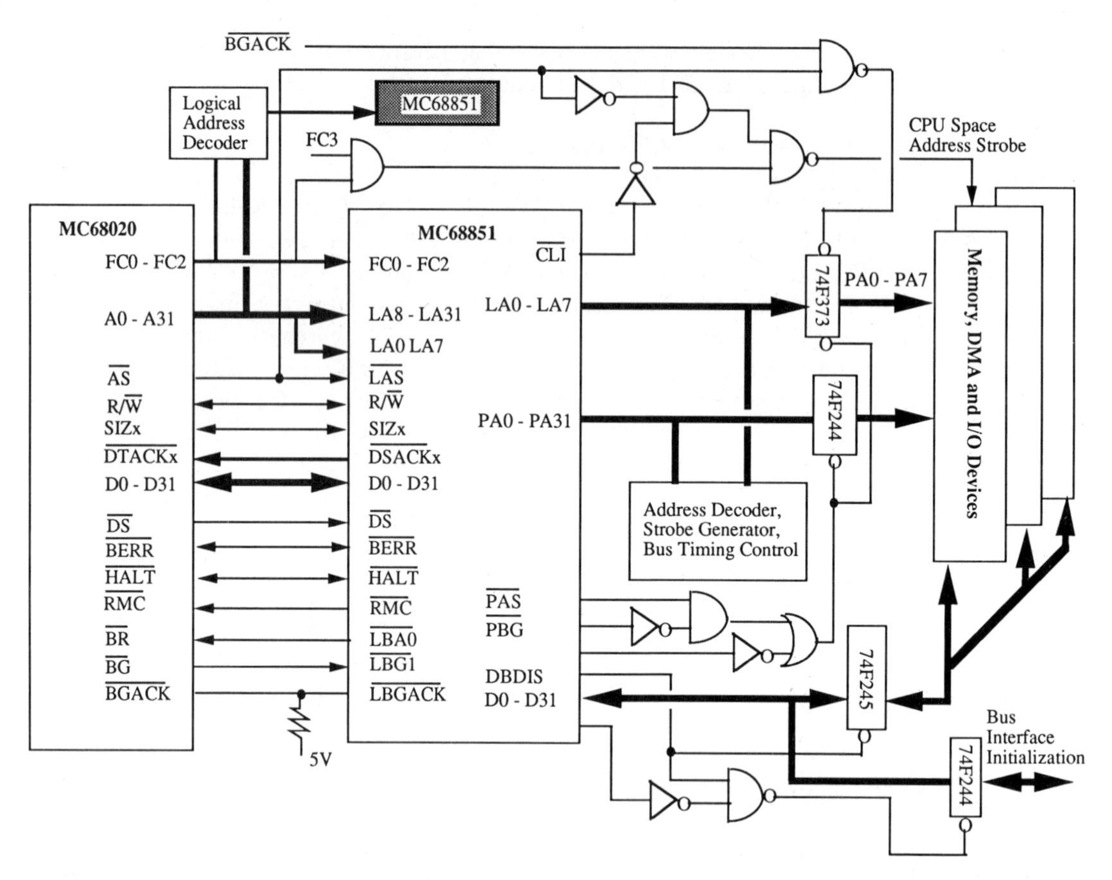

FIGURE 10.18. Hardware configuration for an MC68020-based system, including the MC68851 and MC58881.

The data and address bus drivers are required because the MC68851 cannot provide sufficient power to drive all of the connected devices. The drivers will provide the necessary noise immunity and power. The data and address bus drivers 74F244 and 74F373 are controlled by the signal line (BUS_CONTROL), which is generated from $\overline{\text{PAS}}$, $\overline{\text{PBG}}$, and $\overline{\text{PBGACK}}$. If another bus master requests the physical bus from the MC68851, the BUS_CONTROL will remain valid until $\overline{\text{PBGACK}}$ is not asserted.

The data bus enable signal is generated from DBDIS and $\overline{\text{RESET}}$. It allows the MC68020 to initialize the MC68851 during initialization sequences.

The design of the memory system is fairly straightforward. The designer is required to decide on the memory map, select the device type, and design the appropriate decoding logic.

10.7.2 Software Design

Now, let us assume that the above hardware will be used in an embedded system required to execute a computationally intensive processing task. This task requires a large virtual address space. Since the system is based on the MC68851, we are required to design the organization of the virtual address space.

10.7.2.1 Task Memory Map

To design the virtual address space, we need to decide on the page size, organization of address translation table, separation of the user and operating system (O.S.) routines, etc., and to determine how the MC68851 can be utilized efficiently.

One option is to separate the supervisory (for the O.S.) and user address spaces. To achieve this, we may decide to use the CPU root pointer with the function code look-up as the first index into the translation table. The second option is to use two translation tables and load their starting addresses into the supervisor and CPU root pointers. Using the CPU root pointer and function code look-up we can separate the supervisor and user accesses at the first level of the translation tree, and allow a different supervisor/user mapping for each task in the system.

The MC68851 allows the common sharing of address space among tasks, where the user and O.S. tasks can share the same address space. The advantage is that the O.S. can directly access and use the user task's data area. The general-purpose routines, like the file handler, I/O handler, special system software (e.g., SORT, Find File, etc.) can be shared and made reentrant. We can indicate the sharing status in the translation table entries and load them into the ATC. In this case, the MC68851 will automatically manage entries for all shared tasks.

We can also define a single virtual address space for the entire system, where all tasks run anywhere in the space. This requires only *one translation table* for the entire system. This allows the O.S. to access any item owned by an active task without modification of the root pointer register. To switch a task, the MC68851 needs to update the task's data pointers in the highest-level translation table indexed by the function codes, so that the task can have access to their data.

The other strategy is to maintain the supervisory address space and give each and every user task its own use of the rest of the system's virtual address space. This will require that each user task have its own set of translation tables. In such a case, the supervisor root pointer may or may not be utilized. This will allow a large virtual address space for user tasks. The penalty is that the table management algorithm becomes complex.

10.7.2.2 Protection Issues

The protection mechanism needs to be considered during the implementation of the virtual memory system. We have seen that the MC68020/MC68851 allows partitioning of code and data into eight distinct access levels, where each lower level is accessible by higher levels in such a way that all items owned by a level are available to routines at the higher levels. This requires that all virtual addresses must be distinct for each and every task and data item.

10.7.2.3 Organization of the Translation Table

The MC68851 allows for five levels of address translation tables. A single level is appropriate for numerically intensive real-time tasks, where the overhead of the virtual managed page faults and paging I/O must be minimized. If we select a 32 Kbyte page size for a user program of 64 Mbytes. We may then decide to create entries to be maintained in the ATC. This will allow the user program to map 2 Mbytes using the ATC entries, which may prove sufficient to execute the most complex program.

On the other hand, three or more levels of translation tables are useful when the operating system makes heavy use of shared memory spaces and/or shared page tables. A general-purpose multitasking virtual memory system frequently shares translation tables or program and data areas pointed to at the page table level. A table entry can point to another translation table used by a different task, allowing efficient sharing.

10.7.2.4 Reused Pages

The MC68851 can use the Lock (L) bit feature in a page descriptor to guarantee that once a page is loaded it remains in the memory. And any access to that page will not require the table search, if the SG field is also set. This is useful in implementing an I/O exception handler in a real-time system.

Again, the MC68851 allows the lower level to be partially present. When a limit field is set in the descriptor, the next lower table may have its high- or low-order portions deleted. This will save a considerable amount of memory. The savings are even greater as the number of concurrent tasks present in the system increases.

10.7.2.5 Implementation

In order to minimize the threshing and translation table searches, we decided to use an 8 Kbyte page size. Such a large page size will allow a small number of descriptors to map a large area of virtual space, and will result in fewer descriptor misses in the ATC. However, larger blocks of data will be transferred from disk to memory during a page swap, although only a fraction of a page may be used by the task at any one time.

If we assume that our embedded system has 4 Mbytes of physical memory, then the system could hold as many as 512 pages (based on 8 Kbytes/page) of active virtual memory at any one time. However, part of the operating system, such as the exception handlers, the O.S. kernel, and I/O drivers, will require nonpageable private memory space. That is, these pages can't be removed from the memory to make room for pages required by the application task. Therefore, only a portion of the total number of available pages are available for holding virtual pages.

To manage swappable pages, the O.S. needs to maintain a linked list of all free pages, where a free page contains a pointer to the next free frame on the list within itself. Therefore, whenever a page is required, the first one from the list is taken. If no free page is available, then it will wait until another task releases one to the pool of free pages.

The physical and logical view of the system is shown in Figure 10.19. The physical view represents the actual physical memory map for the MC68020-based embedded system. This is made up of 64 Kbytes for boot and diagnostics, 64

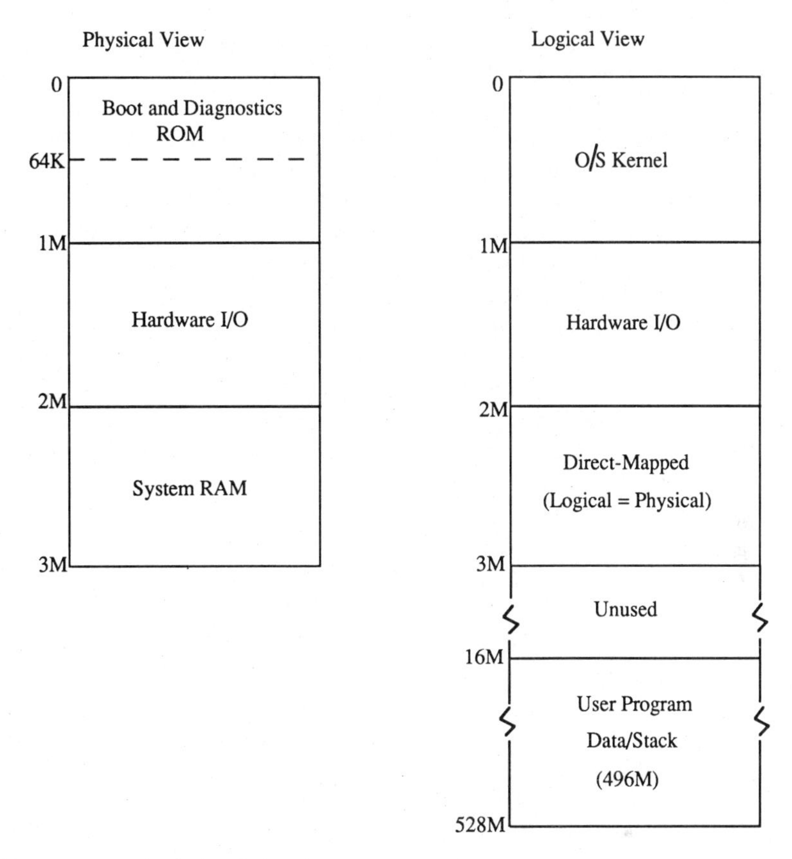

FIGURE 10.19. Physical and virtual memory map of the system.

Kbytes for I/O, and 1 Mbyte for system RAM. The logical view is the virtual memory map that the O.S. and the application task sees after the MC68851 has been initialized. The application task sees the virtual addresses starting at 16 MBytes and higher. This is the area where the code, data, and stack areas for the user tasks are allocated in virtual memory. However, the O.S. sees the entire virtual map that directly accesses the I/O ports, as well as the entire physical memory at untranslated addresses.

In our example, we opted to use the CPU root pointer for address translation. The first entry of the upper-level table for each task (each task has its own translation table) points to the lower-level table. The common lower-level table indicates the supervisor-only access, and maps the entire virtual O.S., physical I/O, and real memory areas. Again, the SG bit of each entry in the common table is set during the initialization of the MC68851, so that task switches do not invalidate any ATC-resident descriptors used by the O.S. Using this scheme, we are able to eliminate the requirements for extra look-up levels or pointer manipulations during task switching.

If the task requires new pages for its execution, it makes a call to the O.S. An O.S. routine GET_PAGE passes a block size in bytes and returns the virtual address of the new block for the task. Another routine GET_PAGE will look for room and, if none is found, then it will return an error. If virtual space for the block is found, then the new page entries are set. The task can then continue with the new present page in place.

In the event of an error, an O.S. routine SWAP_PAGE is called, which points to the MC68851 table entry that contains the disk address of the page to be read in and restored. After reading in the page, SWAP_PAGE replaces the disk address with the physical page address and sets the appropriate flag, so that the entry is available to the MC68851 for execution.

The routines mentioned so far have their corresponding routines. For example, GET_PAGE would have a RETURN_VIRTUAL, SWAP_PAGE would have SWAPOUT_PAGE, etc., and they are generally found in the operating system. If they are not available, then we will have to implement them, and make them part of the O.S.

Chapter 11

The Floating-Point Coprocessor

General-purpose microprocessors provide a level of performance that satisfies a wide range of computer applications. However, its instruction set is not sufficient to effectively satisfy various complex applications involving numerical computations. For such applications, the main processor must be supplemented with a coprocessor that extends the instruction set to allow the necessary computations to be accomplished more efficiently. The coprocessor concept allows the capabilities and performance of a general-purpose processor to be enhanced for a particular application, without unduly encumbering the main processor architecture.

A coprocessor like the floating-point arithmetic coprocessor adds instructions and features to support data types not directly supported by the main processor. The interaction between the main processor and the coprocessors are transparent to the programmer. The programmer can use the floating-point instruction sets without having to know the underlying implementation.

We will describe the capabilities of the MC68881/68882 floating-point arithmetic coprocessor, designed by Motorola to work with the MC68000 family of processors. This coprocessor significantly accelerates binary floating-point operations in adherence with the IEEE floating-point standard. This chapter is organized into several sections. The IEEE floating-point standard is introduced first, followed by the MC68881 coprocessor. The rest of the chapter describes the internal architecture, interfacing technique, instructions, and exception handling.

11.1 THE FLOATING-POINT STANDARD

The application programmer bears the ultimate responsibility for the reliability of numerical computations. To ease that burden, the IEEE-P754 standard for floating-point arithmetic was developed. Since the approval of this standard, semiconductor manufacturers started to market coprocessors compatible with the standard for various families of microprocessors.

11.1.1 The Floating-Point Format

Any nonzero real number may be expressed in "normalized floating-point" form as:

$$\pm (2^e)*f$$

where
e = The signed integer exponent
f = The significant digit field that satisfies $1 \leq f \leq 2$

A normalized nonzero number X has the form:

$$X = [(-1)^s]*[2^{(E\text{-}Bias)}]*(1.F)$$

where
S = Sign bit
E = Exponent
F = Fraction

The logical representation is shown in the following diagram.

Sign	Exponent	Fraction

The standard specifies three binary formats expressly tailored to the needs of the 32-bit architecture. They are:

- Single precision (32 bits)
- Double precision (64 bits)
- Extended precision format (80 bits)

The extended format allows greater precision in intermediate calculations. Each of the three floating-point data formats can represent five unique floating-point data types.

11.1.2 Single Precision Format

The representation is shown in the following diagram

32	30 . . . 23	22 0
Sign	Exponent	Fraction

Sign— 1 bit
Exponent— 8 bits
Fraction— 23 bits

The single precision numbers are represented by the formula:

$$[(-1)^{S}]\ [(1.f1.f2...f23)*2^{(E-127)}]$$

Note: *—multiplication term

11.1.3 Double Precision Format

The double precision format is shown in the following figure:

63	62 . . . 52	51 0
Sign	Exponent	Fraction

Sign— 1 bit
Exponent— 11 bits
Fraction— 52 bits

The double precision numbers are represented by the equation:

$$[(-1)^{S}]\ [(1.f1.f2...f52)*2^{(E-1023)}]$$

11.1.4 Extended Precision Format

This format is illustrated in the following diagram:

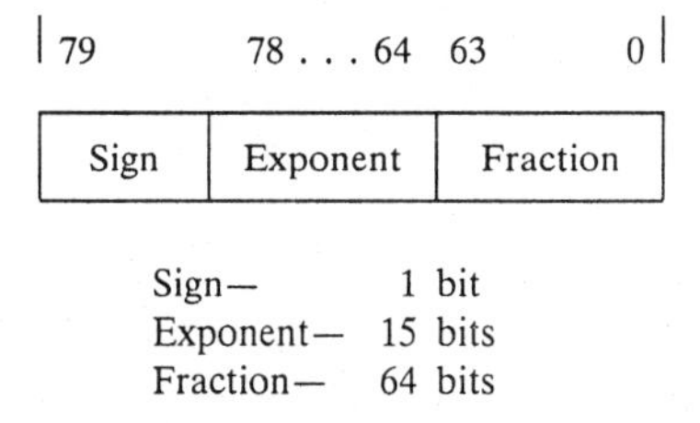

The extended precision numbers are represented by the equation:

$$[(-1)^{S}]\ [(1.f1.f2...f64)*2^{(E-16383)}]$$

Since the extended precision formats are intended only for intermediate results, the details of extended formats are implementation dependent, and the values stored in extended formats are not required to be normalized.

11.1.5 Data Types

The IEEE standard defines five floating-point data types, which permits representation of regular and particular numbers, and of special situations. The data types are:

- Normalized numbers
- Denormalized numbers
- Zeros (+ or − 0)
- Infinity (+ or − infinity)
- NANS (Not a number)

The regular numbers are encoded using three fields:

- A sign bit
- A biased exponent field
- A fraction field

Where redundant representations are possible, the fraction field is left-adjusted as far as possible (using an implicit leftmost bit that is not actually stored), so that the representation used is the one with the smallest possible exponent. If a number is too small to be normalized in this way, it is represented with a reserved exponent, and the fraction field is allowed to have leading zero bits. A number in this form is called subnormal or denormalized.

Normalized numbers. Normalized numbers are those numbers whose exponent lies between the maximum and minimum values. Normalized numbers may be positive or negative. For normalized numbers, the implied integer part bit, in single and double precision, is a one (1). In extended precision, the integer bit is explicitly a one (1). The representation of the normalized number is shown in the following figure:

Sign	Min < Exponent < Max	Mantissa (any bit pattern)

The sign bit is either 0 or 1.

Denormalized numbers. Denormalized numbers are those whose exponent is less than or equal to the minimum exponent value. They may be positive or negative. For denormalized numbers, the implied integer part bit in single and double precision is zero (0). In extended precision, the integer bit is explicitly a one (1). The representation of the denormalized number is shown in the following figure:

Sign	Exponent ≤ 0	Mantissa (any nonzero bit pattern)

The sign bit is either 0 or 1.

The IEEE standard defines a gradual underflow where the resultant mantissa is shifted right (denormalized) while incrementing the resultant exponent, until the resultant exponent reaches the minimum value.

Zero. Zeros are signed (positive or negative), and represent the real values +0.0 and −0.0. The representation is shown in the following figure:

Sign	Exponent = 0	Mantissa = 0

The sign bit is either 0 or 1.

Infinity. Infinities are signed (positive or negative), and represent real values that exceed the overflow threshold. Overflow is detected for a given data format and operation when the resultant exponent is greater than or equal to the maximum exponent value. For extended precision infinities, the most significant bit of the mantissa is a don't care bit.

Sign	Exponent = Max	Mantissa = 0

The sign bit is either 0 or 1.

Not A Number (NAN). Not a Number represents the result of operations that have no mathematical interpretations, such as infinity divided by infinity. All operations involving a NAN operation as an input will return a NAN result.

NANs with the leading fraction bit equal to one are nonsignaling NANs, while NANs with a leading fraction bit equal to zero are signaling NANs (SNAN). The representation of a NAN is illustrated in the following figure:

Sign	Exponent = Max	Mantissa = Any nonzero bit pattern

The sign bit is either 0 or 1.

The NAN provides a way to indicate uninitialized variable or illegal variable access, as well as a means for floating-point routines to communicate error reports through successive calculations.

11.1.6 Arithmetic Operations

Any implementation of the IEEE standard must support the following minimum arithmetic operations.

1. ADD, SUB, MULTIPLY, DIVIDE, and REMAINDER for any two operands of the same format. For each supported format, the destination cannot have smaller exponent range than the operands.
2. COMPARE and MOVE support operands of any, perhaps different, formats.
3. ROUND TO INTEGER and SQUARE ROOT for operands of all formats with results having no less an exponent range than the input operands. In the former operation, rounding shall be to the nearest integer or by truncation toward zero, at the user's option.
4. CONVERSION BETWEEN FLOATING-POINT INTEGERS in all supported formats, including binary integers in the host processor.

5. BINARY and DECIMAL conversions to and from all supported basic formats.

11.1.7 Exceptions

The simplest task can surprise us with exceptions. The standard has recognized that reality and makes provisions so that computations can proceed to their conclusion. The standard provides for exceptions, status flags, and traps. The five exceptions, and the default response for each, are listed in Table 11.1. The exceptions are explained in detail in the following subsubsections.

11.1.7.1 Invalid Operation

Two types of invalid operation exceptions are defined in the P754 standard. The first type occurs if an operand is invalid for the operation to be performed. The second type occurs if the result is invalid for the destination. An operand can be invalid for any of the following reasons:

- Detection of a NAN
- Addition or subtraction of infinity
- Multiplication of zero and infinity
- Division of 0/0, infinity/infinity, or where the divisor is unnormalized and the dividend is not infinite and not a normal 0.
- During a remainder operation X Rem Y, where Y = 0 or unnormalized, or X = infinity.
- During a square root operation, if the operand is less than zero, or is infinity, or is unnormalized.

The result is invalid when the result of any operation is unnormalized, but not denormalized, and has a destination of single or double precision format.

11.1.7.2 Divide-by-Zero

If the dividend is a finite nonzero number, and the divisor is a normal zero, then the divide-by-zero exception will be signaled.

11.1.7.3 Overflow

An overflow will be indicated if a rounded result is finite but not invalid, and the exponent is too large to be represented in the applicable floating-point format.

TABLE 11.1. Default Response for Exceptions.

Exception	Default Response
Invalid Operation	NAN (Not a Number)
Overflow	Infinity
Divide-by-Zero	Infinity Exactly
Underflow	Subnormal Number or Zero
Inexact	Correctly Rounded Number

11.1.7.4 Underflow

An underflow will be indicated when a result is received that is not a normal zero, and has an exponent that is too small to be represented in the applicable floating-point format.

11.1.7.5 Inexact

If there is no invalid operation exception, and the rounded result of an operation is not exact, then the inexact exception will be indicated. This exception will also be signaled when there is no invalid operation exception and the rounded result overflows without the appropriate trap being enabled.

The data type representations of Motorola's MC68881 Floating-Point arithmetic Coprocessor (FPC) and Intel's I80287 FPC are shown in Figure 11.1. Both implementations support the IEEE-P754 standard. It is interesting to note the similarities and differences between them.

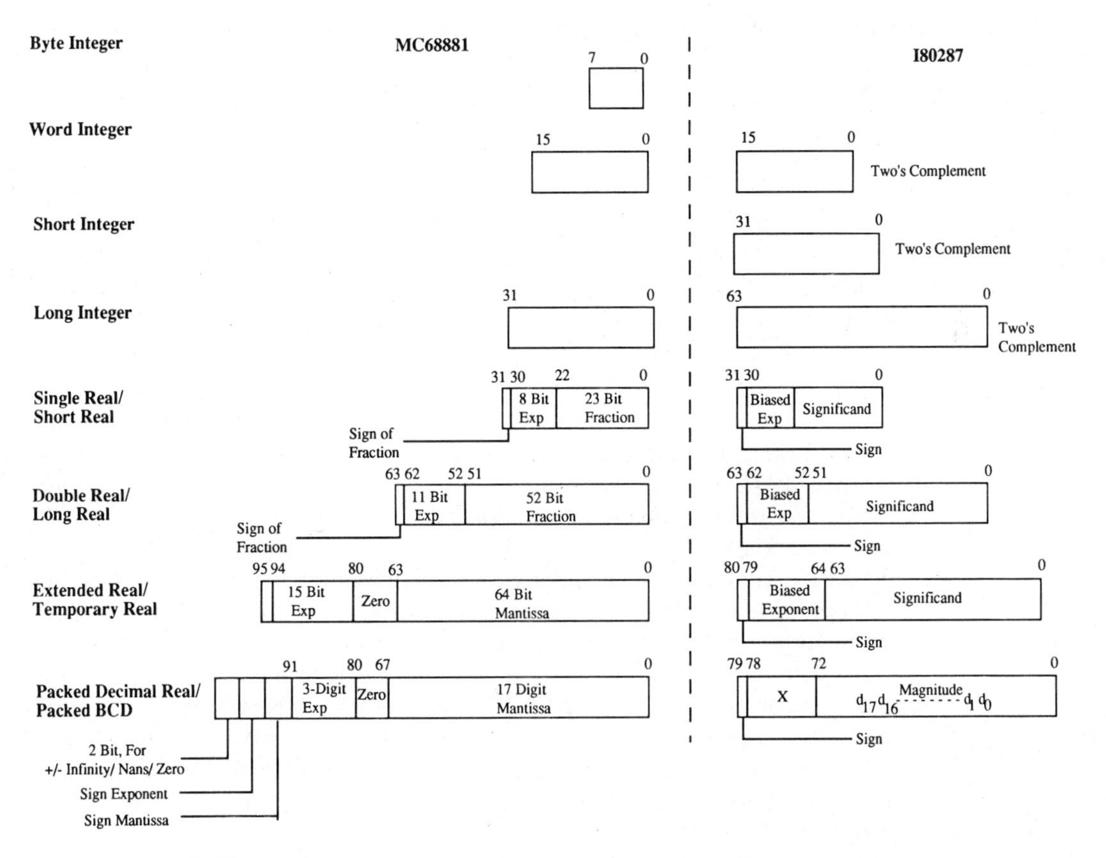

FIGURE 11.1. A comparison of data type representation for the IEEE P754 Standards in the MC68881 and I80287.

11.2 INTRODUCTION TO THE MC68881

The MC68881 floating-point coprocessor is a complete implementation of the IEEE standard P754, for use with the Motorola MC68000 family of microprocessors. It is implemented using VLSI technology, to give systems designers the required functionality in a physically small device. At 5 volts, the MC68881 consumes less than 1 watt of power.

With some performance degradation, it can also be used as a peripheral processor in systems where the MC68020/MC68030 is not the main processor. The configuration of the MC68881 as a peripheral processor or a coprocessor can be completely transparent to user software.

The MC68881 utilizes the MC68000 family coprocessor interface, to provide a logical extension of the CPU's instruction set and register set in a manner that is transparent to the programmer. It can execute concurrently with the MC68020, and usually overlaps its processing with the MC68020's processing, to achieve a high throughput. For applications requiring even more throughput than one MC68881 can deliver, several MC68881's can be used in one system.

The MC68881 is internally divided into three processing elements. The device is also called the FPU (Floating-Point Unit). The internal architecture is shown in Figure 11.2 [Motorola 85b]. The processing elements are:

- Bus Interface Unit (BIU)
- Microcode Control Unit (MCU)
- Execution Control Unit (ECU)

The BIU communicates with the coprocessor unit, while the ECU executes all MC68881 instructions. The MCU contains the clock generator, a two-level microcode sequence that controls the ECU, the microcode ROM, and the self-test logic.

11.2.1 Bus Interface Unit (BIU)

Communication between the main processor and the MC68881 takes place via the standard MC68000 family system bus. This coprocessor is designed to interface with the 8-, 16-, and 32-bit data bus. It is packaged as a 64-pin DIP or 68-pin PGA package.

This FPU contains a number of Coprocessor Interface Registers (CIRs) that are addressed like memory by the main processor. When the MC68020 detects a floating-point coprocessor instruction, it writes the instruction to the memory-mapped command CIR, and reads the response CIR (the handshaking and the protocol are described in Chapter 9). The BIU encodes in this response any additional service required of the main processor on behalf of the MC68881. For example, the response may contain a request to the MC68020 to fetch an operand from the evaluated effective address and to transfer it to the coprocessor interface operand CIR.

The key concern in a coprocessor interface, which allows concurrent instruction execution, is synchronization during main processor and coprocessor com-

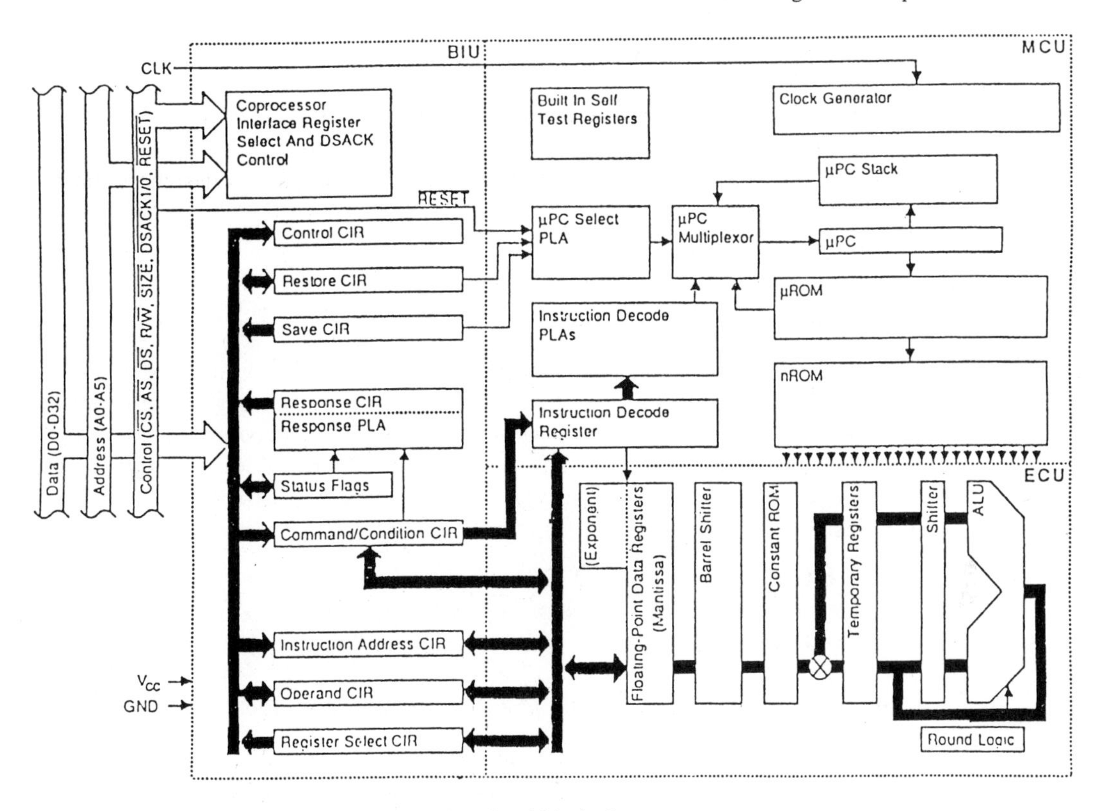

FIGURE 11.2. MC68881 functional block diagram. (Courtesy Motorola Inc.)

munication. If a subsequent instruction is written to the MC68881 before the ECU has completed execution of the previous instruction, then the response is to ask the main processor to wait. Thus, the selection of concurrent or nonconcurrent instruction execution is determined on an instruction-by-instruction basis by the coprocessor. For example, floating-point instructions with a destination in memory or a main processor register do not allow concurrent execution.

Since the coprocessor interface protocol is based on bus transfer in the asynchronous mode, it can be emulated by software when the MC68881 is used as a peripheral device, and memory-mapped over an MC68000-compatible asynchronous bus. Again, the MC68881 need not run at the same speed as the main processor. As a result, optimal cost/performance can be achieved by selecting the appropriate coprocessor and main processor speeds.

When used as a peripheral processor with the 8-bit MC68008 or the 16-bit MC68000/MC68010, all MC68881 instructions are trapped by the main processor to an exception handler at execution time. As a result, the software emulation of the coprocessor interface protocol can be totally transparent to the user. This allows the system to be quickly upgraded by replacing the main processor with an MC68020, without affecting user software.

11.2.2 Microcode Control Unit (MCU)

The MC68881's control unit is implemented as a two-level microcoded machine, much like the MC68000's architecture. The two-level storage structure was chosen primarily to minimize the control store area. The greater part of the unit is implemented as PLAs and ROM, and very little is implemented using random logic.

11.2.3 Execution Control Unit (ECU)

The execution control unit was optimized for extended precision calculations. The complex interconnect problems were minimized by placing a nanostore directly above the execution unit, leaving some space for decoding. Fields in the nanostore are positioned such that control store output lines are close to the corresponding execution unit control points. The special floating-point hardware includes:

- A 76-bit arithmetic unit.
- Various internal 76-bit busses and temporary registers.
- A barrel shifter that can shift right or left from 0 to 67 bits in one microcycle.
- Leading zero detection hardware for normalization.
- Special hardware and a constant ROM used in transcendental calculations.

In addition, special logic has been included to boost the speed of basic functions like ADD, SUB, MULTIPLY, DIVIDE, and MOVES.

11.3 PROGRAMMING MODEL

The programming model of the MC68881 is shown in Figure 11.3. It consists of:

- Eight 80-bit floating-point data registers, FPO through FP7, which are analogous to the integer data registers D0 through D7 in the MC68000 family of processors. The registers FP0–FP7 are the destination for most floating-point operations.
- A 32-bit control register with enable bits for each class of exception trap, and mode bits to select the rounding mode and rounding precision.
- A 32-bit status register that contains the floating-point condition codes, quotient bits (set by remainder and modulo), and exception status bits.
- A 32-bit instruction address register that contains the address (in the main processor memory) of the last floating-point instruction. This address is used during the exception trap handling to locate the instruction that initiated the trap.

11.3.1 Floating-Point Control Register (FPCR)

The 32-bit FPCR contains an exception enable byte, which enables/disables traps for each class of floating-point exception. It also contains a mode byte, which

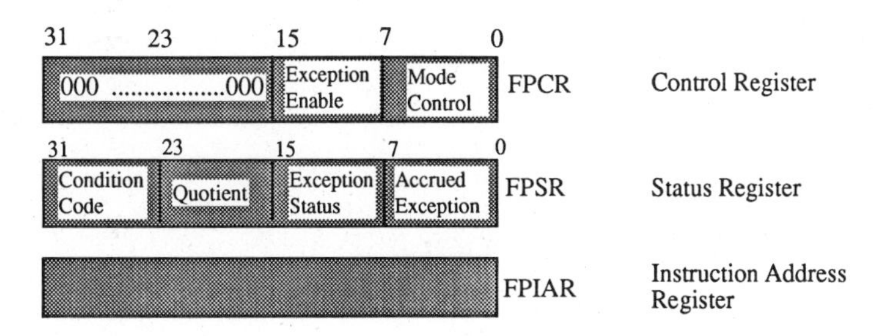

FIGURE 11.3. MC68881 progamming model. (Courtesy Motorola Inc.)

sets the user selectable modes. This register can be read or written to by the main processor. Bits [31:16] within the register are reserved for future use by Motorola. These bits will always be read as zero, and are ignored during write operations. This FPCR provides the default IEEE standard when cleared.

11.3.1.1 FPCR Exception Enable Byte

The exception byte bits [15:8] of the FPCR contains one bit corresponding to each floating-point exception class. The meaning of the bits are illustrated in Figure 11.4. If a bit of this byte is set by the MC68881, and the corresponding *enable* byte in the control register is also set, then an exception will be routed to a specific vector address.

When multiple exceptions occur with traps for more than one exception class, the highest priority exception will be initiated. Priority decreases as we move from left to right in the enable byte. That is, BSUN has the highest priority, and INEX1 has the lowest priority. The most common multiple exceptions are:

- SNAN and INEX1
- OPERR and INEX2

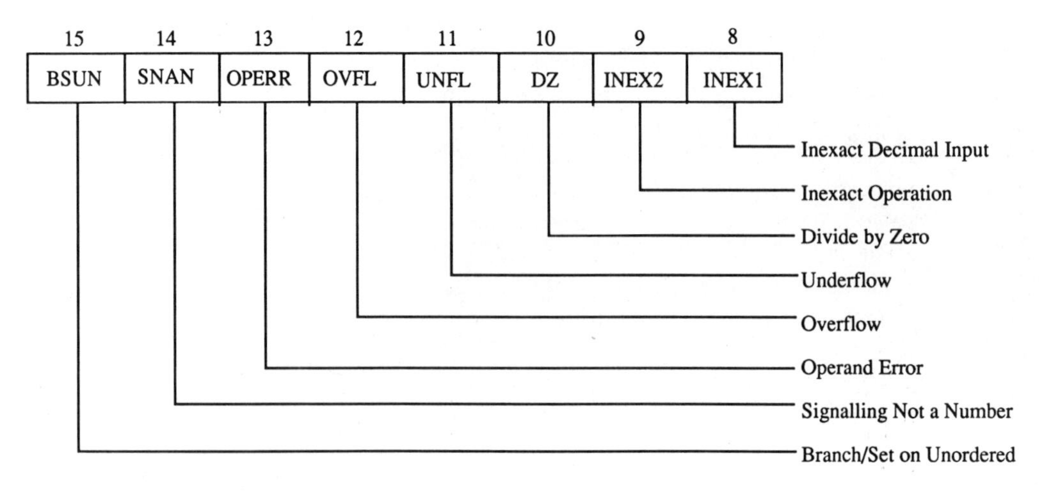

FIGURE 11.4. The exception status/enable byte. (Courtesy Motorola Inc.)

- OPERR and INEX1
- OVFL and INEX2 and/or INEX1
- UNFL and INEX2 and/or INEX1

11.3.1.2 FPCR Mode Control Byte

The *mode* byte bits [7:0] of the FPCR controls the user selectable rounding modes and rounding process. The purpose of the bits is illustrated in Figure 11.5. The bits [3:0] are reserved by Motorola for future use. They are read as zero, and should be altered during write operations, to maintain future software compatibility.

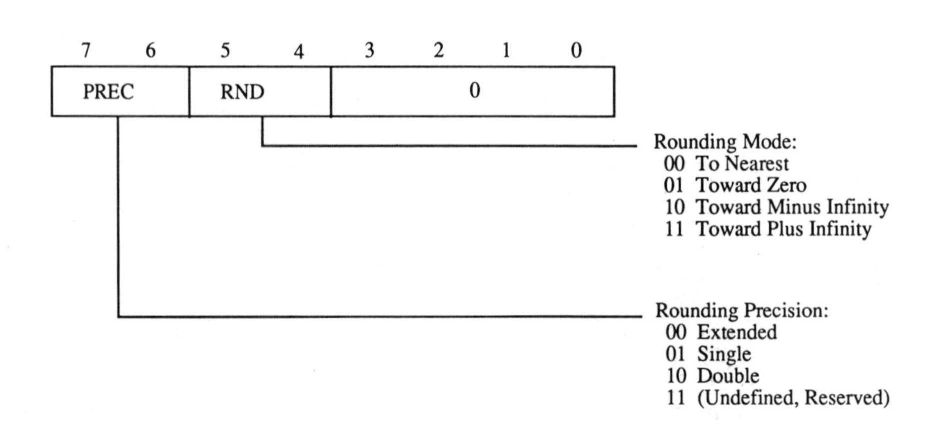

FIGURE 11.5. The mode control byte. (Courtesy Motorola Inc.)

The rounding mode is defined by bits [5:4] and is used to determine rounding precisions. Round to the nearest tells the processor that the nearest number to the infinitely precise result should be selected as the rounded value.

The rounding precision selects where rounding will occur in the mantissa. For extended precision, the result is rounded to a 24-bit boundary. A double precision result is rounded to a 53-bit boundary.

However, the single and double precision rounding modes are considered to be emulation modes. Thus, the execution speed of all instructions is degraded significantly when these modes are used. However, results obtained by the MC68881 will be the same as any other floating-point arithmetic processor that conforms to the IEEE standard, but does not support extended precision calculations.

11.3.2 Floating-Point Status Register (FPSR)

The 32-bit FPSR contains a floating-point condition code byte, a floating-point exceptions status byte, quotient bits, and a floating-point accrued exception byte. Almost all floating-point instructions will modify a part of this register. This register is also cleared by the reset event or a null-state size restore operation.

11.3.2.1 FPSR Floating-Point Condition Code Byte

Bits [31:24] of the FPSR are known as the Floating Point Condition Code (FPCC). Figure 11.6 describes the purpose of each bit. Bits [31:28] are reserved by Motorola for their future use. They are read as zero, and should not be altered. Bits [27:24] are set at the end of all arithmetic instructions. The result of the floating-point operation determines how the four condition code bits are set. Table 11.2 lists the condition code bit settings for each resultant data type.

The MC68881 defined only eight results out of the 16 possible outcomes. The user should be careful during the loading of the FPCC byte, and take the precaution of ensuring that one of the undefined condition code bits is not loaded. Performing a conditional instruction utilizing the undefined combinations may produce an unexpected branch condition.

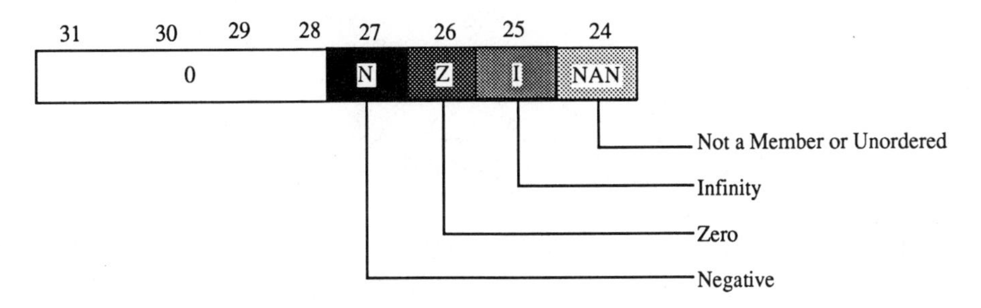

FIGURE 11.6. The condition code byte.

(Courtesy Motorola Inc.)

TABLE 11.2. Results Derived from Setting of Condition Code Bits.

N	Z	I	NAN	Result Data Type
0	0	0	0	+ Normalized or Denormalized
1	0	0	0	− Normalized or Denormalized
0	1	0	0	+ Zero (0)
1	1	0	0	− Zero (0)
0	0	1	0	+ Infinity
1	0	1	0	− Infinity
0	0	0	1	+ NAN
1	0	0	1	− NAN

11.3.2.2 FPSR Quotient Byte

The quotient byte (Figure 11.7), bits [23:16] of the FPSR, is set to reflect the results of the modulo (FMOD) or the IEEE remainder (REM) instruction. This byte contains the least significant bits of the quotient (unsigned), and the sign of the entire quotient. The quotient bits remain set until they are cleared by the user or until another FMOD or FREM instruction is executed. The quotient bits can be utilized in argument reduction for transcendentals and other functions.

11.3.2.3 FPST Accrued Exception Status Byte

Bits [7:0] of the FPSR are called the Accrued Exception (AEXC) status byte. Bits [2:0] of this byte are kept reserved for future use by Motorola, and are read as zero. The remainder of the AEXC contains a bit for each floating-point exception that may occur during the arithmetic instruction or move operation. The definition of the byte is shown in Figure 11.8. The AEXC is cleared by the MC68881 by a reset or a null-state size restore operation.

If a bit is set by the MC68881 in the AEXC byte, and the corresponding bit in the ENABLE byte is also set, then an exception will be signaled to the main processor. When a floating-point exception is detected by the MC68881, the corresponding bit in the AEXC byte will be set, even if the trap for that exception class is disabled.

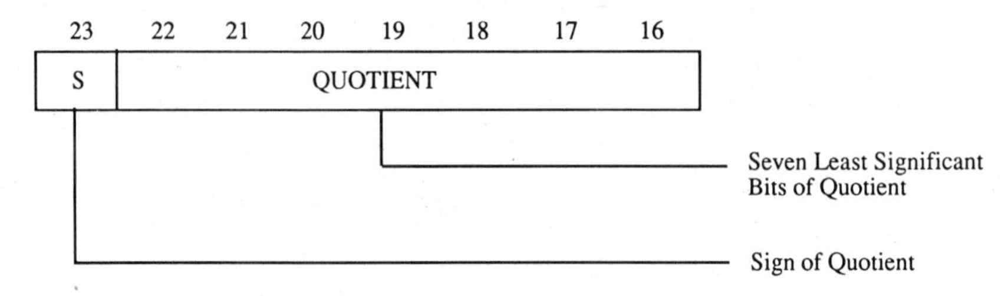

FIGURE 11.7. The quotient byte. (Courtesy Motorola Inc.)

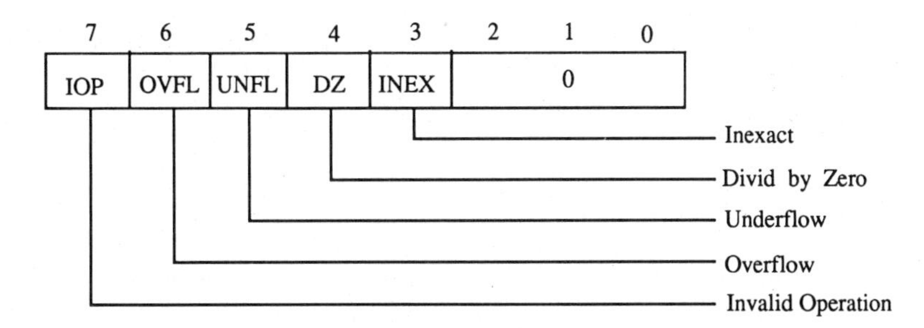

FIGURE 11.8. The accrued exception byte. (Courtesy Motorola Inc.)

11.3.3 Floating-Point Instruction Address Register

The majority of the MC68881 instructions operate concurrently with the main processor, where the main processor (e.g., the MC68020) can execute instructions while MC68881 is executing a floating-point instruction. Due to this concurrency, the PC value stacked by the MC68020 in response to an enabled floating-point exception trap may not point to the instruction that caused the exception. The PC is also cleared by the reset function or by a null-state size restore operation.

Built-in support is provided for the subset of the MC68881 instructions capable of generating floating-point exception traps, where the 32-bit floating-point instruction address register (FPIAR) is loaded with the logical address of an instruction before the instruction is executed. This address can then be used by a floating-point exception handler to locate a floating-point instruction that causes an exception.

Since the MC68881 FMOVE to/from the FPCR, FPSR, or FPIAR and FMOVEM instructions cannot generate floating-point exceptions, they do not modify the FPIAR. Thus, these instructions can be used to read the FPIAR during trap handling without changing the previous values.

11.4 OPERAND DATA FORMATS

The MC68881 supports a number of data formats that implement the IEEE standard in its entirety. The data formats (Figure 11.1) are:

- Byte integer (B)
- Word integer (W)
- Long-word integer (L)
- Single precision binary real (S)
- Double precision binary real (D)
- Extended precision binary real (X)
- Decimal real string—Packed BCD (P)

The letters shown in parentheses are the suffixes added to the Op-code in the assembly language source, to specify the data formats.

11.4.1 Integer Data Format

The integer data format for the MC68881 is identical to the integer data format supported by the MC68000 family of processors. All integer numbers are automatically converted by the FPU into extended precision floating-point numbers before their use. As a result, compilers need not convert the integers into reals at compile time. This saves memory space.

11.4.2 Floating-Point Data Format

Three floating-point data formats are supported by the MC68881, which are:

- Single precision [32 bits]
- Double precision [64 bits]
- Double extended precision [96 bits—80 of which are used, and the other 16 bits are reserved for future expandibility and for long-word alignment of the floating-point data structure]

Table 11.3 lists the sizes of the exponent and mantissa for single, double, and extended precision formats. The exponent is biased, and the mantissa is in sign and magnitude form. Since the single and double precision formats require normalized numbers, the most significant bit of the mantissa is 'implied' as a 1, giving one extra bit of precision.

The extended precision format also conforms to the IEEE standard. However, it is important to note that the IEEE standard does not specify the format to the bit level, as in the case of single and double precision numbers.

By performing all internal computations in the extended precision format, the MC68881 provides protection against unnecessary overflow, underflow, and the loss of significant bits. It also supports the mixing of data formats during evaluation of a single expression. The extended precision format simplifies the accurate computation of elementary functions and protects temporary variables and/or intermediate values.

TABLE 11.3. Representation of Data Formats by the FPU.

Data Format	Exponent Bits	Mantissa Sign	Mantissa Bits
Single	8	1	23 (+1)
Double	11	1	52 (+1)
Extended	15	1	64

11.4.3 Decimal Real String Format

The decimal real string format allows packed BCD strings to be input into, and output from, the MC68881. The strings consist of a 3-digit base 10 mantissa. Both the exponent and mantissa have a sign. All digits are packed BCD so that a whole string fits into 96 bits, equivalent to three long-words. Again, when packed BCD strings are input into the MC68881, they are automatically converted to extended precision real values. This allows packed BCD numbers to be used as input during any operation. For example:

```
FADD.P  #-2.02E+24, FP1
```

The MC68881 can also output BCD numbers in a format desirable to high-level language compilers. For example:

```
FMOVE.P FP1, BUFF[#-5]
```

The instruction asks the MC68881 to convert the value from floating-point data register 1 into a packed BCD string with 5 digits to the right of the decimal point. This is identical to the "F" format used in many Fortran compilers.

The data formats described so far are supported orthogonally by all arithmetic and transcendental operations, and by the appropriate addressing modes of the MC68020/MC68030/MC68040. Examples of floating-point instructions are:

```
FADD.B  #1, FP0
FADD.W  #100, FP1
FADD.L  VALUE, FP2
FADD.S  #3.14159, FP3
FADD.D  (SP)+, FP4
FADD.X  TEMP, FP5
FADD.P  #1.445E20, FP6
```

The MC68881 always performs all computations in full extended precision arithmetic, and converts all operands automatically into extended precision values before their use.

11.5 INSTRUCTION SETS AND ADDRESSING MODES

The instruction set of the MC68881 can be grouped into six major classes:

- Moves between the MC68881 and memory or the main processor
- Move multiples (in and out)
- Monadic operations
- Dyadic operations

- Program control operations
- System control operations

The instructions support all IEEE-defined special values, and return the IEEE-specified results, with accuracy as specified by the standard. The MC68881 added various enhancements to the set of operations defined by the IEEE P754 standard.

The FPU was designed in such a way that a user should not or need not be an expert in floating-point arithmetic in order to use it. As a result, no operations require software envelopes to conform to the standard. Similarly, for transcendentals, all argument reduction is performed by the chip.

11.5.1 Instruction Format

The size of the MC68881's instructions vary from one to eight words. The first word is the operation word, and the next word is the MC68881 command word or conditional predicate. The rest of the words specify the operands or extensions to the effective addressing mode or immediate operands that are part of the instruction. The instruction format is shown in Figure 11.9.

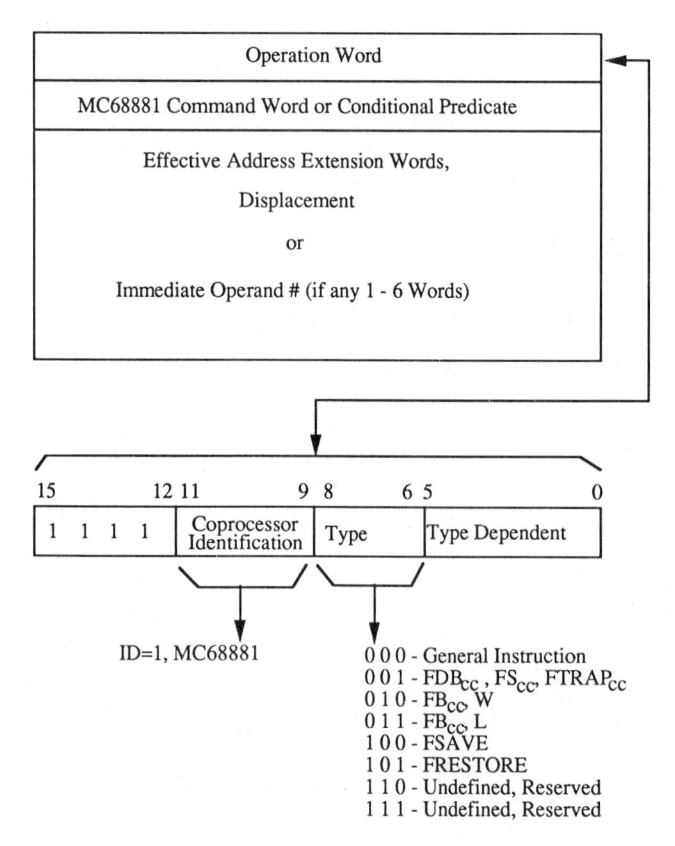

FIGURE 11.9. The coprocessor instructions.

All coprocessor operations are based on the F-line operation codes (i.e., operation words with bits [15:12] = $F), which instruct the MC68020 and the next generation of processors to call upon a coprocessor to execute an instruction.

The coprocessor identification (Cp-ID) field indicates which coprocessor is to be selected. We have already seen that Motorola has reserved Cp-ID positions 000 through 101 for its own use, and left Cp-ID 110 and 111 for a user-defined coprocessor. The MC68020 assembler will utilize the '001' as the Cp-ID for the MC68881. The type field indicates to the main processor the nature of the coprocessor instruction. The type-dependent field normally specifies the effective address or conditional predicate based on the type field.

11.5.2 FPC Instructions

The MC68881 supports 46 instructions, 35 of which are for arithmetic operations. In addition to the floating-point instructions required by the IEEE standard, the MC68881 includes operations for moving data into and out of itself to support branching, and for supporting virtual memory. It also includes a full set of trigonometric and transcendental functions.

11.5.3 MOVES

All MOVES from memory (or from an MC68020 data register) to the MC68881 cause data conversion, from the source data format into the interval extended precision format. On the other hand, all MOVES from the MC68881 to memory or to the MC68020 data register cause conversion from the internal extended precision format to the destination data format. The result is always rounded to the precision specified in the FPCR control byte. The MC68881 always checks for overflow and underflow during the rounding operation.

The syntax for MOVES are:

```
FMOVE.f  <ea>, FPn    ; Move to the MC68881
FMOVE.f  FPn, <ea>    ; Move from the MC68881
FMOVE.X  FPn, FPm     ; Move within the MC68881
```

where

<ea> = MC68020 effective address.
f = Data format size.
FPn,FPm = Floating-point data registers.

11.5.4 MOVE Multiples

The floating-point MOVES are similar to their integer counterparts on the MC68000 family of processors. This instruction allows from 1 to 8 of the MC68881's data registers contents to be moved to or from memory, with one instruction. Specification of which registers are to be moved can be made at assembly time or dynamically specified at execution time. The contents of the registers are moved as 96-bit extended data, with no conversion. The syntax for the move multiple registers are:

FMOVEM <ea>, FP0–FP5/FP7
FMOVECR #PY, FP1

Move multiples are useful during context switches and interrupts, to save and/or restore a program's state. They are also useful at the start and end of a procedure to save and restore the calling program's registers. The various data movement operations are shown in Figure 11.10.

11.5.5 Monadic Operations

Monadic operations are performed on the source operand, and the result is stored in the destination (which is always a floating-point data register). The source operand may either be in a floating-point data register (FPn), a memory location, or an MC68020/MC68030 data register. Examples of monadic instructions are:

FSQRT.f <ea>, FPn
FSQRT.X FPm, FPn

The list of the monadic operations supported by the MC68881 is given in Figure 11.11.

11.5.6 Dyadic Operations

Floating-point dyadic instructions require two input operands. The first operand comes from memory, a floating-point data register, or a main processor data register. The second operand comes from a floating-point data register, where

Instruction	Operand Syntax	Operand Format	Operation
FMOVE	FP_m, FP_n <ea>, FP_n FP_m, <ea> FP_m, <ea>(#K) FP_m, <ea>(D_n) <ea>, FP_{cr} FP_{cr}, <ea>	X B, W, L, S, D, X, P B, W, L, S, D, X P P L L	Source→Destination
FMOVECR	#ccc, FP_n	X	ROM Constant→FP_n
FMOVEM	<ea>, <list> <ea>, D_n <list>, <ea> D_n, <ea>	L, X X L, X X	Listed Register→Destination Source→Listed Register

FIGURE 11.10. Data movement operations.

Non-Transcendental:	
FABS	Absolute Value
FINT	Integer Part
FNEG	Negate
FSQRT	Square Root
FNOP	No Operation (Synchronize)
FGETEXP	Get Exponent
FGETMAN	Get Mantissa
FTST	Test

Transcendental Functions:	
FACOS	Arc Cosine
FASIN	Arc Sine
FATAN	Arc Tangent
FATANH	Hyperbolic Arc Tangent
FCOS	Cosine
FCOSH	Hyperbolic Cosine
FETOX	E to the X Power
FETOXM1	E to the (X-1) Power
FLOG10	Log to the Base 10
FLOG2	Log to the Base 2
FLOGN	Log Base e of X
FLOGNP1	Log Base e of (X+1)
FSIN	Sine
FSINCOS	Simultaneous Sin/Cos
FSINH	Hyperbolic Sine
FTAN	Tangent
FTANH	Hyperbolic Tangent
FTENTOX	Ten to the X Power
FTWOTOX	Two to the X Power

FIGURE 11.11. Monadic Instructions.

the result of the operation is stored. The dyadic function and the instructions are listed in Figure 11.12. Dyadic operations support any data format, and use extended precision arithmetic during execution. The exceptions are single precision multiply and divide operations, which support any precision inputs, but return results accurate only to the single precision format.

Examples of dyadic instructions are:

```
FADD.f <ea>, FPn
FADD.X FPm, FPn
```

11.5.7 Program Control Operations

Floating-point program control operations (e.g., set on condition and trap on condition instructions on the MC68881/MC68882) operate in the same way as their equivalent integer instructions in the MC68000 family processors. However, many more conditions exists for the FPU due to the IEEE floating-point standard. Again, the coprocessor imposes its own set of floating-point condition codes. However, because of the close coupling between the FPU and the main processor, the floating-point branch operations are executed very fast.

Dyadic Functions

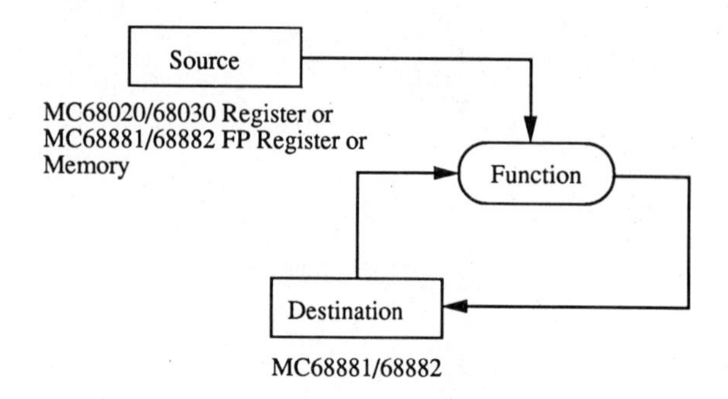

Dyadic Instructions

FADD —Add
FCMP —Compare
FDIV — Divide
FMOD —Modify
FMUL — Multiply
FREM — IEEE Remainder
FSCALE —Scale Exponent
FSUB — Subtract

FIGURE 11.12. Dyadic operations.

Examples of the program control operations are:

FBcc LOOP
FDBcc FP0, LOOP

Where cc is one of 32 floating-point conditional test specifiers. The conditional tests are based on the four condition bits (negative, zero, infinity, and NAN) and confirm to the IEEE's standard conditional test requirements.

The program control operations are shown in Figure 11.13.

11.5.8 System Control Operations

System control operations are listed in Figure 11.14. They are utilized for communication with the operating system via a conditional trap instruction, and are used to save or restore non-user visible information related to the FPU, during context switches (for a virtual memory or other class of multitasking system).

The conditional trap instruction uses the same conditional tests as the program control instructions, and allows an optional 16- or 32-bit immediate operand to be included as part of the instruction for passing parameters to the operating system.

Instruction	Operand Syntax	Operand Size or Format	Operation
FB_{cc}	<Label>	W, L	If condition True Then PC + d → PC
FDB_{cc}	D_n, <Label>	W	If condition True Then no operation; Else D_n - 1 → D_n, If $D_n \neq -1$ Then P C + d → PC
FNOP	None	None	No operation
FS_{cc}	<ea>	B	If condition True Then 1's → Destination Else 0's → Destination
FTST	<ea> FP_n	B, W, L, S, D, X, P X	Set FPSR Condition Codes

FIGURE 11.13. Program control operations.

Instruction	Operand Syntax	Operand Size	Operation
FRESTORE	<ea>	None	State Frame → Internal Registers
FSAVE	<ea>	None	Internal Registers → State Frame
$FTRAP_{cc}$	None #XXX	None W, L	If Condition True Then Take exception

FIGURE 11.14. System control operations.

The list of the instructions that allow the system to be controlled are:

FMOVE <ea>, FPcr — Move to a control register.
FMOVE FPcr, <ea> — Move from a control register.
FSAVE <ea> — Virtual machine state save.
FRESTORE <ea> — Virtual machine state restore.

11.6 FUNCTIONAL SIGNALS OF THE CHIP

The signal lines of the MC68881 can be functionally organized into five groups (Figure 11.15). These are:

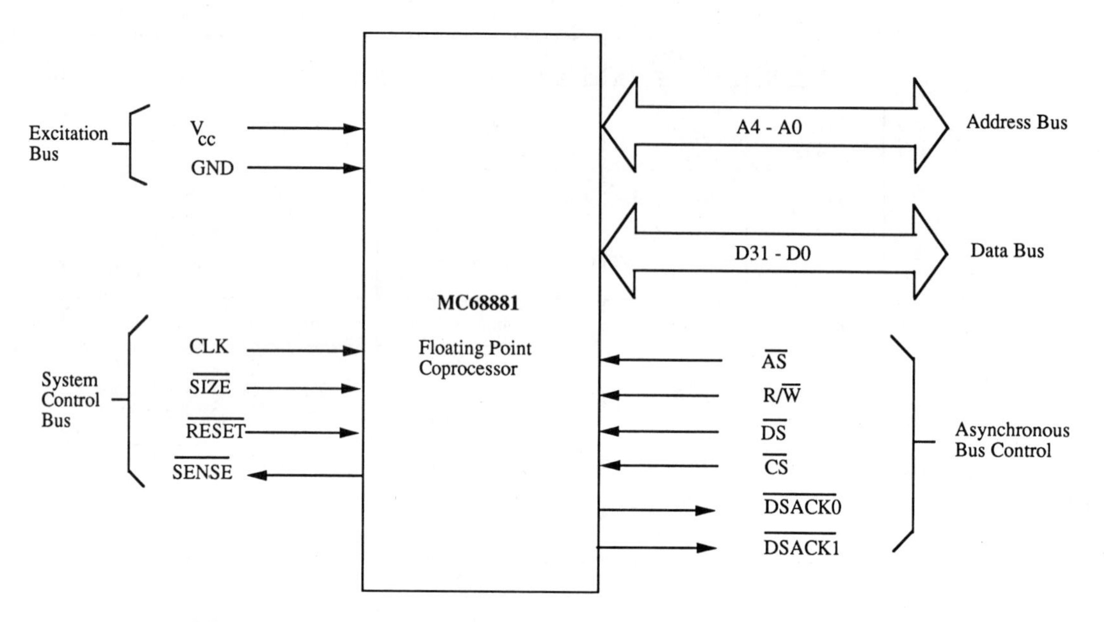

FIGURE 11.15. Pin configuration of the MC68881.

- Address bus [A4–A0]
- Data bus [D31–D0]
- System control bus [CLK, $\overline{\text{SIZE}}$, $\overline{\text{RESET}}$, and $\overline{\text{SENSE}}$]
- Asynchronous bus control [$\overline{\text{AS}}$, R/$\overline{\text{W}}$, $\overline{\text{DS}}$, $\overline{\text{CS}}$, $\overline{\text{DSACK0}}$, and $\overline{\text{DSACK1}}$]
- Excitation bus [Vcc and GND]

11.6.1 Address Bus [A4–A0]

The address lines A4 through A0 are input to the MC68881. They are used by the main processor to select the coprocessor interface registers. The coprocessor interface registers are generally mapped as memory locations within the CPU space of the MC68020.

Table 11.4 illustrates how A0 and the $\overline{\text{SIZE}}$ pins of the MC68881 are utilized to interface with the main processor. When the device is configured to operate over an 8-bit data bus, the A0 pin is used as a byte address signal for the co-

TABLE 11.4. Port Sizing for the MC68881.

A0	Size	Data Bus
—	Low	8-bit
Low	High	16-bit
High	High	32-bit

processor interface registers. However, to interface to a 16- or 32-bit system data bus, both A0 and the $\overline{\text{SIZE}}$ pins are tied to High and/or Low.

11.6.2 Data bus [D31–D0]

The data bus of the MC68881 contains 32-bit, bidirectional, three-state signal lines serving as the data path between the unit and the main processor. The MC68881 will operate over an 8-, 16-, or 32-bit system data bus. Both A0 and the $\overline{\text{SIZE}}$ pins are needed to configure the device's proper operation.

Regardless of whether the MC68881 is interfaced as a coprocessor or as a peripheral device, all interprocessor transfers of instruction information, operand data, status conditions, and requests for service take place as part of a standard MC68000 asynchronous bus access.

11.6.3 System Control Bus

The system control bus is made up of four signals, which are:

- Clock (CLK)
- Size ($\overline{\text{SIZE}}$)
- Reset ($\overline{\text{RESET}}$)
- Sense device ($\overline{\text{SENSE}}$)

11.6.3.1 Clock

This TTL-compatible signal line is internally buffered for a clock signal, to be supplied externally. The clock should be a constant-frequency free-running square wave, and must not be gated at any time. The input clock must conform to minimum and maximum period and pulse width.

The minimum clock frequency for the MC68881 is 4 Mhz, with a 50% duty cycle. The minimum and maximum cycle times for the MC68881RC12 are 80 and 250 nsec. For the MC68881RC16, they are 60 and 250 nsec. The maximum rise time is 5 nsec.

11.6.3.2 Size

The active low input signal $\overline{\text{SIZE}}$ is used in conjunction with A0 to configure the MC68881 for operation over an 8-, 16-, or 32-bit system bus. The configuration is shown in Table 11.4.

11.6.3.3 Reset

The active low input reset signal causes the MC68881 to initialize the internal floating-point data registers to nonsignaling Not-a-Number (NAN). An externally generated reset also clears the floating-point control, status, and instruction address registers.

During a power-up reset, the $\overline{\text{RESET}}$ line must be asserted low for a minimum of four-clock periods after Vcc with tolerance. This is necessary for correct initialization. However, the reset pin should be asserted for 100 milliseconds to maintain compatibility with all of the MC68000 family of processors.

For a rest of reset condition, this line must be asserted low for at least two clock cycles. However, the signal line should be asserted low for 10 clock cycles for compatibility reasons.

11.6.3.4 Sense Device

This output pin declares that a MC68881 is present in the system. A pull-up register is generally connected to this pin, to tell the external hardware that the MC68881 is present in the system. When not used, it acts as an additional GND pin.

11.6.4 Asynchronous Bus Control

We have already seen that the MC68000 family of processors communicates with the outside world in asynchronous mode. The MC68881 utilizes six signal lines to implement the asynchronous protocol. The signal pins are: $\overline{\text{AS}}$, R/$\overline{\text{W}}$, $\overline{\text{DS}}$, $\overline{\text{CS}}$, $\overline{\text{DSACK0}}$, and $\overline{\text{DSACK1}}$.

11.6.4.1 Address Strobe ($\overline{AS}$)

This active low input signal indicates that there is a valid address on the address bus. It also indicates that both the chip select ($\overline{\text{CS}}$) and read/write signal lines are valid.

11.6.4.2 Read/Write (R/$\overline{W}$)

This bidirectional signal indicates the direction of bus transfer (read/write) by the main processor. A signal *high* indicates a read from the MC68881 and a *low* means a write to the MC68881. As indicated, this signal line must be validated using the address strobe signal.

11.6.4.3 Data Strobe ($\overline{DS}$)

This active low input signal indicates that there is valid data on the data bus during a write bus cycle.

11.6.4.4 Chip Select ($\overline{CS}$)

This active low input signal enables the main processor to access the MC68881's interface registers. External logic must generate this signal, so that only one device gets activated in the whole system. The $\overline{\text{CS}}$ must be valid with $\overline{\text{AS}}$.

11.6.4.5 Data and Size Acknowledge ($\overline{DSACK0}$ and $\overline{DSACK1}$)

$\overline{\text{DSACK0}}$ and $\overline{\text{DSACK1}}$ are active three-state output signals. They indicate the completion of a bus cycle to the main processor. The MC68881 asserts these lines to signal the main processor that it has placed valid information on the data bus. During the write cycle on the MC68881, these signal lines are used to acknowledge the acceptance of data by the MC68881. The $\overline{\text{DSACKx}}$ lines are used by the MC68881 to indicate its port size, on a cycle-by-cycle basis, to the MC68020.

Table 11.5 lists the system data bus configurations, including the status of the asynchronous bus pins A4, $\overline{\text{DSACK1}}$, and $\overline{\text{DSACK0}}$.

11.6.5 Excitation Bus [Vcc and GND]

These pins provide the power supply and system reference level for the internal circuitry of the MC68881. The hardware designers should put appropriate capacitive decoupling to reduce the noise level on these pins.

The maximum rating for the supply and input voltages are:

Supply voltage: -0.3volts to $+7.0$ volts.
Input voltage: -0.3volts to $+7.0$ volts

11.7 COPROCESSOR INTERFACE

The MC68000 coprocessor protocol specifies the design of the interface between the MC68881 and MC68020, including functionality of the tasks shared between the two. The tasks are partitioned in such a way that the MC68020 does not have to decode the coprocessor instructions, and the MC68881 does not have to duplicate main processor functions, such as effective address calculations or exception handling.

This clear division provides orthogonal extension to the instruction set, by permitting MC68881 instructions to utilize all MC68020 addressing modes and to generate exception traps. As a result, the programmer is able to view the main processor and coprocessor as a single service provider, where the coprocessor executions may be overlapped with the execution of the MC68020 instructions. The concurrency is completely transparent, including the MC68020's TRACE mode, which is also supported by the MC68881.

Since the coprocessor interface is based solely upon the bus cycle, and the MC68881 is never a bus master, the MC68881 can be placed on either the logical

TABLE 11.5. Location of Valid Data on the Data Bus Placed by the FPU.

Data Bus	A4	$\overline{\text{DSACK1}}$	$\overline{\text{DSACK0}}$	Comments
32-bit	1	Low	High	Valid data on D31–D0
32-bit	0	Low	High	Valid data on D31-D0
16-bit	x	Low	High	Valid data on D31–D16 or D15–D0
8-bit	x	High	Low	Valid data on D31–D24, D23–D16, D15–D8, or D7–D0
32-bit	x	High	High	Insert wait states in current bus cycle

or physical side of the system memory management unit. This adds considerable flexibility to the design of the system.

Again, the virtual machine architecture of the MC68020 is completely supported by the MC68881. As a result, if the main processor detects a page fault and/or a task times out, it can instruct the MC68881 to stop its operation at any time (even in the middle of instruction execution) and save its internal states in memory. The main processor can instruct the MC68881 to reload a previously saved internal state from the memory.

We have already seen in Chapter 9 that the MC68020 communicates with its coprocessor utilizing a well-defined protocol and interface registers forming the mailbox. The MC68881 contains the necessary set of Coprocessor Interface Registers (CIR), separate from the registers described in the programming model. When the MC68881 is used as a coprocessor to the MC68020, the programmer is never required to explicitly access these CIRs. When an MC68020 is not used as the main processor, the MC68881 CIRs are explicitly accessed by the software routine that simulates the behavior of the MC68020 with respect to the coprocessor interface.

11.7.1 Interface Design for the MC68881

The MC68020 does not provide any special bus signal for the MC68881. As a result, the interface logic must look for a specific bit pattern placed by the MC68020 over its pins during the CPU space. This bit pattern is shown in Figure 11.16. The address bits A16–A19 indicate the CPU space function that the main processor is performing. The pattern '0010' over these address lines identifies a coprocessor access. The Cp-ID appears over address lines A15–A13. The A4–A0 selects locations within the coprocessor mailbox.

Therefore, the decoder of the interface must utilize the MC68020's function code outputs (FC0–FC2), the CPU space type field (A19–A16), and the Cp-ID field (A15–A13) to generate the $\overline{\text{CS}}$ signal for the MC68881. An example interface and chip select logic is shown in Figure 11.17. The $\overline{\text{SENSE}}$ signal and the $\overline{\text{CS}}$ are NANDed, and the output is connected to the $\overline{\text{BERR}}$ pin of the MC68020. This informs the MC68020 of a problem with a nonfunctioning MC68881 or chip select decoder. The $\overline{\text{AS}}$, $\overline{\text{DS}}$, R/$\overline{\text{W}}$, $\overline{\text{DSACK0}}$, and $\overline{\text{DSACK1}}$ of the MC68020 are tied to the corresponding pins of the MC68881. The 32-bit data bus is connected to the corresponding 32-bit data pins of the MC68881. This implies that the MC68881 is configured to operate over a 32-bit data bus, where both the A0 and $\overline{\text{SIZE}}$ pins are connected to Vcc. A1 selects the appropriate word location of the

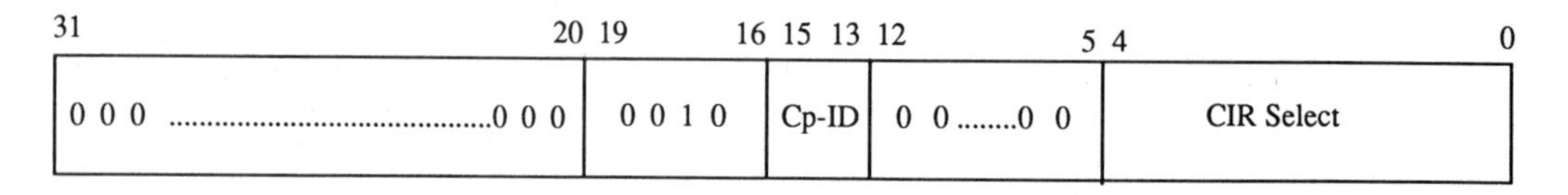

FIGURE 11.16. Address bus encoding for coprocessor access.

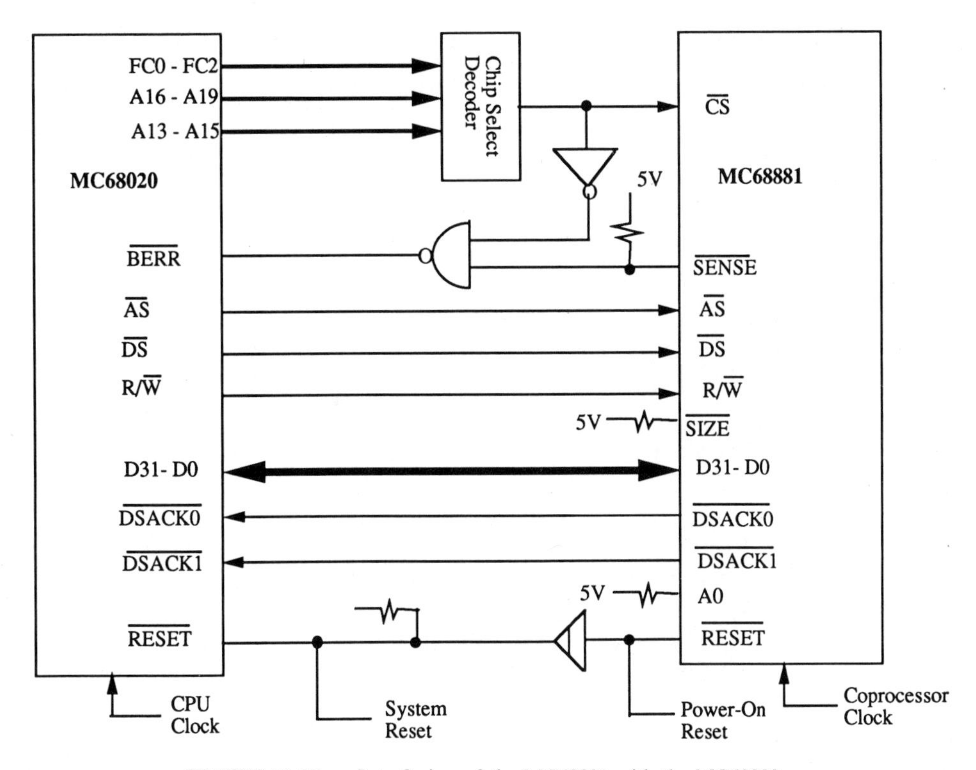

FIGURE 11.17. Interfacing of the MC68881 with the MC68020 processor.

coprocessor interface register. Most CIR access transfers an entire instruction or data item in a single bus cycle. The amount of data transferred with each access to the operand CIR is dependent on the state of instruction dialog, and is determined by the coprocessor, not the MC68020.

11.7.2 Coprocessor Interface Registers

The memory map of the MC68881's interface registers is listed in Table 11.6.

When a chip select is asserted, the MC68881 decodes the CIR select field of the address bus pins (A4–A0) to select the appropriate coprocessor interface register.

11.7.3 16-Bit Data Bus Coprocessor Connection

The connection between the MC68881 and MC68020 over the 16-bit data bus is shown in Figure 11.18. Here the $\overline{\text{SIZE}}$ pin of the coprocessor is tied to Vcc, and A0 to the GND. The 16 least significant data pins of the MC68881 are connected to D31–D16 of the data bus of the main processor. The $\overline{\text{DSACK}}$ pins of the main processor and the coprocessor are also connected to each other.

TABLE 11.6. Internal Register Map of the FPU Unit.

Register	A4–A0	Offset	Width	Type
Response	0000x	$00	16	Read
Control	0001x	$02	16	Write
Save	0010x	$04	16	Read
Restore	0011x	$06	16	Read//Write
Operation word	0100x	$08	16	Read//Write
Command	0101x	$0A	16	Write
(Reserved)	0110x	$0C	16	—
Condition	0111x	$0E	16	Write
Operand	100xx	$10	32	Read//Write
Register select	1010xx	$14	16	Read
(Reserved)	0111x	$16	16	—
Instruction address	110xx	$18	32	Write
Operand address	111xx	$1C	32	Read//Write

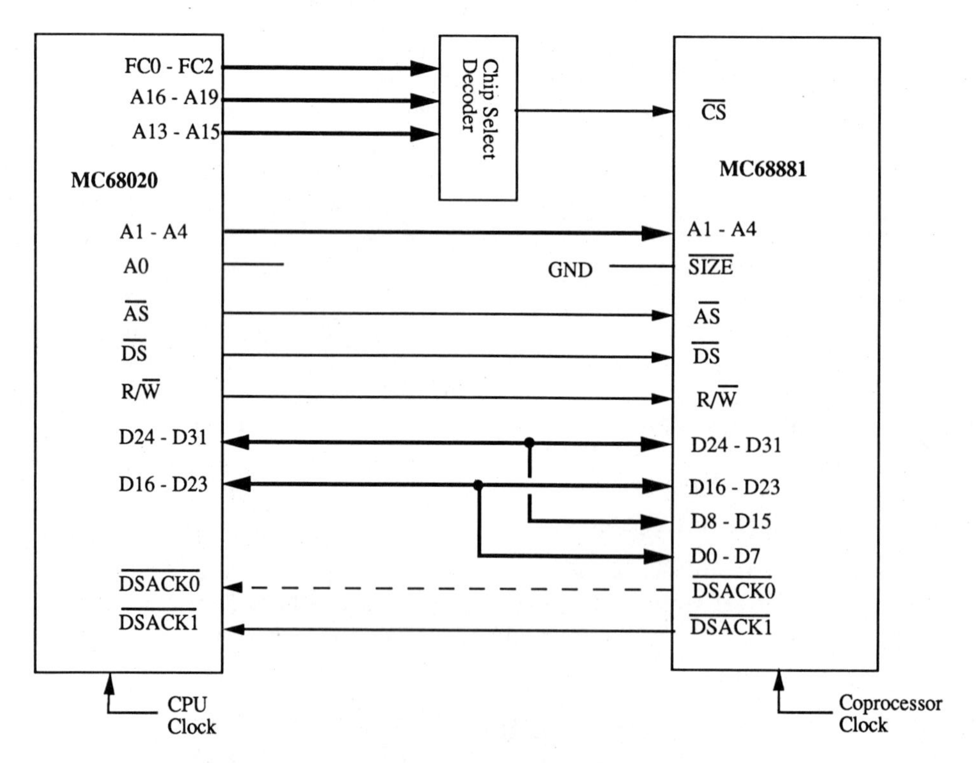

FIGURE 11.18. 16-bit bus coprocessor interconnection.

(Courtesy Motorola Inc.)

11.7.4 8-Bit Data Bus Coprocessor Connection

Figure 11.19 illustrates the connection of the same coprocessor to the MC68020. Here the MC68881 is configured to operate over an 8-bit device where the $\overline{\text{SIZE}}$ pin is connected to the GND. The 24 least significant data pins (D23–D0) of the MC68881 are required to be connected to the 8 most significant data pins (D31–D24) of the main processor. That is, we need to connect D0 of the MC68020 to D8, D16, and D24; D1 to D9, D17, and D25; D7 to D15, D23, and D31, and so on. The $\overline{\text{DSACK}}$ pins of the devices are connected with each other.

The bus-cycle timing requirements are straightforward for the most part, where all signal timings follow the normal MC68020 convention. To detect the start and completion of an access, the MC68881 monitors all three signals $\overline{\text{AS}}$, $\overline{\text{DS}}$, and $\overline{\text{CS}}$. It detects the start of a cycle when all three signals are asserted by the main processor, and a cycle end is detected when the first strobe ($\overline{\text{AS}}$ or $\overline{\text{DS}}$) is negated.

11.7.5 16-Bit Data Bus Peripheral Processor Connection

The MC68881 can be connected as a peripheral device to the previous generation of processors in the MC68000 family. Figure 11.20 illustrates the connection, where the main processor's D8–D15 pins are connected to D24–D31 and D8–D15 of the MC68881, and the main processor's D0–D7 are tied to D0–D7 and D16–

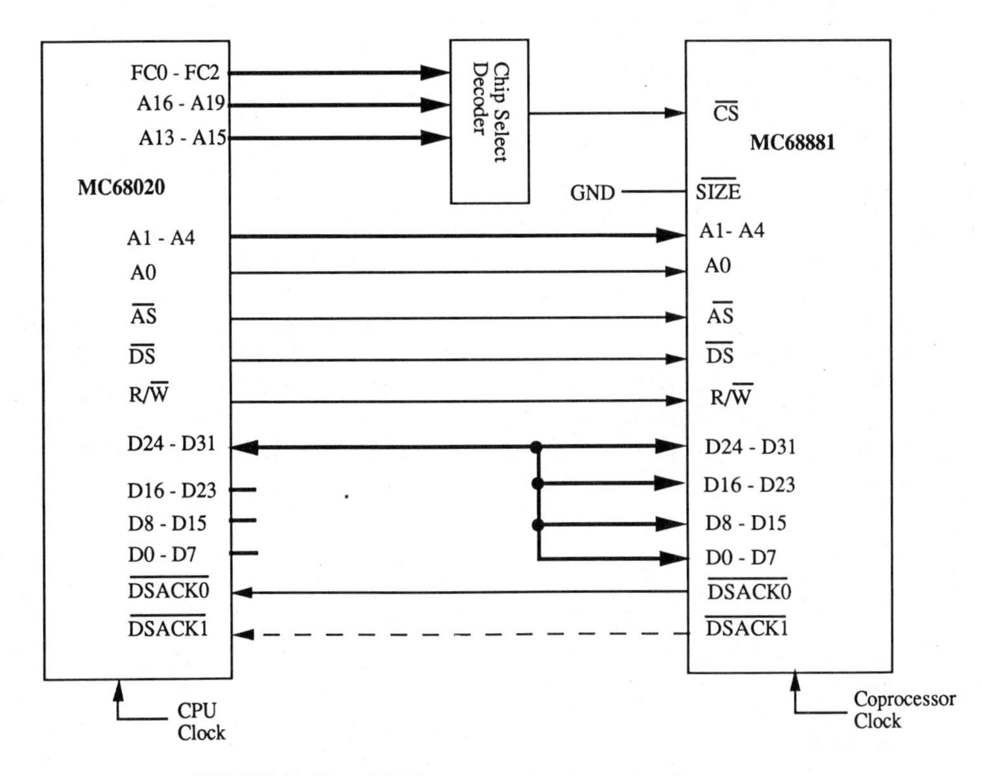

FIGURE 11.19. 8-bit bus coprocessor interconnection.

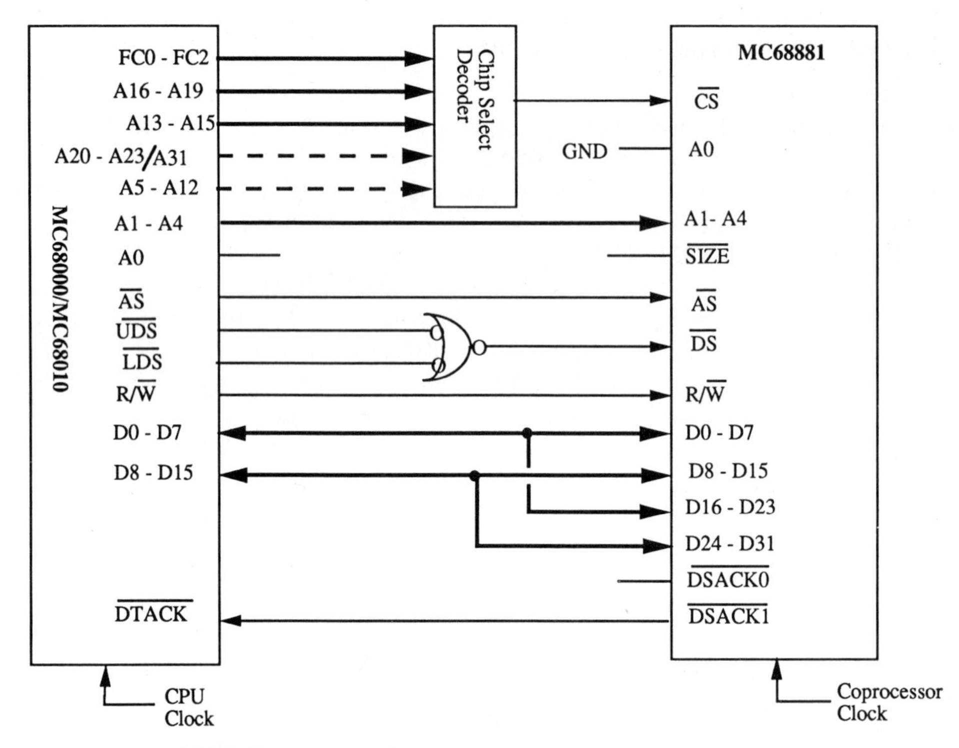

FIGURE 11.20. 16-bit data bus coprocessor interconnection as a peripheral. (Courtesy Motorola Inc.)

D23 of the coprocessor. The $\overline{\text{DSACK1}}$ pin of the MC68881 is connected to the $\overline{\text{DTACK}}$ pin of the main CPU, and the $\overline{\text{DSACK0}}$ pin is not used.

When connected as a peripheral processor, the MC68881 chip select $\overline{\text{CS}}$ decode is system dependent. If the MC68000 is the main CPU, then the MC68881's $\overline{\text{CS}}$ must be decoded in the supervisor or user data spaces. However, if the MC68010 or MC68012 is used as the CPU, the MOVES instruction should be used to emulate any CPU space access that the MC68020 generates for coprocessor communications. Thus, the $\overline{\text{CS}}$ decode logic for such systems may be the same as for an MC68020-based system. Here the MC68881 will not use any part of the data address spaces.

11.8 EXCEPTION PROCESSING

The exception processing architecture of the MC68881 is an extension of the MC68020's exception processing when they work together as a system, where the main processor coordinates the exception processing for the system, irrespective of whether they are detected during the execution of an instruction related to the main processor or during a coprocessor instruction.

The basis steps involved during the execution of MC68881-related instructions are [Motorola 85b]:

1. Detect the exception.
2. Determine the exception vector number and forward that to the main processor, in case it was detected by the MC68881.
3. Change processing states, if required.
4. Save the old context of the main processor.
5. Load the new context from the location pointed to by the exception vector.
6. Execute the exception handler routine.
7. Restore the previous context.

11.8.1 MC68881-Detected Exceptions

MC68881-detected exceptions fall into two categories:

- Traps and protocol violations.
- Floating-point instruction computational errors (e.g., divide-by-zero or instructions designed to cause a trap, such as the FTRAPcc instruction).

The above exceptions are processed by the MC68020 using the following algorithm:

1. The MC68881 encodes the appropriate take exception primitive (pre- or mid-instruction), along with the vector number in the response CIR to be read by the MC68020.
2. The MC68881 reads the response CIR (usually in an attempt to initiate the next instruction), and receives the take exception request from the MC68020.
3. The MC68020 acknowledges the request and writes an exception acknowledgment to the control CIR. It then stores the appropriate stack frame in the memory and transfers control to the exception handler routine.
4. In response to the exception acknowledgment, the MC68881 aborts any active internal operations (this only applies to protocol violations), clears all pending exceptions, and enters into an *idle state.*

11.8.2 Exception Vectors

We have already seen that the MC68000 family of processors uses the least significant 1024 bytes of memory locations for the vector table. Each entry in the vector table is a pointer to the routine that is to take over the control in response to a specific exception. Figure 7.2 listed the exception vector table for the MC68020. The MC68881 utilizes three vector entries defined by the MC68020, and defines seven additional vectors for the support of floating-point exceptions. The vectors defined by the MC68881 are listed in Table 11.7.

The vector number shown in Table 11.7 is encoded in the take exception response primitive specified by the MC68881. The exception is FTRAPcc, which is generated by the MC68020. The MC68020 will always add the value of the vector

TABLE 11.7. Exception Vector Table Related to FPU Operations.

Vector Number (Decimal)	Vector Offset (Hex)	Assignment
07	$01C	FTRAPcc instruction
11	$02C	F-line emulator
13	$034	Coprocessor protocol violation
48	$0C0	Branch or set on unordered condition
49	$0C4	Inexact result
50	$0C8	Floating-point divide-by-zero
51	$0CC	Underflow
52	$0D0	Operand error
53	$0D4	Overflow
54	$0D8	Signaling NAN

base register to the vector offset, to calculate the absolute address of the vector. However, it is the responsibility of system designers to put the address of the appropriate service routines at those addresses, so that the MC68020 can find the desired service routines and execute them.

11.8.3 Instruction Exception and Traps

The MC68881 can generate one of eight exceptions during the execution of a floating-point instruction. The location of the exception bits in the MC68881's status register is shown in Figure 11.4. The priority decreases from left to right, where BSUN has the highest priority and INEX1 has the lowest priority. It is also possible that an instruction can cause more than one exception. In the event of multiple exceptions, the MC68881 services the highest priority exception that is enabled. The lowest priority exception cannot trigger a second exception. Thus, it is the responsibility of the programmer to check whether the exception bits that have a lower priority than the exception taken are set properly.

During the execution of a floating-point algorithm, the execution control unit of the MC68881 performs all computations using floating-point numbers. In the ECU, a floating-point number contains a 67-bit mantissa, for the purpose of rounding, and a 17-bit exponent to ensure that an overflow or underflow can never occur. At the end of computation, the intermediate results are stored into a floating-point data register, in the MC68020 data register, or into memory. The MC68881 then checks the intermediate results for underflow, overflow, rounding, etc., before generating the final result.

11.8.4 The Branch Set or Unordered (BSUN) Exception

The BSUN exceptions may occur during the execution of the MC68881's conditional instructions, with the following branch condition predicates. These predicates are not a part of the IEEE P754 standard, but they are implemented for performance enhancement. The predicates are:

GT—Greater Than
NGT—Not Greater Than
GE—Greater Than or Equal To
NGE—Not Greater Than or Equal
LT—Less Than
NLT—Not Less Than
LE—Less Than or Equal To
NLE—Not Less Than or Equal To
GL—Greater or Less Than
NGL—Not Greater or Less Than
GLE—Greater or Less or Equal
NGLE—Not Greater or Less or Equal
SF—Signaling False
ST—Signaling True
SEQ—Signaling Equal To
NSE—Signaling Not Equal To

During the execution of MC68881 conditional instructions, the MC68020 writes the conditional predicates to the condition CIR, and reads the response from the response CIR.

The MC68881 detects a BSUN exception if the conditional predicate is one of the IEEE nonaware branches, and sets the NAN condition code bit. When this exception is detected, the BSUN bit in the FPSR is set by the FPU.

11.8.5 Signaling Not-a-Number (SNAN)

The SNAN are used for user-defined non-IEEE data types. The MC68881 will never create a SNAN as a result of an operation. However, the NANs created by operand error exceptions are always nonsignaling NANs.

11.8.6 Operand Error

Operand errors occur when an operation has no mathematical interpretations for the given operand. When an operand error occurs, the OPERR bit is set in the FPSR exception status byte.

11.8.7 Overflow

An overflow is detected when the resultant exponent is greater than or equal to the maximum exponent value of the format. Overflow can occur only when the destination is in the S, D, or X formats. However, overflows during data conversion to the B, W, or L integer and packed decimal formats are labeled as operand errors.

At the end of any operation, if a potential overflow condition exists, the intermediate result is checked first for underflow, then it is rounded and checked for overflow, before the result is stored to the destination. If an overflow occurs, then the OVFL bit in the FPSR exception byte is set.

11.8.8 Underflow

Underflow is detected for a given operation and data format when the exponent of an intermediate result is less than or equal to the minimum exponent value of the destination format. Underflow can only occur when the destination format is S, D, or X. However, if the destination format is packed decimal, then underflows are included as operand errors. However, if they are B, W, or L integer, then the conversion underflows to zero, without causing either an underflow or an operand error.

11.8.9 The MC68881-Detected Protocol Violation

The MC68881 will initiate a midinstruction exception if it detects an interprocess communication failure between the MC68881 and MC68020. Coprocessor-detected protocol violations occurs when:

- It is expecting a write to the command or condition CIR, and instead an access to the register select or operand CIR occurs.
- The MC68881 is expecting a read of the register select or operand CIR, and instead a write to the command, condition, or operand CIR occurs.
- It is expecting a write to the operand CIR, and instead either a write to the command or condition CIR or a read of the register select or operand CIR occurs.

In response, the coprocessor will generate a coprocessor protocol violation. For the above three situations, the MC68881 maps the 16-bit register select CIR onto the upper word of the 32-bit operand register. Thus, the MC68020 reads inconsistent data from the operand CIR, and write cycles cannot store the correct values. This will have a minimum effect on the coprocessor and main processor, since the protocol violation will invalidate any operation being attempted.

A protocol violation has the highest priority exception detected by the MC68881. It is also considered to be a fatal exception, since the MC68020 acknowledgment of the protocol violation exception clears any pending MC68881 instruction exceptions or illegal instructions.

11.9 AN EXAMPLE OF INTERFACING WITH THE MC68881

Figure 11.21 illustrates a simple technique by which the MC68881 can be placed in the logical address space on an MC68020-based system. Here, the chip select

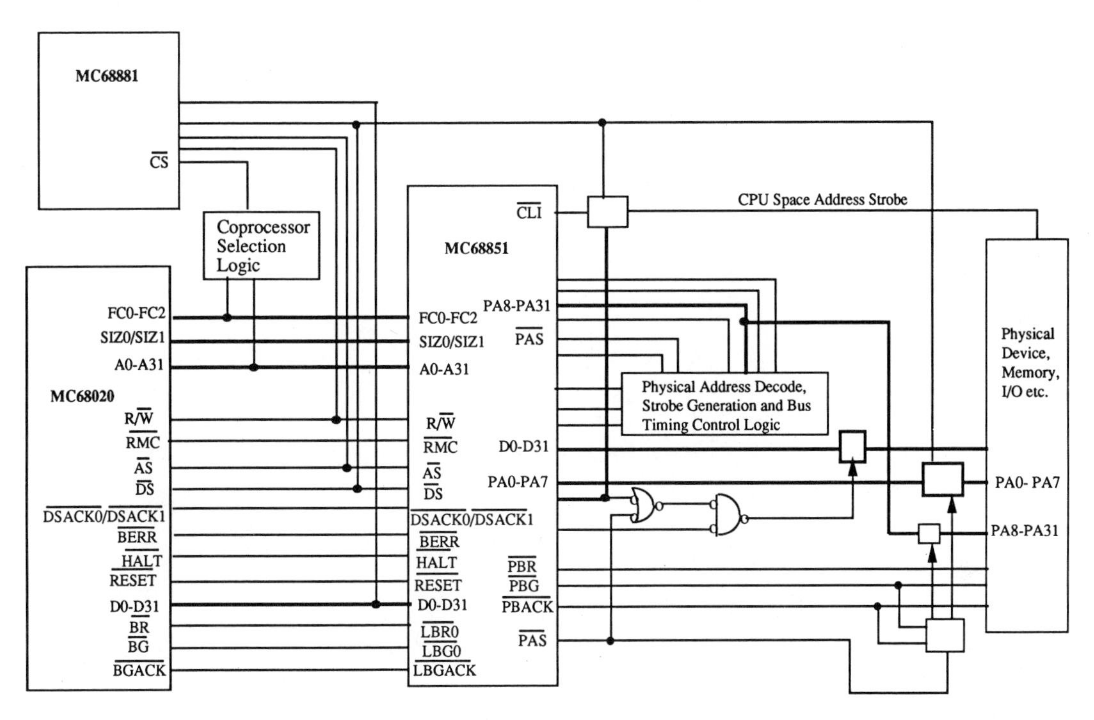

FIGURE 11.21. Interconnection of an MC68020-based system composed of an MC68851 and an MC68881.

(Courtesy Motorola Inc.)

is generated by decoding the logical address bus and the control strobes of the MC68020. The corresponding signal lines of the MC68020 are connected to the corresponding lines of the MC68881. The floating-point coprocessor is referenced using Cp-ID = \$2 (A15:A13 = 010). The gate *no-block* is used to block passage of the CPU space access strobe to other devices during access of the MC68881.

The system configuration shown in Figure 11.21 can be coupled to the system in Figure 10.18, and develops a simple system composed of the MC68020, the memory management unit MC68851, and the MC68881, and connects them to the physical memory, to create a complete virtual system. One can port an operating system, and then this will be ready to run one or more applications.

Chapter 12

The MC68030 Microprocessor

The 32-bit MC68030 microprocessor is the second-generation member of the MC68000 family of processors. It is object-code compatible with the previous members of the family. It extended the capabilities of the MC68020 by adding an on-chip memory management unit and a data cache, and included an enhanced bus interface. The Harvard architecture (discussed in Chapter 1) was the motivation behind the design of this processor. A smart controller is integrated, to support asynchronous and synchronous bus cycles, including burst data transfers. The controller also supports dynamic port sizing on a cycle-by-cycle basis, as the CPU transfers operands to and from external devices.

12.1 GENERAL FEATURES OF THE MC68030 PROCESSOR

One can model the MC68030 as combinations of the MC68020, MC68851, and data and instruction caches packaged on one chip which operates as fact as 40 MHz. The features of this processor are:

- Object-code compatibility with the MC68020 and other members of the MC86000 processor family.
- Two 32-bit supervisor stack pointers
- Ten special-purpose control registers.
- 256-byte instruction and data caches, which can be accessed simultaneously.
- A paged memory management unit capable of translating addresses in parallel with instruction execution and internal cache access.

- Two transparent segments to allow untranslated access to physical memory, and to transfer large blocks of data between predefined physical addresses.
- Pipelined architecture with enhanced parallelism allows accesses to internal caches in parallel with bus transfers, including the concurrent execution of multiple instructions.
- 4 Gbytes of direct addressing range.

Figure 12.1 illustrates the various components the MC68030. The CPU, MMU, and bus controller operate autonomously. The bus controller consists of the address buffer, data buffer, the multiplexers to support dynamic bus sizing, and a macro bus controller, which schedules bus cycles on the basis of priority.

The micromachine is made up of an execution unit and control logic. The bilevel microROM and nanoROM contained in the micromachine generate the microcode control. However, hardware logic is utilized for instruction decode and sequencing. The instruction pipeline units and the control sections perform the secondary decoding of instructions and generate the necessary control signals. These control signals are then utilized in the decoding and interpretation of nanoROM and microROM information.

The instruction and data caches operate independently, and retain information for future usage. The caches operate at the same speed as the CPU, and allow for simultaneous access via separate address and data buses. The caches are organized as 64 long-word entries with a block size of 4 long-words. The data cache coherency is maintained via a write-through policy with no write allocation on cache miss.

The programmable memory management unit supports demand paging for

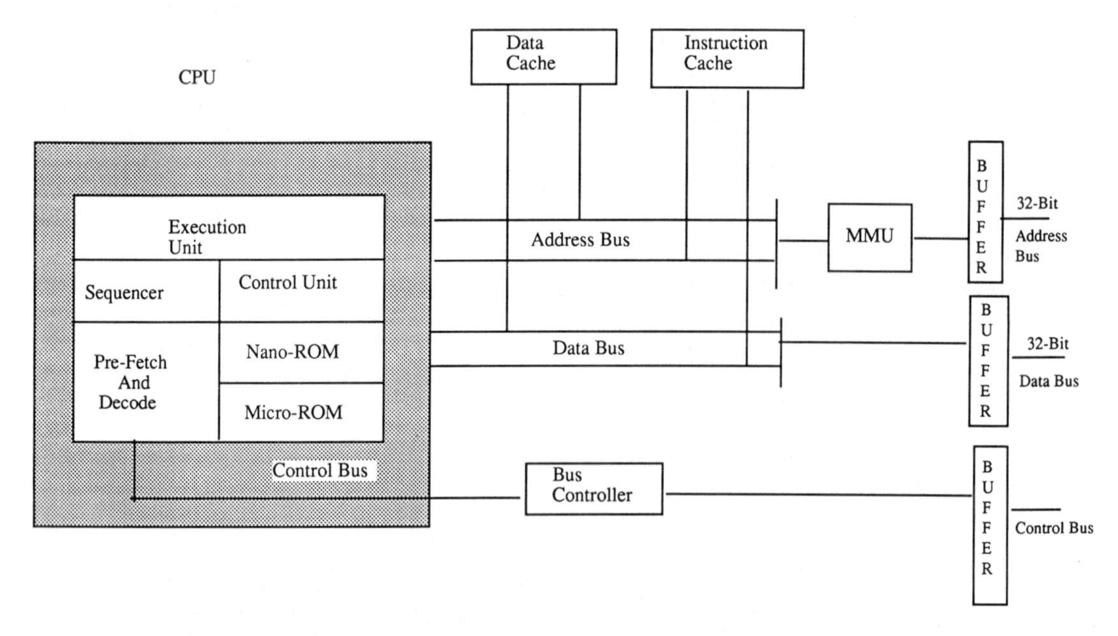

FIGURE 12.1. Architectural partitioning of the MC68030 processor.

pages that may vary from 256 bytes to 32 Kbytes. The translation table in memory stores address mapping information, and is automatically fetched by the MC68030 on demand. Recently-used descriptors are maintained in a 22 entry fully associative content-addressable Address Translation Cache (ATC).

12.1.1 Programming Model

The MC68030's programming model is the same as the MC68020's, except for the added memory management registers. The user programming model is the same as that for the MC68020. The resources available within the user programming model are:

- 16 general-purpose 32-bit registers (D0–D7, A0–A7).
- 32-bit Program Counter (PC).
- 8-bit Condition Code Register (CCR).

In the supervisory mode, the programs have access to the resources available in the user mode, including the following:

- Two 32-bit supervisory Stack Pointers (ISP and MSP)
- 16-bit Status Register (SR)
- 32-bit Vector Base Register (VBR)

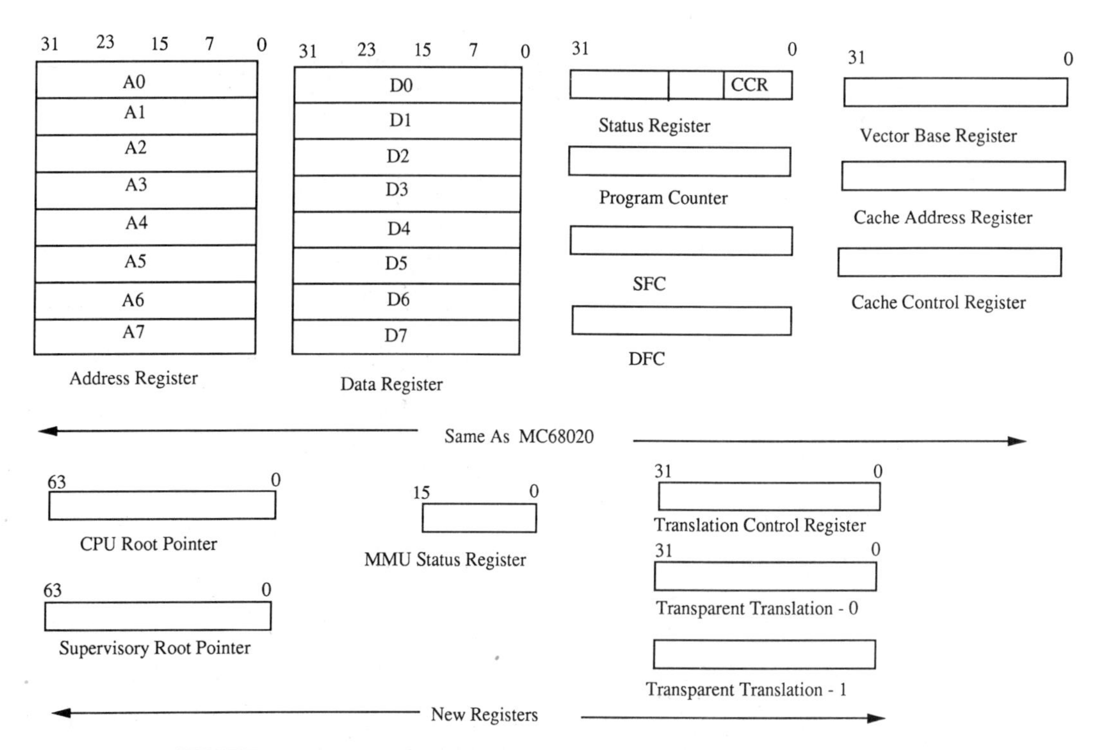

FIGURE 12.2. The MC68030 programming model.

- Two 32-bit alternate Function Code registers (SFC and DFC)
- 32-bit Cache Control Register (CACR)
- 32-bit Cache Address Register (CAAR)
- 64-bit CPU Root Pointer (CRP)
- 64-bit Supervisor Root Pointer (SRP)
- 32-bit Translation Control register (TC)
- 32-bit Transparent Translation registers (TT0 and TT1)
- 16-bit MMU Status Register (MMUSR)

Other than the MMU-related registers, the model is the same as the MC68020's supervisory model, with the exception of the cache control register, where various control bits are added. These new control bits manage the operations of the data and instruction caches added to MC68030.

The supervisory programming model is used exclusively by MC68030 system programmers who utilize the supervisory privilege level to implement operating system functions, I/O control, and memory management subsystems.

12.1.2 Instruction Set

The MC68030's instruction sets are downward compatible with the MC68020 and other members of the MC68000 family. The instructions (with some exceptions) generally operate on bytes, words, and long-words, and most of them can utilize the 18 addressing modes supported by the MC68020. The memory management instructions included in the MC68030 are a subset of the instructions found in the MC68851. Table 12.1 lists the new instructions added to the MC68030.

TABLE 12.1. New Instruction Sets Included in the MC68030.

Instruction	Meaning
	MMU Instructions
PMOVE	-Move to or from MMU Register
PLOAD	-Load Page Descriptor into ATC
PTEST	-Test Translation
PFLUSH	-Flush Selected ATC Entries
PFLUSH	-Flush All ATC Entries
	Coprocessor Instructions
cpBCC	-Branch Conditionally
cpDBcc	-Test Coprocessor Condition, Decrement, and Branch
cpGEN	-Coprocessor General Instruction
cpRESTORE	-Restore Internal State of the Coprocessor
cpSAVE	-Save Internal State of the Coprocessor
cpScc	-Set Conditionally
cpTRAPcc	-Trap Conditionally

12.2 FUNCTIONAL PIN DIAGRAM

The functional pin diagram of the MC68030 is shown in Figure 12.3. The functions of various pins are identical to the MC68020, with several exceptions. The new pins can be grouped into three classes, namely: the cache control bus, synchronous bus control, and the emulator support bus. The pin functional groups and their members are listed in Table 12.2.

12.2.1 Cache Control Bus

The cache control bus is composed of 4 pins, listed in the following paragraphs.

Cache Inhibit Input ($\overline{CIIN}$). This input signal, when asserted, prevents data from being loaded into the MC68030's instruction and data caches. It is a synchronous input signal. The status of this pin needs to be interpreted during each bus cycle. During the write cycle, the status of this pin is ignored.

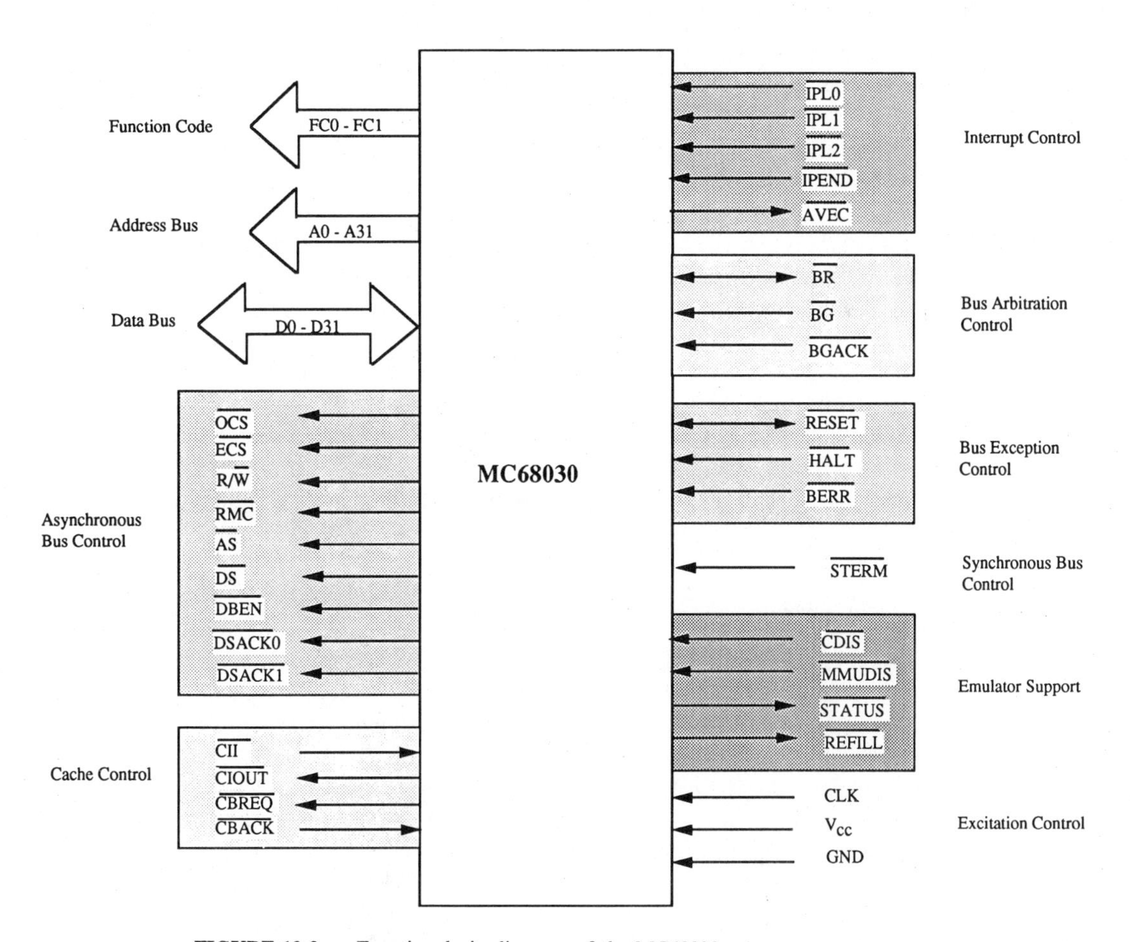

FIGURE 12.3. Functional pin diagram of the MC68030.

TABLE 12.2. Functional Grouping of the MC68030's Signals.

Functional Code	Membership
Function Code	FC0–FC2
Address Bus	A0–A31
Data Bus	D0–D31
Asynchronous Bus Control	$\overline{\text{OCS}}$, $\overline{\text{ECS}}$, R/$\overline{\text{W}}$, $\overline{\text{RMC}}$, $\overline{\text{AS}}$, $\overline{\text{DS}}$, $\overline{\text{DBEN}}$, $\overline{\text{DSACK0}}$, $\overline{\text{DSACK1}}$
Cache Control	$\overline{\text{CIIN}}$, $\overline{\text{CIOUT}}$, $\overline{\text{CBREQ}}$, $\overline{\text{CBACK}}$
Interrupt Control	$\overline{\text{IPL0}}$, $\overline{\text{IPL1}}$, $\overline{\text{IPL2}}$, $\overline{\text{IPEND}}$, $\overline{\text{AVEC}}$
Bus Arbitration Control	BR, $\overline{\text{BG}}$, $\overline{\text{BGACK}}$
Bus Exception Control	$\overline{\text{RESET}}$, $\overline{\text{HALT}}$, $\overline{\text{BERR}}$
Synchronous Bus Control	$\overline{\text{STERM}}$
Emulator Control	$\overline{\text{CDIS}}$, $\overline{\text{MMUDIS}}$, $\overline{\text{STATUS}}$, $\overline{\text{REFILL}}$
Excitation Control	CLK, Vcc, GND

Cache Inhibit Out ($\overline{\textit{CIOUT}}$). This is a three-state output signal for the processor. This signal reflects the status of the CI bit in the ATC entry for each referenced logical address, asking the external cache to ignore the bus transfer.

Whenever the CI bit is set, an external bus cycle is required for either a read or a write, and $\overline{\text{CIOUT}}$ is asserted by the processor. It is also used by external hardware logic to inhibit caching from the external caches.

Cache Burst Request ($\overline{\textit{CBREQ}}$). This is also a three-state signal for the MC68030. The bus controller uses this signal to request the external device to fill a line of data into the instruction or data cache using burst mode. The responding device may sequentially supply one to four long-words of cachable data.

Cache Burst Acknowledge ($\overline{\textit{CBACK}}$). This input signals the completion of the burst mode operation. The referenced device tells the processor that it can operate in burst mode and can supply at least one more long-word of data to the instruction or data cache.

However, if the responding device can't support the burst mode and terminates the activity with $\overline{\text{STERM}}$, then it should not assert a $\overline{\text{CBACK}}$ signal, because the MC68030 ignores the $\overline{\text{CBACK}}$ signal whenever the bus cycle is terminated by $\overline{\text{DSACKx}}$.

12.2.2 Emulator Control

Debugging real-time software is a complex task, because, in such a system, one can't take advantage of the trace exception mechanism (trace takes processing time away from the real-time event). The designers of the MC68030 have acknowledged those difficulties and provided several pins to allow real-time visibility into the MC68030. Four pins together form the emulation control bus. These pins can disable on-chip caches and the memory management unit, and force the processor to place all logical address on the address bus. Tracing capability can be added by decoding the processor's control signals and capturing important events. The signal lines are listed in the following paragraphs.

Cache disable ($\overline{\textit{CDIS}}$). When asserted, this disables the internal cache operation. $\overline{CDIS}$ does not flush cache entries. Rather, it simply disables them and they become available again once the signal is negated.

MMU disable ($\overline{MMUDIS}$). This input signal, when asserted, disables the address translation capabilities of the MMU. When $\overline{MMUDIS}$ is asserted, no further descriptor fetches or address translation occurs, and the internal state of the MMU is preserved. This is valuable for emulators that need to have as little effect as possible on the target system.

Again, by asserting $\overline{MMUDIS}$ when leaving the emulation state and negating $\overline{MMUDIS}$ when reentering the emulation state, the external system is unaffected by the fact that the emulation state has never been exited.

Pipeline refill ($\overline{REFILL}$). This output signal, when asserted, indicates that the MC68030 is beginning to refill the internal instruction pipeline. Refill requests are a result of a break in instruction sequence (e.g., branches, jumps, traps, etc.) execution to process nonsequential events.

Internal microsequencer status ($\overline{STATUS}$). This output line indicates the state of the internal microsequencer. This signal allows trace hardware to examine the progress of the execution unit as it accesses the program memory operands and allows some exceptions. The varying number of clocks for which this signal is asserted indicate instruction boundaries, pending exceptions, and the halted condition.

12.2.3 Synchronous Bus Control

Two modes of bus transfers are supported in the MC68030. They are:

- Asynchronous bus transfer
- Synchronous bus transfer

The asynchronous transfer is identical to the MC68020's asynchronous transfer, where external devices connected to the bus can operate at any speed. This protocol uses the following signals: $\overline{\mathrm{AS}}$, $\overline{\mathrm{DS}}$, $\overline{\mathrm{DSACKx}}$, $\overline{\mathrm{BERR}}$, and $\overline{\mathrm{HALT}}$. The $\overline{\mathrm{AS}}$ and $\overline{\mathrm{DS}}$ indicate the start of a bus cycle, and are used as a condition for valid data during the write cycle. The slave device then places the data or reads the data from the data bus and terminates the cycle using $\overline{\mathrm{DSACKx}}$. If no slave device responds, or an invalid access is initiated, then the timeout control logic must terminate the cycle by asserting $\overline{\mathrm{BERR}}$ or/and $\overline{\mathrm{HALT}}$. The asynchronous bus cycle takes as little as three clock cycles, and accesses data from any size port.

The synchronous transfer reduces data-transfer time from a three-cycle bus to two cycles. The synchronous transfers are characterized by:

- A two-clock cycle access.
- A 32-bit port.
- $\overline{\mathrm{STERM}}$ must be asserted on the rising edge of S2.

Thus, whenever $\overline{\mathrm{STERM}}$ (an input signal to the MC68030) is used instead of the data transfer and size acknowledge signals $\overline{\mathrm{DSACKx}}$ in any bus cycle, it makes the cycle synchronous. Wait cycles can also be inserted in this case by delaying the assertion of $\overline{\mathrm{STERM}}$ appropriately. However if $\overline{\mathrm{STERM}}$ and

$\overline{\text{DSACKx}}$ are asserted during the same bus cycle, then the results become unedictable. The processor will then make a decision internally to resolve the conflict.

Burst mode used for a cache fill, another variation of synchronous transfer, is also controlled by $\overline{\text{STERM}}$. The first cycle of any burst transfer must be a synchronous cycle with a minimum cycle time of two clocks. Thus, if a burst mode is initiated and allowed to terminate normally, then the second, the third, and the fourth cycles may latch data on successive falling edges of the clock. Again, the exact timing of these subsequent cycles is controlled by the timing of $\overline{\text{STERM}}$ for each these cycles, and wait cycles can be inserted as necessary.

The burst fill will be aborted under the following conditions:

- Cache Inhibit In ($\overline{\text{CIIN}}$) asserted.
- Bus Error ($\overline{\text{BERR}}$) asserted.
- $\overline{\text{CBACK}}$ negated prematurely.

When the processor recognizes a bus error condition, it terminates the current bus cycle in the normal way and initiates exception processing. The exception processing of the MC68030 is described later in this chapter.

12.3 INTERNAL CACHES

The MC68030 contains two separate instruction and data caches. Each is 256 bytes long, and they are located on the logical side of the MMU, having their own address and data buses, allowing them to be accessed simultaneously by the MMU. They are also accessible by the execution unit without going through the MMU. The logical location means that when an O.S. makes a context switch, the caches must be flushed or cleaned, and the flushing can be done within one instruction.

12.3.1 Instruction Cache

The instruction cache of the MC68030 is also 64 long-words in length, as found in the MC68020. However, they are organized differently. The instruction cache is organized as 16 lines with 4 entries per line. Each entry is a long-word, and the columns in the entries are labeled as LW0–LW3 (Figure 12.4). Whenever an instruction access is initiated, the addresses are driven to the instruction cache's address bus. If the cache is enabled in the CACR (discussed in subsection 12.3.8), then the index bit of the addresses A4–A7 select one of the 16 lines. The tag field of the cache contains 24 address bits, including a function code bit. The tag field of the selected line is compared against FC2, including the access address bits A8–A31. In the event of a match, and indicating a valid entry, then an entry hit is asserted. This determines that the data that resides in this cache entry is same as the entry in the memory. Thus, an external memory access is not required for this logical address.

Now, the information gets released from the cache location and routed to the instruction cache output register. This register is not accessible by the user. From

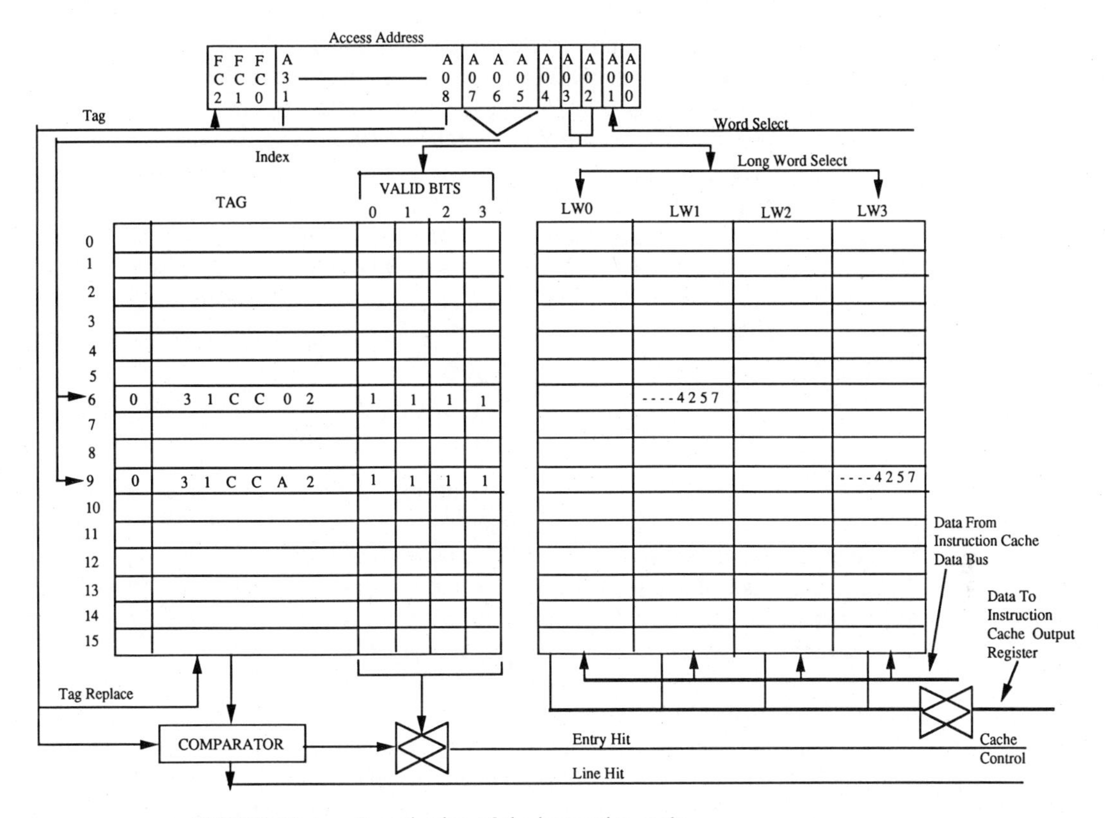

FIGURE 12.4. Organization of the instruction cache.

this register, the word select bit A1 of the address selects the appropriate word to be placed on the cache data bus. When an entry miss occurs, the cache entry can be updated, utilizing the information returned from the external access via the instruction cache data bus. If the cache is disabled or frozen in the CACR, updating of the cache will not take place [MTTA3 87].

12.3.2 Instruction Cache Entry Hit and Miss

The instruction cache entry hit and miss is illustrated via two examples. Let us assume that the processor is executing an instruction:

> CLR.W (A7) with PC pointed to the address $31CCA29E, and that location contains a value of $4257.

The address $31CCA29E is placed on the instruction cache address bus. The index value, A4–A7, selects the tenth tag in the cache. The first part of the tag value, with FC2 = 0, indicates a user space. In the second part of the address, A8–A31 = $31CCA2. Since this matches the access address, a line hit has occurred. The long-word select value being equal to $3 selects the associated valid bit entry illustrated in Figure 12.4. The valid bit indicates that the entry is valid. Therefore, an entry hit has occurred. The entry selected by the index and long-word select is enabled to the instruction cache output register. The word select value binary '1' will release the value $4257. The information (Op-code for the clear word) gets placed into the instruction cache data bus.

Let us examine what happens when the processor can't find a unit of data in the cache. This is explored using the same example, where the program counter points to a different location.

CLR.W (A7)
where PC→$376F1A64

The PC points to $376F1A64, and this address is placed over the address lines of the instruction cache. The index value A4–A7 = $6 selects the seventh tag in the cache. In the first part, FC2 = 0 (i.e., user space), the tag location contains $31CC02. This is not equal to the corresponding bit of the accessed address, and a miss occurs. Because of the miss, the instruction must be fetched from the external memory. The instruction size is one word. However, when the access is executed, the instruction from the long-word location $376F1A64 is returned to the CPU, and the value $——4256 is placed into the cache.

The tag must be updated at the same time that the entry is updated. The tag value is changed to $376F1A. This change invalidates the other three entries in the line and their valid bit is changed to 0. In this case, the MC68030 will attempt to update these entries, using a burst fill, only if the instruction cache burst fill bit is set to 1 in the CACR. A bus error can occur when the external access is executed. If it does, then the valid bit for the entry will be cleared and the MC68030 will recognize the bus error and will initiate exception processing.

12.3.3 Data Cache

The organization of the data cache shown in Figure 12.5 is similar to the instruction cache. The operation is also similar to the instruction cache with the exception of address comparison and cache update policy. In the case of a data cache, all three FC bits, including the address bits A8–A31, must be equal to the tag field for a line hit to occur. The control logic selects the cache tag, using the address bits A4–A7. The address bits A2–A3 are used to locate the long-word stored in the data cache. The bits A0–A1 indicate byte or bytes to be placed on the data lines of the data cache. The data cache can be updated during write as well as read operations. If the data cache is disabled in the CACR register, it will not be updated during each read or write operation. If it is enabled but frozen in the CACR, the cache will not be updated during read operations. However, updates will occur during write operations. If data cache is enabled and not frozen, then it will update during both read and write operations.

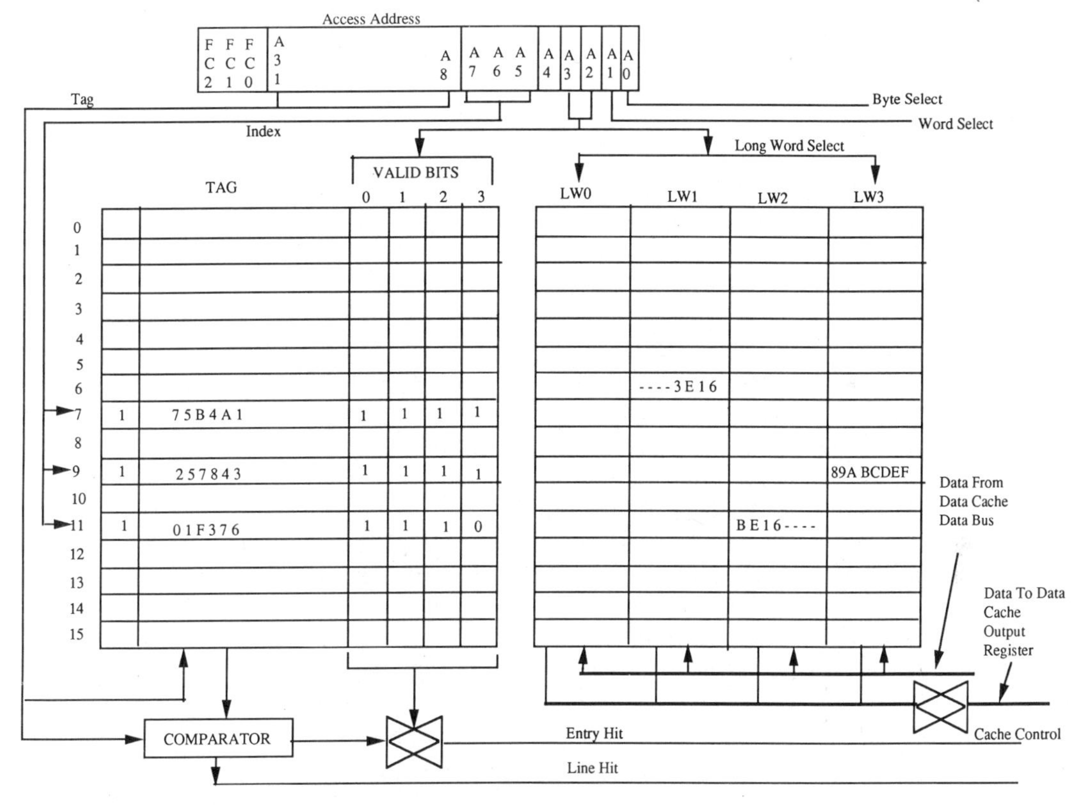

FIGURE 12.5. Organization of the data cache.

12.3.4 Data Cache Read Example

The activities that take place during a read entry hit in the data cache are illustrated by an example. The instruction is:

MOVE.W (A6),D7
where A6 → $75B4A176

When the CPU needs the data, the address $75B4A176 is placed over the data cache address bus. The index value A4–A7 = $7 selects a tag from the cache, which is shown in Figure 12.5. The first part of the tag value identifies the function code FC2 = $1 (i.e., user data space), and the second part, A8–A31 = $75B4A1. Both are equal to the access address. Therefore, a line hit occurs. The long-word select value, which is '01' in this case, selects the associated valid bit and entry in the cache shown in the diagram. The valid bit indicates that this entry valid bit is valid. Therefore, an entry hit occurs in the cache. The highlighted entry selected by the index and long-word select is enabled and the value is latched into the data cache output register. The word select value, which is 1, will release the contents, i.e., \$3E16, and the desired data is placed onto the data cache data bus.

12.3.5 Data Cache Read Miss

The data cache read miss operation is illustrated via an example that is similar to a miss in the instruction cache. The instruction executed is:

MOVE.W (A6), D7
where A6→\$1F376B8

When the CPU needs the data, it places the address \$1F376B8 on the address bus of the data cache. The index value derived from A4–A7 (which is \$B) selects the tag in the cache, which is highlighted in Figure 12.5. The first part of the tag value indicates the function code of 1, which means that the processor is accessing the user data space. The second part is derived from A8–A31 (which is \$01F376). The access address function code value is 5, indicating a supervisory data space, and does not equal the cache tag function code value. Therefore, no line hit occurs. Since there is a miss, the data must be fetched externally. The data size is one word. However, when the access is made, the entire long-word from address location \$1F376B8 is returned to the CPU, and is also put into the cache. At the same time, this entry status is updated. The tag will remain the same because the address is the same, but the function code part will be updated to 5. This change invalidates the entry in the other three lines, and the valid bits are changed to zero.

The MC68030 will attempt to update the entries using burst fill mode if the data cache burst fill bit is set to 1 in the CACR. If the burst fill is successful, then all three entries will be updated and the valid bits will be changed to 1. If not successful, or if not enabled, the valid bits remain zero. If a bus error occurs during an external access, then the valid bit for the entry will be cleared. The CPU will recognize the bus error within the bus cycle in which it occurs.

12.3.6 Data Cache Write

An example of the occurrence of a write entry hit in the data cache is illustrated below. The instruction is:

MOVE.W D7, (A6) is executed with A6 pointing to \$2578439C

When the CPU writes the data, the address \$2578439C is placed on the address bus of the data cache. The index value derived from the address bits A4–A7 (which is \$9 in this case) selects a tag in cache (Figure 12.5). In the first part of the tag, indicating the presence of the function code, is 1, which identifies a user data space. The second part, derived from A8–A31 (which is \$257843), is compared against the location entry and is found to be equal to the access address. Therefore, a line hit occurs. Since the long-word select value is '11', the processor selects the associated valid bit in the entry, which is shown in the diagram. The valid bit indicates a valid entry. Therefore, an entry hit occurs. The data cache uses a write-through policy (i.e., entry write is extended if there is an entry hit) to update the cache entry. When the hit occurs, an entry hit is asserted. Here, the word select bit is zero. Therefore, the upper word of the entry is updated with the word being written from the data register D7. If a bus error occurs during the execution of the external write, then the updated cache entry will remain valid. The processor will recognize the bus error during the bus cycle in which it occurs.

12.3.7 Cache Inhibit

The MC68030 makes extensive use of internal and external caches in support of a large performance enhancement. In most systems, certain blocks of instructions or data information are not cached. For example, I/O device data normally would not be cached, and common data in a multiprocessor system would certainly not be cached. The MC68030 has two pins that can be used to manage such cases. Pin $\overline{\text{CIIN}}$ is input and, when asserted, inhibits the updating of both of the internal caches.

The output in $\overline{\text{CIOUT}}$ may be asserted during the synchronous and asynchronous transfer. The processor asserts it whenever the internal MMU encounters a page designated as noncachable. Certain pages are marked as noncachable because they are the locations of I/O devices, shareable memory, or other similar areas. $\overline{\text{CIOUT}}$ indicates to the external caches that the information on the data buses are unlikely to be accessed or, if accessed, it is unlikely to be the same. Therefore, it should not be cached. If cache inhibit out is asserted, then updating of the internal caches is inhibited also.

12.3.8 Burst Fill Mode

Burst mode filling is based on the concept of locality of reference, where all entries of the line associated with a tag are filled when a miss occurs. A burst fill is executed if it has been enabled in the CACR. If it is requested and if the responding device has a 32-bit port, then it performs synchronous transfer. The MC68030 has two pins to facilitate burst fills. Cache burst request $\overline{\text{CBREQ}}$ is an output pin that indicates "another long-word is needed" for a fill. The responding device may sequentially supply one to four long-words of cachable data. If the responding device does not support burst fill and terminates the cycle with $\overline{\text{STERM}}$, then it should assert $\overline{\text{CBACK}}$. MC68030 ignores the $\overline{\text{CBACK}}$ signal status during the burst fill cycle if it is terminated with $\overline{\text{DSACKx}}$.

12.3.9 Cache Control Mechanism

Depending on application and need, the user may want the ability to control and affect the operation of the cache. For example, an operating system needs to invalidate all cache entries on a context switch, or when an emulator is used.

From time to time the user may want to freeze the cache, to minimize the performance side-effects from an emulator.

Both instruction and data caches are subject to various controls, namely:

- Enable: Both hit and fill can occur
- Freeze: Only a hit can occur
- Clear: Invalidates the cache entries
- Clear Entry: Invalidates a specific cache entry

These controls are activated by turning on the appropriate bits in the cache control register. The bit positions in the CACR, including their functions, are shown Figure 12.6. On reset, both caches are disabled. If the use of either or both caches is desired, then the user enables one or both in the CACR. Most often, the user will want the caches to be enabled. However, there may be situations when the caches are disabled, such as debugging, when one wants to see every access made by the program. That visibility exists only on the address bus. So by disabling the caches, one forces all accesses to go external.

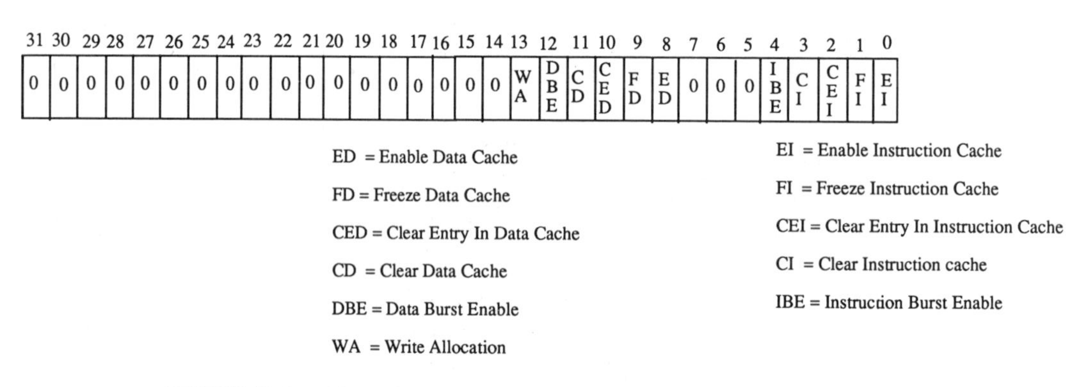

FIGURE 12.6. The cache control register.

Another control is freeze. If freeze is turned on, then the contents of the caches are saved and are available for hit. However, no update occurs in the queue except for writes in the data cache. This feature is valuable when some critical code must be executed as soon as possible, and during emulation, when the content of the caches are not allowed to be changed by the emulator. In this case the emulator is kept maximally accurate.

Clear a cache means to invalidate all cache entries. This happens when an operating system switches context and the data in the caches are no longer accurate for the new context. The clear can be executed with an instruction and, following the clear, the caches operate normally. There may be times when one may want to clear a particularly entry. This can happen in a multiprocessor system, when one processor changes a location in a sharable memory and communicates the change to a second processor. The second processor could clear the associated entry in its cache, thus eliminating the possibility of stale data.

12.3.10 Cache Control and Address Registers

All of the preceding features are controlled via the address and cache control registers. Both registers are 32 bits in size. In the CACR (Fig. 12.6), bits 5 through 7 and 14 through 31 are unused, and read as zeros. A write to those bit positions has no effect. Bits 0 and 8 in the CACR are EI and ED, which enable and disable the instruction and data caches. The flags are cleared during reset and disable the caches, which forces the processor to access the external memory and suppresses the updating of cache entries. The caches can be enabled by writing 1 into either or both of the EI and ED flags. Bits 1 and 9 of the CACR are the FI and FD flags, and freeze the respective caches. If the cache is enabled and the freeze bit is a zero, then the caches operate normally, hits are allowed, and fill is allowed during misses. If the freeze bit is 1, then the caches are frozen, meaning that hits can occur, but no fills occur during misses. Bits 2 and 10 of the CACR are the CED and CEI flags, which allow entries in the instruction and data caches to be invalidated. The cache address register is used in conjunction with the setting of these flags. When an entry in the cache must be invalidated, the memory address of the instruction or data is written to the CAAR first, then 1 is written to the clear entry flag. Writing a 1 to these bits causes the CAAR index field bits 4 through 7 and long-word select bits 2 through 3 to be used to select and invalidate an entry in the cache (Figure 12.7).

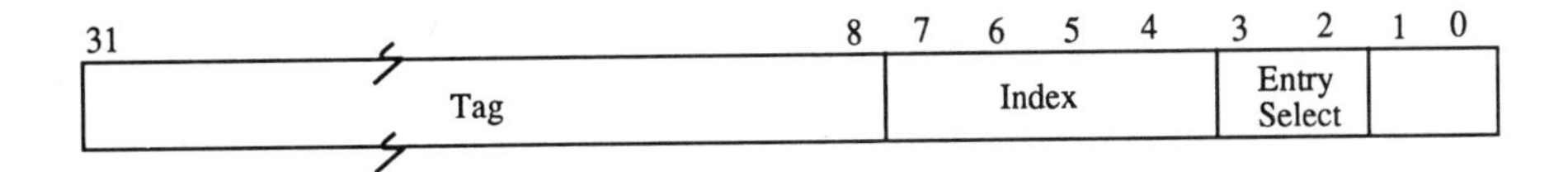

FIGURE 12.7. The cache address register.

The clear entry bits are always read as zeros. Writing a 1 for the clear entry bit performs the desired operation. However, the 1 is not saved in the clear entry bit. An entry can be invalidated, regardless of the state of the E or F bits or the cache disable pin. Writing a 1 to bits 3 or 11 of the CACR, that is the CI and CD flags, causes all entries in the instruction or data cache to be invalidated. Like the clear entry bits, these bits are always read as zero. Bits 4 and 12 of the CACR are the IBE and DBE flags, and are used to control burst fills for both caches. However, there may be instances when the processor may follow a special locality of reference. In such cases, the user may prefer the burst fill to remain disabled for either or both caches. Bit 13 selects the write policy to be used in managing the data cache. If the flag is set to 1, then the write allocation policy is activated.

The CAAR is only used with the clear entry function implemented in the CACR. Although the index portion address (i.e., bits 4 through 7) and the long-word select key (designed by address bits 2 and 3) are used to address the specific cache entry, for compatibility a complete 32-bit address should always be written into the CAAR. This will ensure that current software will work in future versions of the MC68000 family of processors.

The initialization of the CACR is illustrated via an example. In this example, both caches are enabled, including their corresponding burst fill capabilities.

```
MOVE.L  #1111, D7
MOVEC   D7, CACR
```

The following example illustrates how the write-allocation policy of the cache can be changed.

```
MOVEC  CACR, D7   ; get the status of the CACR into D7
OR.L  #$2000, DO  ; invoke the write-allocation policy
MOVEC  D7, CACR   ; update the CACR register
```

In this example, the current value of the CACR is saved in D7 before it is 'ORed', and then the result is finally updated, using MOVEC.

12.4 MEMORY MANAGEMENT UNIT (MMU)

The Memory Management Unit (MMU) is an on-chip module of the MC68030 that translates logical addresses to physical addresses. The MMU is made up of three major blocks:

- The registers.
- The address translation cache.
- Table search logic.

12.4.1 The MMU Registers

The MMU registers (Figure 12.2) are visible only in the supervisory mode, and are referenced using MMU instruction sets.

The two 64-bit CPU Root Pointer CRP registers contain a pointer to the translation table of the current task. It is initialized as a part of the context switch. The MMU uses this as a base address into the task's translation table. The least significant long-word of the CRP is the pointer, and the most significant 32 bits contain status and limit information.

The 64-bit Supervisory Root Pointer (SRP) functions like the CRP, except that it is used during supervisory access. The user may invoke this register by selecting the appropriate bits in the translation control register. It is normally initialized as a part of the general initialization procedure following a system reset.

The 32-bit Translation Control (TC) register controls the active functions of the MMU. They are: MMU enable, root pointer enable, page size selection, and the number of levels to be searched by the table search algorithm.

The two 32-bit Transparent Translation (TT0 and TT1) registers are used to allow two logical address ranges, also called *windows,* to be output as physical addresses. If a logical address matches the contents of any of the registers, then it is placed on the address bus as a physical address. The windows enable users to allocate contiguous areas in the logical address maps that do not require paging and/or protection. The windows may overlap, and have a minimum and maximum size of 16 Mbytes and 4 Gbytes, respectively. Windows allow for memory-mapped devices such as video display RAMs, shared library, and common operating system resources, without additional programming overhead.

The 16-bit MMU Status Register (MMUSR) indicates the results of the MMU instruction PTEST, where results could reflect a bus error or an attempt by a user program to access supervisory space, or an attempt to write into protected memory. MMUSR can be accessed using the instructions associated with the MMU, all of which can only be executed from the supervisory mode.

12.4.2 Address Translation Cache (ATC)

The address translation cache provides a fast method for address translation by minimizing the overhead associated with accessing the address translation table. The ATC is a 22-entry, fully associative cache that allows any data item to be stored in any tag location. To enhance the performance, all tags are compared in parallel during an access. The address translation cache operates in parallel with on-chip virtual caches for instruction and operands. Since the cache access is highly pipelined, a physical bus transfer can be achieved in two clock cycles [Vegesna 86]. Thus, for back-to-back bus cycles without translation overhead, the translation for the second cycle overlaps the first cycle. However, during write accesses, the address translation cache is accessed in parallel with an on-chip operand cache. Write protect flag in the address translation cache help complete the write cycles [Holden 87].

The organization of the ATC is shown in Figure 12.8. An entry consists of two parts, a virtual part and a physical part. The virtual part is made up of 24

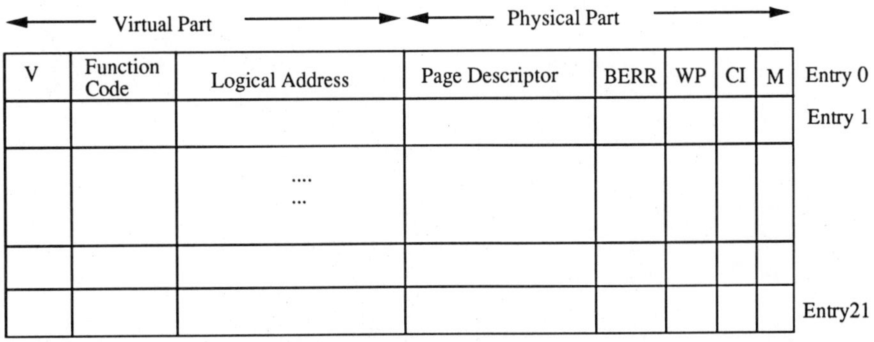

← Virtual Part →			← Physical Part →					
V	Function Code	Logical Address	Page Descriptor	BERR	WP	CI	M	Entry 0
								Entry 1
		....						
								Entry21

FIGURE 12.8. Organization of the address translation cache.

bits of virtual address, three bits of function code, and a valid bit to indicate whether entries have been changed or not. During the translation process, function codes are passed unchanged to FC0, FC1, and FC2.

The logical part of the entry also consists of two sections: a 24-bit page descriptor, and a 4-bit status field containing protection flags. The protection flags are: BERR (Bus error), WP (Write-protect), CI (Cache-inhibit), and M (Modify) bits. The flags provide status information and run-time system protection on a page-by-page basis.

The WP bit, when set in a table descriptor, write-protects all pages accessed with that descriptor (i.e., these pages can only be accessed in read-only mode).

The CI bit indicates protection for this page (i.e., whether the page is allowed to be cached or not). If $\overline{CI}$ is set, then the processor asserts $\overline{CIOUT}$ when this page is accessed.

The M bit, indicates whether data in a page has been modified. In a demand page system, this is important because if the page is not modified, the operating system does not need to write the page back onto the disk. Again, in a demand page system, the valid bit can be checked as part of the decision algorithm, in determining which page should be replaced from memory so that another can be read-in from disk.

The basic set of activities that occur in an address translation is shown in Figure 12.9. Upon receiving a logical address from the CPU, the MMU checks the ATC for a match. If a match is found, then the associated page frame number is placed on the address bus. An access involving this type of translation is executed quickly, as if the MMU is not used. If a match is not found, then the

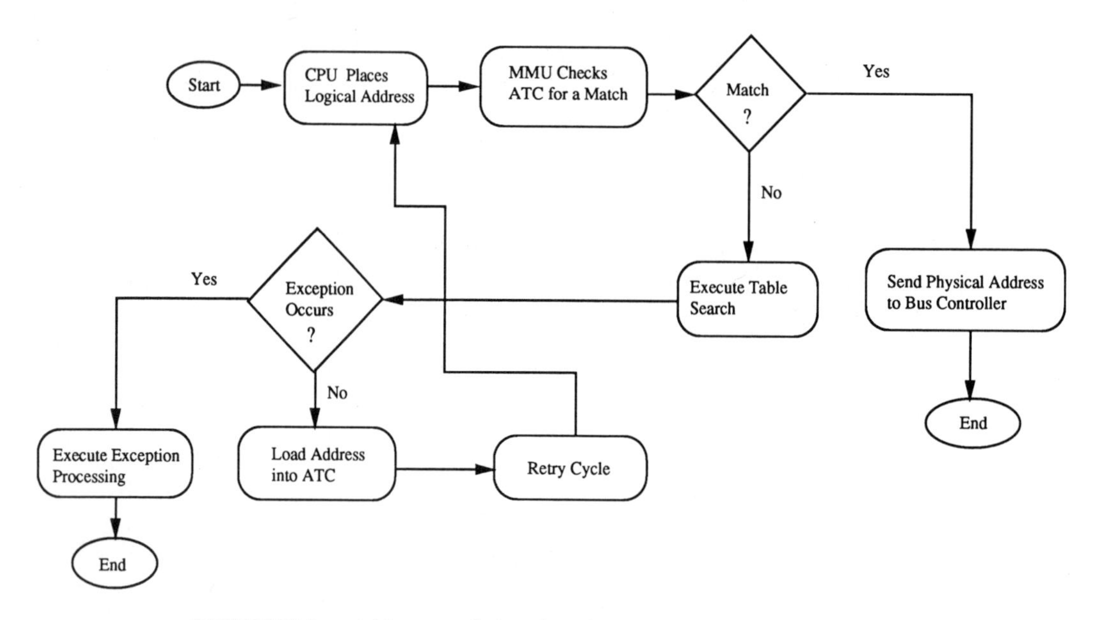

FIGURE 12.9. Address translation flow chart.

MMU must search the translation tables, and a search is initiated. If the search is successful, then a page descriptor is loaded into the ATC and the original cycle is retried. However, if a search is not successful, then a bus error exception occurs and control is returned to the operating system.

12.4.3 Address Translation Cache Miss

The MC68030 suspends its normal execution and enters into the address translation miss state whenever the address translation cache contains invalid virtual to physical mapping entry or when it encounters an unmodified bit in the cache, during a write cycle, to a valid entry. This detour allows the processor to update the translation tables. A miss causes the processor to save its internal state, including the bus state information required to restore the machine to its original state prior to the detour.

12.4.4 Table Search

The MC68030 organizes the logical-to-physical translation tables into a tree structure. The architecture of the table determines the complete logical view of the memory. It is built as part of the system software and generally resides in the physical memory for fast operation. The processor can support zero to six levels of translation tree. The levels are based on the trade-off between table-walk time and the memory required to hold the mapping tables.

The organization of the translation table is shown in Figure 12.8. The nodes of the translation tree contain descriptors, which are data structures to describe the translation and relevant access information. The root pointer for a table is a descriptor that contains the base address of the top-level table for the tree. The base address is placed in one of the two root-pointer registers, CRP or SRP. If the function code look-up is used, then they are also used to index into the first-level descriptor. The pointer descriptor provides the base address for the next-level table. Each field is then used to index into a table that gives a base address for the next level's table, and the table search process continues. At the final level, it must reach a page descriptor. The page descriptor mapping is then transferred to the ATC, and the bits that were not used as table indexes specify an offset into the selected page. If the ATC is full, then an entry is selected and updated, using a replacement algorithm. Since the processor uses a fully associative organization, each address can be mapped into any location of the ATC.

12.4.5 Translation Control

The Translation Control (TC) register defines the translation table's architecture. It essentially turns the MMU on or off, and selects various configuration variables, such as SRP enable and disable, page size, and the translation table index variable.

The IS (Initial Shift) field of this register defines the size of the logical address page. For example, if the IS = 15, then the logical address space contains only (2^{32}-2^{15}) bytes.

The PS (Page Size) field identifies the page size of the system. The valid page sizes are: 256 bytes, 512 bytes, 1 Kbyte, 2 Kbytes, 4 Kbytes, 8 Kbytes, 16 Kbytes, and 32 Kbytes. The total number of pages for the system is obtained by the logical address space divided by the page size.

The proper page size selection is important during the design of the system, and an optimal size should be adopted to allow a fine-enough granularity for the application's requirement. For example, if the system will have small adjacent regions that need to be mapped differently, then the page size will have to be small enough to fit within those regions. Here, larger pages will waste memory space. However, large pages mean smaller translation tables, because fewer logical bits are needed to index down to page level. However, criteria like swap size to disk may also affect the size of the page.

The E (Enable) bit enables and disables the address translation mechanism. The SRE (Supervisory Root Pointer Enable) controls the use of the SRP. When the SRP is disabled, then the MMU uses a single translation table, otherwise it uses two translation tables (one during user mode and the other during supervisory mode).

The table indexes TIA, TIB, TIC, and TID specify the sizes of logical address bits used to index the four possible levels of translation tables.

12.4.6 MMU Instructions

The MMU instructions are listed in Table 12.1. The instructions are:

PTEST: This causes the MMU to search the ATC or translation tables for a function code, address, and read/write operand. The result of the search is reflected in the MMUSR. This instruction is generally used to obtain the status of a page (i.e., whether it is modified or not).

PLOAD: This instruction asks the MMU to search the translation table for the specified function code and address operands. Upon completion of the search, a descriptor is loaded into the ATC. However, the MMUSR is not changed. It is generally used during a context switch, to initialize the ATC with one or more page descriptors related to the current task.

PFLUSH: When executed, it flushes a part of or the entire ATC. The ATC can be flushed completely by function code alone or by effective address and function code.

12.5 PROGRAMMING EXAMPLE FOR THE MMU

The use of the MMU is illustrated through an example. Motorola's MC68030-based MVME147 VMEbus board is used as the base hardware. The memory map of an application that runs on this hardware is listed in the following table [Howard 91b].

Function	Memory Address	Access Classification
Local I/O	$FFFE5000	Supervisor Read and Write
Blocked Out	$FFFE0000	
EPROM Data	$FFC00000	Supervisor Read-Only
EPROM Code	$FFA00000	Supervisor Read-Only
Blocked Out	$FF800000	
Master Stack	$00400000	Supervisor Read and Write
Stack Limit Page	$003FF000	Not Accessible
Interrupt Stack	$003FE000	Supervisor Read and Write
Stack Limit Page	$003FD000	Not Accessible
User Stack	$003FC000	User-Accessible Read and Write
Stack Limit Page	$003F1000	Not Accessible
Heap	$00F00000	User-Accessible Read and Write
Limit Page	$00301000	Not Accessible
Application Data	$00300000	User-Accessible Read and Write
Application Code/IData	$00200000	User-Accessible Read-Only
System Code/IData	$00100000	Supervisor Read-Only
System Data	$00080000	Supervisor Read and Write

Accessible and nonaccessible address spaces are utilized to eliminate the possibility of stack overflow and the underflow of the heap. The MMU will signal a bus error if the nonaccessible areas are accessed by a program.

Programming the MMU has two parts: configuration of the translation table and initialization of the MMU control registers.

12.5.1 Configuration of the Translation Table

The logical address space in this example is organized into a four-level translation table, with some early termination descriptors and several complete branches. A design decision is made to keep the page size of 4 Kbytes. The early-termination descriptors are used to keep the translation table to a manageable size. It also makes the design flexible and allows future modification of the hardware through the additions of physical memory and I/O areas.

Figure 12.10 illustrates the tree structure with the mapping. We disabled the FCL and SRC bits in the TC register, so that the MMU can use the CRP to access the translation table, and points to the first-level table. We also set IS = 0 and PS = 12. Now, we are left with 20 bits to be accounted for by the table index registers. By properly selecting the values of table index, we can map regions to align with the physical layout of memory and I/O devices. In the example: 'a' marks a bit that is used to index the A Level table, 'b' for indexing B level table, and so on. An 'o' bit is used as an offset within the page.

```
1098 7654 3210 9876 5432 1098 7654 3210
aaaa bbbb cccc dddd dddd oooo oooo oooo
```

The upper four bits (TIA) of the 32-bit address divides the address space into sixteen 256 Mbyte regions, giving a first-level translation table size of 16. The next four bits (TIB) of the 32-bit address index the B Level table. They divide a

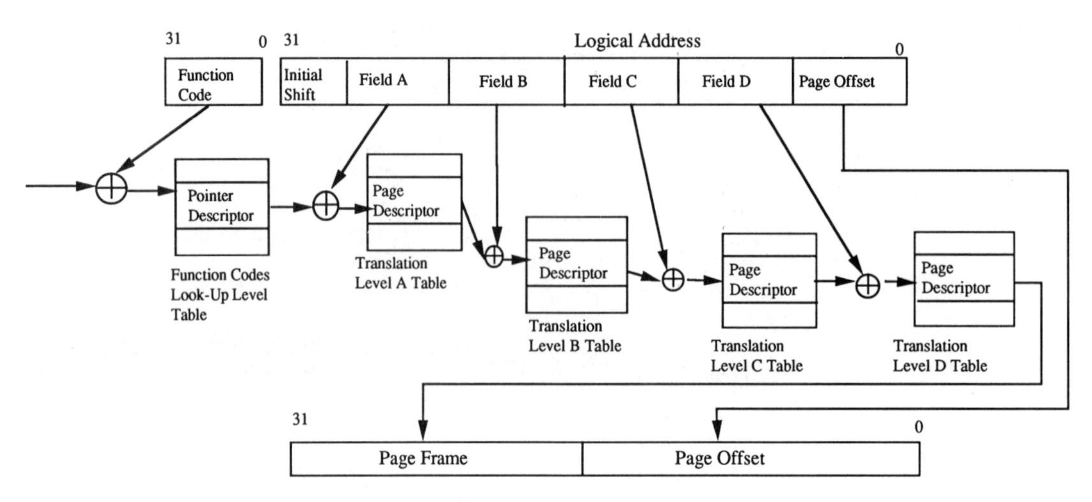

FIGURE 12.10. Logical-to-physical translation table walk.

256-Mbyte region into 16 regions of 16 Mbytes each. This method works similar to the A Level table, further dividing the address space. The next four bits (TIC) of the 32-bit address index the C-level table, dividing each 16-Mbyte region into 16, 1-Mbyte regions. The C Level table maps the DRAM area into four entries. The EPROM and I/O space will have 16 entries.

Now, the table mapping the DRAM region will have two early-termination page descriptors (LEFT) for the second and third megabytes, which align evenly with the application's memory map, described earlier. A D-Level page table will not be utilized in our example. Instead, it will map the entire megabyte using a single descriptor and related access control. This strategy limits the overall page-table size, and can be used whenever a memory area aligns evenly (top and/or bottom) with the region mapped by a specific table descriptor. Again, the table mapping the high-memory area will use early-termination descriptors to map the EPROM and I/O space. The EPROM space is 4 Mbytes and utilizes 4 C-Level LEFTs. The I/O space is less than a megabyte, but has a full megabyte region mapping it as supervisory only with CI (cache inhibit).

TID is the next higher 8 bits, and it indexes the D-Level page table for tree branches that require a D-Level. 4 Kbyte granularity in mapping the region is available. That is, the tables on this level have 256 entries, unless limited. The remaining 12 bits are used as offsets within a page. They are used directly.

It is good practice to conceptualize the translation table and document it on paper. This will allow us to experiment with TIx sizes and evaluate early termination. Figure 12.11 is an example of such a methodology. This is created to get a clear picture of the tree. This also maps very well to the software used to implement the translation table. Here, the address mapped by a particular branch is progressively defined at each successive level, until a final descriptor is reached.

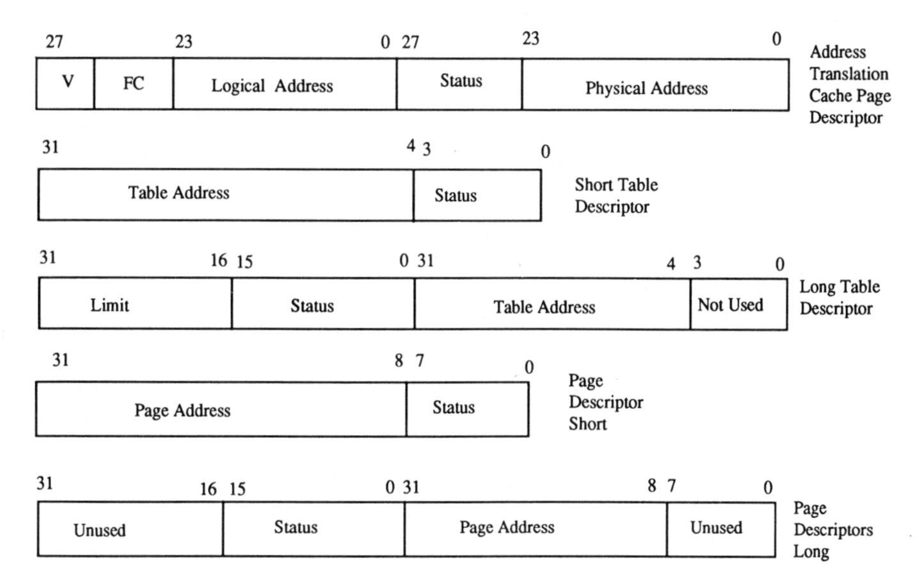

FIGURE 12.11. Summary of page descriptors.

12.5.2 Creation of a Data Structure

This design utilizes a directly mapped logical-to-physical address scheme. With this type of static mapping, it is possible to design a scheme that performs real translations. Additionally, it would enable the user to move memory blocks and rearrange the address space to fit the need. This is also called remapping. Remapping is primarily used in situations where we have several discontiguous memory regions, and want to remap the regions so that they appear more logical and contiguous to the applications.

Most operating systems set the table's organization at run time, and the table changes continually. For embedded systems, the table organization is created at compile time. As a result, the remapping strategy has been found effective.

To facilitate the table construction, we can use macros to define the individual descriptor types, and use a systematic labeling convention to define tables and their addresses. To set up the table, we are required to set up the root pointer, which has an 8-byte long structure. The root pointer points to the A-Level table. Once the root pointer is set up, each successive level is constructed. In this example, macro definitions and predefined status-bit symbols are used. The labeling convention utilizes a two-letter prefix, which defines the table's level and type. AT is prefixed to the label on the A Level table. BT is prefixed to all B Level table labels. An eight-digit hex address is used to define the memory region's base address. The label on each table is then used as the TADDRESS field in the next-level table. PADDRESS values specify the true physical address of the page that is mapped.

An example of the use of the labeling convention and macros to define a four-entry A-Level table and the next B-Level tree is given in the following, for completeness. Care should be taken to ensure that the translation table tree's structure has the same number of index bits in each level as the table index in the TC register.

```
        /* Base of the region for A Table is always $00000000 */

    ALIGN16: TABLE_SECTION AT00000000
        LFTD   MMU_LLIM, 4, BT00000000     ; points to the B
                                             table
        LFTD   MMU_LLIM, 4, BT40000000     ; points to the B
                                             table
        LFID                               ; invalid
    LFID                                   ; invalid
    BT00000000: *                          ; early termination
        LFET MMU_LLIM, 8, MMU_SUPV,
        MMU_CENA, MMU_RDWR, $00000000
                    :
                    :
    BT40000000:
                    LFTD MMU_LLIM, 8, CT40000000
                    LFTD MMU_LLIM, 8, CT50000000
                    :
                    :
```

12.5.3 MMU Initialization Code

The MMU is initialized once the translation table is set up. The initialization routine assigns the proper values to the MMU registers. Here, the CPU root pointer register is loaded with a pointer to AT00000000, the two transparent translation registers are disabled, and the MMU is enabled with the TC register.

The MMU registers are loaded with the MC68030's unique PMOVE instruction. This instruction moves the appropriate data from the memory to the MMU. For example, the CRP is initialized from a data structure in memory that is 64 bits long and similar to a table descriptor. This instruction has no immediate mode, so the programmer must set up the values in memory, then PMOVE is used with the proper addressing mode. In this case, the programmer may use address register A0 indirectly, to point to the memory.

If, at any time, the program tries to load the MMU registers with some sort of invalid value, then MMU configuration exception will take place. As a result, we should design an appropriate exception handler routine to address this exception.

12.5.4 Bus Error Handler

Before the design of an application, we must decide how to handle bus errors. If we have an operating system or a monitor, we can forward the bus error abort to the monitor, for which a bus-error handler is required. For this type of system, the handler is relatively simple to develop if we consider bus error to be a fatal exception. If we adopt the strategy that a bus error is a fatal error, then we can use the monitor screen or other output device to display any type of error message and take appropriate action.

12.6 EXCEPTION PROCESSING

Like the MC68020, an exception can originate internally or from external sources for the MC68030. The externally generated exceptions are caused by an interrupt, bus error, and reset. Internally generated exceptions are triggered by instructions, tracing, or breakpoints. The coprocessor-generated exceptions, CKH, CKH2, TRAPVcc, TRAP, etc. are also considered to be internally generated exceptions. Those detected by the MMU of the MC68030 are considered to be internally generated exceptions.

The exception vector tables of the MC68020 and MC68030 are identical, with the exceptions of MMU-related vectors. The exception vectors related to MMU are listed in the following table.

Vector Number	Assignment
56	MMU Configuration Error
57	Defined for the MC68851, But Not Used by the MC68030
58	Defined for the MC68851, But Not Used by the MC68030

Again, through the $\overline{\text{STATUS}}$ signal, the MC68030 identifies instruction boundaries and some exceptions [MC68030]. This signal also indicates when an address translation cache miss happened within the MMU, and when the processor is about to begin a table search for the logical address that caused the miss.

The exception-processing algorithm of the MC68030 involves the same four steps as found in the MC68020. Six different stack frames are used to save relevant data during exception processing. The stacks are: the normal 4-word and 6-word stack frames, the throwaway 4-word stack frame, 16-word coprocessor midinstruction exception stack frame, and the short and long bus fault stack frames. The format and the purpose of the exception stack frames are described in Chapter 7.

The MC68030 uses long-word operand transfers during writes to or reads from the stack frame. However, it does not necessarily read from or write to the stack frame data in sequential order. The system software should not also depend on a particular exception generating a particular stack frame. For the purpose of portability with future devices and hardware configuration, the software should be created in such a way that it is able to manage any type of stack frame related to any type of exception.

12.7 ADAPTATION OF THE MC68030 TO AN MC68020-BASED DESIGN

It is relatively simple to use the MC68030 in a MC68020 or MC68020/MC68851 system. One can place an adapter board containing the MC68030 into the existing socket of the MC68020 processor. However, in such a configuration, the MC68030 will operate in asynchronous mode. The signals that do not have a compatible signal on the MC68020 need to be pulled up or left unconnected, and the common signals are routed to the corresponding signal of the other processor. The following signals to the MC68030 should be handled as:

Pulled-Up	No Connect
$\overline{\text{STERM}}$	$\overline{\text{STATUS}}$
$\overline{\text{CBACK}}$	$\overline{\text{REFILL}}$
$\overline{\text{CIIN}}$	$\overline{\text{CBREQ}}$
$\overline{\text{MMUDIS}}$	$\overline{\text{CIOUT}}$

If the MC68851 is removed from its socket and replaced with a jumper header, then special attention should be given to a number of signals. They are: $\overline{\text{CLI}}$, $\overline{\text{RMC}}$, $\overline{\text{LBRO}}$, $\overline{\text{LBG}}$, $\overline{\text{LBGACK}}$, and $\overline{\text{LBGI}}$. For the MC68020/MC68851, without any logical cache and bus arbitration logic, the following MC68851 signals should be connected together via jumper:

$\overline{\text{LAS}}$ ↔ $\overline{\text{PAS}}$
$\overline{\text{LBRO}}$ ↔ $\overline{\text{PBR}}$
$\overline{\text{LBGI}}$ ↔ $\overline{\text{PBG}}$
$\overline{\text{LBGACK}}$ ↔ $\overline{\text{PBGACK}}$
LA8–LA31 ↔ PA8–PA31
$\overline{\text{CLI}}$ ↔ $\overline{\text{LAS}}$

The software differences that exists between the MC68851 and the MC68030 need to be considered whenever we upgrade a system originally designed for the MC68020/MC68851 pair. The following features are not supported by the MC68030's MMU unit.

- No access level (i.e., CALLM and RTM instructions are not supported).
- Breakpoint registers are missing.
- DMA is not supported (i.e., no DMA root pointer is available).
- Task aliasing is not supported.
- 22-entry address translation cache, instead of the 64 found in the MC68851.
- CAM entries can't be locked.
- Sharing of global entries is not supported.
- The following instructions are not supported: F-line trap, PVALID, PFLUSHR, PFLUSHS, PBcc, PDBcc, PScc, PTRAPcc, PSAVE, and PRESTORE.
- Control-alterable effective addresses are only used for MMU instructions.

This indicates that we need to make several modifications to the software that is executed on the MC68020 and MC68851 configuration.

Chapter 13

The MC68040 Microprocessor

The MC68040 is the third-generation member of the MC6800 family of processors. It integrates an MC68030-compatible CPU core, an MC68881-compatible floating-point core, a dual-demand paged memory management unit like the MC68851, and two independent 4-Kbyte instruction and data caches, on one VLSI device. The processor is fabricated utilizing a less than 1 micron HCMOS process, and operates at 25 MHz. The MC68040 implements full Harvard architecture. The high degree of instruction execution parallelism is achieved via multiple independent pipelines, multiple internal buses, and individual physical caches for instructions and data. This CPU is downward object-code compatible with the MC68020 and its predecessors.

This highly integrated processor makes system design very convenient. Here, designers do not have to interface the MC68020, MC68851, MC68881, and caches to build a system. They are replaced by a single chip, which operates at high speed, consumes less power, provides high reliability, and conserves valuable board space. Most previously written software can be directly ported on the hardware.

13.1 INTERNAL ARCHITECTURE AND FEATURES

The features of the MC68040 include:

- A 13.5 Mips, MC68030-compatible integer execution unit.
- A 3.6 MFlops, MC68881/MC68882-compatible floating-point unit.

- Independent instruction and data memory management units.
- Concurrently accessible 4-Kbyte instruction and data caches.
- Multimaster/multiprocessor support via bus snooping.
- 32-bit Nonmultiplexed external address and data buses with synchronous interface.
- 4 Gbytes of direct addressing capability.
- Object-code compatibility with previous members of the MC68000 family of processors.
- Concurrent integer unit, FPU, MMU, and bus controller operation that maximizes throughput.
- Software support, including an optimizing C compiler and UNIX system support.

Figure 13.1 illustrates the functional units of this processor, and they are:

- The floating-point unit.
- The integer execution unit.
- The memory units.
- The bus interface unit.

The units execute concurrently. Memory controllers reside on separate buses and fetch instructions and operands simultaneously. Each MMU stores recently used address maps in separate 64-entry ATCs. To minimize the search time of the ATCs, hardware logic is utilized in its implementation rather than microprogramming techniques. Each MMU is equipped with a transparent translation register, which defines a one-to-one mapping for address space segments ranging from 16 Mbytes to 4 Gbytes.

The bus controller is designed to support high-speed nonmultiplexed synchronous bus operation. It supports high data transfer to and from the caches during read and write operations. Additional bus signals are generated by the bus controller to support bus snooping and maintenance of the external cache tags.

The data cache supports write-through or copy-back write modes, and can be configured to operate on a page-by-page basis. The caches are physically mapped, and support external bus snooping to maintain cache coherency in multitasking systems.

13.2 PROGRAMMING MODEL

Like the previous members of the MC68000 family, the MC68040 also supports two programming modes. They are the *user programming mode* and the *supervisory programming mode.*

13.2.1 User Programming Model

The user program utilizes a model shown in Figure 13.2. This consists of:

- Eight 32-bit Data Registers (D7:D0)
- Eight 32-bit Address Registers (A7:A0)

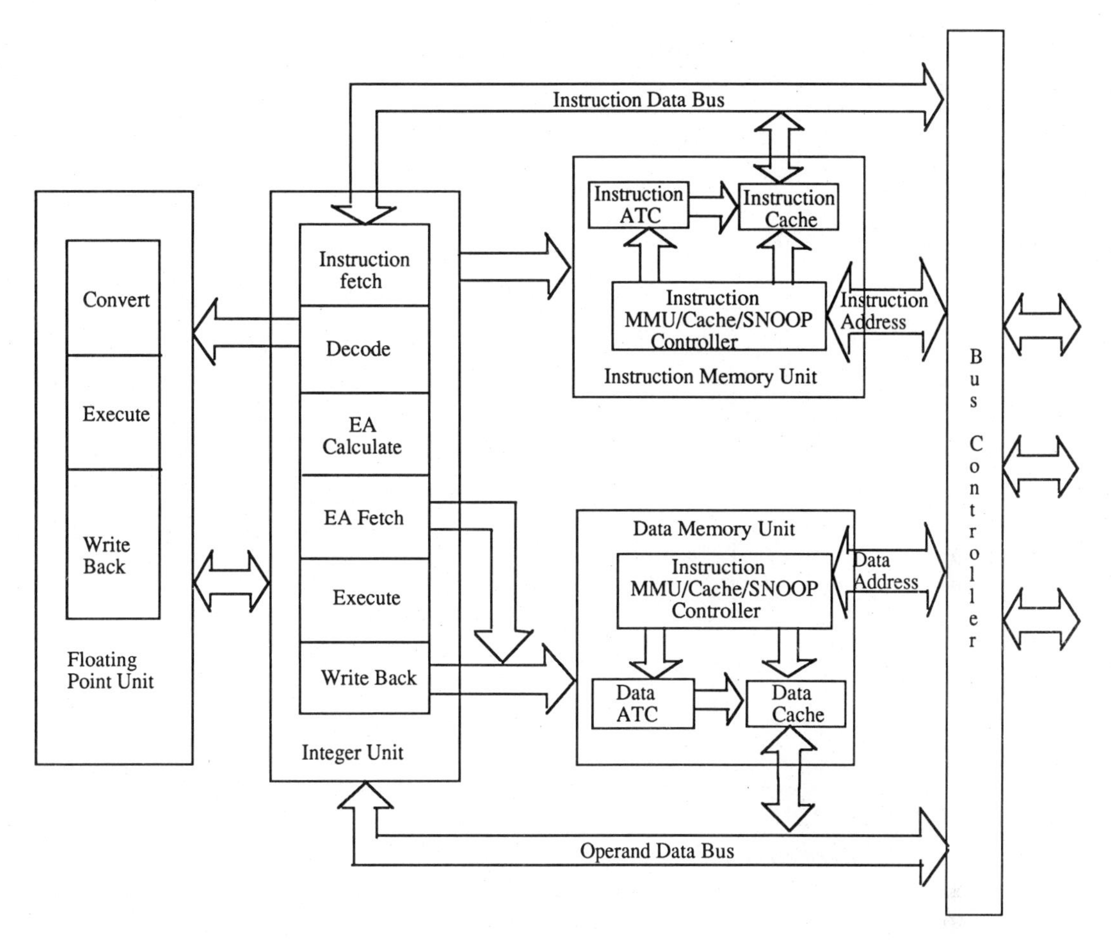

FIGURE 13.1. Block diagram of the MC68040. (Courtesy Motorola Inc.)

- A 32-bit Program Counter (PC)
- An eight-bit Condition Code Register (CCR)
- Eight 80-bit Floating-Point Data Registers (FP8:FP0)
- A 32-bit Floating-Point Control Register (FPCR)
- A 32-bit Floating-Point Status Register (FPSR)
- A 32-bit Floating-Point Instruction Address Register (FPIAR)

13.2.2 Supervisory Programming Model

The supervisory programming model of the MC68040 is designed for the system programmers to implement the operating system, I/O driver, and memory man-

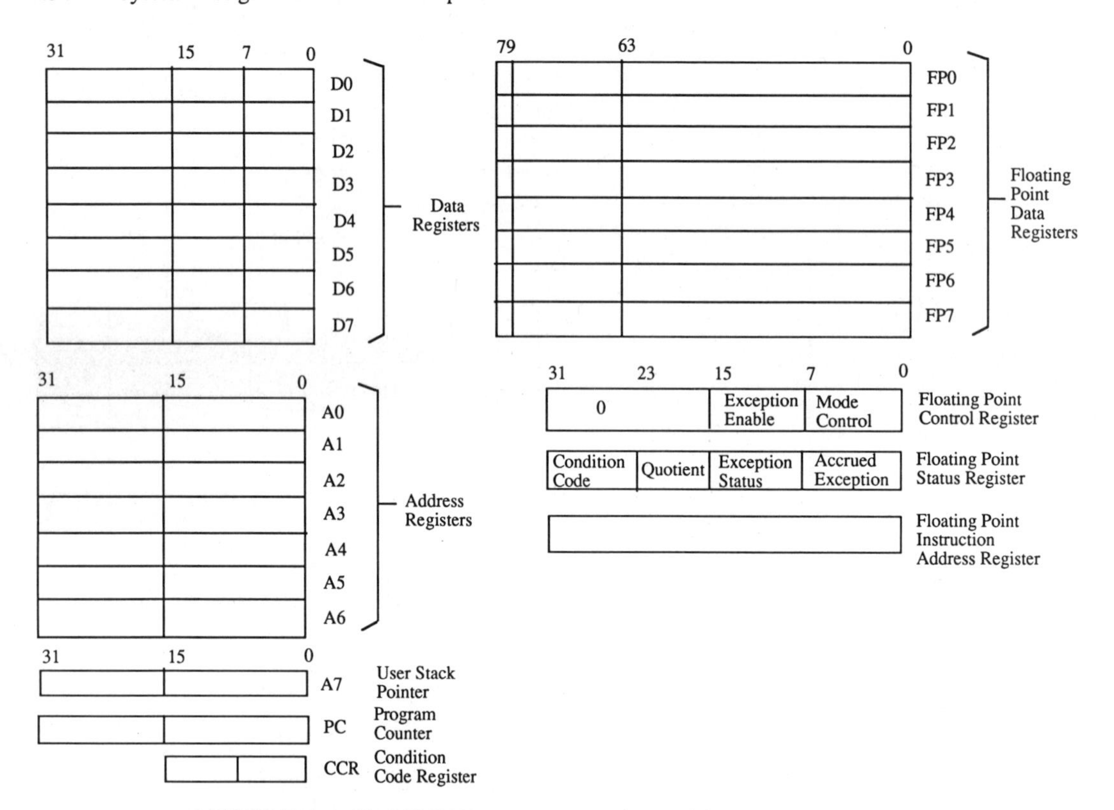

FIGURE 13.2. The MC68040 user programming model.

(Courtesy Motorola Inc.)

agement functions. Resources available under this mode are shown in Figure 13.3, and include [MC68040 89]:

- A 16-bit Status Register (SR)
- A 32-bit Interrupt Stack Pointer (ISP)
- A 32-bit Master Stack Pointer (MSP)
- A 32-bit Vector Base Register (VBR)
- A 32-bit Source Function Code (SFC) Register
- A 32-bit Destination Function Code (DFC) Register
- A 32-bit Cache Control Register (CACR)
- A 32-bit User Root Point (URP) Register
- A 32-bit Supervisor Root Pointer (SRP) Register
- A 16-bit Translation Control (TC) Register
- A 32-bit Data Transparent Translation Register 0 (DTT0)
- A 32-bit Data Transparent Translation Register 1 (DTT1)
- A 32-bit Instruction Transparent Translation Register 0 (ITT0)
- A 32-bit Instruction Transparent Translation Register 1 (ITT1)
- A 16-bit Memory Management Unit Status Register (MMUSR)

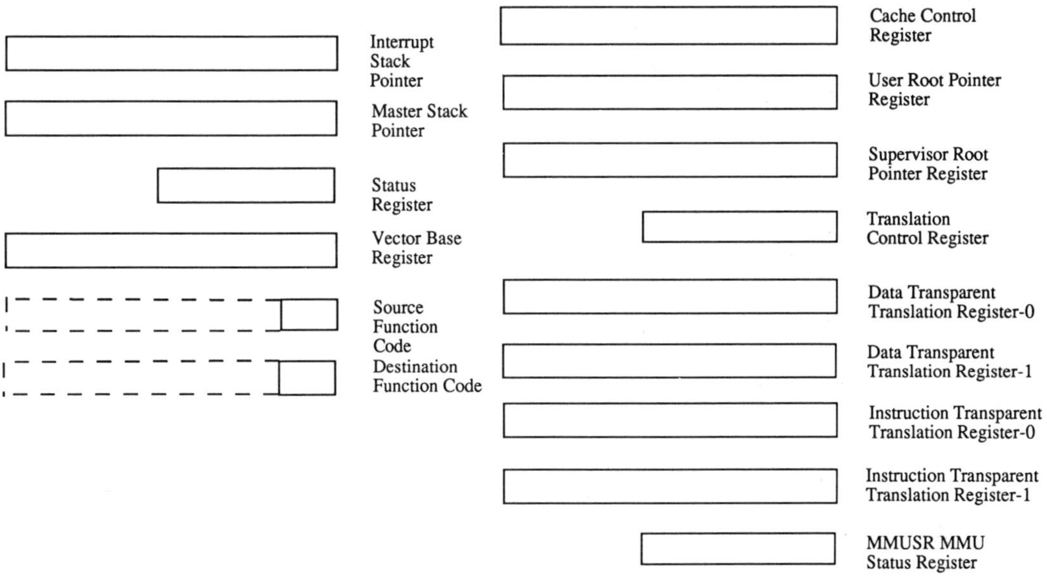

Interrupt Stack Pointer

Master Stack Pointer

Status Register

Vector Base Register

Source Function Code

Destination Function Code

Cache Control Register

User Root Pointer Register

Supervisor Root Pointer Register

Translation Control Register

Data Transparent Translation Register-0

Data Transparent Translation Register-1

Instruction Transparent Translation Register-0

Instruction Transparent Translation Register-1

MMUSR MMU Status Register

FIGURE 13.3. The MC68040 supervisory progamming model.

The contents of the TC selects the logical-to-physical address translation mechanism, and selects either a 4 Kbyte or 8 Kbyte page size. The four translation registers, DTT0, DTT1, ITT0, and ITT1, allow portions of the logical address space to be transparently mapped and accessed without the resident descriptors in the ATC.

The results of the PTEST instruction are reflected in the MMUSR, which contains various status information. The PTEST instruction examines the translation tables for the logical address, as indicated in its effective address field, including the status within the DFC. It then returns status information corresponding to the translation.

13.2.3 Data Types

Table 13.1 lists the data types supported by the MC68040. The two arithmetic units, the IU (Integer Unit) and FPU (Floating-Point Unit) allow the processor to support a wide range of data types. Again, the instruction set supports operations on variable data types (not supported directly) via memory addresses.

The integer data formats supported by both the IU and FPU follow the two's complement arithmetic found in the MC68000 family architecture. However, integer data formats are converted into extended-precision floating-point numbers by the FPU before their usage.

The single-double-precision, and extended precision floating-point data formats implemented in the FPU comply with the IEEE standard (described in Chapter 11). The extended precision format for FPU is 96 bits, out of which 80

TABLE 13.1. Data Types Supported by the M68040.

Operand Data Type	Size (Bits)	Execution Unit	Explanations
Bit	1	IU	
Bit Field	1-32	IU	Field of Consecutive Bits
BCD	8	IU	Packed and Unpacked
Byte Integer	8	IU and FPU	
Word Integer	16	IU and FPU	
Long-Word Integer	32	IU and FPU	
Quad-Word Integer	64	IU	Any Two Data Registers
16-Byte	128	IU	Memory Only, Aligned to 16-Byte Boundary
Single Precision Real	32	FPU	1-Bit Sign, 8-Bit Exp., 23-Bit Mantissa
Double Precision Real	64	FPU	1-Bit Sign, 11-Bit Exp., 52-Bit Mantissa
Extended Precision Real	80	FPU	1-Bit Sign, 15-Bit Exp., 64-Bit Mantissa

bits are currently defined, and the rest are reserved for future expansion. The extended precision format is used primarily for temporary variables, intermediate values, and for extra precisions.

13.2.4 Addressing Modes

The addressing modes of the MC68040 are identical to the MC68020. The MC68040 calculates the effective address for all modes within one processor cycle. The data is fetched within one cycle from the data cache into the *effective address fetch stage* of the integer pipeline. The instruction is then processed in the execution unit. When all stages are working on different instructions, the concurrency in the integer pipeline hides the effective address calculation and fetch for instructions with optimized addressing modes. Table 13.2 compares the

TABLE 13.2. Comparison of Processor Cycles Between the MC68040 and MC68020.

Instruction	Source	Dest.	Processor Cycles	
			MC68040	MC68020
MOVE	Rn	Rn	1	1
MOVE	<ea>	Rn	1	3
MOVE	Rn	<ea>	1	3
MOVE	<ea>	<ea>	2	4
Arithmetic/Logic(Simple)	Rn	Rn	1	1
Arithmetic/Logic(Simple)	<ea>	Rn	1	3
Arithmetic/Logic(Simple)	Rn	<ea>	1	3
Arithmetic/Logic(Simple)	Immediate	Rn	1	2
Shift	Rn		2	2
Shift	<ea>		2	4
Bcc Taken			2	3
Bcc Not-Taken			3	2

execution time for common optimized instructions between the MC68040 and MC68020. The information is based on the assumption that data fetches hit in their respective TLB and cache.

The processor cycles listed in the table refer to the number of times the execution unit worked per instruction. A processor cycle is equal to one bus cycle for the MC68040 and two bus cycles for the MC68020. Therefore, the MC68040 cycles twice as often as the MC68020 for the same input clock speed.

The MC68040 supports a subset of the MC68881/68882's floating-point instructions. Several new arithmetic instructions are added to explicitly select single- or double-precision rounding. However, the nonsupported MC68881/68882's instructions can be emulated in software.

A new user mode instruction, MOVE16, has been added to the MC68040. This allows for high-speed transfer of 16-byte blocks of data between devices, namely memory-to-memory, coprocessor-to-memory, or slave-device-to-memory.

13.3 SIGNAL DESCRIPTION

The signal lines of the MC68040 can be assigned to 12 functional groups. Figure 13.4 shows the groups, along with their members. The groups are [MC68040 89]:

- Address bus (A31:A0)
- Data bus (D31:D0)
- Transfer attributes
- Master transfer control
- Slave transfer control
- Bus snoop control and response
- Bus arbitration
- Processor control
- Interrupt control
- Status and clocks
- Test
- Excitation control

13.3.1 Address Bus

This 32-line three-state address bus is identical to the MC68020's address bus.

13.3.2 Data Bus

These three-state bidirectional signals provide the general-purpose data path between the MC68040 and all external devices. Like the MC68020/68030, this bus can transfer 8, 16, or 32 bits of data during a bus access. The burst transfer is a new feature, where the data lines are time-multiplexed to carry 128 bits of burst request, utilizing four 32-bit transfers.

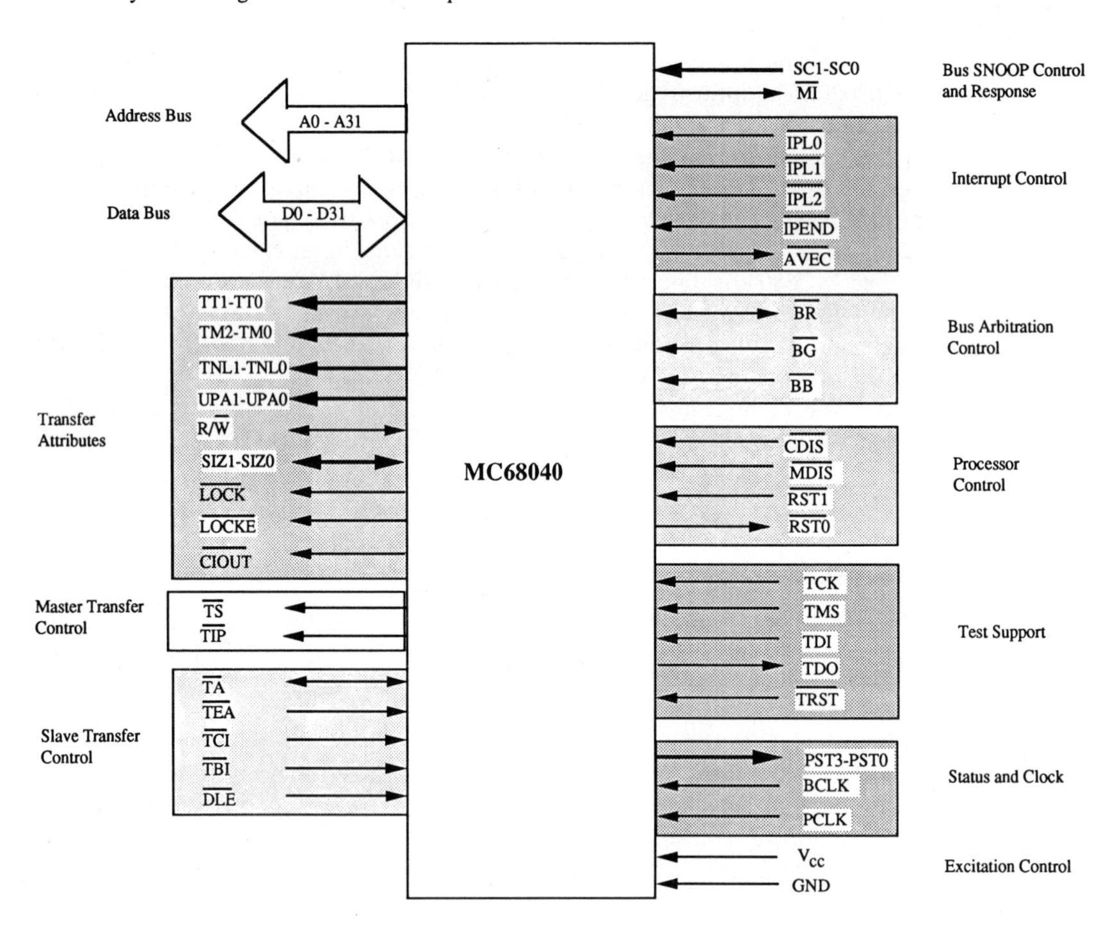

FIGURE 13.4. Functional signal groups of the MC68040.

13.3.3 Transfer Attribute Signals

The data transfer over the data bus is controlled by 15 signal lines, which are:

- Transfer Type (TT1, TT0)
- Transfer Modifier (TM2, TM1, and TM0)
- Transfer Line Number (TLN1 and TLN0)
- User Programmable Attributes (UPA1 and UPA0)
- Transfer Size (SIZ1 and SIZ0)
- Read/Write (R/$\overline{W}$)
- Lock ($\overline{LOCK}$)
- Lock End ($\overline{LOCKE}$)
- Cache Inhibit Out ($\overline{CIOUT}$)

The MC68040 bus interface supports synchronous data transfer between the processor and other devices in the system. When the processor becomes the bus master, it uses the bus to access instructions and data from memory, in case of a cache miss, and updates data into the memory.

Transfer Types (TT1 and TT0) are asserted by the MC68040 to declare the purpose of the current bus transfer. During the bus transfer by an alternate bus master, these signals are sampled by the MC68040 to determine whether it needs to snoop (monitor) the transfer. It snoops during the normal and MOVE16 bus access. Table 13.3 lists the encoded values of the transfer type.

TABLE 13.3. Data Transfer Configuration.

TT1	TT0	Transfer Type
0	0	Normal Access
0	1	MOVE16 Access
1	0	Alternate Logical Function Code Access
1	1	Acknowledge Access

Transfer Modifiers (TM2–TM0) are three-state output signals. They supplement the information provided by the transfer type signals. Table 13.4 lists the encodings for normal and MOVE16 transfers, and Table 13.5 lists the encodings for alternate access transfers. TMx signals are asserted by the MC68040 only when it is the bus master.

TABLE 13.4. Transfer Modifiers.

TM2	TM1	TM0	Transfer Type
0	0	0	Data Cache PUSH Access
0	0	1	User Data Access (MOVE16)
0	1	0	User Code Access
0	1	1	MMU Table Search Data Access
1	0	0	MMU Table Search Code Access
1	0	1	Supervisor Data Access (MOVE16)
1	1	0	Supervisor Code Access
1	1	1	Reserved for Future Use

TABLE 13.5. Encodings for Alternate Access Transfers.

TM2	TM1	TM0	Transfer Modifier
0	0	0	Logical Function Code 0
0	0	1	Reserved
0	1	0	Reserved
0	1	1	Logical Function Code 3
1	0	0	Logical Function Code 4
1	0	1	Reserved
1	1	0	Reserved
1	1	1	Logical Function Code 7

Transfer Line (TLN1–TNL0) signals are also three-state outputs for the processor. They indicate which entry within the set of four instruction or data cache lines is being accessed. These lines do not carry information the rest of the time. The transfer lines are driven to the high-impedance state as the MC68040 releases the control of the bus. Table 13.6 lists the encoded meaning of the TLNx signals.

TABLE 13.6. Encoding for Transfer Modifiers.

TLN1	TLN0	Transfer Modifier
0	0	Zero
0	1	One
1	0	Two
1	1	Three

User Programmable Attributes (UPA1–UPA0) are also three-state outputs for the MC68040. These signals are defined for normal code and data access, including MOVE16 accesses, and are zero for the other accesses.

The R/$\overline{W}$ bidirectional three-state signal defines the data transfer direction for the current bus cycle.

The bidirectional three-state SIZx are identical to the MC68020's SIZx.

The $\overline{LOCK}$ is a three-state output for the MC68040. It is equivalent to the $\overline{RMC}$ signal for the MC68020. Although $\overline{LOCK}$ indicates that the bus will be locked as a part of an indivisible read-modify-write operation, the processor will release the bus if the bus grant ($\overline{BG}$) signal is negated by the external arbitrator.

The three-state output Lock End ($\overline{LOCKE}$) indicates that the current transfer is the last in a locked sequence of transfers for the read-modify-write operation. This signal helps the MC68040 to implement multiple indivisible operations.

The Cache Inhibit Out ($\overline{CIOUT}$) is a three-state output signal for MC68040. The signal is asserted by the processor during the access to the noncachable pages, and directs the external cache to ignore this operation.

13.3.4 Bus Transfer Control Signals

The MC68040 utilizes transfer control signals to control the bus transfer operation.

Transfer Start ($\overline{TS}$). This is a bidirectional line, and is asserted by the processor for one clock period during the start of each transfer. The processor monitors this signal during the bus access performed by an alternate bus master. The processor uses this information in order to detect the start of each transfer to be snooped.

Transfer In Progress ($\overline{TIP}$). This is a three-state output signal for the processor. When asserted, it indicates the progress of a bus transfer. It negates $\overline{TIP}$ after the completion of the current transfer, and then transitions to a high-impedance state.

Transfer Acknowledge ($\overline{TA}$). This is also a bidirectional signal for the

MC68040. It signals the completion of a requested data transfer operation. During the processor-controlled bus transfer, $\overline{TA}$ is driven by the slave device, indicating the completion of the transfer (using this signal, we can implement an asynchronous protocol during bus access). During the alternate master accesses, $\overline{TA}$ is normally three-stated, to allow the referenced slave device to respond, and is sampled by the MC68040 to determine the completion of the bus transfer. $\overline{TA}$ is also asserted by the MC68040 to acknowledge the data transfer while an alternate master accesses its internal caches.

Transfer Error Acknowledgment ($\overline{TEA}$). This is an input signal to the processor, and is driven by the current slave to signal an error condition during the bus transfer. When $\overline{TEA}$ is asserted along with $\overline{TA}$, it indicates that the processor should retry the access. The $\overline{TEA}$ is sampled by the MC68040 during the access of the system bus by the alternate bus master.

Transfer Cache Inhibit ($\overline{TCI}$). This input signal inhibits the incoming data from being loaded into the MC68040 instruction or data caches. The processor ignores the $\overline{TCI}$ status during all writes, including the first data transfer for both burst line reads and burst-inhibited line reads.

Transfer Burst Inhibit ($\overline{TBI}$). This is also an input signal to the MC68040 processor. The referenced device asserts this signal *low* to declare that it cannot support burst mode accesses, and the requested transfer should be broken up into several long-word transfers. During the bus access by an alternate master, the MC68040 samples $\overline{TBI}$ to detect the completion of each bus access.

Data Latch Enable ($\overline{DLE}$). This input signal is utilized by the processor to latch data from the data bus during $\overline{DLE}$ mode. This signal can be utilized in support of asynchronous data transfer to/from the MC68040.

13.3.5 Snoop Control Signals

The process of bus monitoring and intervention by the MC68040 is known as snooping. During the bus access by another bus master (which has been granted control of the bus), the MC68040 continues to snoop the bus. This is done in case the current bus master requests the MC68040 to intervene during the access. The control input signal lines to the processor allow an external entity to specify the required snoop operation during each bus transfer by an alternate master.

Several signal lines are provided in the MC68040 to control the snooping operation. These signals are:

Snoop Controls (SC1 and SC0) are input signals. They specify the snoop operation to be performed by the processor for an alternate bus master. During the snooping of a read operation performed by an alternate bus master, the MC68040 can intervene during the access by supplying data from its data cache when the memory copy is stale, ensuring that proper data gets to the alternate bus master. In the case of a write, the alternate bus master can be snooped to either update the MC68040's internal data cache with the new data, or to invalidate the matching cache lines. This ensures valid data during subsequent reads made by the processor. The encoded meaning of the snoop control lines are listed in Table 13.7.

TABLE 13.7. Encoding of Snoop Control Signals.

SC1	SC0	Requested Snoop Operation	
		Read Access	Write Access
0	0	Inhibit Snooping	Inhibit Snooping
0	1	Supply Dirty Data and Leave Dirty	Sink Byte/Word/Long-Word Data
1	0	Supply Dirty Data and Make Line Invalid	Invalid Line
1	1	Reserved (No Snoop)	Reserved (No Snoop)

Memory Inhibit ($\overline{\text{MI}}$) is an output signal from the MC68040. It inhibits the memory from responding to an alternate master's access, when the processor is snooping the access. The processor keeps $\overline{\text{MI}}$ asserted until it determines whether intervention is warranted. If no intervention is required, $\overline{\text{MI}}$ asserted until it determines whether intervention is warranted. If no intervention is required, $\overline{\text{MI}}$ is negated, and this allows the memory to respond to an alternate bus master's request. If an intervention is required, $\overline{\text{MI}}$ remains asserted, and the MC68040 completes the transfer as a slave. The processor updates its cache during a write, or supplies data to the alternate master during a read operation.

13.3.6 Bus Arbitration Signals

The MC68040 uses three signals, $\overline{\text{BR}}$, $\overline{\text{BG}}$, and $\overline{\text{BB}}$, to negotiate the control of system bus with another potential bus master. The functions of the signals are described in the following list:

Bus Request ($\overline{\text{BR}}$) signal line is used by the potential bus master to request the system bus.

Bus Grant ($\overline{\text{BG}}$) indicates that an external device may assume bus mastership.

Bus Busy ($\overline{\text{BB}}$) is a three-state bidirectional signal, and indicates that the bus is currently owned. The MC68040 must detect $\overline{\text{BG}}$ asserted and $\overline{\text{BB}}$ negated (indicating that the bus is free) before it asserts $\overline{\text{BB}}$ as an output in order to assume the bus ownership. To release the bus, the processor negates $\overline{\text{BB}}$, and then allows it to be set to the high-impedance state to be used again as an input.

13.3.7 Processor Control Signals

The processor control signals are utilized to manage the caches and related memory management units, including the initialization of the external associated devices.

Cache Disable ($\overline{\text{CDIS}}$), when asserted, inhibits the operation of the MC68040's cache. When asserted, the processor does not flush the data and instruction caches. Entries remain unchanged and become available after $\overline{\text{CDIS}}$ is negated.

MMU Disable ($\overline{\text{MDIS}}$), when asserted, disables the address translation features of the MMUs, internal to the processor. The address translation cache entries become reactivated once the $\overline{\text{MDIS}}$ is negated.

Reset In ($\overline{\text{RSTI}}$) causes the MC68040 to enter into reset exception processing. $\overline{\text{RSTI}}$ is an asynchronous input. However, it is synchronized internally before being acted upon by the processor.

Reset Out ($\overline{\text{RSTO}}$) is asserted by the processor when it executes the RESET instruction. It is used to initialize external devices.

13.3.8 Interrupt Control Signals

These signals, and their functions, are identical to their counterparts in the MC68020 processor.

Interrupt Priority Level ($\overline{\text{IPL2}}$: $\overline{\text{IPL0}}$) provides an encoded interrupt level to the processor.

Interrupt Pending ($\overline{\text{IPEND}}$) indicates that an interrupt is pending.

Autovector ($\overline{\text{AVEC}}$), when asserted by an external device, requests that the processor generate the interrupt vector internally.

13.3.9 Status and Clocks

The complexity and synchronous data transfer features of the MC68040 require that it provides more comprehensive information to the outside world, so that external devices are aware of the processor's status and act accordingly. The MC68040 has also extended the notion of a single clock, which was prevalent in its predecessors.

Processor Status (PST3:PST0) are output signals for the MC68040. They display the internal status of the MC68040, and are synchronous with Bus Clock (BCLK). Table 13.8 lists the encoded messages placed over the processor status lines.

Bus Clock (BCLK) is a input to the MC68040. It is used as a reference for all bus timing. This is a TTL-compatible clock and has a 50% duty cycle.

Processor Clock (PCLK) is used by the processor to drive all internal circuitry. This is also a TTL-compatible signal, and can not be gated off.

13.3.10 Test Signals

The five test pins provide an interface in support of the IEEE-P1149.1 (Test Access Port for Boundary Scan Testing of Board Interconnect) standard. These signals support boundary SCAN only [Abramovici 90].

TABLE 13.8. Encoded Representation of the Processor Status.

PST3	PST2	PST1	PST0	Internal Status
0	0	0	0	User Start/Continue Current Instruction
0	0	0	1	User End Current Instruction
0	0	1	0	User Branch Not Taken, End Current Instruction
0	0	1	1	User Branch Taken, End Current Instruction
0	1	0	0	User Table Search
0	1	0	1	Halted State (Double Bus Fault)
0	1	1	0	Reserved
0	1	1	1	Reserved
1	0	0	0	Supervisor Start/Continue Current Instruction
1	0	0	1	Supervisor End Current Instruction
1	0	1	0	Supervisor Branch Not Taken, End Current Instruction
1	0	1	1	Supervisor Branch Taken, End Current Instruction
1	1	0	0	Supervisor Table Search
1	1	0	1	Stopped State (Supervisor Instruction)
1	1	1	0	RTE Executed
1	1	1	1	Exception Stacking

Test Clock (TCK) is used as a dedicated clock for test logic.

Test Mode Select (TMS) is an input signal for the *Test Access Port* (TAP) controller.

Test Data In (TDI) provides a serial input to the TAP.

Test Data Out (TDO) is a three-state output signal. It provides a serial data output from the TAP controller. It allows parallel connection of board-level test data paths.

Test Reset ($\overline{\text{TRST}}$) is also an input signal to the processor. It provides an asynchronous reset of the TAP controller.

13.4 BUS OPERATION

The MC68040 supports synchronous data transfer with slave devices. This is different from the previous members of the MC68000 family of processors. All operations are synchronized with the rising edge of the bus clock. The processor does not support dynamic bus sizing. As a result, memory should be 32 bits wide, and devices with less than 32-bit data ports must be placed on long-word boundaries. This organization is adopted to optimize high-speed transfer.

We have seen that the processor uses two clocks; a Bus Clock (BCLK) and a Processor Clock (PCLK) to generate timing information. The PCLK operates twice as fast as the BCLK. Such dual clock inputs allows the bus to operate at half the speed of the internal operation, and places less stringent bus interface requirements. All inputs to the processor, with the exceptions of $\overline{\text{IPL2}}$–$\overline{\text{IPL0}}$ and $\overline{\text{RSTI}}$, are sampled synchronously and must be stable during the sample window for the input setup and hold time, to guarantee proper operation. The signals $\overline{\text{IPLn}}$ and $\overline{\text{RSTI}}$ are also sampled on the rising edge of the BCLK, and are synchronized internally to resolve all levels before their usage.

13.4.1 Bus Transfer Protocol

To initiate a bus transfer protocol, the MC68040 places the address on the address bus, asserts the transfer size pins SIZ2–SIZ0 with the proper values, specifies the type of transfer to be performed over TT1–TT0, and drives the related control signals. During a write operation, the processor places the data over the data bus. For a read operation, the selected device responds by supplying the data on the data bus, utilizing a transfer termination ($\overline{\text{TA}}$) signal. The slave device can control the rate of the data transfer operation utilizing $\overline{\text{TA}}$ and $\overline{\text{TEA}}$ signals.

During burst transfers, the address and transfer attributes remain unaltered. The value placed on the address bus will indicate the starting address of the required long-words, and the memory system is required to supply data from locations at an offset of 0, 1, 2, and 3 long-words.

The transfer sequence may be terminated in various ways. For a byte, word, or long-word bus request, the transfer is terminated on the first cycle via a $\overline{\text{TA}}$ or $\overline{\text{TEA}}$ signal. For cache line transfers, the bus transfer is terminated on the last assertion of $\overline{\text{TA}}$ for a normal acknowledgment. For abnormal acknowledgment, the slave device utilizes $\overline{\text{TA}}$ and $\overline{\text{TEA}}$ to specify the reason for the termination. Table 13.9 lists the reasons for termination.

TABLE 13.9. Bus Transfer Termination.

$\overline{\text{TA}}$	$\overline{\text{TEA}}$	$\overline{\text{TBI}}$	$\overline{\text{TCI}}$	Encoded Meaning
N	N	X	X	Insert wait states in current bus transfer
A	N	N	N	Normal transfer acknowledgment
A	N	N	A	Transfer ACK with cache inhibit
N	A	X	X	Transfer terminate with bus error trap
A	A	X	X	Transfer teminate with immediate entry
A	N	A	N	Transfer ACK with no burst
A	N	A	A	Transfer ACK with no burst, cache inhibit

N = Negated, A = Asserted, X = Don't Care.

13.4.2 Snoop Protocol

The MC68040 is capable of monitoring the bus activity of another bus master. Snooping begins as soon as the bus is granted to another bus master and the processor detects the assertion of $\overline{\text{TS}}$ signal. The processor latches the status of the address bus, SIZ1–SIZ0, TT1–TT0, R/$\overline{\text{W}}$, and SC1–SC0 signals on the rising edge of BCLK for which $\overline{\text{TS}}$ is detected. It then examines the snoop control and transfer type, to decide whether this access should be snooped or not. The MC68040 inhibits the memory from responding by asserting an $\overline{\text{MI}}$ signal, while it checks the internal cache for a match. However, if the cache contains a dirty line corresponding to this access, then the processor intervenes and supplies the data directly to the bus master. If no match is found in the cache, then the processor will allow the memory to respond to the request by negating $\overline{\text{MI}}$. In a system with multiple bus masters, the memory must wait for each snooping bus master to negate their $\overline{\text{MI}}$ signals before responding to an access.

If snooping with sink data is enabled for a byte, word, or long-word write access, and the corresponding data cache line contains dirty data, then the MC68040 inhibits the memory, and responds to the access as a slave device to read the data from the bus and update the data cache line. The dirty bit is set for the long-word changed in the cache line.

The MC68040 can be configured to respond to snooping operation on a page-by-page basis during its operation as a bus master. In this mode, the processor also monitors the levels of $\overline{TA}$, $\overline{TEA}$, and $\overline{TBI}$ to identify normal, bus error, retry, and burst inhibit terminations.

13.4.3 Bus Arbitration Protocol

The MC68040 bus protocol also supports the notion of a single bus master at a given instance. It supports an arbitration mechanism, where an external arbiter controls bus arbitration and the processor acts as a slave device in requesting ownership of the bus from the arbiter.

Three primary signals, $\overline{BR}$ (bus request), $\overline{BG}$ (bus grant), and $\overline{BB}$ (bus busy), are used by the MC68040 during this bus arbitration. The bus arbitration unit operates synchronously, and the state transition takes place on the rising edge of the BCLK. The bus arbitration logic, including the relevant signals, is shown in Figure 13.5.

The MC68040 will assert $\overline{BR}$ to initiate the bus request. It continues to assert

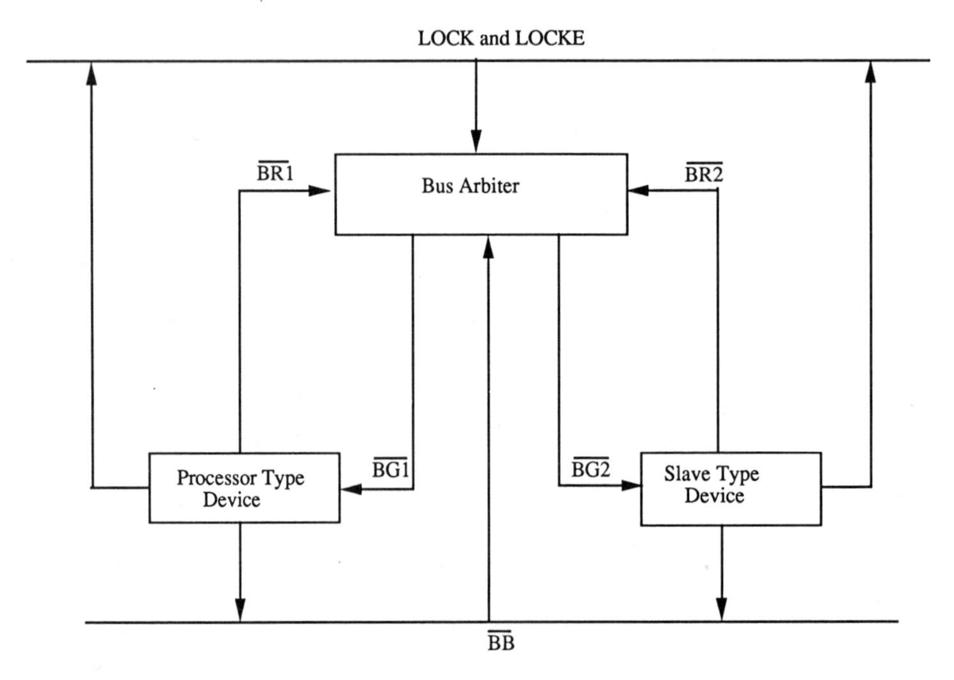

FIGURE 13.5. Bus arbitration logic.

$\overline{\text{BR}}$ until it receives $\overline{\text{BG}}$ from the arbiter, allowing the processor to begin the bus cycle. If $\overline{\text{BG}}$ is found to be negated while a bus cycle is in progress, then the processor relinquishes the bus at the completion of the current bus cycle. The read and write portions of a locked read-modify-write sequence are divisible for the MC68040, allowing the bus to be arbitrated away during the locked sequence. This last portion is different from the previous members of the processor family.

The processor monitors the $\overline{\text{BB}}$ (bus busy) signal to identify when the bus cycle of the previous master has completed after receiving $\overline{\text{BG}}$. After $\overline{\text{BB}}$ is negated by the alternate bus master, the processor asserts $\overline{\text{BB}}$ to declare the ownership of the bus, and begins the bus cycle. At the completion of the bus transaction, the processor negates $\overline{\text{BR}}$ and hands over the bus to the external arbiter, to allow another master to acquire the bus.

For applications that cannot allow locked sequences to be broken, the arbiter can use the $\overline{\text{LOCK}}$ signal to detect locked accesses and prevent the negation of $\overline{\text{BG}}$ during these sequences. The $\overline{\text{LOCKE}}$ signal is also asserted by the processor to indicate the last write cycle of a locked sequence. This allows arbitration between back-to-back locked sequences. An external arbiter can use the $\overline{\text{LOCKE}}$ status to determine when to remove $\overline{\text{BG}}$ without breaking a locked sequence. This allows arbitration to be overlapped during the last transfer in a locked sequence.

13.5 FLOATING-POINT (FP) UNIT

The MC68040 incorporated an on-chip floating-point unit to enhance its performance. The FP unit conforms fully to the IEEE754 standard. Again, it maintains compatibility with the MC68881/MC68882 coprocessors. The on-chip floating-point unit increased the MC68040's performance to 3.6 MFLOPS (Millions of Floating-Point Instructions per Second).

The on-chip FP unit operates concurrently with the Integer Unit (IU). However, the instruction prefetch, effective address calculation, and operand prefetch are performed by IU for both the integer and FP operations. The FP unit is composed of three-stage pipes. They are:

- Convert/tag unit
- FP arithmetic unit
- Write back unit

The convert/tag pipe stage converts single, double, extended, and integer data format memory operands to the internal extended precision format, and tags the operands by data type. This unit executes instructions, namely: FMOVE, FADD, FMUL, etc.

The FP arithmetic unit pipe stage is configured as a single, nonpipelined functional unit, with no concurrency during the execution of ADD, SUB, MUL, and DIV operations.

The FP write-back pipe stage contains dedicated hardware to perform rounding, overflow/underflow detection, condition code and exception setting, and write-back into the FP data register.

13.5.1 Floating-Point Instruction Set

Table 13.10 lists the FP instructions supported by the MC68040's on-chip FP unit. Table 13.11 lists the data formats and data types supported by the FP unit for the instructions listed in the Table 13.10. The instructions, data formats, and the data types that are not supported by the MC68040 need to be implemented by software, in order to maintain software compatibility with the MC68881/ MC68882.

TABLE 13.10. Floating-Point Instruction Set Available in the MC68040.

FADD	FMOVE	FNEG	FSUB	FCMP
FBcc	FMUL	FTST	FDBcc	FScc
FDIV	FABS	FTRAPcc		
FSAVE	FRESTORE			
FMOVEM of FP data registers (In/Out)				
FMOVE and FMOVEM of FPCR, FPSR, and FPIAR registers (In/Out)				

Note: S/W = Data format/type subset needs to be supported by software emulation; Y = Data format/type subset supported;— = Not applicable.

The floating-point instructions of the MC68040 support the following binary data format types:

- Single
- Double
- Extended Precision

The above formats are not supported when the operand data type is denormalized or unnormalized. However, the conversions (integer → float) and (float → integer) are supported by the on-chip unit.

TABLE 13.11. Available Data Types with the FPU of the MC68040.

Data Type	Data Format						
	SGL	DBL	EXT	Decimal	Byte	Word	LWord
Norm	Y	Y	Y	S/W	Y	Y	Y
Zero	Y	Y	Y	S/W	Y	Y	Y
Infinity	Y	Y	Y	S/W	—	—	—
NAN	Y	Y	Y	S/W	—	—	—
DeNorm	S/W	S/W	S/W	S/W	—	—	—
UnNorm	—	—	S/W	S/W	—	—	—

13.6 ON-CHIP MEMORY MANAGEMENT UNIT (MMU)

We have seen that the MC68040 includes independent instruction and data Memory Management Units (MMUs), and support demand-paged virtual memory environment. The MMU assumes that the physical memory is paged, the logical address space is divided into pages of the same size, and the operating system assigns pages to page frames to reflect the requirements of the programs.

The MMU supports the following features:

- A 32-bit logical address
- A 32-bit physical address with a 2-bit extension
- A selectable 4/8 Kbyte page size
- A three-level page table hierarchy
- A 64-entry four-way set-associative ATC for each MMU.
- A global bit, which allows flushing of all nonglobal entries from the ATCs.
- Descriptor tables aligned to natural boundaries
- Support for individual supervisor and user translation trees
- Cache maintenance support
- Supervisor and write protection
- An external translation disable input signal ($\overline{\text{MDIS}}$)

13.6.1 MMU Programming Model

Figure 13.6 shows the programming model for the MMU. The MMU resources can only be accessed by the supervisor programs. The operations of MMUs are supported by the following registers, namely:

- Two root pointer registers.
- Four transparent translation registers.
- A status register.
- A control register.

The root pointer registers point to the address translation tree in the memory. The translation tree describes the logical-to-physical mapping for user and supervisor accesses. Both root pointers can point to the same address translation tree. A common tree structure may be utilized in support of a merged supervisor and user address space.

Each transparent translation register can identify a block of logical addresses to be used as physical addresses, that require no translations.

The status register contains resultant status information for a translation performed as a part of a PTEST instruction.

The translation control register contains two bits: one to enable/disable page address translation, and the other bit to select the page size.

13.6.2 Translation Table Organization

The MC68040 supports a three hierarchical level translation tree. The address translation tree consists of tables of descriptors. The first and second levels are known as pointer table descriptors, and can be either resident or invalid. The third level, called the page table descriptor, can be either resident, invalid, or

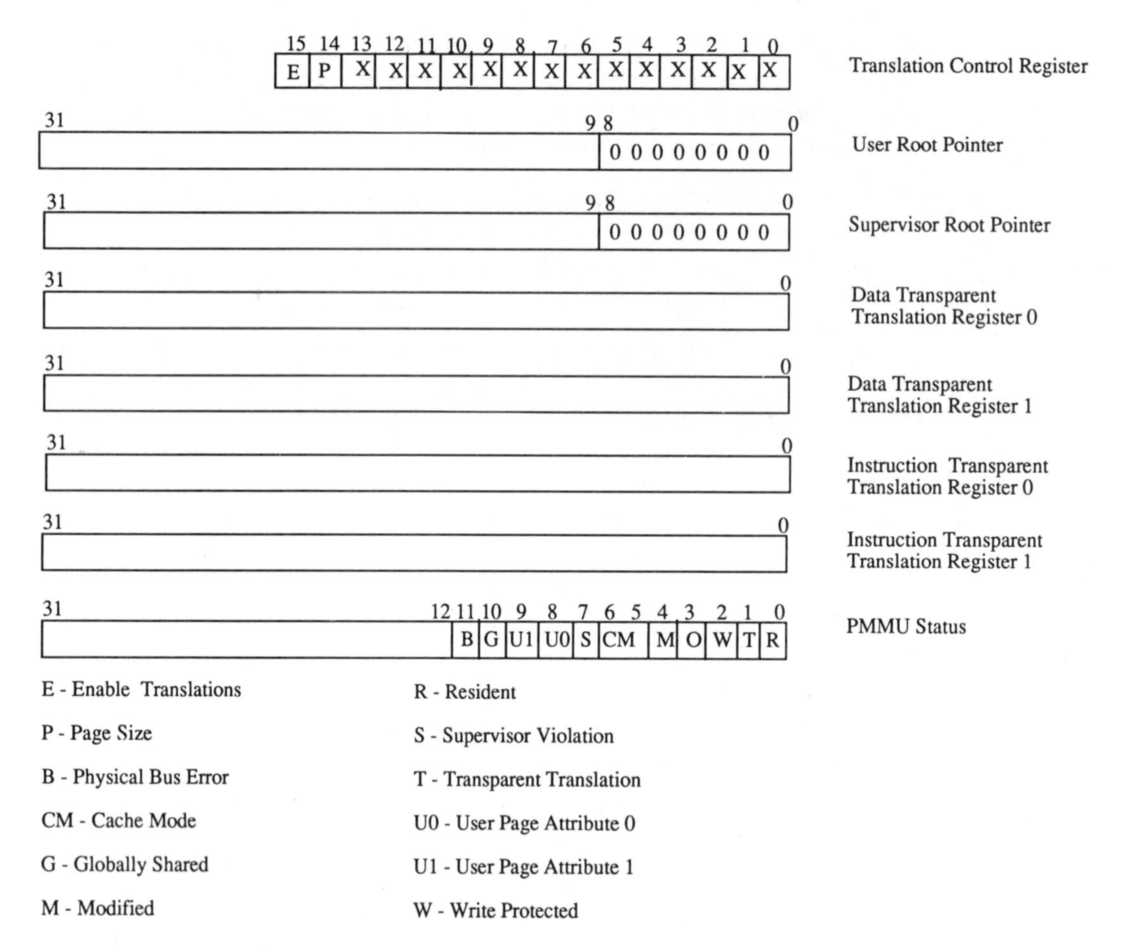

FIGURE 13.6. Memory management unit programming model.

indirect. A page descriptor specifies the physical address of a page frame in memory, which corresponds to the logical address of a page.

The current privilege mode is used to select the supervisor or user root pointer to access the translation tree. Each root pointer contains the base address of the first-level table of an address translation tree. The base address of each table is indexed by a field obtained from the logical address. The organization of the table index fields extracted from the logical address, including the translation table tree, are illustrated in Figure 13.7. The 7-bit TIA (Table Index A) field is used to index into the first-level pointer table. The first-level pointer table contains 128 pointer descriptors, to point to the second-level table. The next 7-bit TIB (Table Index B) selects one of the 128 pointer descriptors that points to a page table in the third-level table. The TIC (Table Index C) selects one of either 32 (for 8 Kbyte pages) or 64 (for 4 Kbyte pages) descriptors for the third-level page table. The descriptors in the page tables contain either a page descriptor for

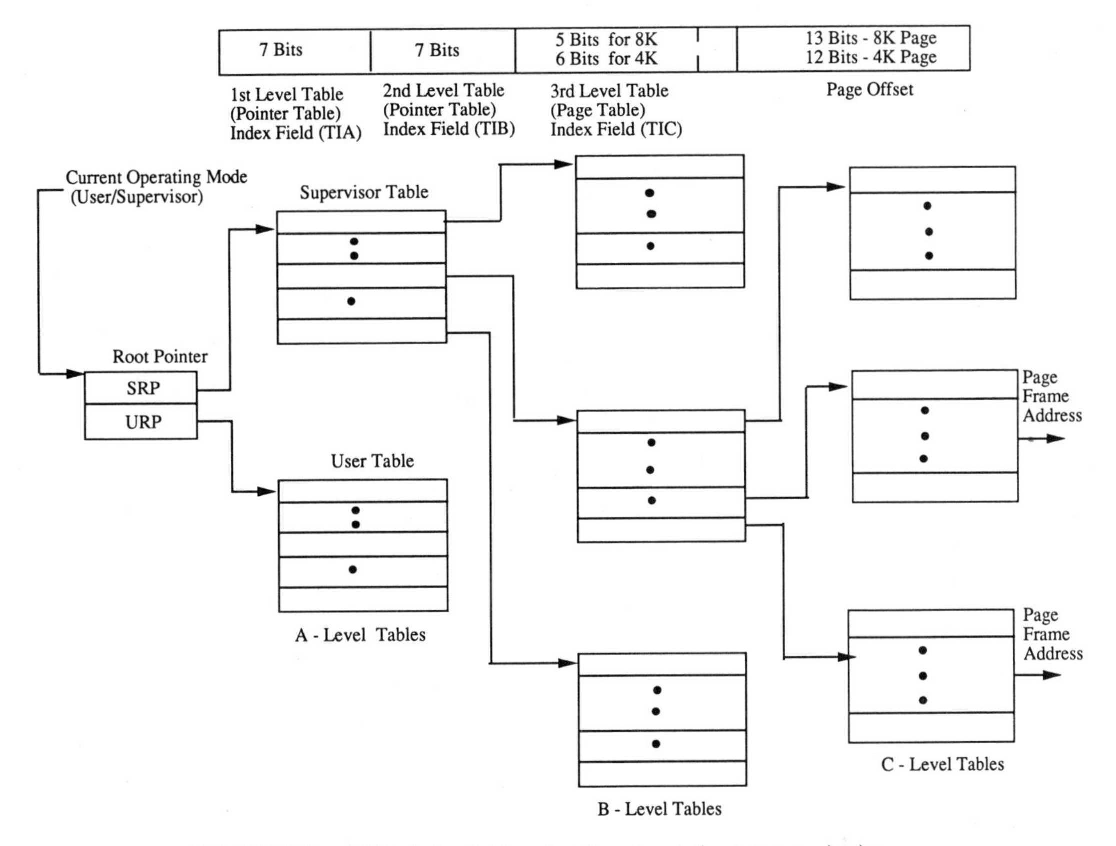

FIGURE 13.7. Table index fields and address translation tree organization.

the translation, or an indirect descriptor that points to a memory location containing the page descriptor.

13.6.3 Address Translation

The MMU translates the logical address to the physical address. The translation process is guided by the control information found in the MMU registers and in the translation table tree resident in the memory. In general, the operating system initializes the MMU registers and sets up the translation table as it loads a program for execution.

During normal program execution the address translation algorithm proceeds as follows:

1. It compares the logical address and the associated privilege mode against the parameters found in the transparent translation registers. If a match is found in one of the transparent translation registers, then it uses the logical address as a physical address for the access.

2. It compares the logical address and privilege mode contained in the tag portions of the entries in the ATC. If a match occurs, then it uses the corresponding physical address to access the memory.
3. When no transparent translation register is found, and no valid ATC entry is found, then it initiates a table search operation to obtain the corresponding physical address from the translation tree. If the required item is found, then it creates a valid ATC entry for the logical address and repeats the second step.

The general flow of the address translation algorithm is illustrated in Figure 13.8. The top branch of the flowchart applies to transparent translation. The bottom three branches are applicable to ATC (Address Translation Cache) translation. If a requested access is missed in the ATC, then a table search operation is initiated by the MMU. An ATC entry is created after the table search is finished, and the access is repeated. If the processor encounters an error after the hit in the ATC, or detects an invalid descriptor during the table search, then it aborts that access and initiates bus error exception processing.

However, if a write or indivisible read-modify-write access results during an ATC hit, but the page is found to be write-protected, then the MMU aborts the access and initiates exception processing. However, if the page is not write-protected, and the modify bit in the ATC is cleared, then the table search proceeds, and the modified bit is set for both the page descriptor in the memory and the ATC. The incomplete access is then reinitiated.

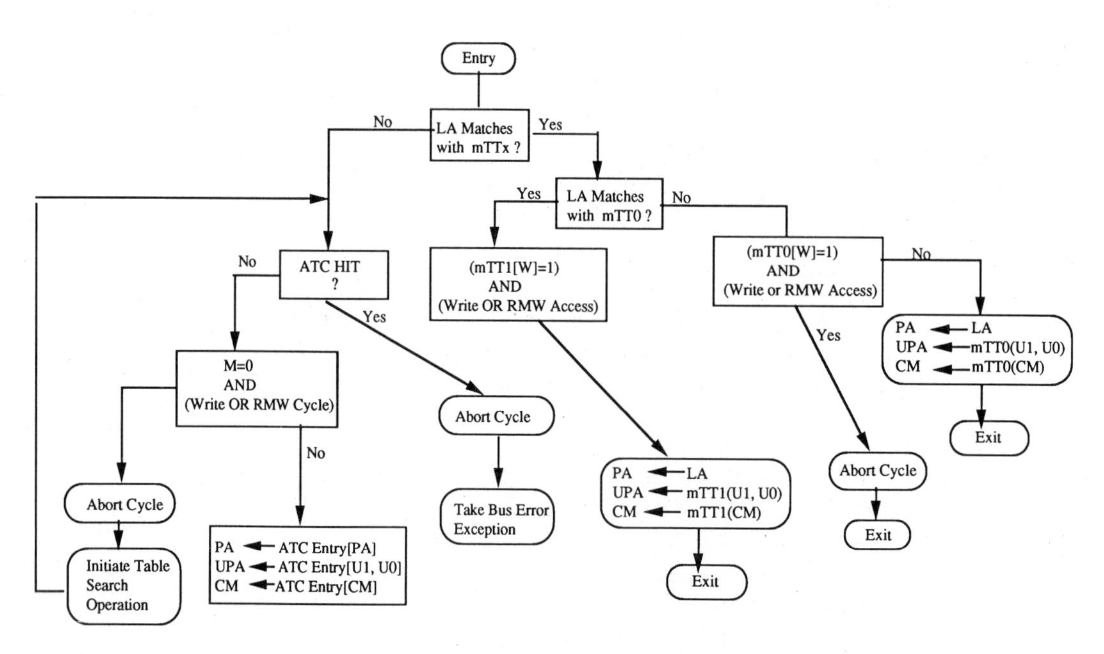

FIGURE 13.8. General address translation mechanism.

If, during the translation phase, the modified bit is set for a write and read-modify-write access to an unprotected page, and the table search for an entry is completed successfully, and none of the entries in the mTTx registers match, then the ATC provides the address translation for the access.

13.7 ADDRESS TRANSLATION CACHES (ATCS)

The MC68040 uses two ATCs, one for the instruction MMU and the other for the data MMU. ATCs are made up of four-way set-associative caches. They each preserve the 64 most recently utilized logical-to-physical address translation entries and associated page information. However, the actual tables are stored in the memory, and the translation tables for a program are loaded into memory by the operating system.

Figure 13.9 illustrates the organization of the ATC. The four immediate higher-order bits of the logical address, just above the page offset, are used as an index into the TLB's RAM. The tags are compared against the remaining upper bits of the logical address, including the FC2. If the tag matches, and the

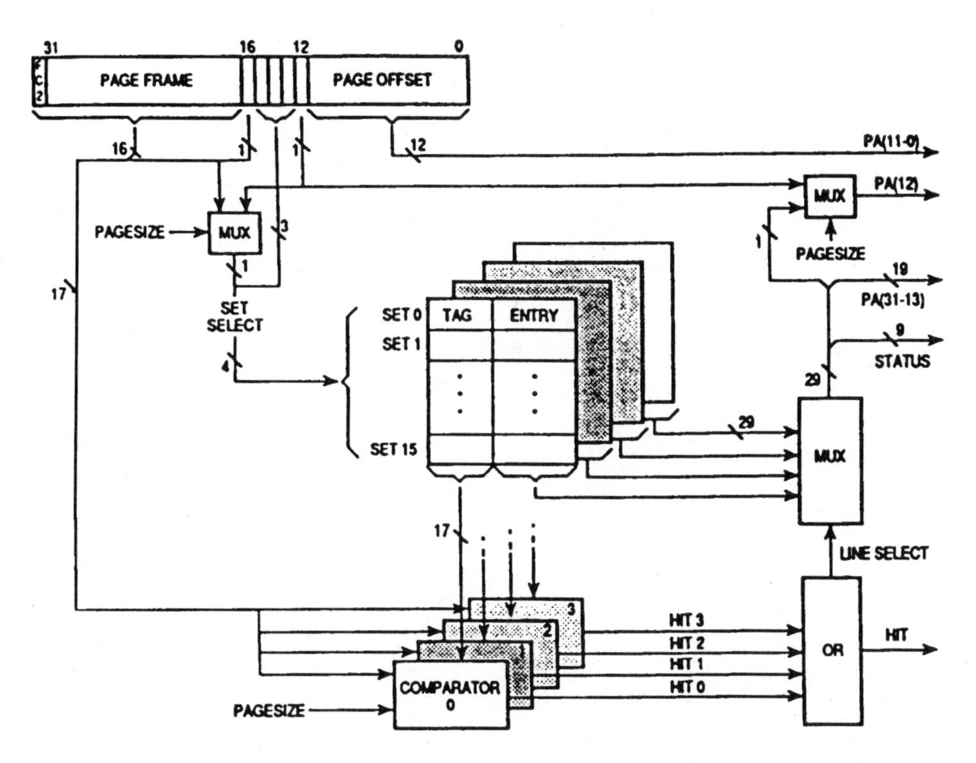

FIGURE 13.9. ATC organization. (Courtesy Motorola Inc.)

matched entry is found to be valid, then the corresponding entry is selected by the multiplexer to generate the physical address and related status information. If no match is found during the compare operation, then no mapping for this logical address exists in the TLB and a table-walk algorithm is initiated.

The logical-to-physical mapping will be guided by the selected page size. If a 4K page size is selected, then bit 12 of the logical address is used to access the ATC, and bit 16 is used by the tag comparators. The least significant 12 bits of the logical address are routed as ATC output. However, for a page size of 8K, the logical address bit 16 is used to access the ATC's memory, and is ignored by the tag comparators. In such a case, the physical address bit 12 is driven by the logical address bit 12.

The organization of the set within the ATC is shown in Figure 13.10. A set is composed of two fields, *tag* and *entry*.

- A tag contains a logical address and status information.
- An entry contains a physical address and related attribute information from a corresponding page descriptor.

When the ATC stores a new address translation, it replaces a valid entry. When all entries in an ATC set are valid, then the ATC selects a valid entry for replacement, using a pseudorandom replacement algorithm.

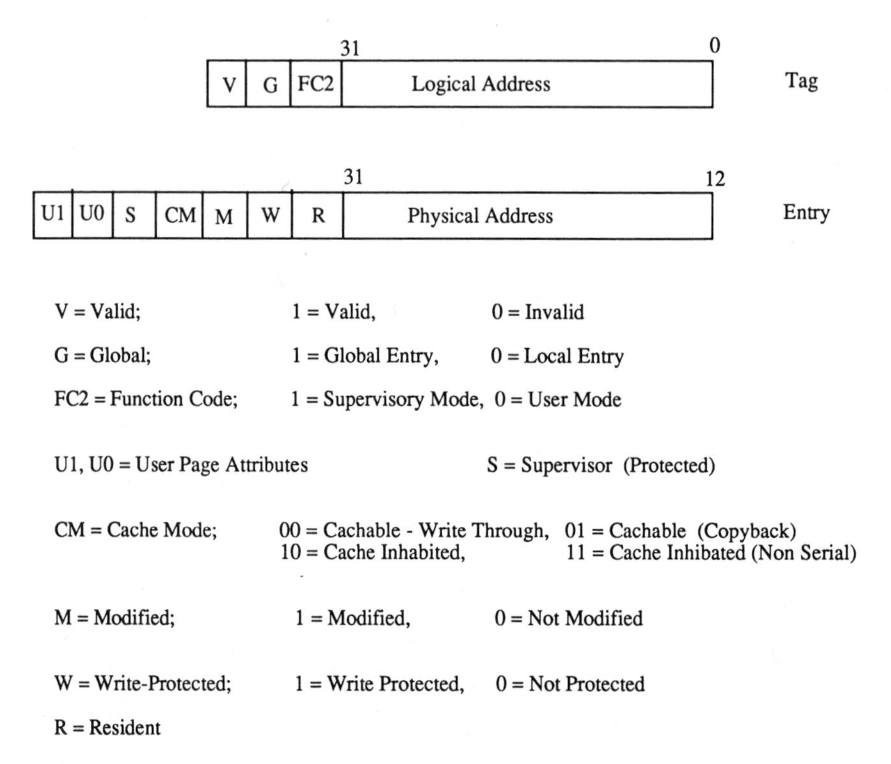

FIGURE 13.10. Organization of the set within the ATC.

13.7.1 Table Walking

A table-walk algorithm is invoked whenever one of the two TLBs (instruction or data) detects a miss. If both misses occur at the same time, then the data TLB gets priority over the instruction TLB. This is done because it is more efficient to complete existing instructions in the pipeline than to prefetch new ones. The general table-walk algorithm for a page size of 8K is shown in Figure 13.11. Here, the upper 23 bits of the appropriate root pointer are concatenated with the upper 7 bits of the logical address, and then multiplied by 4, to generate the physical address of the first-level descriptor. The second-level descriptor is then fetched and the upper 25 bits are then concatenated with the next 5 bits of the logical address. The result is then multiplied by 4, to form the page descriptor address. The page descriptor is then fetched. The status from the previous descriptor is merged with it, and the resulting entry is loaded into the appropriate TLB. The used and modified bits are then updated in the tables as necessary.

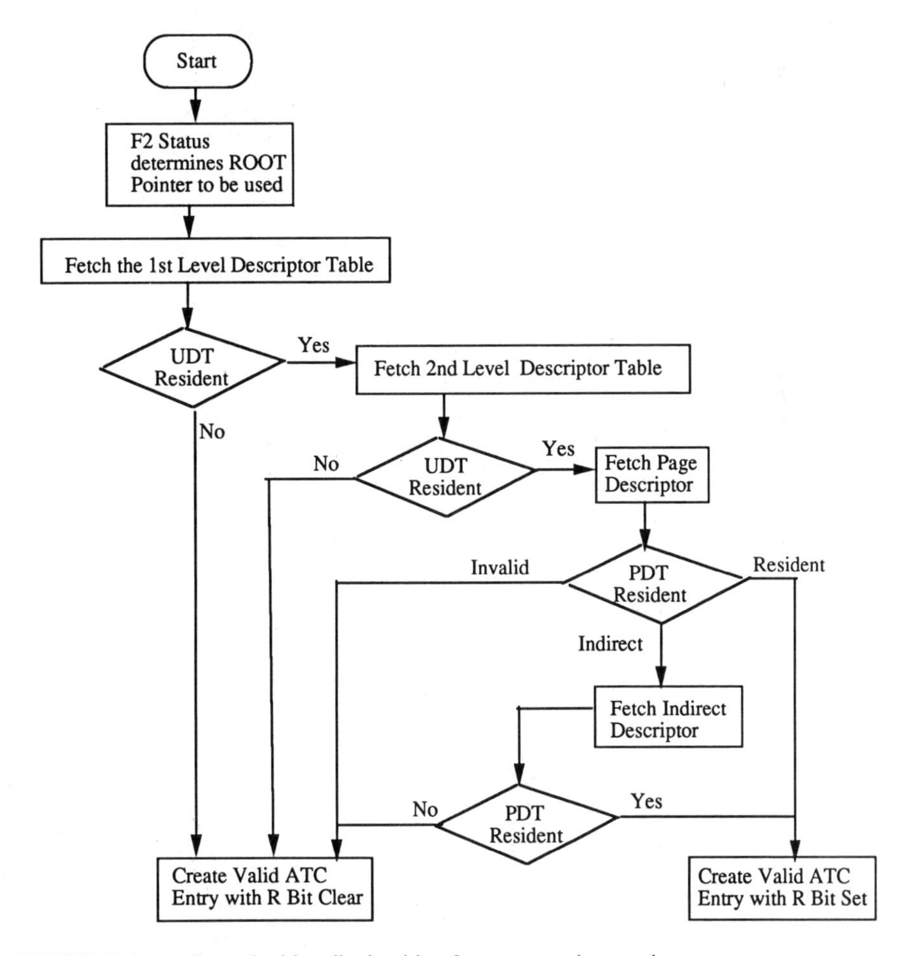

FIGURE 13.11. General tablewalk algorithm for page-search operation.

The table-walk algorithm for a 4K page is identical, with the exception that the page table is twice as large, and one more bit is concatenated to the second-level descriptor to form the address of the page descriptor. The descriptor tables must be aligned to their natural boundaries. This simplifies the hardware design of the table-walk logic, which relieves the operating system, and allows it to manage other functions.

13.7.2 Organization of the Translation Table

The MC68040 allows for various flexibilities to organize the translation tree. The processor allows the replacement of an entry in a page table with a pointer to an alternate entry. This capability allows the page frame to appear at various addresses within the logical address space of each task.

The MC68040 also allows a page or pointer table to be shared among tasks. This is achieved by placing the pointer to the shared table in the address translation tables of more than one task. For example, the upper table can have different write-protection settings, thus allowing different tasks to use the same memory area, with different write protection.

To optimize the memory utilization, the processor allows the paging of tables. Here, the tables that describe the resident set of pages needs to reside in main memory. Thus, when a task attempts to use an address that requires a nonresident table for address translation, then the processor creates a bus error exception. As the execution unit retries the bus access (after the table is loaded into the memory), it causes a table search to be initiated.

13.8 INTERNAL CACHES

We have noted that the MC68040 contains a 4 Kbyte on-chip instruction cache and a 4 Kbyte data cache. The caches are located in the physical address space. To facilitate data coherency, write-through or copy-back write features are supported on a page-by-page basis. Writes to memory are either buffered or deferred, based on the update policy. These caches are optimized for a uniprocessor environment, where the primary access into each cache is made from the CPU. DMA and shared memory support is provided with a bus snooper, which maintains cache-memory coherency, but does not guarantee cache-to-cache coherency.

13.8.1 Cache Organization

The cache organization of the MC68040 is shown in Figure 13.12. Each cache is organized to be a four-way set-associative, with 64 entries. Each set is organized as a group of four 16-byte lines, where a cache line contains:

- An address tag (TAG)
- Status field
- Four long-words of data

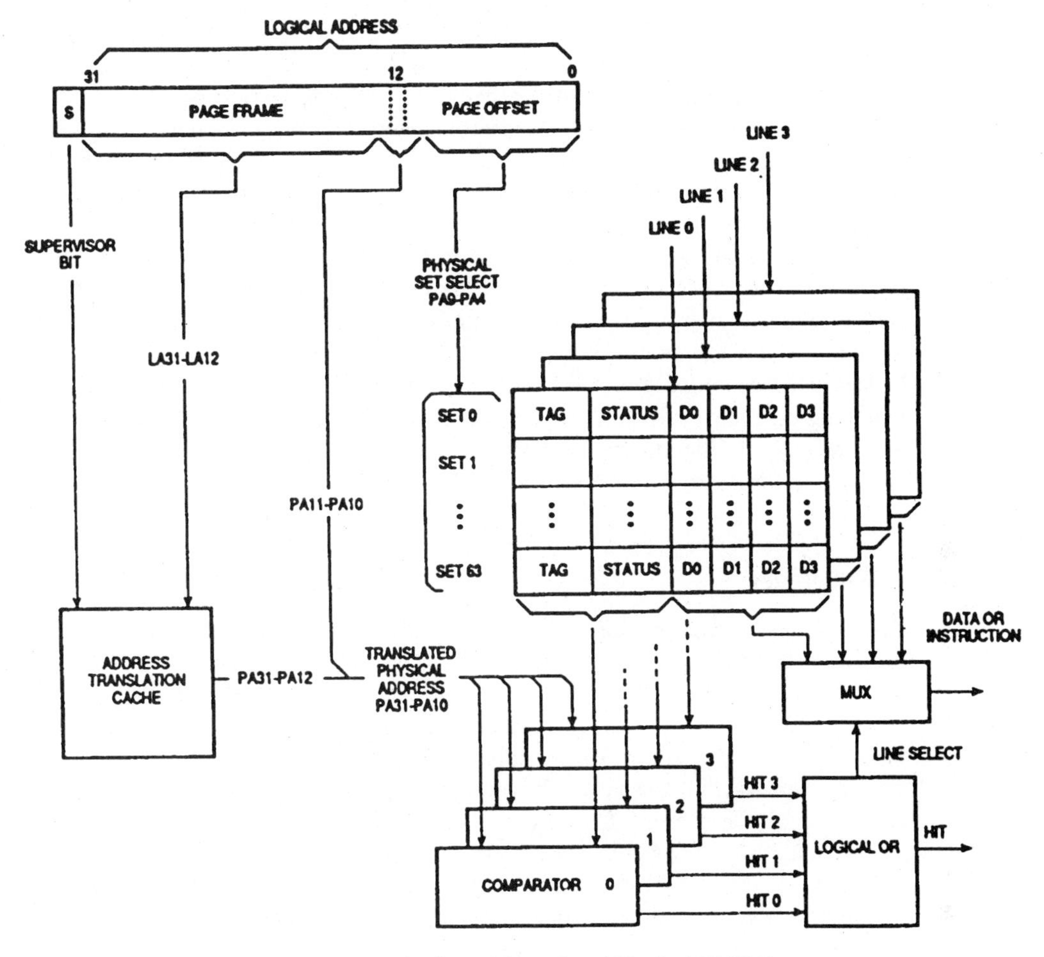

FIGURE 13.12. Organization of the cache within the MC68040. (Courtesy Motorola Inc.)

The tag consists of the upper 22 bits of the physical address.

The status field contains a *valid bit,* plus four additional bits that specify the *dirty status* for each long-word in the line.

Since entry validity is indicated on a line-by-line basis, an entry line must be loaded from system memory in order for the cache to update an entry. However, the burst mode access that successfully reads four long-words can also be cached.

The caches are associated with a physical address, where the set is indexed using the physical address bits PA9–PA4, generated by the MMU. These bits are used to index into the cache and select one of the 64 sets of four lines. The matched four tags are compared against the translated physical address bits PA31–PA12 from the MMU, including the two bits PA11–PA10. A hit is said to have occurred if any of the four tags matches, and the tag status is found to be valid and/or dirty.

If a hit takes place during read access, then the corresponding half-line from the selected *cache line* is multiplexed into the internal bus. However, for a write access, the data gets written into the cache line regardless of its size. Therefore, for a write operation, crossing line boundaries will require two accesses into the cache.

The address marked as noncachable in the TLB will bypass the cache (push matching dirty entries, invalidate clean entries) to allow for the support of I/O, DMA, etc. The operations performed under the TAS instruction that misses the cache are treated as noncachable, and will also bypass the cache. However, for locked instructions, a read will invalidate a matching valid entry and will push a matching dirty entry. In such a case, the read will bypass the cache. However, a write will update the matched entry and perform a write transfer to the memory, in order to maintain cache coherency.

13.8.2 Cache Coherency

The cache coherency problem arises if an alternate bus master references data that is already cached by the MC68040. The processor supports both write-through and copy-back memory update policies to solve this problem.

The MC68040 can monitor the system bus during bus transfers by other masters. The processor can update its internal caches if a write access hit occurs, or can intervene during the access by supplying dirty data, in the event of a read access hit. The watching of bus master's activity is known as bus snooping. The snooping is controlled by the external bus master via snoop control signals. Table 13.7 lists the meanings of requested snoop operations. It is quite possible that the processor and the bus snooper may access the cache simultaneously. In such a situation, the snoop controller will have priority.

Under the copy-back strategy, only one processor can cache data at one time. Thus, if one processor accesses shared data cached by another processor, then the slave can source data to the master without invalidating its own copy. However, the master must invalidate the slave processor's copy before caching the same data. As a result, the memory controller of the MC68040 must monitor the data transfer between the processors, and update memory with the transferred data. The memory update is necessary because the master processor is unaware of the source of data, and initially creates a valid cache line and loses its dirty status if the data was supplied by a snooping processor.

The coherency between the instruction cache and the data cache must be maintained by software, since the instruction cache does not monitor data accesses. Processor writes modify the code segments and access memory through the data memory unit. Because these data accesses are not monitored by the instruction cache, stale data occurs in the instruction cache if the corresponding data in memory is modified. This coherency problem can be resolved by invalidating any line in the instruction cache before writing to the corresponding line in memory.

13.8.3 Cache Operation

In the MC68040, the instruction and data caches operate independently in processing access requests from the on-chip integer unit. The instruction and data caches operate in different ways.

13.8.3.1 Instruction Cache Protocol

Instructions are prefetched by the MC68040. They are accessed sequentially from the memory until a change takes place in the program flow, or until the SR register reflects a status change (as a result of instruction execution). In such cases, instruction pipes are flushed and refilled automatically. The structure of the instruction cache lines are shown in Figure 13.13. The elements are:

- A tag
- A single valid bit
- Four long-words of data (128 bits)

Tag	V	LW3	LW2	LW1	LW0

Tag = 22-Bit Physical Address Tag Information

V = Line Valid Bit

LW_n = 32-Bit Data Entry

FIGURE 13.13. Organization of the instruction cache line.

During program execution, if a hit takes place during instruction prefetch, then the cache half-line selected by address bit PA3 is multiplexed into the internal instruction data bus. When an access is missed in the cache, the cache controller requests the required data from the memory. When the required data arrives into the data buffer, the cache controller copies the line into the cache and sets the valid bit. If all line entries are full in the cache, then a pseudorandom technique is used to select one of the four lines to be replaced. The cache controller then replaces the tag and data contents of the line with the updated information.

13.8.3.2 Data Cache Protocol

The data cache protocol is more complex than the instruction cache. The data cache controller supports two modes of operation, namely, write-through and copy-back. Both maintain memory coherency. Both modes are selectable on a page-by-page basis. The organization of the entries in the data cache are shown in Figure 13.14, where an entry consists of:

- The tag.
- Status information.
- Data information.

Tag	V	LW3	D3	LW2	D2	LW1	D1	LW0	D0

Tag = 22-Bit Physical Address Tag Information

D_n = Dirty Bit for Long-Word n

Invalid = $\overline{V}$

Valid = V AND $(\overline{D3 + D2 + D1 + D0})$

Dirty = V AND (D3 + D2 + D1 + D0)

LW_n = 32-Bit Data Entry

FIGURE 13.14. Organization of the data cache line.

- An encoded algorithm for defining the line state for the status bit combinations.

The cache controller fetches a new cache line from memory in the event of a read-miss and/or write-miss. The controller first attempts to select an invalid line within the selected set. In the event of a match, the selected set is updated with the tag address, with the status bit set to *valid* for the data memory. However, if all lines in the set are found to be *valid* or *dirty,* a pseudorandom replacement algorithm is invoked to select one of the four lines. The cache controller then updates the tag and data contents of the line with the new information.

In order to resolve cache coherency problems and to support DMA, including the shared memory environments, two simple cache coherency protocols are defined for data cache. The write-through protocol only uses the VALID and INVALID states, because all writes go to the bus as well as to the cache. However, the copy-back protocol uses all three states, since the state of each cache line is based on the requests from the processor and actions seen on the bus by the bus snooper.

The state transition diagram in Figure 13.15 illustrates the three possible states for a data cache line, including the transitions caused by either processor accesses or snooped accesses. Transitions are labeled with a capital letter, indicating the previous state, followed by a number indicating the specific case. The cases are described below:

Read miss. During a read miss, the cache will supply the data during a read operation initiated by the processor. However, no bus transaction will be performed and the state of the cache line will not change. Note that a read miss caused by a snooped external bus transfer will have no effect on the cache.

Write miss. A write to a copy-back page initiated by the processor is managed according to the selected caching mode. A miss will force the processor to perform a bus transaction, and bring in the required cache line to its cache and update the line. The processor then sets the status bit to the dirty state to reflect

Cache Activities	Current State		
	If Invalid Then	If Valid Then	If Dirty Then
CPU Read Miss	Read data line from memory, supply data to CPU and update cache	Read data line from memory, supply data to CPU and update the cache Remain in Current State	Buffer dirty Cache Line, Read new data line from memory, supply data to CPU and update cache, write the buffer's content to memory and move to the Valid State.
CPU read HIT	Not applicable	Supply data to CPU and remain in the current mode	Supply data to CPU and remain in the current mode
CPU Write Miss Cache Mode is Copyback	Read data line from memory into cache, write data to cache, set the dirty bits to modified long-word	Read data line memory into cache, write data to cache and set dirty bits Make a transition to Dirty mode	Buffer the dirty cache line, read new line from memory, write data to cache and modify the dirty bits, write buffered dirty data to memory Remain in current mode
CPU write Miss Copy mode = Copyback	Write data into memory Remain in the current mode	Write data into memory Remain in the current mode	Write data into memory Reamin in the current mode
CPU write HIT Copy mode = Copyback	Not applicable	Write data line into cache, set dirty bits of modified long-words Make a transition to Dirty mode	Write data line into cache, set dirty bits of modified long-words Remain in the current mode
CPU write HIT Copy mode = Writethrough	Not applicable	Write data into cache and Memory	Write data into cache, set dirty bits of modified long-words Remain in the current mode
Cache Invalidated	Remain in the current mode	Enter into the Invalid mode	Enter into the Invalid mode
Cache push	Remain in the current mode	Enter into the Invalid mode	Write dirty data into memory Enter into the Invalid mode
Alternate Master read HIT (Snoop Control = 01 - leave dirty)	Not applicable	Remain in the current mode	Inhibit memory and source data Remain in the current mode
Alternate Master read HIT (Snoop Control = 10 - Invalidated)	Not applicable	Enter into the Invalid mode	Inhibit memory and source data Enter into the Invalid mode
Alternate Master write HIT (Snoop Control = 10 - Invalidate/Size = Line)	Not applicable	Enter into the Invalid mode	Enter into the Invalid mode
Alternate Master write HIT (Snoop Control = 01 - Sink data and size ≠ Line)	Not applicable	Enter into the Invalid mode	Inhibit memory and sink data, set dirty bits of modified long-words Remain in the current mode

FIGURE 13.15. Status of the data cache transitions. (Courtesy Motorola Inc.)

the change. However, a write miss for a write-through page will cause a write to the bus, but will not allocate the line in the cache.

Read hit. In the event of a hit, the cache supplies data to the processor regardless of its mode of operation during the write (i.e., write-through or copy-back). In such cases, bus transactions are not performed and the status of the cache lines remains unaltered. However, a read hit caused by a snooped external bus transfer will have no effect on the cache.

Write hit. For a page selected as a write-through class, a write initiated by the processor (which hits in the cache) will cause the processor to update the cache line, and perform a memory write transfer to enforce coherency. Here, the cache line will end up in the dirty state.

On the other hand, if the page access is marked as a copy-back, the cache controller updates the cache line and sets the dirty bit of the appropriate long-words in the cache line. However, the processor will not execute a bus write cycle to update the memory, and the cache line will be marked as the dirty state.

An alternate bus master can drive the snoop control signals for a write, with an indication to the MC68040 that it should sink the data, if the access caused a hit in the cache. A snooped line write or a snooped write that hits a valid line always causes the corresponding cache line to be invalidated. For a snoop write of size byte, word, or long-word that hits a dirty line, the processor inhibits the memory and responds to the alternate bus master as a slave, sinking the write data. Data received from the alternate bus master is then written to the appropriate long-word in the cache line, and the dirty bit is set for that entry. However, if the snoop control pins indicate that a matching cache line should be marked as invalid, then the line will be invalidated.

13.9 EXCEPTION PROCESSING

The MC68040's exception model is different from the rest of the MC68000 family of processor's exception models. This processor does not support instruction continuation. The rest of the model remains the same, and is composed of four steps. The steps are the same as for the MC68020 and MC68030.

The MC68040 supports a 1024-byte vector table, made up of 256 exception vectors. Vector number 2 is called the Access Fault. Vector number 13 is not defined for this processor. Vector number 55 has been added to the table and is called the FP Unimplemented data type. Vector numbers 56 through 58 are not used by the MC68040.

All exception vectors are located in the supervisory address space, and are accessed via data references. However, only the initial reset vector is memory mapped during initialization. After the initialization, the vector table can be placed in any areas of memory using the VBR as a base address. It can even be dynamically relocated (a prerequisite for a virtual operating system) for each task executed by an operating system.

13.9.1 Stack Frames

We have seen that stack frames vary from processor to processor. The MC68040 supports five stack frames during exception processing. They are:

- Four-word stack frame
- Six-word stack frame
- Four-word throwaway stack frame
- Floating-point postinstruction stack frame
- Access error stack frame

The format and the usage of the first three stack frames are common to all members of the MC68000 family of processors. The other two are specific to the MC68040.

The floating-point postinstruction stack frame is generated as a part of an illegal or unimplemented instruction. When the MC68040 has identified an illegal or unimplemented instruction, it initiates exception processing instead of attempting to execute the instruction. The processor saves the illegal or unimplemented instruction vector offset, current PC, and a copy of the SR on the stack.

The saved value of the PC is the address of the illegal or unimplemented instruction. After the execution of the RTE, instruction execution resumes at the address contained in the instruction vector.

Exception processing for unimplemented floating-point instructions is slightly different. A long stack frame is created that contains the effective address of the memory operand. The stacked PC points to the next instruction to be executed after the unimplemented floating-point instruction, and the address of the unimplemented floating-point instruction is available to the exception handler in the FSAVE instruction stack frame.

An FSAVE instruction is executed to save the current floating-point internal state for context switches and floating-point exception handling. When an FSAVE is executed, the processor waits until the FPU either completes execution of all current instructions or is unable to proceed due to a pending exception.

An access fault exception frame is generated during the access fault. The access fault occurs when a data access or instruction prefetch access faults due to either an external bus error or an internal address translation fault. External logic can abort a bus cycle and signal a bus error by asserting the $\overline{\text{TEA}}$ input signal. An internally generated access fault is detected when the data MMU or instruction MMU detects that an address translation is not possible because the page is write-protected, supervisor-only, or nonresident. During the processing of the access fault, MC68040 saves its foot-print into the access error stack frame. The information saved on the stack is sufficient to identify the source of the fault, complete pending write-backs, and recover from errors.

13.9.2 Exception Priorities

In the event of multiple exceptions, the MC68040 processes them according to fixed priorities. The priorities are listed in Table 13.12, and are arranged as groups. Members within a group are assigned priorities from 0 through 7, with 0 as the highest priority. We have already indicated that the exception processing sequence of the MC68040 is different from its predecessors, due to the notion of the *restart* exception model.

The access error exception type is incorporated within the processor to support the on-chip MMU units. This group of exceptions creates a type $7 stack frame to hold status information that can indicate a pending trace, namely a floating-point post-instruction or unimplemented floating-point instruction. During the execution of RTE instruction, the processor checks the status bits for one of these pending exceptions. If so, then the RTE changes the access error stack frame to match against the pending exceptions and fetches the related exception vector. Instruction execution then resumes in the new exception handler. However, the floating-point trace exception handler must check the trace bits on the stack and must invoke the trace handler directly.

Similarly for group-3 and group-4 traces, exceptions are not reported and the exception handler for these exception conditions must check for the trace condition.

TABLE 13.12. Exception Priority for the MC68040.

Group/ Priority	Exception and Relative Priority	Attributes
0	Reset	Aborts all processing without saving the old context.
1	Data Access error (ATC fault or Bus error)	Aborts current instructions—can have pending trace. Floating Point Unit—post-instruction and unimplemented instructions.
2	Floating-point pre-instruction	Exception processing begins before current floating-point instruction is completed. Instruction is restarted upon return from exception.
3	BKPT #n, CHK, CHK2, Divide-by-Zero, FTRAPcc, RTE, TRAP #n, TRAPV	Exception processing is an integral part of the instruction execution.
	Illegal instructions: Unimplemented line-A and line-F, Privilege violation	Exception processing begins before instruction is executed.
	Unimplemented floating-point-instruction	Exception processing is initiated after the memory operands are fetched and before instruction is executed.
4	Floating-point post-instruction	Exception processing is initiated when FMOVE instruction and previous exception processing is completed.
5	Address error	Exception is registered after all previous instructions and asociated exceptions are completed.
6	Trace	Exception processing is initiated when current instruction or previous exception processing is completed.
7	Instruction access error due to ATC fault or bus error	After all previously reported instructions and associated exceptions are completed.
8	Interrupt	Exception processing is initiated after the current instruction or previous exception processing is completed.

The remaining exceptions perform like the exceptions of the MC68020 and MC68030.

13.10 PROGRAMMING CHANGES FOR MC68040 TO MAINTAIN DOWNWARD COMPATIBILITY

Systems designers must incorporate certain modifications in their operating systems and application programs to run them under the MC68040. This is necessary because the processor does not implement all of the features of the MC68851 and MC68881/68882. Again, the MC68040 has added two larger-size caches, which are absent in previous processors. From a software point of view, systems composed of the MC68020/MC68030, MC68881/68882, and MC68851 are not identical.

The O.S. designed to run on the MC68030 and MC68020 must be modified to run successfully on the MC68040. The main impact will show up in a virtual memory system because of the dual internal MMUs and dual caches resident on the processor.

The bus error recovery routine must be changed to reflect the new architectural enhancement. In addition, the software emulation routine must be included for the unsupported floating-point and data types.

13.10.1 Modifications to the MMU Software

The MMU software designed for the MC68030 must be changed to fit the requirements of the MC68040. The O.S. must use a 4 Kbyte- or 8 Kbyte-size page. The page size must be modified in virtual memory software, I/O, file system routines, and various subsystems designed to operate under a page size other than 4/8 Kbytes.

The O.S. needs to configure the MMU for a three-level address translation table. The software needs to reflect the new MMU's instruction and register formats. One should also change the routines that flush entries from ATC, bus error handler, etc. The system should also modify various exception routines to reflect the various changes incorporated within the MC68040.

13.10.2 On-Chip Cache Management

It is the O.S. that enables the caches and maintains cache coherency. The system needs to make changes to the physical rather than logical caches. For optimal performance, the O.S. should enable the copy-back mode without bus snooping [Lindhorst 91].

The new supervisory instruction CINV invalidates cache entries and selectively invalidates information in the cache. The cache entries can be saved in memory utilizing the CPUSH' instruction, which updates the dirty entries into memory and labels them invalid. This is applicable for data caching during the copy-back mode. However, enabling the copy-back mode of the data cache becomes very difficult, especially if the system crashes when it gets into a cache overrun.

The physical caches of the MC68040 are more easily manageable than the logical caches of the MC68030. On the MC68040, the only time that the O.S. needs to push the cache is after writing instructions or while managing the DMA. Rather than updating the data cache after writing instructions, the O.S. can be configured to place the data cache in write-though mode for pages containing instructions.

13.10.3 Page Fault Exception Processing

The MC68040 has incorporated various changes to the exception processing models of the MC68020 and MC68030. Thus the O.S. must be modified to manage the new exception model of the MC68040. The access-error/page-fault exception frame defines three writes that may have to be completed by system software after the cause of the fault has been corrected. However, the implementation of write-backs can be tricky, because they may themselves cause page faults. One solution is to use the PTEST instruction to check the write-back addresses prior to the execution of actual write-back. Here, if a fault is detected, then the fault handler must be invoked to repair it before the actual write-back takes place.

The other approach is to execute the write-back, and if an error occurs, then re-execute the access-error handler to correct the problem. As a precaution, we can use NOP after a write instruction, to manage the pipeline and guarantee that the write will be in the pipe alone when it is executed. This will guarantee that there are no other pending write-backs in the event of a fault.

13.10.4 Floating-Point Emulation

The floating-point operation of the MC68040 is object-code compatible with the MC68882. This is now supported by on-chip hardware for various common operations, and off-chip emulation software for the remaining operations. The hardware-implemented instructions include data movement, comparison and branching, addition/substraction, multiplication/division, square root, compilation of absolute values, and negations. The denormalized and packed-decimal floating-point format are not supported by the MC68040. The unsupported instructions must be emulated in software.

The floating-point software emulator is invoked whenever unsupported instructions are encountered. Since no separate exception vectors are available, the illegal instruction exception processing routine is generally invoked to manage such situations. The software emulator needs to be invoked for the underflow exception and the new unsupported-data-type exception that handles denormalized and packed-decimal representations.

The required emulation software can be procured from Motorola, or written independently. In the case of a UNIX system, such emulation software generally resides entirely in kernel.

13.10.5 Application Software

The MC68040 has been designed to be object-code compatible with the previous members of the MC68000 family. However, the application programmer should be aware of the implications of various new enhancements made to this processor.

Self-modifying code can create serious problems when it is used with the cache enabled. Specifically, the separation of the data and instruction caches have all but ensured that the older self-modifying MC68000 codes will fail, because all data writes flow through the data cache, and if instructions are already in the instruction cache at the same address as the new instructions, then the execution stream will not see all or part of the new instructions. Even without the instruction cache, the copy-back data cache may not write the new instructions to memory before they are fetched for execution. This cache coherency problem is important for dynamic loaders and hardware emulators. A possible solution to this problem is to push and invalidate the cache at the particular addresses where they affect instruction.

The MOVE16 is an important new instruction added to the user mode. It moves 16 bytes of data at a time. It is efficient in operating between aligned 16-byte memory locations. The list, table, and I/O moves can benefit from this instruction.

13.10.6 Memory Utilization

In the MC68040, the data alignment is a concern, due to the new caches and their management. For the purpose of efficiency, the data should be contained within one cache line, so that the entire piece of data can be procured within one bus access instead of two (which is expensive, performance-wise). For this reason, we are to move to natural alignment of that data block.

Branches are optimized for 'branch taken,' which is implemented in two clocks, over the three clocks found in the previous members. So, the loops and other control structure should be replaced by the 'branch taken.' Again, branches to instructions on 8-byte boundaries that miss the instruction cache benefit from a one-cycle overhead over branches to misaligned instructions. Thus, for maximum efficiency, branch targets should be aligned on an 8-byte boundary.

To take full advantage of the elegant design of the MC68040, O.S. and compiler programmers need to take great care in using the various features of this processor. It even helps application programmers to be aware of the details of the MC68040.

Appendix A

Number and Character System

Numbering is a fundamental concept. It is learned in childhood, and from then on it is considered physical law. However, numbering is not a physical law, it is a concept. We are most familiar with the *decimal* numbering system because it is generally taught in kindergarten and we use it in our daily lives. The decimal number system is based on ten using the digits 0 to 9. However, computers are designed to operate with *binary numbers.* Due to the cumbersome nature of binary numbers, humans generally use *hexadecimal* numbers when dealing with computers.

In this appendix, we will summarize the fundamentals behind the numbering systems, and cover some basic concepts.

THE DECIMAL SYSTEM

The base ten system represents a numbering system using powers of ten. Each power is weighted by an integer between 0 and 9. For example

425 = 4 (10) **2 + 2 (10) **1 + 5 (10) **0

The general format for an integer number N is given by

$$\begin{aligned} N_i &= a_i \quad a_{i-1} \qquad a_2\, a_1\, a_0 \\ &= a_i\,(10)^i + a_{i-1}\,(10)^{i-1} \ldots + a_2\,(10)^2 + a_1\,(10)^1 + a_0\,(10)^0 \end{aligned}$$

where $0 \leq a_k \leq 9$

Similarly, a fraction can be derived by using negative powers of ten. Thus

$$0.526 = 5\,(10)^{-1} + 2\,(10)^{-2} + 6\,(10)^{-3}$$

In general, the decimal fraction, Nf, is given by

$$\begin{aligned} Nf &= a_{-1}\, a_{-2} \quad\; a_{-j+1}\, a_{-j} \\ &= a_{-1}\,(10)^{-1} + a_{-2}\,(10)^{-2} \qquad + a_{-j+1}\,(10)^{-j+1} + a_{-j}\,(10)^{-j} \end{aligned}$$

The general format for a decimal number N is given by

$$N = N_I + N_F = \sum_{k=-j}^{i} a_k\,(10)^k \quad ;\, 0 \le a_k \le 9$$

BINARY NUMBER SYSTEM

Microprocessors, and in fact all digital systems, operate as binary devices. The binary number system has a base of 2, and relies on two digits, 0 and 1. The formulas previously shown can be used to represent binary numbers, with the following changes: the base is 2 instead of 10, and a is either 0 or 1. The addition and multiplication tables for them are as follows:

$0 + 0 = 0$
$0 + 1 = 1 + 0 = 1$
$1 + 1 = 10$
$0 \cdot 0 = 0$
$0 \cdot 1 = 0$
$1 \cdot 0 = 0$
$1 \cdot 1 = 1$

To be able to represent positive and negative numbers, the Most Significant Bit (MSB) (the initial numeral in a sequence) in a byte, word, or long word is reserved for the sign bit. 1 in the MSB denotes a negative number. In the case of a byte:

The number 127 would be represented as	0111 1111
The number −127 is represented as	1000 0001

We can move from positive to negative and from negative to positive by using the Complement and Increment technique.

HEXADECIMAL NUMBERS

The base of a hexadecimal number system is 16, and utilizes the following sixteen symbols: 0, 1, 2, 3, 4, 5, 6, 7, 8, 9, A, B, C, D, E, and F. The relationships among the decimal, binary, and the hexadecimal systems are listed in the following table.

Decimal	Binary	Hexadecimal	Decimal	Binary	Hexadecimal
0	0	0	8	1000	8
1	1	1	9	1001	9
2	10	2	10	1010	A
3	11	3	11	1011	B
4	100	4	12	1100	C
5	101	5	13	1101	D
6	110	6	14	1110	E
7	110	7	15	1111	F

CONVERSION FROM A DECIMAL NUMBER TO A BINARY OR HEXADECIMAL NUMBER

It is often necessary to convert decimal numbers to binary or hexadecimal numbers. To convert a decimal number to its equivalent binary number, the following procedure may be used:

1. Get N, the decimal number to be converted.
2. If N is odd, then subtract 1 from it and write 1, else write 0. Go to Step 3.
3. Get a new value of N by dividing it by 2.
4. If N is greater than 1, then go back to Step 2.
5. If $N = 1$, then write 1. The digit written is the binary equivalent of the decimal number, where the first digit written is the least significant digit, and the last written digit is the most significant digit.

To convert the binary number to its hexadecimal equivalent, group the digits into 4 bits each and convert them into their hex equivalent utilizing the above table.

ASCII CHARACTER CODE

The standard binary code for alphanumeric characters is the ASCII code. The seven bits of the code are identified by b1 through b7. Representations of 128 characters are listed below.

b4b3b2b1	b7b6b5							
	000	001	010	011	100	101	110	111
0000	NULL	DEL	SP	0	@	P		p
0001	SOH	DC1	!	1	A	Q	a	q
0010	STX	DC2	"	2	B	R	b	r
0011	ETX	DC3	#	3	C	S	c	s
0100	EOT	DC4	$	4	D	T	d	t
0101	ENQ	NAK	%	5	E	U	e	u
0110	ACK	SYN	&	6	F	V	f	v

(continued)

b4b3b2b1	b7b6b5							
	000	001	010	011	100	101	110	111
0111	BEL	ETB	’	7	G	W	g	w
1000	BS	CAN	(	8	H	X	h	x
1001	HT	EM	)	9	I	Y	i	y
1010	LF	SUB	*	:	J	Z	j	z
1011	VT	ESC	+	;	K	[	k	{
1100	FF	FS	,	<	L	\	l	|
1101	CR	GS	-	=	M	]	m	}
1110	SO	RS	.	>	N		n	~
1111	SI	US	/	?	O	_	o	DEL

Appendix B

Summary of Instruction Set

1. Notation for operands:

PC—Program counter
SR—Status register
V—Overflow condition code
Immediate Data—Immediate data from the instruction
Source—Source contents
Destination—Destination contents
Vector—Location of exception vector
+ inf—Positive infinity
– inf—Negative infinity
<fmt>—Operand data format: byte (B), word (W), long (L), single (S), double (D), extended (X), or packed (P)
FPm—One of eight floating-point data registers (always specifies the source register)
FPn—One of eight floating-point data registers (always specifies the destination register)

2. Notation for subfields and qualifiers:

<bit> of <operand>—Selects a single bit of the operand
<ea>{offset:width}—Selects a bit field
(<operand>)—The contents of the referenced location
<operand>10—The operand is a binary coded decimal; operations are performed in decimal
(<address register>)—The register indirect operator
–(<address register>)—Indicates that the operand register points to the memory

(<address register>)+—Location of the instruction operand—the optional mode qualifiers are −, +, (d), and (d,ix)

#xxx or #<data>—Immediate data that follows the instruction word(s)

3. Notations for operations that have two operands, written <operand> <op> <operand>, where <op> is one of the following:

➡—The source operand is moved to the destination operand

⬅➡—The two operands are exchanged

+—The operands are added

−—The destination operand is subtracted from the source operand

×—The operands are multiplied

÷—The source operand is divided by the destination operand

<—Relational test, true if source operand is less than destination operand

>—Relational test, true if source operand is greater than destination operand

V—Logical OR

⊕—Logical exclusive OR

Λ—Logical AND

shifted by, rotated by—The source operand is shifted or rotated by the number of positions specified by the second operand

4. Notation for single-operand operations:

~<operand>—The operand is logically complemented

<operand>sign-extended—The operand is sign-extended—all bits of the upper portion are made equal to the high order bit of the lower portion

<operand>tested—The operand is compared to zero and the condition codes are set appropriately

5. Notation for other operations:

TRAP—Equivalent to Format/Offset Word ➡ (SSP); SSP–2 ➡ SSP; PC ➡ (SSP); (SSP–4 ➡ SSP; SR ➡ (SSP); SSP–2 ➡ SSP; (vector) ➡ PC

STOP—Enter the stopped state, waiting for interrupts

If <condition> then—tions <operations> else <operations> The condition is tested. If true, the operations after "then" are performed. If the condition is false and the optional "else" clause is present, the operations after "else" are performed. If the condition is false and else is omitted, the instruction performs no operation. Refer to the Bcc instruction description as an example.

TABLE B.1.

MC68020, MC68030, and MC68040 Instruction Set Extensions		Applies To		
Instruction	Notes	MC68020	MC68030	MC68040
Bcc	Supports 32-Bit Displacements	✓	✓	✓
BFxxxx	Bit Field Instructions (BCHG, BFCLR, BFEXTS, BFEXTU, BFFFO, BFINS, BFSET, BFTST)	✓	✓	✓
BKPT	New Instruction Functionally	✓	✓	
BRA	Supports 32-Bit Displacements	✓	✓	✓
BSR	Supports 32-Bit Displacement	✓	✓	✓
CALLM	New Instruction	✓		
CAS, CAS2	New Instructions	✓	✓	✓
CHK	Supports 32-Bit Operands	✓	✓	✓
CHK2	New Instruction	✓	✓	✓
CINV	Cache Maintenance Instruction			✓
CMPI	Supports Program Counter Relative Addressing Modes	✓	✓	✓
CMP2	New Instruction	✓	✓	✓
CPUSH	Cache Maintenance Instruction			✓
cp	Coprocessor Instructions	✓	✓	
DIVS/DIVU	Supports 32-Bit and 64-Bit Operands	✓	✓	✓
EXTB	Supports 8-Bit Extend to 32-Bits	✓	✓	✓
FABS	New Instruction			✓
FADD	New Instruction			✓
FBcc	New Instruction			✓
FCMP	New Instruction			✓
FDBcc	New Instruction			✓
FDIV	New Instruction			✓
FMOVE	New Instruction			✓
FMOVEM	New Instruction			✓
FMUL	New Instruction			✓
FNEG	New Instruction			✓
FRESTORE	New Instruction			✓
FSAVE	New Instruction			✓
FScc	New Instruction			✓
FSQRT	New Instruction			✓
FSUB	New Instruction			✓
FTRAPcc	New Instruction			✓
FTST	New Instruction			✓
LINK	Supports 32-Bit Displacement	✓	✓	✓
MOVE16	New Instruction			✓
MOVEC	Supports New Control Registers	✓	✓	✓
MULS/MULU	Supports 32-Bit Operands	✓	✓	✓
PACK	New Instruction	✓	✓	✓
PFLUSH	MMU Instruction		✓	✓
PLOAD	MMU Instruction		✓	
PMOVE	MMU Instruction		✓	
PTEST	MMU Instruction		✓	✓
RTM	New Instruction	✓		
TST	Supports Program Counter Relative Addressing Modes	✓	✓	✓
TRAPcc	New Instruction	✓	✓	✓
UNPK	New Instruction	✓	✓	✓

TABLE B.2.

Opcode	Operation	Syntax
ABCD	Source$_{10}$ Destination$_{10}$ X ➧ Destination	ABCD Dy, Dx ABCD – (Ay), – (Ax)
ADD	Source Destination ➧ Destination	ADD <ea>,Dn ADD Dn,<ea>
ADDA	Source + Destination ➧ Destination	ADDA <ea>,An
ADDI	Immediate Data + Destination ➧ Destination	ADDI #<data>,<ea>
ADDQ	Immediate Data + Destination ➧ Destination	ADDQ #<data>,<ea>
ADDX	Source + Destination + X ➧ Destination	ADDX Dy,Dx ADDX –(Ay),–(Ax)
AND	SourceΛDestination ➧ Destination	AND <ea>,Dn AND Dn,<ea>
ANDI	Immediate DataΛDestination ➧ Destination	ANDI #<data>,<ea>
ANDI to CCR	SourceΛCCR ➧ CCR	ANDI #<data>,CCR
ANDI to SR	If supervisor state the Source. SR ➧ SR else TRAP	ANDI #<data>,SR
ASL,ASR	Destination Shifted by <count> ➧ Destination	ASd Dx,Dy ASd #<data>,Dy ASd <ea>
Bcc	If (condition true) then PC + d ➧ PC	Bcc <label>
BCHG	~(<number> of Destination) ➧ Z; ~(<number> of Destination) ➧ <bit number> of Destination	BCHG Dn,<ea> BCHG #<data>,<ea>
BCLR	~(<bit number> of Destination) ➧ Z; 0 ➧ <bit number> of Destination	BCLR DN,<ea> BCLR #<data>,<ea>
BFCHGR	~(<bit field> of Destination) ➧ <bit field> of Destination	BFCHG <ea>{offset:width}
BFCLR	0 ➧ <bit field> of Destination	BFCLR <ea>{offset:width}
BFEXTS	<bit field> of Source ➧ Dn	BFEXTS <ea>{offset:width},Dn
BFEXTU	<bit offset> of Source ➧ Dn	BFEXTU <ea>{offset:width},Dn
BFFFO	<bit offset> of Source Bit Scan ➧ Dn	BFFFO <ea>{offset:width},Dn
BFINS	Dn ➧ <bit field> of Destination	BFINS Dn,<ea>{offset:width}
BFSET	1s ➧ <bit field> of Destination	BFSET <ea>{offset:width}
BFTST	<bit field> of Destination	BFTST <ea>{offset:width}
BKPT	Run breakpoint acknowledge cycle; TRAP as illegal instruction	BKPT #<data>
BRA	PC + d ➧ PC	BRA <label>
BSET	~(<bit number> of Destination ➧ Z; 1 ➧ <bit number> of Destination	BSET Dn,<ea> BSET #<data>,<ea>
BSR	SP – 4 ➧ SP; PC + d ➧ PC	BSR <label>
BTST	–(<bit number> of Destination) ➧ Z;	BTST Dn,<ea> BTST #<data>,<ea>
CAS CAS2	CAS Destination—Compare Operand ➧ cc; if Z, Update Operand ➧ Destination else Destination ➧ Compare Operand CAS2 Destination 1—Compare 1 ➧ cc; if Z, Destination 2—Compare ➧ cc; if Z, Update 1 ➧ Destination 1; Update 2 ➧ Destination 2 else Destination 1 ➧ Compare 1; Destination 2 ➧ Compare 2	CAS Dc,Du,<ea> CAS2 Dc1:Dc2,Du1:Du2,(Rn1):(Rn2)
CHK	If Dn < 0 or Dn > Source then TRAP	CHK <ea>,Dn
CHK2	If Rn < lower bound or Rn > upper bound then TRAP	CHK2 <ea>,Rn

(continued)

TABLE B.2. (continued)

Opcode	Operation	Syntax
CINV	If supervisor state then invalidate selected cache lines else TRAP	CINVL <caches>[1] (An) CINVP <caches>[1] (An) CINVA <caches>[1]
CLR	0 → Destination	CLR <ea>
CMP	Destination—Source → cc	CMP <ea>,Dn
CMPA	Destination—Source	CMPA <ea>,An
CMPI	Destination—Immediate Data	CMPI #<data>,<ea>
CMPM	Destination—Source → cc	CMPM (Ay) +,(Ax) +
CMP2	Compare Rn < lower-bound or Rn > upper-bound and Set Condition Codes	CMP2 <ea>,Rn
CPUSH	If supervisor state then if data cache then push selected dirty data cache lines invalidate selected cache lines else TRAP	CPUSHL <caches>[1],(An) CPUSHP <caches>[1],(An) CPUSHA <caches>[1]
DBcc	If condition false then (Dn – 1 → Dn; If Dn ≠ – 1 then PC + d → PC)	DBcc Dn,<label>
DIVS DIVSL	Destination/Source → Destination	DIVS.W <ea>,Dn 32/16 → 16r:16q DIVS.L <ea>,Dq 32/32 → 32q DIVS.L <ea>,Dr:Dq 64/32 → 32r:32q DIVSL.L <ea>,Dr:Dq 32/32 → 32r:32q
DIVU DIVUL	Destination/Source → Destination	DIVU.W <ea>,Dn 32/16 → 16r:16q DIVU.L <ea>,Dq 32/32 → 32q DIVU.L <ea>,Dr:Dq 64/32 → 32r:32q DIVUL.L <ea>,Dr:Dq 32/32 → 32r:32q
EOR	Source ⊕ Destination → Destination	EOR Dn,<ea>
EORI	Immediate Data ⊕ Destination → Destination	EORI #<data>,<ea>
EORI to CCR	Source ⊕ CCR → CCR	EORI #<data>,CCR
EORI to SR	If supervisor state the Source ⊕ SR → SR else TRAP	EORI #<data>,SR
EXG	Rx ↔ Ry	EXG Dx,Dy EXG Ax,Ay EXG, Dx,Ay EXG Ay,Dx
EXT EXTB	Destination Sign-Extended → Destination	EXT.W Dn extend byte to word EXT.L L Dn extend word to long word EXTB.L Dn extend byte to long word
FABS	Absolute Value of Source → FPn	FABS.<fmt> <ea>,FPn FABS.X FPm,FPn FABS.X FPn FrABS.<fmt>;2 <ea>,FPn FrABS.X[2] FPm,FPn FrABS.X[2] FPn
FADD	Source + FPn → FPn	FADD.<fmt> <ea>,FPn FADD.X FPm,FPn FrADD.<fmt>[2] <ea>,FPn FrADD.X[2] FPm,FPn
FBcc	If condition true, then PC + d → PC	FBcc.<size> <label>

TABLE B.2. (continued)

FCMP	FPn—Source	FCMP.<fmt> FCMP.X	<ea>,FPn FPm,FPn
FDBcc	If condition true then no operation else Dn – 1 → Dn if Dn ≠ –1 then PC + d → PC else execute next instruction	FDBcc	Dn,<label>
FDIV	FPn (÷) Source → FPn	FDIV.<fmt> FDIV.X FrDIV.<fmt>[2] FrDIV.X[2]	<ea>,FPn FPm,FPn <ea>,FPn FPm,FPn
FMOVE	Source → Destination	FMOVE.<fmt> FMOVE.<fmt> FMOVE.P FMOVE.P FrMOVE.<fmt>[2]	<ea>,FPn FPM,<ea> FPm,<ea>{Dn} FPm,<ea>{#k} <ea>,FPn
FMOVE	Source → Destination	FMOVE.L <ea>,FPcr FMOVE.L FMcr,<ea>	
FMOVEM	Register List → Destination Source → Register List	FMOVEM.X <list>[3],<ea> FMOVEM.X Dn,<ea> FMOVEM.X <ea>,<list>[3] FMOVEM.X <ea>,Dn	
FMOVEM	Register List → Destination Source → Register List	FMOVEM.L <list>[4],<ea> FMOVEM.L <ea>,<list>[4]	
FMUL	Source × FPn → FPn	FMUL.<fmt> FMUL.X FrMUL<fmt>[2] FrMUL.X[2]	<ea>,FPn FPm,FPn <ea>,FPn FPm,FPn
FNEG	–(Source) → FPn	FNEG.<fmt> FNEG.X FNEG.X FrNEG.<fmt>[2] FrNEG.X[2] FrNEG.X[2]	<ea>,FPn FPm,FPn FPn <ea>,FPn FPm,FPn FPn
FNOP	None	FNOP	
FRESTORE	If in supervisor state then FPU State Frame → Internal State else TRAP	FRESTORE <ea>	
FSAVE	If in supervisor state then FPU Internal State → State Frame else TRAP	FSAVE <ea>	
FScc	If condition true then 1s → Destination else Os → Destination	FScc.<size> <ea>	
FSQRT	Square Root of Source → FPn	FSQRT.<fmt> FSQRT.X FSQRT.X FrSQRT <fmt>[2] FrSQRT[2] FrSQRT[2]	<ea>,FPn FPm,FPn FPn <ea>,FPn FPm,FPn FPn

(continued)

TABLE B.2. (continued)

Opcode	Operation	Syntax
FSUB	FPn-Source ➧ FPN	FSUB.<fmt> <ea>,FPn FSUB.X FPm,FPn FrSUB.<fmt> <ea>,FPn FrSUB.X[2] FPm,FPn
FTRAPcc	If condition true, then TRAP	FTRAPcc FTRAPcc.W #<data> FTRAPcc.L #<data>
FTST	Condition Codes for Operand ➧ FPCC	FTST.<fmt> <ea> FTST.X FPm
ILLEGAL	SSP–2 ➧ SSP; Vector Offset ➧ (SSP); SSP–4 ➧ SSP; PC ➧ (SSP); SSp–2 ➧ SSP; SR ➧ (SSP); Illegal Instruction Vector Address ➧ PC	ILLEGAL
JMP	Destination Address ➧ PC	JMP <ea>
JSR	SP–4 ➧ SP; PC ➧ (SP) Destination Addres ➧ PC	JSR <ea>
LEA	<ea> ➧ An	LEA <ea>,An
LINK	SP–4 ➧ SP; An ➧ (SP) SP ➧ An, SP + d ➧ SP	LINK An,#<displacement>
LSL,LSR	Destination Shifted by <count> ➧ Destination	LSd[5] Dx,Dy LSd[5] #<data>,Dy LSd[5] <ea>
MOVE	Source ➧ Destination	MOVE <ea>,<ea>
MOVEA	Source ➧ Destination	MOVEA <ea>,An
MOVE to CCR	CCR ➧ Destination	MOVE CCR,<ea>
MOVE to CCR	Source ➧ CCR	MOVE <ea>,CCR
MOVE from SR	If supervisor state then SR ➧ Destination else TRAP	MOVE SR,<ea>
MOVE to SR	If supervisor state then Source ➧ SR else TRAP	MOVE <ea>,SR
MOVE USP	If supervisor state then USP ➧ An or An ➧ USP else TRAP	MOVE USP,An MOVE An,USP
MOVE16	Source block > Destination block	MOVE 16 (Ax) +,(Ay)+ MOVE 16 xxx.L.(An) MOVE16 xxx.L,(An) + MOVE16 (An),xxx.L MOVE16 (An) +,xxx.L
MOVEC	If supervisor state then Rc ➧ Rn or Rn ➧ Rc else TRAP	MOVEC Rc,Rn MOVEC Rn,Rc
MOVEM	Registers ➧ Destination Source ➧ Registers	MOVEM register list,<ea> MOVEM <ea>,register list
MOVEP	Source ➧ Destination	MOVEP Dx,(d,Ay) MOVEP (d,Ay),Dx
MOVEQ	Immediate Data ➧ Destination	MOVEQ #<data>,Dn

TABLE B.2. (continued)

MOVES	If supervisor state then Rn ➧ Destination [DFC] or Source [SFC] ➧ Rn else TRAP	MOVES Rn,<ea> MOVES <ea>,Rn	
MULS	Source × Destination ➧ Destination	MULS.W <ea>,Dn MULS.L <ea>,DI MULS.L <ea>,Dh:DI	16 × 16 ➧ 32 32 × 32 ➧ 32 32 × 32 ➧ 64
MULU	Source × Destination ➧ Destination	MULU.W <ea>,Dn MULU.L <ea>,DI MULU.L <ea>,Dh:DI	16 × 16 ➧ 32 32 × 32 ➧ 32 32 × 32 ➧ 64
NBCD	0–(Destination$_{10}$)–X ➧ Destination	NBCD <ea>	
NEG	0–(Destination) ➧ Destination	NEG <ea>	
NEGX	0–(Destination)– X ➧ Destination	NEGX <ea>	
NOP	None	NOP	
NOT	~Destination ➧ Destination	NOT <ea>	
OR	Source V Destination ➧ Destination	OR <ea>,Dn OR Dn,<ea>	
ORI	Immediate Data V Destination ➧ Destination	ORI #<data>,<ea>	
ORI or CCR	Source V CCR ➧ CCR	ORI #<data>,CCR	
ORI to SR	If supervisor state Then Source V SR ➧ SR else TRAP	ORI #<data>,SR	
PACK	Source (Unpacked BCD) + adjustment ➧ Destination (Packed BCD)	PACK – (Ax), – (Ay), # (adjustment) PACK Dx,Dy,#<adjustment>	
PEA	Sp–4 ➧ SP; <ea> ➧ (SP)	PEA <ea>	
PFLUSH	If supervisor state then invalidate instruction and data ATC entries for destination address else TRAP	PFLUSH (An) PFLUSHN (An) PFLUSHA PFLUSHAN	
PTEST	If supervisor state then logical address status ➧ MMUSR; entry ➧ ATC else TRAP	PTESTER (An) PTESTW (An)	
RESET	If supervisor state then Assert $\overline{RSTO}$ Line else TRAP	RESET	
ROL,ROR	Destination Rotated by <count> ➧ Destination	ROd[5] Rx,Dy ROd[5] #<data>,Dy ROd[5]<ea>	
ROXL, ROXR	Destination Rotated with X by <count> ➧ Destination	ROXd[5]Dx,Dy ROXd[5]#<data>,Dy ROXd[5]<ea>	
RTD	(SP) ➧ PC; SP + 4 + d ➧ SP	RTD #<displacement>	
RTE	If supervisor state the (SP) ➧ SR; SP + 2 ➧ SP; (SP) ➧ PC; SP + 4 ➧ SP; restore state and deallocate stack according to (SP) else TRAP	RTE	
RTR	(SP) ➧ CCR; SP + 2 ➧ SP; (SP) ➧ PC; SP + 4 ➧ SP	RTR	
RTS	(SP) ➧ PC; SP + 4 ➧ SP	RTS	
SBCD	Destination$_{10}$–Source$_{10}$–X ➧ Destination	SBCD Dx,Dy SBCD – (Ax), – (Ay)	

(continued)

TABLE B.2. continued

Opcode	Operation	Syntax
Scc	If Condition True then 1s ➧ Destination else 0s ➧ Destination	Scc <ea>
STOP	If supervisor state then Immediate Data ➧ SR; STOP else TRAP	STOP #<data>
SUB	Destination–Source ➧ Destination	SUB <ea>,Dn SUB Dn,<ea>
SUBA	Destination–Source ➧ Destination	SUBA <ea>,An
SUBI	Destination–Immediate Data ➧ Destination	SUBI #<data>,<ea>
SUBQ	Destination–Immediate Data ➧ Destination	SUBQ #<data>,<ea>
SUBX	Destination–Source–X ➧ Destination	SUBX Dx,Dy SUBX–(Ax),–(Ay)
SWAP	Register [31:16] ⬌ Register [15:0]	SWAP Dn
TAS	Destination Tested ➧ Condition Codes; 1 ➧ bit 7 of Destination	TAS <ea>
TRAP	SSP–2 ➧ SSP; Format/Offset ➧ (SSP); SSP–4 ➧ SSP; PC ➧ (SSP); SSP–2 ➧ SSP; SR ➧ (SSP); Vector Address ➧ PC	TRAP #<vector>
TRAPcc	If cc then TRAP	TRAPcc TRAPcc.W #<data> TRAPcc.L #<data>
TRAPV	If V then TRAP	TRAPV
TST	Destination Tested ➧ Condition Codes	TST<ea>
UNLK	An ➧ SP; (SP) ➧ An; SP + 4 ➧ SP	UNLK An
UNPK	Source (Packed BCD) + adjustment ➧ Destination (Unpacked BCD)	UNPACK – (Ax),–(Ay), #<adjustment> UNPACK Dx,Dy,#<adjustment>

NOTES:

1. Specifies either the instruction (IC), data (DC), or IC/DC caches.
2. Where r is rounding precision, S or D.
3. A list of any combination of the eight floating-point data registers, with individual register names separated by a slash (/); and/or contiguous blocks of registers specified by the first and last register names separated by a dash(–).
4. A list of any combination of the three floating-point system control registers (FPCR, FPSR, and FPIAR) with individual register names separated by a slash(/).
5. where d is direction, L or R.

Appendix C

Instruction Timing Diagrams

TABLE C.1. Timing Diagram for MC68020

Num	Characteristic	Symbol	MC68020RC12 Min	MC68020RC12 Max	MC68020RC16 Min	MC68020RC16 Max	Unit
6	Clock High to Address/FC/Size/$\overline{RMC}$ Valid	^{t}CHAV	0	40	0	30	ns
6A	Clock High to $\overline{ECS}$, $\overline{OCS}$ Asserted	^{t}CHEV	0	30	0	20	ns
7	Clock High to Address, Data, FC, $\overline{RMC}$, Size High Impedance	^{t}CHAZx	0	80	0	60	ns
8	Clock High to Address/FC/Size/$\overline{RMC}$ Invalid	^{t}CHAZn	0	—	0	—	ns
9	Clock Low to $\overline{AS}$, $\overline{DS}$ Asserted	^{t}CLSA	3	40	3	30	ns
9A[1]	$\overline{AS}$ to $\overline{DS}$ Assertion (Read) (Skew)	^{t}STSA	–20	20	–15	15	ns
10	$\overline{ECS}$ Width Asserted	^{t}ECSA	25	—	20	—	ns
10A	$\overline{OCS}$ Width Asserted	^{t}OCSA	25	—	20	—	ns
11[6]	Address/FC/Size/$\overline{RMC}$ Valid to $\overline{AS}$ (and $\overline{DS}$ Asserted Read)	^{t}AVSA	20	—	15	—	ns
12	Clock Low to $\overline{AS}$, $\overline{DS}$ Negated	^{t}CLSN	0	40	0	30	ns
12A	Clock Low to $\overline{ECS}$/$\overline{OCS}$ Negated	^{t}CLEN	0	40	0	30	ns
13	$\overline{AS}$, $\overline{DS}$ Negated to Address, FC, Size Invalid	^{t}SNAI	20	—	15	—	ns
14	$\overline{AS}$ (and $\overline{DS}$ Read) Width Asserted	^{t}SWA	120	—	100	—	ns
14A	$\overline{DS}$ Width Asserted Write	^{t}SWAW	50	—	40	—	ns
15	$\overline{AS}$, $\overline{DS}$ Width Negated	^{t}SN	50	—	40	—	ns
16	Clock High to $\overline{AS}$, $\overline{DS}$, R/$\overline{W}$, $\overline{DBEN}$ High Impedance	^{t}CSZ	—	80	—	60	ns
17[6]	$\overline{AS}$, $\overline{DS}$ Negated to R/$\overline{W}$ High	^{t}SNRN	20	—	15	—	ns
18	Clock High to R/$\overline{W}$ High	^{t}CHRH	0	40	0	30	ns
20	Clock High to R/$\overline{W}$ Low	^{t}CHRL	0	40	0	30	ns
21[6]	R/$\overline{W}$ High to $\overline{AS}$ Asserted	^{t}RAAA	20	—	15	—	ns
22[6]	R/$\overline{W}$ Low to $\overline{DS}$ Asserted (Write)	^{t}RASA	90	—	70	—	ns
23	Clock High to Data Out Valid	^{t}CHDO	—	40	—	30	ns
25[6]	$\overline{DS}$ Negated to Data Out Invalid	^{t}SNDI	20	—	15	—	ns
26[6]	Data Out Valid to $\overline{DS}$ Asserted (Write)	^{t}DVSA	20	—	15	—	ns
27	Data-In Valid to Clock Low (Data Setup)	^{t}DICL	10	—	5	—	ns
27A	Late $\overline{BERR}$/$\overline{HALT}$ Asserted to Clock Low Setup Time	^{t}BELCL	25	—	20	—	ns
28	$\overline{AS}$, $\overline{DS}$ Negated to $\overline{DSACKx}$, $\overline{BERR}$, $\overline{HALT}$, $\overline{AVEC}$ Negated	^{t}SNDN	0	110	0	80	ns

(continued)

TABLE C.1 (continued)

			MC68020RC12		MC68020RC16		
Num	Characteristic	Symbol	Min	Max	Min	Max	Unit
29	DS Negated to Data-In Invalid (Data-In Hold Time)	tSNDI	0	—	0	—	ns
29A	DS Negated to Data-In (High Impedance)	tSNDI	—	80	—	60	ns
31[2]	DSACKx Asserted to Data-In Valid	tDADI	—	60	—	50	ns
31A[3]	DSACKx Asserted to DSACKx Valid (DSACK Asserted Skew)	tDADV	—	20	—	15	ns
32	RESET Input Transition Time	tHRrf	—	2.5	—	2.5	Clk Per
33	Clock Low to BG Asserted	tCLBA	0	40	0	30	ns
34	Clock Low to BG Negated	tCLBN	0	40	0	30	ns
35	BR Asserted to BG Asserted (RMC Not Asserted)	tBRAGA	1.5	3.5	1.5	3.5	Clk Per
37	BGACK Asserted to BG Negated	tGAGN	1.5	3.5	1.5	3.5	Clk Per
39	BG Width Negated	tGN	120	—	90	—	ns
39A	BG Width Asserted	tGA	120	—	90	—	ns
40	Clock High to DBEN Asserted (Read)	tCHDAR	0	40	0	30	ns
41	Clock Low to DBEN Negated (Read)	tCLDNR	0	40	0	30	ns
42	Clock Low to DBEN Asserted (Write)	tCLDAW	0	40	0	30	ns
43	Clock High to DBEN Negated (Write)	tCHDNW	0	40	0	30	ns
44[6]	R/W Low to DBEN Asserted (Write)	tRADA	20	—	15	—	ns
45[5]	DBEN Width Asserted Read		80	—	60	—	ns
	Write	tDA	160	—	120	—	ns
46	R/W Width Asserted (Write or Read)	tRWA	180	—	150	—	ns
47a	Asynchronous Input Setup Time	tAIST	10	—	5	—	ns
47b	Asynchronous Input Hold Time	tAIHT	20	—	15	—	ns
48[4]	DSACKx Asserted to BERR, HALT Asserted	tDABA	—	35	—	30	ns
53	Data Out Hold from Clock High	tDOCH	0	—	0	—	ns
55	R/W Asserted to Data Bus Impedance Change	tRADC	40	—	40	—	ns
56	RESET Pulse Width (Reset Instruction)	tHRPW	512	—	512	—	Clks
57	BERR Negated to HALT Negated (Rerun)	tBNHN	0	—	0	—	ns

NOTES:

1. This number can be reduced to 5 nanoseconds if strobes have equal loads.
2. If the asynchronous setup time (#47) requirements are satisfied, the DSACKx low to data setup time (#31) and DSACKx low to BERR low setup time (#48) can be ignored. The data must only satisfy the data-in to clock low setup time (#27) for the following clock cycle, BERR must only satisfy the late BERR low to clock low setup time (#27A) for the following clock cycle.
3. This parameter specifies the maximum allowable skew between DSACK0 to DSACK1 asserted or DSACK1 to DSACK0 asserted, specification #47 must be met by DSACK0 or DSACK1.
4. In the absence of DSACKx, BERR is an asynchronous input using the asynchronous input setup time (#47).
5. DBEN may stay asserted on consecutive write cycles.
6. Actual value depends on the clock input waveform.

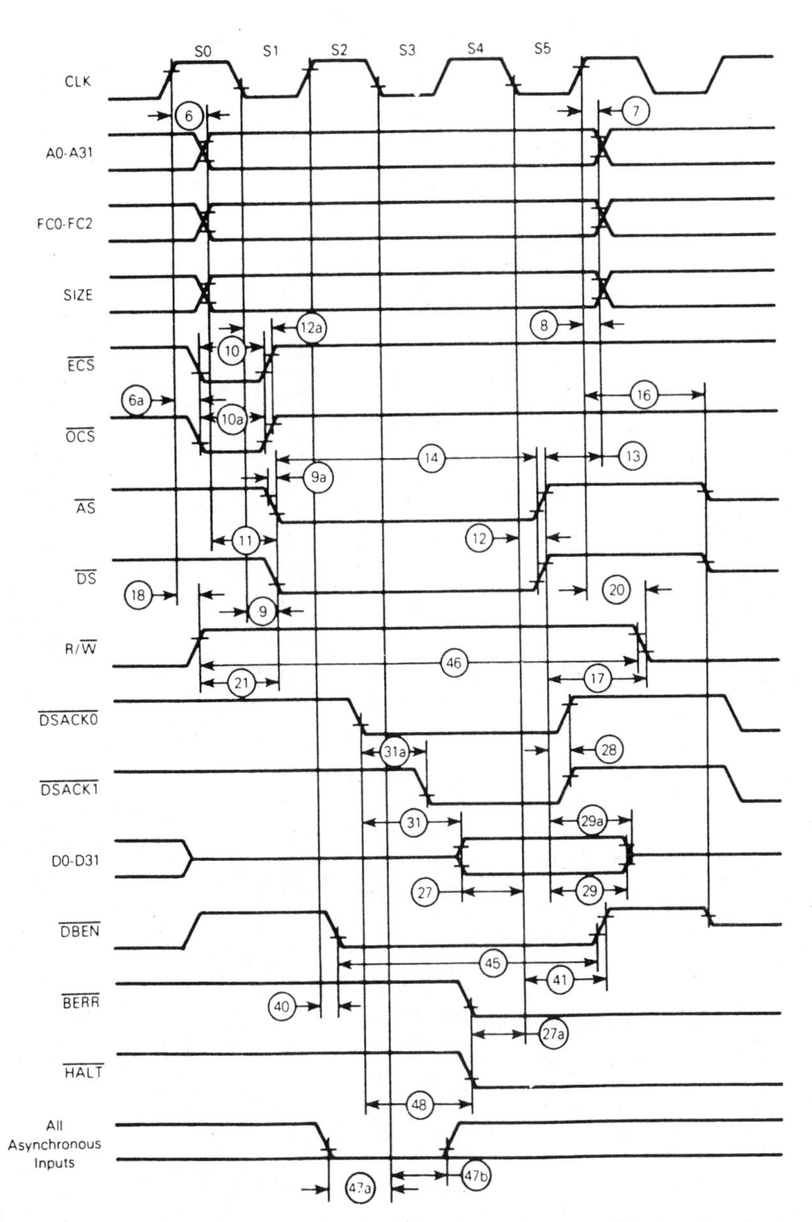

NOTE:
Timing measurements are referenced to and from a low voltage of 0.8 volt and a high voltage of 2.0 volts, unless otherwise noted. The voltage swing through this range should start outside and pass through the range such that the rise or fall will be linear between 0.8 volt and 2.0 volts.

Figure C.1. Read Cycle Timing Diagram for MC68020

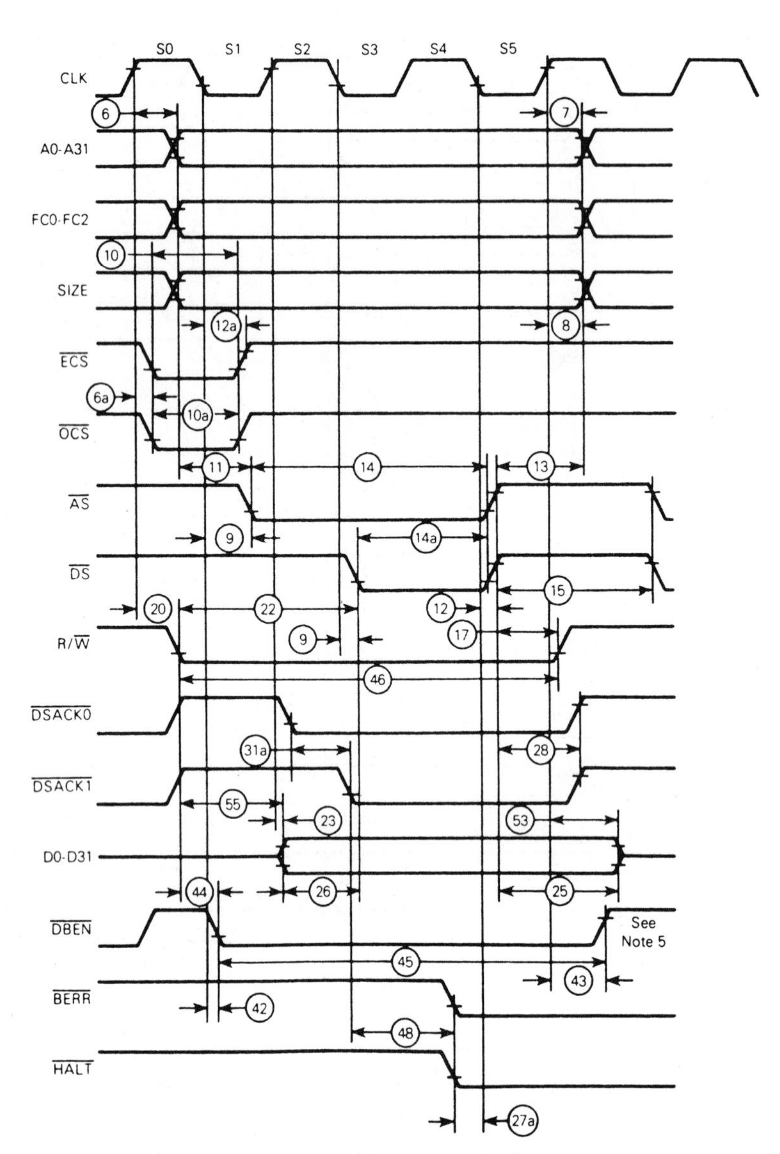

NOTE:
Timing measurements are referenced to and from a low voltage of 0.8 volt and a high voltage of 2.0 volts, unless otherwise noted. The voltage swing through this range should start outside and pass through the range such that the rise or fall will be linear between 0.8 volt and 2.0 volts.

Figure C.2. Write Cycle Timing Diagram for MC68020

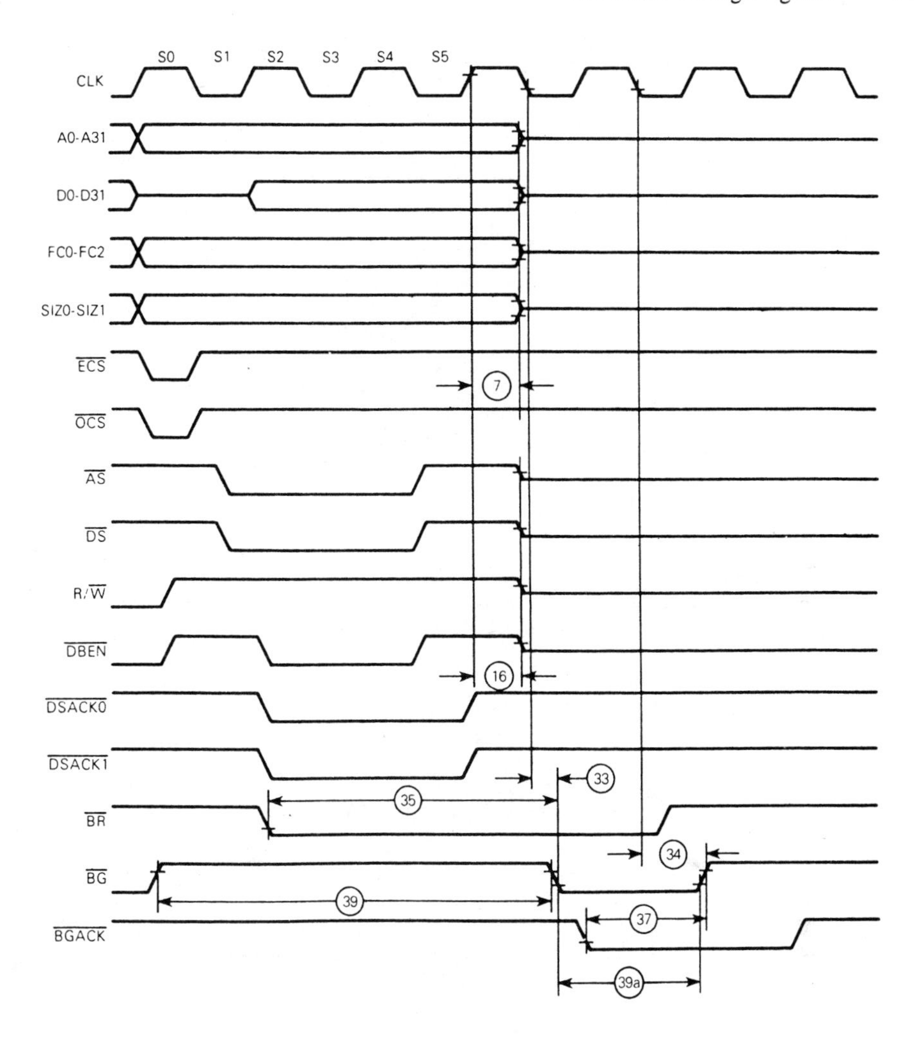

NOTE: Timing measurements are referenced to and from a low voltage of 0.8 volt and a high voltage of 2.0 volts, unless otherwise noted. The voltage swing through this range should start outside and pass through the range such that the rise or fall will be linear between 0.8 volt and 2.0 volts.

Figure C.3. Bus Arbitration Timing Diagram for MC68020

TABLE C.2. Read Write Cycles for MC68030

		20 MHz		25 MHz		33.33 MHz		
Num.	Characteristic	Min	Max	Min	Max	Min	Max	Unit
6	Clock High to Function Code, Size, $\overline{RMC}$, $\overline{IPEND}$, $\overline{CIOUT}$, Address Valid	0	25	0	20	0	14	ns
6A	Clock High to $\overline{ECS}$, $\overline{OCS}$ Asserted	0	15	0	15	0	12	ns
6B	Function Code, Size, $\overline{RMC}$, $\overline{IPEND}$, $\overline{CIOUT}$, Address valid to Negating Edge of $\overline{ECS}$	4	—	3	—	3	—	ns
7	Clock High to Function Code, Size, $\overline{RMC}$, $\overline{CIOUT}$, Address, Data High Impedance	0	50	0	40	0	30	ns
8	Clock High to Function Code, Size, $\overline{RMC}$, $\overline{IPEND}$, $\overline{CIOUT}$, Address Invalid	0	—	0	—	0	—	ns
9	Clock Low to $\overline{AS}$, $\overline{DS}$ Asserted, $\overline{CBREQ}$ Valid	3	20	3	18	2	10	ns
9A[1]	$\overline{AS}$ to $\overline{DS}$ Assertion Skew (Read)	−10	10	−10	10	−8	8	ns
9B[14]	$\overline{AS}$ Asserted to $\overline{DS}$ Asserted (Write)	32	—	27	—	22	—	ns
10	$\overline{ECS}$ Width Asserted	15	—	10	—	8	—	ns
10A	$\overline{OCS}$ Width Asserted	15	—	10	—	8	—	ns
10B[7]	$\overline{ECS}$, $\overline{OCS}$ Width Negated	10	—	5	—	5	—	ns
11	Function Code, Size, $\overline{RMC}$, $\overline{CIOUT}$, Address Valid to $\overline{AS}$ Asserted (and $\overline{DS}$ Asserted, Read)	10	—	7	—	5	—	ns
12	Clock Low to $\overline{AS}$, $\overline{DS}$, $\overline{CBREQ}$ Negated	0	20	0	18	0	10	ns
12A	Clock Low to $\overline{ECS}$ $\overline{OCS}$ Negated	0	20	0	18	0	15	ns
13	$\overline{AS}$, $\overline{DS}$ Negated to Function Code, Size, $\overline{RMC}$ $\overline{CIOUT}$, Address Invalid	10	—	7	—	5	—	ns
14	$\overline{AS}$ (and $\overline{DS}$ Read) Width Asserted (Asynchronous Cycle)	85	—	70	—	45	—	ns
14A[11]	$\overline{DS}$ Width Asserted (Write)	38	—	30	—	23	—	ns
14B	$\overline{AS}$ (and $\overline{DS}$, Read) Width Asserted (Synchronous Cycle)	35	—	30	—	23	—	ns
15	$\overline{AS}$, $\overline{DS}$ Width Negated	38	—	30	—	23	—	ns
15A[8]	$\overline{DS}$ Negated to $\overline{AS}$ Asserted	30	—	25	—	18	—	ns
16	Clock High to $\overline{AS}$, $\overline{DS}$, R/$\overline{W}$, $\overline{DBEN}$, $\overline{CBREQ}$ High Impedance	—	50	—	40	—	30	ns
17	$\overline{AS}$, $\overline{DS}$ Negated to R/$\overline{W}$ Invalid	10	—	7	—	5	—	ns
18	Clock High to R/$\overline{W}$ High	0	25	0	20	0	15	ns
20	Clock High to R/$\overline{W}$ Low	0	25	0	20	0	15	ns
21	R/$\overline{W}$ High to $\overline{AS}$ Asserted	10	—	7	—	5	—	ns
22	R/$\overline{W}$ Low to $\overline{DS}$ Asserted (Write)	60	—	47	—	35	—	ns
23	Clock High to Data-Out Valid	—	25	—	20	—	14	ns
24	Data-Out Valid to Negating Edge of $\overline{AS}$	8	—	5	—	3	—	ns
25[11]	$\overline{AS}$, $\overline{DS}$ Negated to Data-Out Invalid	10	—	7	—	5	—	ns
25A[9,11]	$\overline{DS}$ Negated to $\overline{DBEN}$ Negated (Write)	10	—	7	—	5	—	ns
26[11]	Data-Out Valid to $\overline{DS}$ Asserted (Write)	10	—	7	—	5	—	ns
27	Data-In Valid to Clock Low (Setup)	4	—	2	—	1	—	ns
27A	Late $\overline{BERR}$ $\overline{HALT}$ Asserted to Clock Low (Setup)	10	—	5	—	3	—	ns
28[12]	$\overline{AS}$, $\overline{DS}$ Negated to $\overline{DSACKx}$, $\overline{BERR}$, $\overline{HALT}$, $\overline{AVEC}$ Negated (Asynchronous Hold)	0	50	0	40	0	30	ns

TABLE C.2 (continued)

Num.	Characteristic	20 MHz		25 MHz		33.33 MHz		Unit
		Min	Max	Min	Max	Min	Max	
28A[12]	Clock Low to DSACKx, BERR, HALT, AVEC Negated (Synchronous Hold)	12	85	8	70	6	50	ns
29[12]	AS, DS Negated to Data-In Invalid (Asynchronous Hold)	0	—	0	—	0	—	ns
29A[12]	AS, DS Negated to Data-In High Impedance	—	50	—	40	—	30	ns
30[12]	Clock Low to Data-In Invalid (Synchronous Hold)	12	—	8	—	6	—	ns
30A[12]	Clock Low to Data-In High Impedance (Read followed by Write)	—	75	—	60	—	45	ns
31[2]	DSACKx Asserted to Data-In Valid (Asynchronous Data Setup)	—	43	—	28	—	20	ns
31A[3]	DSACKx Asserted to DSACKx Valid (Skew)	—	10	—	7	—	5	ns
32	RESET Input Transition Time	—	1.5	—	1.5	—	1.5	Clks
33	Clock Low to BG Asserted	0	25	0	20	0	15	ns
34	Clock Low to BG Negated	0	25	0	20	0	15	ns
35	BR Asserted to BG Asserted (RMC Not Asserted)	1.5	3.5	1.5	3.5	1.5	3.5	Clks
37	BGACK Asserted to BG Negated	1.5	3.5	1.5	3.5	1.5	3.5	Clks
37A	BGACK Asserted to BR Negated	0	1.5	0	1.5	0	1.5	Clks
39[6]	BG Width Negated	75	—	60	—	45	—	ns
39A	BG Width Asserted	75	—	60	—	45	—	ns
40	Clock High to DBEN Asserted (Read)	0	25	0	20	0	18	ns
41	Clock Low to DBEN Negated (Read)	0	25	0	20	0	18	ns
42	Clock Low to DBEN Asserted (Write)	0	25	0	20	0	18	ns
43	Clock High to DBEN Negated (Write)	0	25	0	20	0	18	ns
44	R/W Low to DBEN Asserted (Write)	10	—	7	—	5	—	ns
45[5]	DBEN Width Asserted Asynchronous Read	50	—	40	—	30	—	
	Asynchronous Write	100	—	80	—	60	—	ns
45A[9]	DBEN Width Asserted Synchronous Read	10	—	5	—	5	—	
	Synchronous Write	50	—	40	—	30	—	ns
46	R/W Width Asserted (Asynchronous Write or Read)	125	—	100	—	75	—	ns
46A	R/W Width Asserted (Synchronous Write or Read)	75	—	60	—	45	—	ns
47A	Asynchronous Input Setup Time to Clock Low	4	—	2	—	2	—	ns
47B	Asynchronous Input Hold Time from Clock Low	12	—	8	—	6	—	ns
48[4]	DSACKx Asserted to BERR, HALT Asserted	—	20	—	25	—	18	ns
53	Data-Out Hold from Clock High	3	—	3	—	2	—	ns
55	R/W Asserted to Data Bus Impedance Change	25	—	20	—	15	—	ns
56	RESET Pulse Width (Reset Instruction)	512	—	512	—	512	—	Clks
57	BERR Negated to HALT Negated (Rerun)	0	—	0	—	0	—	ns

(continued)

TABLE C.2 Read Write Cycles for MC68030

Num.	Characteristic	20 MHz		25 MHz		33.33 MHz		Unit
		Min	Max	Min	Max	Min	Max	
58[10]	$\overline{\text{BGACK}}$ Negated to Bus Driven	1	—	1	—	1	—	Clks
59[10]	$\overline{\text{BG}}$ Negated to Bus Driven	1	—	1	—	1	—	Clks
60[13]	Synchronous Input Valid to Clock High (Setup Time)	4	—	2	—	2	—	ns
61[13]	Clock High to Synchronous Input Invalid (Hold Time)	12	—	8	—	6	—	ns
62	Clock Low to $\overline{\text{STATUS}}$, $\overline{\text{REFILL}}$ Asserted	0	25	0	20	0	15	ns
63	Clock Low to $\overline{\text{STATUS}}$, $\overline{\text{REFILL}}$ Negated	0	25	0	20	0	15	ns

NOTES:

1. This number can be reduced to 5 nanoseconds if strobes have equal loads.
2. If the asynchronous setup time (#47A) requirements are satisfied, the $\overline{\text{DSACKx}}$ low to data setup time (#31) and $\overline{\text{DSACKx}}$ low to $\overline{\text{BERR}}$ low setup time (#48) can be ignored. The data must only satisfy the data-in clock low setup time (#27) for the following clock cycle and $\overline{\text{BERR}}$ must only satisfy the late $\overline{\text{BERR}}$ low to clock low setup time (#27A) for the following clock cycle.
3. This parameter specifies the maximum allowable skew between $\overline{\text{DSACK0}}$ to $\overline{\text{DSACK1}}$ asserted or $\overline{\text{DSACK1}}$ to $\overline{\text{DSACK0}}$ asserted; specification #47A must be met by $\overline{\text{DSACK0}}$ or $\overline{\text{DSACK1}}$.

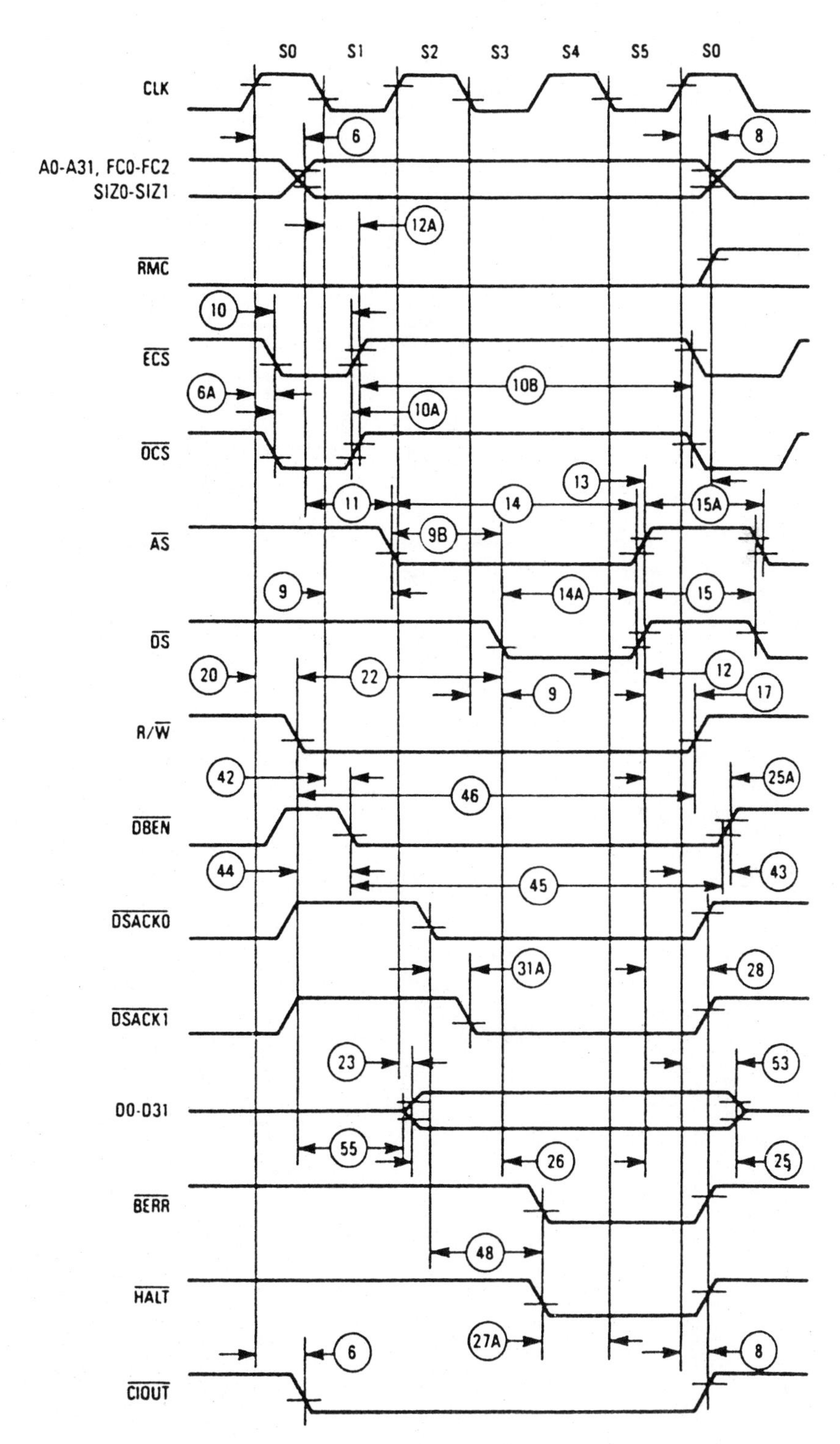

Figure C.4. Asynchronous Write Cycle Timing Diagram for MC68030

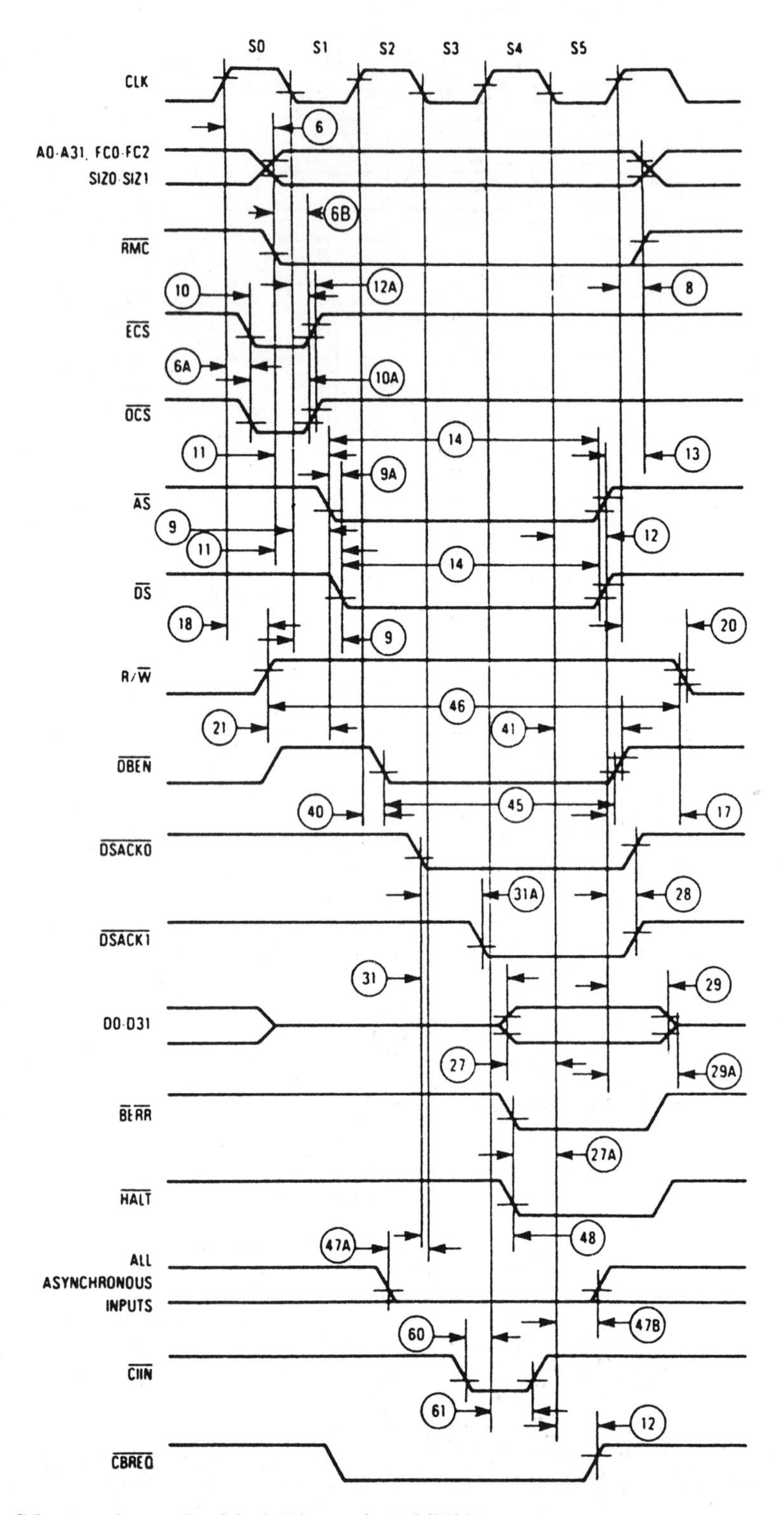

Figure C.5. Asynchronous Read Cycle Diagram for MC68030

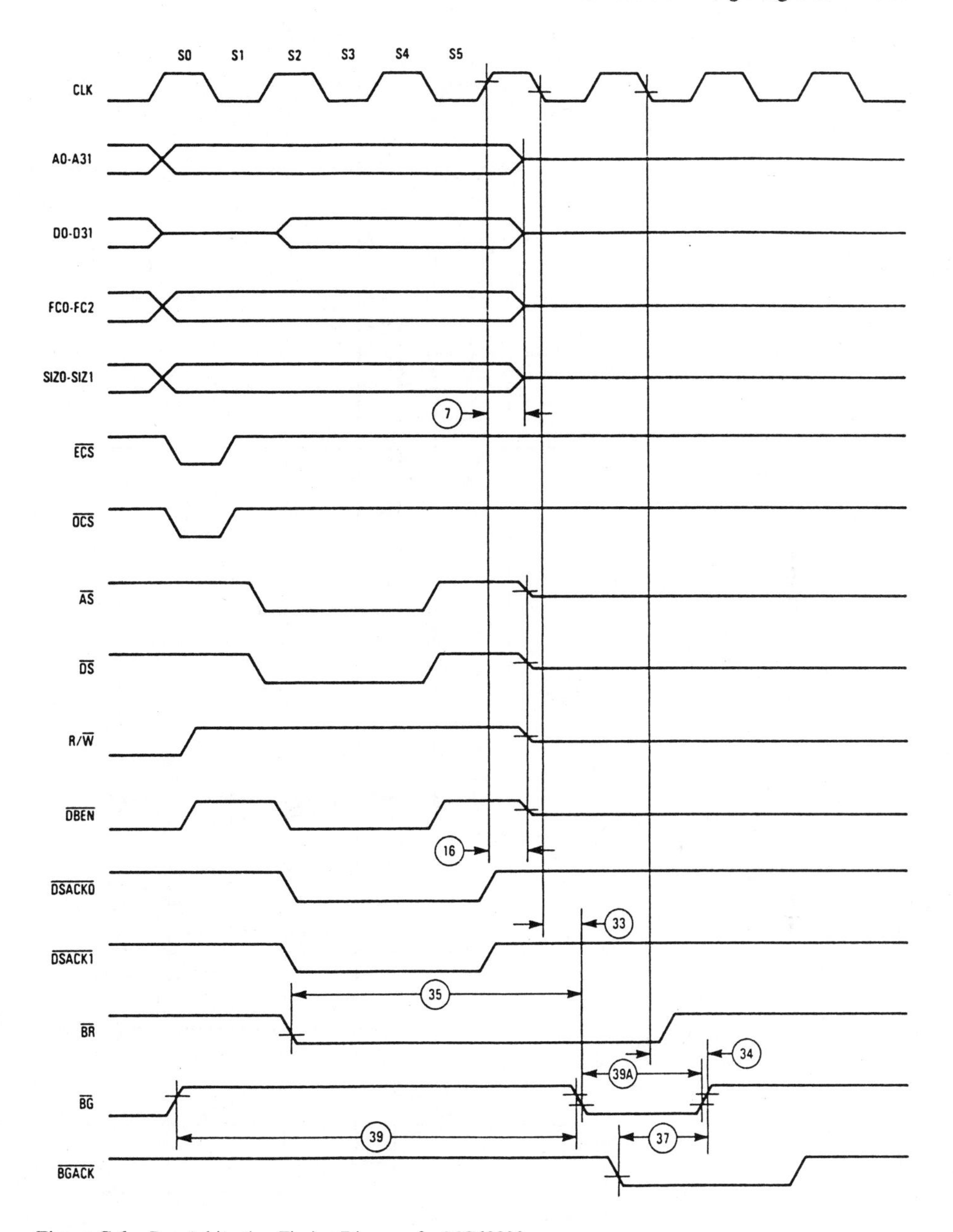

Figure C.6. Bus Arbitration Timing Diagram for MC68030

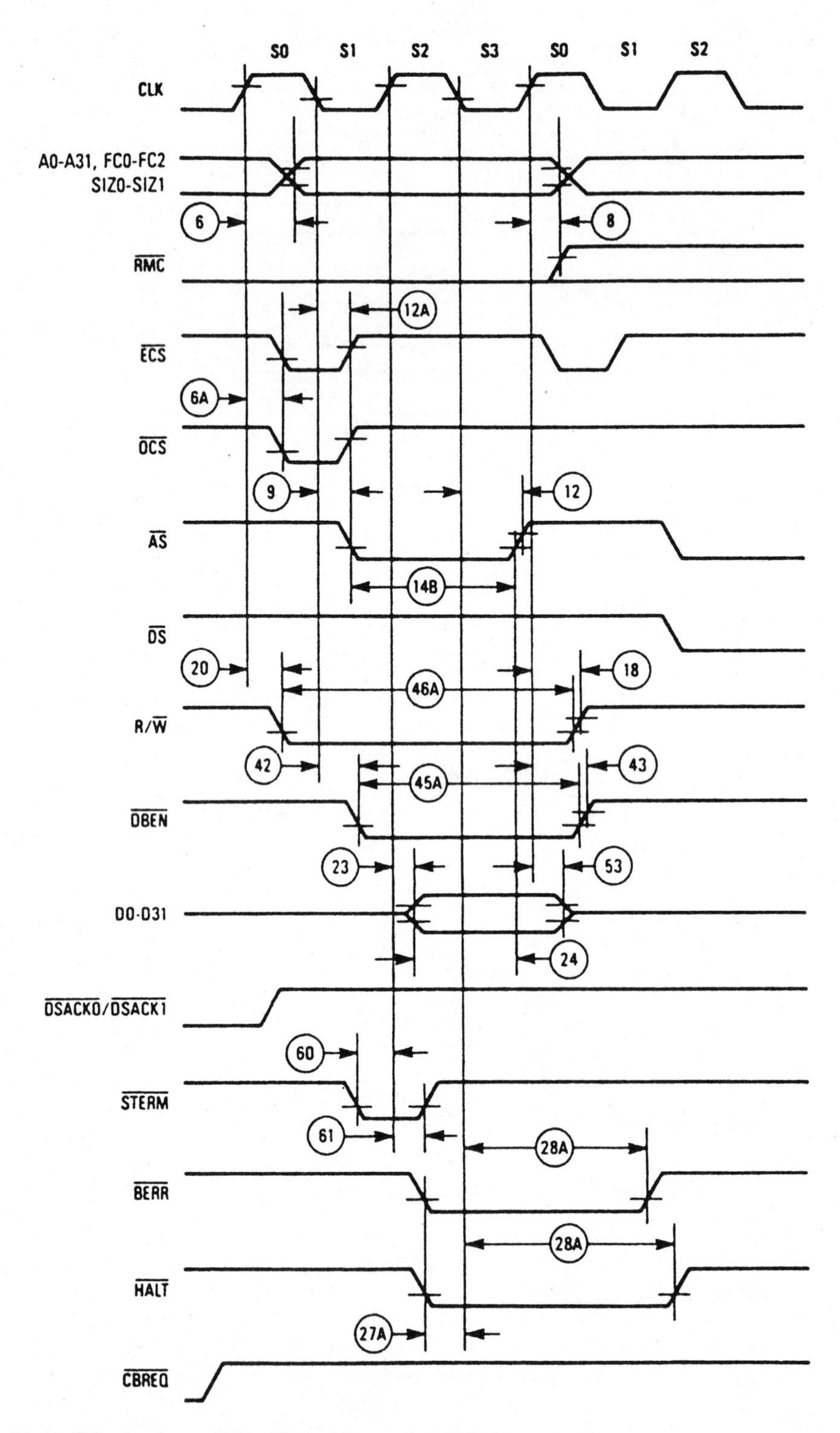

Figure C.7. Synchronous Write Timing Diagram for MC68030

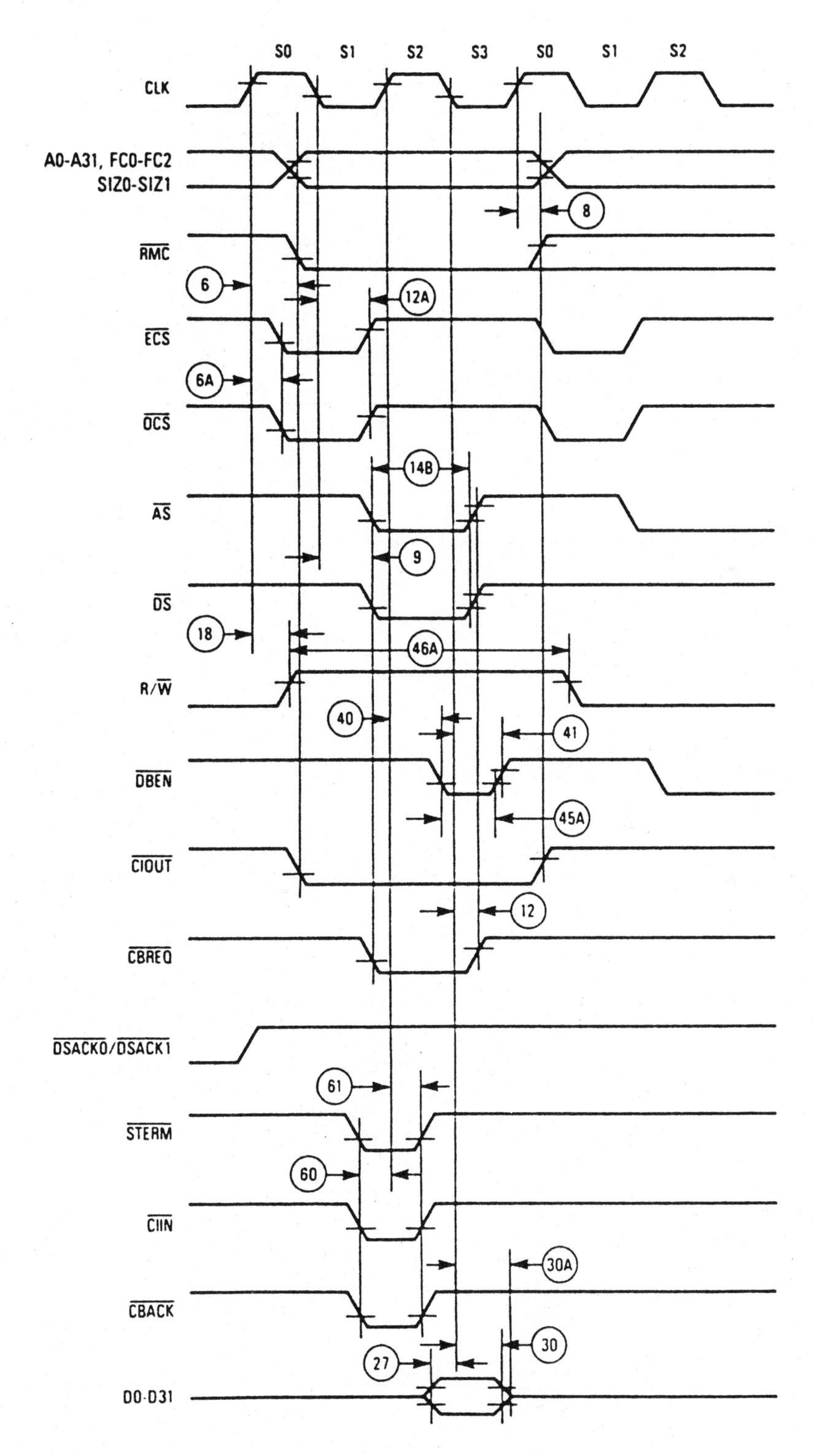

Figure C.8. Synchronous Read Cycle Timing Diagram for MC68030

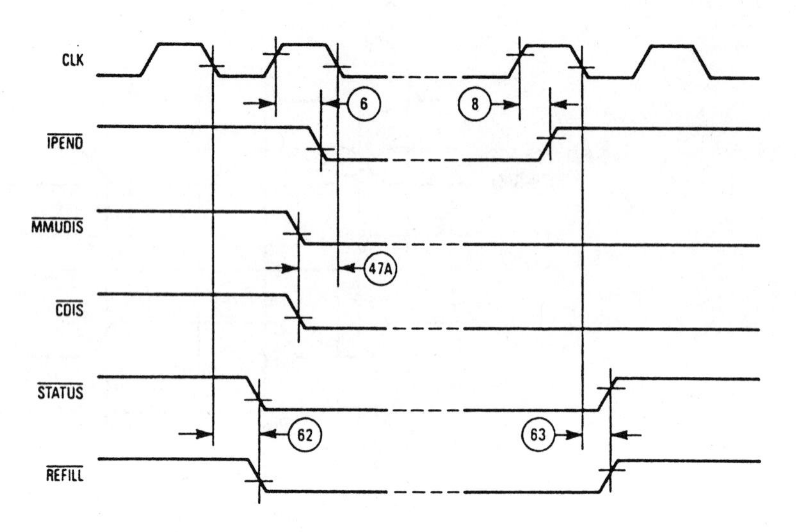

Figure C.9. Additional Timing Diagram for MC68030

Appendix D

Instruction Execution Timing Tables

INSTRUCTION TIMING TABLES

This appendix deals with the instruction timings for MC68020, MC68030, and MC68040. The timing information is extracted directly from Motorola's data books.

Instruction Timing Table for MC68020

The instruction times are based on the following assumptions:

All operands are long-word aligned as is the stack
32-bit data bus
3 cycle read/write (no wait state memory)

Three values are specified for each instruction and addressing mode namely:

Best case (BC) which reflects the time in clock cycles when the instruction is in the cache and benefits from maximum overlap due to other instructions.
Cache-only case (CC) when the instruction is in the cache but has no overlap.
Worst Case (WC), when the instruction is not in cache or the cache is disabled and there is no instruction overlap.

Within each set or column of instruction timings four sets of numbers exist for the processor, three of which are enclosed in parentheses.

The outer number is the total number of operand read cycles performed by the instruction.
The second value inside the parentheses is the number of the instruction accesses performed by the instruction, including all prefetches to keep the instruction pipe filled.
The third value within parentheses is the number of write cycles performed by the instruction.

The first tables deal exclusively with fetching and calculating effective addresses and immediate operands. The tables are arranged in such a manner because some instructions do not require effective address calculation or fetching. For example, the instruction CLR<ea> found in the table of Single Operand Instruction, only needs to have a calculated EA time added to its table entry because no fetch of an operand is required. This instruction only writes into a register or memory location. Some instructions use specific addressing modes which exclude timing for calculation or fetching of an operand. When these instructions arise, they are footnoted to indicate which other tables are needed in the timing calculation.

The first five timing tables deal exclusively with fetching and calculating effective addresses and immediate operands for MC68030. The remaining tables are instruction and operation timings. Some instructions use addressing modes that are not included in the corresponding instruction timings. These cases refer to footnotes that indicate the additional table needed for the timing calculation. All read and write accesses take two clock periods.

The only instructions for which the size of the operand has any effect are the instructions with immediate operands. Again, unless specified otherwise, immediate byte and word operands have identical execution times.

It is important to understand that the timing tables listed below cannot accurately predict the instruction timing that would be observed when executing an instruction stream on the MC68020, the tables can be used to calculate best case and worst case bounds for instruction timing. Absolute instruction timing must be measured by using the microprocessor itself, to execute the target instruction stream.

Fetch Effective Address

The fetch effective address table indicates the number of clock periods needed for the processor to calculate and fetch the specified effective address. The total number of clock cycles is outside the parentheses, the number of read, prefetch, and write cycles are given inside the parentheses as (r/p/w). They are included in the total clock cycle number.

TABLE D.1.

Address Mode	Best Case	Cache Case	Worst Case
Dn	0 (0/0/0)	0 (0/0/0)	0 (0/0/0)
An	0 (0/0/0)	0 (0/0/0)	0 (0/0/0)
(An)	3 (1/0/0)	4 (1/0/0)	4 (1/0/0)
(An)+	4 (1/0/0)	4 (1/0/0)	4 (1/0/0)
– (An)	3 (1/0/0)	5 (1/0/0)	5 (1/0/0)

TABLE D.1. (continued)

Address Mode	Best Case	Cache Case	Worst Case
(d_{16},An) of (d_{16},PC)	3 (1/0/0)	5 (1/0/0)	6 (1/1/0)
(xxx).W	3 (1/0/0)	4 (1/0/0)	6 (1/1/0)
(xxx).L	3 (1/0/0)	4 (1/0/0)	7 (1/1/0)
#<data>.B	0 (0/0/0)	2 (0/0/0)	3 (0/1/0)
#<data>.W	0 (0/0/0)	2 (0/0/0)	3 (0/1/0)
#<data>.L	0 (0/0/0)	4 (0/0/0)	5 (0/1/0)
(d_8,An,Xn) or (d_8, PC, Xn)	4 (1/0/0)	7 (1/0/0)	8 (1/1/0)
(d_{16},An,Xn) or (d_{16},PC,Xn)	4 (1/0/0)	7 (1/0/0)	9 (1/1/0)
(B)	4 (1/0/0)	7 (1/0/0)	9 (1/1/0)
(d_{16},B)	6 (1/0/0)	9 (1/0/0)	12 (1/1/0)
(d_{32},B)	10 (1/0/0)	13 (1/0/0)	16 (1/2/0)
([B],I)	9 (2/0/0)	12 (2/0/0)	13 (2/1/0)
([B],I,d_{16})	11 (2/0/0)	14 (2/0/0)	16 (2/1/0)
([B],I,d_{32})	11 (2/0/0)	14 (2/0/0)	17 (2/2/0)
([d_{16},B],I)	11 (2/0/0)	14 (2/0/0)	16 (2/1/0)
([d_{16},B],I,d_{16})	13 (2/0/0)	16 (2/0/0)	19 (2/2/0)
([d_{16},B],I,d_{32})	13 (2/0/0)	16 (2/0/0)	20 (2/2/0)
([d_{32},B],I)	15 (2/0/0)	18 (2/0/0)	20 (2/2/0)
([d_{32},B],I,d_{16})	17 (2/0/0)	20 (2/0/0)	22 (2/2/0)
([d_{32},B],I,d_{32})	17 (2/0/0)	20 (2/0/0)	24 (2/3/0)

B = Base address; 0, An, PC, Xn, An + Xn, PC + Xn. Form does not affect timing.
I = Index; 0, Xn
NOTE: Xn cannot be in B and I at the same time. Scaling and size of Xn does not affect timing.

Fetch Immediate Effective Address

The fetch immediate effective address table indicates the number of clock periods needed for the processor to fetch the immediate source operand, and calculate and fetch the specified destination operand. The total number of clock cycles is outside the parentheses, the number of read, prefetch, and write cycles are given inside the parentheses as (r/p/w). They are included in the total clock cycle number.

TABLE D.2.

Address Mode	Best Case	Cache Case	Worst Case
#<data>.W, Dn	0 (0/0/0)	2 (0/0/0)	3 (0/1/0)
#<data>.L, Dn	1 (0/0/0)	4 (0/0/0)	5 (0/1/0)
#<data>.W, (An)	3 (1/0/0)	4 (1/0/0)	4 (1/1/0)
#<data>.L, (An)	3 (1/0/0)	4 (1/0/0)	7 (1/1/0)
#<data>.W, (An) +	4 (1/0/0)	6 (1/0/0)	7 (1/1/0)
#<data>.L, (An) +	5 (1/0/0)	8 (1/0/0)	9 (1/1/0)
#<data>.W, – (An)	3 (1/0/0)	5 (1/0/0)	6 (1/1/0)
#<data>.L, – (An)	4 (1/0/0)	7 (1/0/0)	8 (1/1/0)
#<data>.W, (bd,An)	3 (1/0/0)	5 (1/0/0)	7 (1/1/0)
#<data>.L, (bd,An)	4 (1/0/0)	7 (1/0/0)	10 (1/2/0)
#<data>.W, xxx.W	3 (1/0/0)	5 (1/0/0)	7 (1/1/0)
#<data>.L,xxx.W	4 (1/0/0)	7 (1/0/0)	10 (1/2/0)

(continued)

TABLE D.2. (continued)

Address Mode	Best Case	Cache Case	Worst Case
#<data>.W,xxx.L	3 (1/0/0)	6 (1/0/0)	10 (1/2/0)
#<data>.L,xxx.L	4 (1/0/0)	8 (1/0/0)	12 (1/2/0)
#<data>.W, #<data>.B,W	0 (0/0/0)	4 (0/0/0)	6 (0/2/0)
#<data>.W, #<data>.B,W	1 (0/0/0)	6 (0/0/0)	8 (0/2/0)
#<data>.W, #<data>.L	0 (0/0/0)	6 (0/0/0)	8 (0/2/0)
#<data>.L, #<data>.L	1 (0/0/0)	8 (0/0/0)	10 (0/2/0)
#<data>.W, (d_8,An,Xn) or (d_8,PC,Xn)	4 (1/0/0)	9 (1/0/0)	11 (1/2/0)
#<data>.L, (d_8,An,Xn) or (d_8,PC,Xn)	5 (1/0/0)	11 (1/0/0)	13 (1/2/0)
#<data>.W, (d_{16},An,Xn) or (d_{16},PC,Xn)	4 (1/0/0)	9 (1/0/0)	12 (1/2/0)
#<data>.L, (d_{16},An,Xn) or (d_{16},PC,Xn)	5 (1/0/0)	11 (1/0/0)	15 (1/2/0)
#<data>.W, (B)	4 (1/0/0)	9 (1/0/0)	12 (1/2/0)
#<data>.L, (B)	5 (1/0/0)	11 (1/0/0)	14 (1/2/0)
#<data>.W, (bd,PC)	10 (1/0/0)	15 (1/0/0)	19 (1/3/0)
#<data>.L, (bd,PC)	11 (1/0/0)	17 (1/0/0)	21 (1/3/0)
#<data>.W, (d_{16},B)	6 (1/0/0)	11 (1/0/0)	15 (1/2/0)
#<data>.L, (d_{16},B)	7 (1/0/0)	13 (1/0/0)	17 (1/2/0)
#<data>.W, (d_{32}, B)	10 (1/0/0)	15 (1/0/0)	19 (1/3/0)
#<data>.L, (d_{32}, B)	11 (1/0/0)	17 (1/0/0)	21 (1/3/0)
#<data>.W, ([B],I)	9 (2/0/0)	14 (2/0/0)	16 (2/2/0)
#<data>.L, ([B],I)	10 (2/0/0)	16 (2/0/0)	18 (2/2/0)
#<data>.W, ([B],I,d_{16})	11 (2/0/0)	16 (2/0/0)	19 (2/2/0)
#<data>.L, ([B],I,d_{16})	12 (2/0/0)	18 (2/0/0)	21 (2/2/0)
#<data>.W, ([B],I,d_{32})	11 (2/0/0)	16 (2/0/0)	20 (2/3/0)
#<data>.L, ([d_{16},B],I,d_{32})	12 (2/0/0)	18 (2/0/0)	22 (2/3/0)
#<data>.W,([d_{16},B],I)	11 (2/0/0)	16 (2/0/0)	19 (2/2/0)
#<data>.L,([d_{16},B],I)	12 (2/0/0)	18 (2/0/0)	21 (2/2/0)
#<data>.W, ([d_{16},B],I,d_{16})	13 (2/0/0)	18 (2/0/0)	22 (2/3/0)
#<data>.L, ([d_{16},B],I,d_{16})	14 (2/0/0)	20 (2/0/0)	24 (2/3/0)
#<data>.W, ([d_{32},B],I)	15 (2/0/0	20 (2/0/0)	23 (2/3/0)
#<data>.L, ([d_{32},B],I)	16 (2/0/0)	22 (2/0/0)	25 (2/3/0)
#<data>.W, ([d_{32},B],I,d_{16})	17 (2/0/0)	22 (2/0/0)	25 (2/3/0)
#<data>.L, ([d_{32},B],I,d_{16})	18 (2/0/0)	24 (2/0/0)	27 (2/3/0)
#<data>.W, ([d_{32},B],I,d_{32})	17 (2/0/0)	22 (2/0/0)	27 (2/4/0)
#<data>.L, ([d_{32},B],I,d_{32})	18 (2/0/0)	24 (2/0/0)	29 (2/4/0)

B = Base address; 0, An, PC, Xn, An + Xn, Pc + Xn. Form does not affect timing.
I = Index 0, Xn
NOTE: Xn cannot be in B and I at the same time. Scaling and size of Xn does not affect timing.

Calculate Effective Address

The calculate effective address table indicates the number of clock periods needed for the processor to calculate the specified effective address. Fetch time is only included for the first level of indirection on memory indirect addressing modes. The total number of clock cycles is outside the parentheses, the number of read, prefetch, and write cycles are given inside the parentheses as (r/p/w). They are included in the total clock cycle number.

TABLE D.3.

Address Mode	Best Case	Cache Case	Worst Case
Dn	0 (0/0/0)	0 (0/0/0)	0 (0/0/0)
An	0 (0/0/0)	0 (0/0/0)	0 (0/0/0)
(An)	2 (0/0/0)	2 (0/0/0)	2 (0/0/0)
(An) +	2 (0/0/0)	2 (0/0/0)	2 (0/0/0)
– (An)	2 (0/0/0)	2 (0/0/0)	2 (0/0/0)
(d_{16},An) or (d_{16},PC)	2 (0/0/0)	2 (0/0/0)	3 (0/1/0)
<data>.W	2 (0/0/0)	2 (0/0/0)	3 (0/1/0)
<data>.L	1 (0/0/0)	4 (0/0/0)	5 (0/1/0)
(d_8,An,Xn) or (d_8,PC,Xn)	1 (0/0/0)	4 (0/0/0)	5 (0/1/0)
(d_{16},An,Xn) or (d_{16},PC,Xn)	3 (0/0/0)	6 (0/0/0)	7 (0/1/0)
(B)	3 (0/0/0)	6 (0/0/0)	7 (0/1/0)
(d_{16},B)	5 (0/0/0)	8 (0/0/0)	10 (0/1/0)
(d_{32},B)	9 (0/0/0)	12 (0/0/0)	15 (0/2/0)
([B],I)	8 (1/0/0)	11 (1/0/0)	12 (1/1/0)
([B],I,d_{16})	10 (1/0/0)	13 (1/0/0)	15 (1/1/0)
([B],I,d_{32})	10 (1/0/0)	13 (1/0/0)	16 (1/2/0)
([d_{16},B],I)	10 (1/0/0)	13 (1/0/0)	15 (1/1/0)
([d_{16},B],I,d_{16})	12 (1/0/0)	15 (1/0/0)	18 (1/2/0)
([d_{16},B],I,d_{32})	12 (1/0/0)	15 (1/0/0)	19 (1/2/0)
([d_{32},B],I)	14 (1/0/0)	17 (1/0/0)	19 (1/2/0)
([d_{32},B],I,d_{16})	16 (1/0/0)	19 (1/0/0)	21 (1/2/0)
([d_{32},B],I,d_{32})	16 (1/0/0)	19 (1/0/0)	24 (1/3/0)

B = Base address; 0, An, PC, Xn, An + Xn, PC + Xn. Form does not affect timing.
I = Index; 0, Xn
NOTE: Xn cannot be in B and I at the same time. Scaling and size of Xn does not affect timing.

Calculate Immediate Effective Address

The calculate immediate effective address table indicates the number of clock periods needed for the processor to fetch the immediate source operand and calculate the specified destination effective address. Fetch time is only included for the first level of indirection on memory indirect addressing modes. The total number of clock cycles is outside the parentheses, the number of read, prefetch, and write cycles are given inside the parentheses as (r/p/w). They are included in the total clock cycle number.

TABLE D.4.

Address Mode	Best Case	Cache Case	Worst Case
#<data>.W, Dn	0 (0/0/0)	2 (0/0/0)	3 (0/1/0)
#<data>.L, Dn	1 (0/0/0)	4 (0/0/0)	5 (0/1/0)
#<data>.W, (An)	0 (0/0/0)	2 (0/0/0)	3 (0/1/0)
#<data>.L, (An)	1 (0/0/0)	4 (0/0/0)	5 (0/1/0)
#<data>.W, (An) +	2 (0/0/0)	4 (0/0/0)	5 (0/1/0)
#<data>.L, (An) +	3 (0/0/0)	6 (0/0/0)	7 (0/1/0)
#<data>.W, (bd,An)	1 (0/0/0)	4 (0/0/0)	5 (0/1/0)
#<data>.L, (bd,An)	3 (0/0/0)	6 (0/0/0)	8 (0/2/0)
#<data>.W, (xxx).W	1 (0/0/0)	4 (0/0/0)	5 (0/1/0)

(continued)

TABLE D.4. (continued)

Address Mode	Best Case	Cache Case	Worst Case
#<data>.L,(xxx).W	3 (0/0/0)	6 (0/0/0)	8 (0/2/0)
#<data>.W,(xxx).L	2 (0/0/0)	4 (0/0/0)	6 (0/2/0)
#<data>.L,(xxx).L	3 (0/0/0)	8 (0/0/0)	10 (0/2/0)
#<data>.W, (d_8,An,Xn) or (d_8,PC,Xn)	0 (0/0/0)	6 (0/0/0)	8 (0/2/0)
#<data>.L, (d_8,An,Xn) or (d_8,PC,Xn)	2 (0/0/0)	8 (0/0/0)	10 (0/2/0)
#<data>.W, (d_{16},An,Xn) or (d_{16},PC,Xn)	3 (0/0/0)	8 (0/0/0)	10 (0/2/0)
#<data>.L, (d_{16},An,Xn) or (d_{16},PC,Xn)	4 (0/0/0)	10 (0/0/0)	12 (0/2/0)
#<data>.W, (B)	3 (0/0/0)	8 (0/0/0)	10 (0/2/0)
#<data>.L, (B)	4 (0/0/0)	10 (0/0/0)	12 (0/2/0)
#<data>.W, (bd,PC)	9 (0/0/0)	14 (0/0/0)	18 (0/3/0)
#<data>.L, (bd,PC)	10 (0/0/0)	16 (0/0/0)	20 (0/3/0)
#<data>.W, (d_{16},B)	5 (0/0/0)	10 (0/0/0)	13 (0/2/0)
#<data>.L, (d_{16},B)	6 (0/0/0)	12 (0/0/0)	15 (0/2/0)
#<data>.W, (d_{32},B)	9 (0/0/0)	14 (0/0/0)	18 (0/3/0)
#<data>.L, (d_{32},B)	10 (0/0/0)	16 (0/0/0)	20 (0/3/0)
#<data>.W, ([B],I)	8 (1/0/0)	13 (1/0/0)	15 (1/2/0)
#<data>.L, ([B],I)	9 (1/0/0)	15 (1/0/0)	17 (1/2/0)
#<data>.W, ([B],I,d_{16})	10 (1/0/0)	15 (1/0/0)	18 (1/2/0)
#<data>.L, ([B],I,d_{16})	11 (1/0/0)	17 (1/0/0)	20 (1/2/0)
#<data>.W, ([B],I,d_{32})	10 (1/0/0)	15 (1/0/0)	19 (1/3/0)
#<data>.L, ([B],I,d_{32})	11 (1/0/0)	17 (1/0/0)	21 (1/3/0)
#<data>.W,([d_{16},B],I)	10 (1/0/0)	15 (1/0/0)	18 (1/2/0)
#<data>.L,([d_{16},B],I)	11 (1/0/0)	17 (1/0/0)	20 (1/2/0)
#<data>.W, ([d_{16},B],I,d_{16})	12 (1/0/0)	17 (1/0/0)	21 (1/3/0)
#<data>.L, ([d_{16},B],I,d_{16})	13 (1/0/0)	19 (1/0/0)	23 (1/3/0)
#<data>.([d_{16},B],I,d_{32})	12 (1/0/0)	17 (1/0/0)	22 (1/3/0)
#<data>.([d_{16},B],I,d_{32})	13 (1/0/0)	19 (1/0/0)	24 (1/3/0)
#<data>.W,([d_{32},B],I)	14 (1/0/0)	19 (1/0/0)	22 (1/3/0)
#<data>.L,([d_{32},B],I)	15 (1/0/0)	21 (1/0/0)	24 (1/3/0)
#<data>.W, ([d_{32},B],I,d_{16})	16 (1/0/0)	21 (1/0/0)	24 (1/3/0)
#<data>.L, ([d_{32},B],I,d_{16})	17 (1/0/0)	23 (1/0/0)	26 (1/3/0)
#<data>.W,([d_{32},B],I,d_{32})	16 (1/0/0)	21 (1/0/0)	24 (1/3/0)
#<data>.L,([d_{32},B],I,d_{32})	17 (1/0/0)	23 (1/0/0)	29 (1/3/0)

B = Base address; 0, An, PC, Xn, An + Xn, PC + Xn. Form does not affect timing.
I = Index; 0, Xn
NOTE: Xn cannot be in B and I at the same time. Scaling and size of Xn does not affect timing.

Jump Effective Address

The jump effective address table indicates the number of clock periods needed for the processor to calculate and jump to the specified effective address. Fetch time is only included for the first level of indirection on memory indirect addressing modes. The total number of clock cycles is outside the parentheses, the number of read, prefetch, and write cycles are given inside the parentheses as (r/p/w). They are included in the total clock cycle number.

TABLE D.5.

Address Mode	Best Case	Cache Case	Worst Case
(An)	0 (0/0/0)	2 (0/0/0)	2 (0/0/0)
(d_{16},An)	1 (0/0/0)	4 (0/0/0)	4 (0/0/0)
(xxx).W	0 (0/0/0)	2 (0/0/0)	2 (0/0/0)
(xxx).L	0 (0/0/0)	2 (0/0/0)	2 (0/0/0)
(d_8,An,Xn) or (d_8,PC,Xn)	3 (0/0/0)	6 (0/0/0)	6 (0/0/0)
(d_{16},An,Xn) or (d_{16},PC,Xn)	3 (0/0/0)	6 (0/0/0)	6 (0/0/0)
(B)	3 (0/0/0)	6 (0/0/0)	6 (0/0/0)
(B,d_{16})	5 (0/0/0)	8 (0/0/0)	8 (0/1/0)
(B,d_{32})	9 (0/0/0)	12 (0/0/0)	12 (0/1/0)
([B],I)	8 (1/0/0)	11 (1/0/0)	11 (1/1/0)
([B],I,d_{16})	10 (1/0/0)	13 (1/0/0)	14 (1/1/0)
([B],I,d_{32})	10 (1/0/0)	13 (1/0/0)	14 (1/1/0)
([d_{16},B],I)	10 (1/0/0)	13 (1/0/0)	14 (1/1/0)
([d_{16},B],I,d_{16})	12 (1/0/0)	15 (1/0/0)	17 (1/1/0)
([d_{16},B],I,d_{32})	12 (1/0/0)	15 (1/0/0)	17 (1/1/0)
([d_{32},B],I)	14 (1/0/0)	17 (1/0/0)	19 (1/2/0)
([d_{32},B],I,d_{16})	16 (1/0/0)	19 (1/0/0)	21 (1/2/0)
([d_{32},B],I,d_{32})	16 (1/0/0)	19 (1/0/0)	23 (1/3/0)

B = Base address; 0, An, PC, Xn, An + Xn, PC + Xn. Form does not affect timing.
I = Index; 0, Xn
NOTE: Xn cannot be in B and I at the same time. Scaling and size of Xn does not affect timing.

MOVE Instruction

The MOVE instruction timing table indicates the number of clock periods needed for the processor to fetch, calculate, and perform the MOVE or MOVEA with the specified source and destination effective addresses, including both levels of indirection on memory indirect addressing modes. No additional tables are needed to calculate the total effective execution time for the MOVE or MOVEA instruction. The total number of clock cycles is outside the parentheses, the number of read, prefetch, and write cycles are given inside the parentheses as (r/p/w). They are included in the total clock cycle number.

TABLE D.6.

Best Case

	Destination							
Source Address Mode	An	Dn	(An)	(An) +	− (An)	(d_{16},An)	(xxx).W	(xxx).L
Rn	0 (0/0/0)	0 (0/0/0)	3 (0/0/1)	4 (0/0/1)	3 (0/0/1)	3 (0/0/1)	3 (0/0/1)	5 (0/0/1)
#<data>.B,W	0 (0/0/0)	0 (0/0/0)	3 (0/0/1)	4 (0/0/1)	3 (0/0/1)	3 (0/0/1)	3 (0/0/1)	5 (0/0/1)
#<data>.L	0 (0/0/0)	0 (0/0/0)	3 (0/0/1)	4 (0/0/1)	3 (0/0/1)	3 (0/0/1)	3 (0/0/1)	5 (0/0/1)
(An)	3 (1/0/0)	3 (1/0/0)	6 (1/0/1)	6 (1/0/1)	6 (1/0/1)	6 (1/0/1)	6 (1/0/1)	8 (1/0/1)
(An) +	4 (1/0/0)	4 (1/0/0)	7 (1/0/1)	7 (1/0/1)	7 (1/0/1)	7 (1/0/1)	7 (1/0/1)	9 (1/0/1)
− (An)	3 (1/0/0)	3 (1/0/0)	6 (1/0/1)	6 (1/0/1)	6 (1/0/1)	6 (1/0/1)	6 (1/0/1)	8 (1/0/1)
(d_{16},An) or (d_{16},PC)	3 (1/0/0)	3 (1/0/0)	6 (1/0/1)	6 (1/0/1)	6 (1/0/1)	6 (1/0/1)	6 (1/0/1)	8 (1/0/1)
(xxx).W	3 (1/0/0)	3 (1/0/0)	6 (1/0/1)	6 (1/0/1)	6 (1/0/1)	6 (1/0/1)	6 (1/0/1)	8 (1/0/1)

(continued)

TABLE D.6. (continued)

Best Case

Source Address Mode	Destination An	Dn	(An)	(An) +	– (An)	(d_{16},An)	(xxx).W	(xxx).L
(d_8,An,Xn) or (d_8,PC,Xn)	4 (1/0/0)	4 (1/0/0)	7 (1/0/1)	7 (1/0/1)	7 (1/0/1)	7 (1/0/1)	7 (1/0/1)	9 (1/0/1)
(d_{16},An,Xn) or (d_{16},PC,Xn)	4 (1/0/0)	4 (1/0/0)	7 (1/0/1)	7 (1/0/1)	7 (1/0/1)	7 (1/0/1)	7 (1/0/1)	9 (1/0/1)
(B)	4 (1/0/0)	4 (1/0/0)	7 (1/0/1)	7 (1/0/1)	7 (1/0/1)	7 (1/0/1)	7 (1/0/1)	9 (1/0/1)
(d_{16},B)	6 (1/0/0)	6 (1/0/0)	9 (1/0/1)	9 (1/0/1)	9 (1/0/1)	9 (1/0/1)	9 (1/0/1)	11 (1/0/1)
(d_{32},B)	10 (1/0/0)	10 (1/0/0)	13 (1/0/1)	13 (1/0/1)	13 (1/0/1)	13 (1/0/1)	13 (1/0/1)	15 (1/0/1)
([B],I)	9 (2/0/0)	9 (2/0/0)	12 (2/0/1)	12 (2/0/1)	12 (2/0/1)	12 (2/0/1)	12 (2/0/1)	14 (2/0/1)
([B],I,d_{16})	11 (2/0/0)	11 (2/0/0)	14 (2/0/1)	14 (2/0/1)	14 (2/0/1)	14 (2/0/1)	14 (2/0/1)	16 (2/0/1)
([B],I,d_{32})	11 (2/0/0)	11 (2/0/0)	14 (2/0/1)	14 (2/0/1)	14 (2/0/1)	14 (2/0/1)	14 (2/0/1)	16 (2/0/1)
([d_{16},B],I)	11 (2/0/0)	11 (2/0/0)	14 (2/0/1)	14 (2/0/1)	14 (2/0/1)	14 (2/0/1)	14 (2/0/1)	16 (2/0/1)
([d_{16},B],I,d_{16})	13 (2/0/0)	13 (2/0/0)	16 (2/0/1)	16 (2/0/1)	16 (2/0/1)	16 (2/0/1)	16 (2/0/1)	18 (2/0/1)
([d_{16},B],I,d_{32})	13 (2/0/0)	13 (2/0/0)	16 (2/0/1)	16 (2/0/1)	16 (2/0/1)	16 (2/0/1)	16 (2/0/1)	18 (2/0/1)
([d_{32},B],I)	15 (2/0/0)	15 (2/0/0)	18 (2/0/1)	18 (2/0/1)	18 (2/0/1)	18 (2/0/1)	18 (2/0/1)	20 (2/0/1)
([d_{32},B],I,d_{16})	17 (2/0/0)	17 (2/0/0)	20 (2/0/1)	20 (2/0/1)	20 (2/0/1)	20 (2/0/1)	20 (2/0/1)	22 (2/0/1)
([d_{32},B],I,d_{32})	17 (2/0/0)	17 (2/0/0)	20 (2/0/1)	20 (2/0/1)	20 (2/0/1)	20 (2/0/1)	20 (2/0/1)	22 (2/0/1)

Source Address Mode	Destination (d_8,An,Xn)	(d_{16},An,Xn)	(B)	(d_{16},B)	(d_{32},B)	([B],I)	([B],I,d_{16})	([B],I,d_{32})
Rn	4 (0/0/1)	6 (0/0/1)	5 (0/0/1)	7 (0/0/1)	11 (0/0/1)	9 (1/0/1)	11 (1/0/1)	12 (1/0/1)
#<data>.B,W	4 (0/0/1)	6 (0/0/1)	5 (0/0/1)	7 (0/0/1)	11 (0/0/1)	9 (1/0/1)	11 (1/0/1)	12 (1/0/1)
#<data>.L	4 (0/0/1)	6 (0/0/1)	5 (0/0/1)	7 (0/0/1)	11 (0/0/1)	9 (1/0/1)	11 (1/0/1)	12 (1/0/1)
(An)	8 (1/0/1)	10 (1/0/1)	9 (1/0/1)	11 (1/0/1)	15 (1/0/1)	13 (2/0/1)	15 (2/0/1)	16 (2/0/1)
(An) +	9 (1/0/1)	11 (1/0/1)	10 (1/0/1)	12 (1/0/1)	16 (1/0/1)	14 (2/0/1)	16 (2/0/1)	17 (2/0/1)
– (An)	8 (1/0/1)	10 (1/0/1)	9 (1/0/1)	11 (1/0/1)	15 (1/0/1)	13 (2/0/1)	15 (2/0/1)	16 (2/0/1)
(d_{16},An) or (d_{16},PC)	8 (1/0/1)	10 (1/0/1)	9 (1/0/1)	11 (1/0/1)	15 (1/0/1)	13 (2/0/1)	15 (2/0/1)	16 (2/0/1)
(xxx).W	8 (1/0/1)	10 (1/0/1)	9 (1/0/1)	11 (1/0/1)	15 (1/0/1)	13 (2/0/1)	15 (2/0/1)	16 (2/0/1)
(xxx).L	8 (1/0/1)	10 (1/0/1)	9 (1/0/1)	11 (1/0/1)	15 (1/0/1)	13 (2/0/1)	15 (2/0/1)	16 (2/0/1)
(d_8,An,Xn) or (d_8,PC,Xn)	9 (1/0/1)	10 (1/0/1)	10 (1/0/1)	12 (1/0/1)	16 (1/0/1)	14 (2/0/1)	16 (2/0/1)	17 (2/0/1)
(d_{16},An,Xn) or (d_{16},PC,Xn)	9 (1/0/1)	11 (1/0/1)	10 (1/0/1)	12 (1/0/1)	16 (1/0/1)	14 (2/0/1)	16 (2/0/1)	17 (2/0/1)
(B)	9 (1/0/1)	11 (1/0/1)	10 (1/0/1)	12 (1/0/1)	16 (1/0/1)	14 (2/0/1)	16 (2/0/1)	17 (2/0/1)
(d_{16},B)	11 (1/0/1)	13 (1/0/1)	12 (1/0/1)	14 (1/0/1)	18 (1/0/1)	16 (2/0/1)	18 (2/0/1)	19 (2/0/1)
(d_{32},B)	15 (1/0/1)	17 (1/0/1)	18 (1/0/1)	18 (1/0/1)	22 (1/0/1)	20 (2/0/1)	22 (2/0/1)	23 (2/0/1)
([B],I)	14 (2/0/1)	16 (2/0/1)	17 (2/0/1)	17 (2/0/1)	21 (2/0/1)	19 (3/0/1)	21 (3/0/1)	22 (3/0/1)
([B],I,d_{16})	16 (2/0/1)	18 (2/0/1)	19 (2/0/1)	19 (2/0/1)	23 (2/0/1)	21 (3/0/1)	23 (3/0/1)	24 (3/0/1)
([B],I,d_{32})	16 (2/0/1)	18 (2/0/1)	19 (2/0/1)	19 (2/0/1)	23 (2/0/1)	21 (3/0/1)	23 (3/0/1)	24 (3/0/1)
([d_{16},B],I)	16 (2/0/1)	18 (2/0/1)	19 (2/0/1)	19 (2/0/1)	23 (2/0/1)	21 (3/0/1)	23 (3/0/1)	24 (3/0/1)
([d_{16},B],I,d_{16})	18 (2/0/1)	20 (2/0/1)	21 (2/0/1)	21 (2/0/1)	25 (2/0/1)	23 (3/0/1)	25 (3/0/1)	26 (3/0/1)
([d_{16},B],I,d_{32})	18 (2/0/1)	20 (2/0/1)	21 (2/0/1)	21 (2/0/1)	25 (2/0/1)	23 (3/0/1)	25 (3/0/1)	26 (3/0/1)

Source Address Mode	Destination ([d_{16},B],I)	([d_{16},B],I,d_{16})	([d_{16},B],I,d_{32})	([d_{32},B],I)	([d_{32},B],I,d_{16})	([d_{32},B],I,d_{32})
Rn	11 (1/0/1)	13 (1/0/1)	14 (1/0/1)	15 (1/0/1)	17 (1/0/1)	18 (1/0/1)
#<data>.B,W	11 (1/0/1)	13 (1/0/1)	14 (1/0/1)	15 (1/0/1)	17 (1/0/1)	18 (1/0/1)
#<data>.L	11 (1/0/1)	13 (1/0/1)	14 (1/0/1)	15 (1/0/1)	17 (1/0/1)	18 (1/0/1)
(An)	15 (2/0/1)	17 (2/0/1)	18 (2/0/1)	19 (2/0/1)	21 (2/0/1)	22 (2/0/1)
(An) +	16 (2/0/1)	18 (2/0/1)	19 (2/0/1)	20 (2/0/1)	22 (2/0/1)	23 (2/0/1)
– (An)	15 (2/0/1)	17 (2/0/1)	18 (2/0/1)	19 (2/0/1)	21 (2/0/1)	22 (2/0/1)
(d_{16},An) or (d_{16},PC)	15 (2/0/1)	17 (2/0/1)	18 (2/0/1)	19 (2/0/1)	21 (2/0/1)	22 (2/0/1)
(xxx).W	15 (2/0/1)	17 (2/0/1)	18 (2/0/1)	19 (2/0/1)	21 (2/0/1)	22 (2/0/1)
(xxx).L	15 (2/0/1)	17 (2/0/1)	18 (2/0/1)	19 (2/0/1)	21 (2/0/1)	22 (2/0/1)

TABLE D.6. (continued)

Source Address Mode	Destination					
	([d_{16},B],I)	([d_{16},B],I,d_{16})	([d_{16},B],I,d_{32})	([d_{32},B],I)	([d_{32},B],I,d_{16})	([d_{32},B],I,d_{32})
(d_8,An,Xn) or (d_8,PC,Xn)	16 (2/0/1)	18 (2/0/1)	19 (2/0/1)	20 (2/0/1)	22 (2/0/1)	23 (2/0/1)
(d_{16},An,Xn) or (d_{16},PC,Xn)	16 (2/0/1)	18 (2/0/1)	19 (2/0/1)	20 (2/0/1)	22 (2/0/1)	23 (2/0/1)
(B)	16 (2/0/1)	18 (2/0/1)	19 (2/0/1)	20 (2/0/1)	22 (2/0/1)	23 (2/0/1)
(d_{16},B)	18 (2/0/1)	20 (2/0/1)	21 (2/0/1)	22 (2/0/1)	24 (2/0/1)	25 (2/0/1)
(d_{32},B)	22 (2/0/1)	24 (2/0/1)	25 (2/0/1)	26 (2/0/1)	28 (2/0/1)	29 (2/0/1)
([B],I)	21 (3/0/1)	23 (3/0/1)	24 (3/0/1)	25 (3/0/1)	27 (3/0/1)	28 (3/0/1)
([B],I,d_{16})	23 (3/0/1)	25 (3/0/1)	26 (3/0/1)	27 (3/0/1)	29 (3/0/1)	30 (3/0/1)
([B],I,d_{32})	23 (3/0/1)	25 (3/0/1)	26 (3/0/1)	27 (3/0/1)	29 (3/0/1)	30 (3/0/1)
([d_{16},B],I)	23 (3/0/1)	25 (3/0/1)	26 (3/0/1)	27 (3/0/1)	29 (3/0/1)	30 (3/0/1)
([d_{16},B],I,d_{16})	25 (3/0/1)	27 (3/0/1)	28 (3/0/1)	29 (3/0/1)	31 (3/0/1)	32 (3/0/1)
([d_{16},B],I,d_{32})	25 (3/0/1)	27 (3/0/1)	28 (3/0/1)	29 (3/0/1)	31 (3/0/1)	32 (3/0/1)
([d_{32},B],I)	27 (3/0/1)	29 (3/0/1)	30 (3/0/1)	31 (3/0/1)	33 (3/0/1)	34 (3/0/1)
([d_{32},B],I,d_{16})	29 (3/0/1)	31 (3/0/1)	32 (3/0/1)	33 (3/0/1)	35 (3/0/1)	36 (3/0/1)
([d_{32},B],I,d_{32})	29 (3/0/1)	31 (3/0/1)	32 (3/0/1)	33 (3/0/1)	35 (3/0/1)	36 (3/0/1)

TABLE D.7.

Cache Case

Source Address Mode	Destination							
	An	Dn	(An)	(An) +	– (An)	(d_{16},An)	(xxx).W	(xxx).L
Rn	2 (0/0/0)	2 (0/0/0)	4 (0/0/1)	4 (0/0/1)	5 (0/0/1)	5 (0/0/1)	4 (0/0/1)	6 (0/0/1)
#<data>.B,W	4 (0/0/0)	4 (0/0/0)	6 (0/0/1)	6 (0/0/1)	7 (0/0/1)	7 (0/0/1)	6 (0/0/1)	8 (0/0/1)
#<data>.L	6 (0/0/0)	6 (0/0/0)	8 (0/0/1)	8 (0/0/1)	9 (0/0/1)	9 (0/0/1)	8 (0/0/1)	10 (0/0/1)
(An)	6 (1/0/0)	6 (1/0/0)	7 (1/0/1)	7 (1/0/1)	7 (1/0/1)	7 (1/0/1)	7 (1/0/1)	9 (1/0/1)
(An) +	6 (1/0/0)	6 (1/0/0)	7 (1/0/1)	7 (1/0/1)	7 (1/0/1)	7 (1/0/1)	7 (1/0/1)	9 (1/0/1)
– (An)	7 (1/0/0)	7 (1/0/0)	8 (1/0/1)	8 (1/0/1)	8 (1/0/1)	8 (1/0/1)	8 (1/0/1)	10 (1/0/1)
(d_{16},An) or (d_{16},PC)	7 (1/0/0)	7 (1/0/0)	8 (1/0/1)	8 (1/0/1)	8 (1/0/1)	8 (1/0/1)	8 (1/0/1)	10 (1/0/1)
(xxx).W	6 (1/0/0)	6 (1/0/0)	7 (1/0/1)	7 (1/0/1)	7 (1/0/1)	7 (1/0/1)	7 (1/0/1)	9 (1/0/1)
(xxx).L	6 (1/0/0)	6 (1/0/0)	7 (1/0/1)	7 (1/0/1)	7 (1/0/1)	7 (1/0/1)	7 (1/0/1)	9 (1/0/1)
(d_8,An,Xn) or (d_8,PC,Xn)	9 (1/0/0)	9 (1/0/0)	10 (1/0/1)	10 (1/0/1)	10 (1/0/1)	10 (1/0/1)	10 (1/0/1)	12 (1/0/1)
(d_{16},An,Xn) or (d_{16},PC,Xn)	9 (1/0/0)	9 (1/0/0)	10 (1/0/1)	10 (1/0/1)	10 (1/0/1)	10 (1/0/1)	10 (1/0/1)	12 (1/0/1)
(B)	9 (1/0/0)	9 (1/0/0)	10 (1/0/1)	10 (1/0/1)	10 (1/0/1)	10 (1/0/1)	10 (1/0/1)	12 (1/0/1)
(d_{16},B)	11 (1/0/0)	11 (1/0/0)	12 (1/0/1)	12 (1/0/1)	12 (1/0/1)	12 (1/0/1)	12 (1/0/1)	14 (1/0/1)
(d_{32},B)	15 (1/0/0)	15 (1/0/0)	16 (1/0/1)	16 (1/0/1)	16 (1/0/1)	16 (1/0/1)	16 (1/0/1)	18 (1/0/1)
([B],I)	14 (2/0/0)	14 (2/0/0)	15 (2/0/1)	15 (2/0/1)	15 (2/0/1)	15 (2/0/1)	15 (2/0/1)	17 (2/0/1)
([B],I,d_{16})	16 (2/0/0)	16 (2/0/0)	17 (2/0/1)	17 (2/0/1)	17 (2/0/1)	17 (2/0/1)	17 (2/0/1)	19 (2/0/1)
([B],I,d_{32})	16 (2/0/0)	16 (2/0/0)	17 (2/0/1)	17 (2/0/1)	17 (2/0/1)	17 (2/0/1)	17 (2/0/1)	19 (2/0/1)
([d_{16},B],I)	16 (2/0/0)	16 (2/0/0)	17 (2/0/1)	17 (2/0/1)	17 (2/0/1)	17 (2/0/1)	17 (2/0/1)	19 (2/0/1)
([d_{16},B],I,d_{16})	18 (2/0/0)	18 (2/0/0)	19 (2/0/1)	19 (2/0/1)	19 (2/0/1)	19 (2/0/1)	19 (2/0/1)	21 (2/0/1)
([d_{16},B],I,d_{32})	18 (2/0/0)	18 (2/0/0)	19 (2/0/1)	19 (2/0/1)	19 (2/0/1)	19 (2/0/1)	19 (2/0/1)	21 (2/0/1)
([d_{32},B],I)	20 (2/0/0)	20 (2/0/0)	21 (2/0/1)	21 (2/0/1)	21 (2/0/1)	21 (2/0/1)	21 (2/0/1)	23 (2/0/1)
([d_{32},B],I,d_{16})	22 (2/0/0)	22 (2/0/0)	23 (2/0/1)	23 (2/0/1)	23 (2/0/1)	23 (2/0/1)	23 (2/0/1)	25 (2/0/1)
([d_{32},B],I,d_{32})	22 (2/0/0)	22 (2/0/0)	23 (2/0/1)	23 (2/0/1)	23 (2/0/1)	23 (2/0/1)	23 (2/0/1)	25 (2/0/1)

(continued)

TABLE D.7. (continued)

	Destination							
Source Address Mode	(d_8,An,Xn)	(d_{16},An,Xn)	(B)	(d_{16},B)	(d_{32},B)	([B],I)	([B],I,d_{16})	([B],I,d_{32})
Rn	7 (0/0/1)	9 (0/0/1)	8 (0/0/1)	10 (0/0/1)	14 (0/0/1)	12 (1/0/1)	14 (1/0/1)	15 (1/0/1)
#<data>.B,W	7 (0/0/1)	9 (0/0/1)	8 (0/0/1)	10 (0/0/1)	14 (0/0/1)	12 (1/0/1)	14 (1/0/1)	15 (1/0/1)
(An)	9 (1/0/1)	11 (1/0/1)	10 (1/0/1)	12 (1/0/1)	16 (1/0/1)	14 (2/0/1)	16 (2/0/1)	17 (2/0/1)
(An) +	9 (1/0/1)	11 (1/0/1)	10 (1/0/1)	12 (1/0/1)	16 (1/0/1)	14 (2/0/1)	16 (2/0/1)	17 (2/0/1)
– (An)	10 (1/0/1)	12 (2/0/1)	11 (1/0/1)	13 (1/0/1)	17 (1/0/1)	15 (2/0/1)	17 (2/0/1)	18 (2/0/1)
(d_{16},An) or (d_{16},PC)	10 (1/0/1)	12 (1/0/1)	11 (1/0/1)	13 (1/0/1)	17 (1/0/1)	15 (2/0/1)	17 (2/0/1)	18 (2/0/1)
(xxx).W	9 (1/0/1)	11 (1/0/1)	10 (1/0/1)	12 (1/0/1)	16 (1/0/1)	14 (2/0/1)	16 (2/0/1)	17 (2/0/1)
(xxx).L	9 (1/0/1)	11 (1/0/1)	10 (1/0/1)	12 (1/0/1)	16 (1/0/1)	14 (2/0/1)	16 (2/0/1)	17 (2/0/1)
(d_8,An,Xn) or (d_8,PC,Xn)	12 (1/0/1)	14 (1/0/1)	13 (1/0/1)	15 (1/0/1)	19 (1/0/1)	17 (2/0/1)	19 (2/0/1)	20 (2/0/1)
(d_{16},An,Xn) or (d_{16},PC,Xn)	12 (1/0/1)	14 (1/0/1)	13 (1/0/1)	15 (1/0/1)	19 (1/0/1)	17 (2/0/1)	19 (2/0/1)	20 (2/0/1)
(B)	12 (1/0/1)	14 (1/0/1)	13 (1/0/1)	15 (1/0/1)	19 (1/0/1)	17 (2/0/1)	19 (2/0/1)	20 (2/0/1)
(d_{16},B)	14 (1/0/1)	16 (1/0/1)	15 (1/0/1)	17 (1/0/1)	21 (1/0/1)	19 (2/0/1)	21 (2/0/1)	22 (2/0/1)
(d_{32},B)	18 (1/0/1)	20 (1/0/1)	19 (1/0/1)	21 (1/0/1)	25 (1/0/1)	23 (2/0/1)	25 (2/0/1)	26 (2/0/1)
([B],I)	17 (2/0/1)	19 (2/0/1)	18 (2/0/1)	20 (2/0/1)	24 (2/0/1)	22 (3/0/1)	24 (3/0/1)	25 (3/0/1)
([B],I,d_{16})	19 (2/0/1)	21 (2/0/1)	20 (2/0/1)	22 (2/0/1)	26 (2/0/1)	24 (3/0/1)	26 (3/0/1)	27 (3/0/1)
([B],I,d_{32})	19 (2/0/1)	21 (2/0/1)	20 (2/0/1)	22 (2/0/1)	26 (2/0/1)	24 (3/0/1)	26 (3/0/1)	27 (3/0/1)
([d_{16},B],I)	19 (2/0/1)	21 (2/0/1)	20 (2/0/1)	22 (2/0/1)	26 (2/0/1)	24 (3/0/1)	26 (3/0/1)	27 (3/0/1)
([d_{16},B],I,d_{16})	21 (2/0/1)	23 (2/0/1)	22 (2/0/1)	24 (2/0/1)	28 (2/0/1)	26 (3/0/1)	28 (3/0/1)	29 (3/0/1)
([d_{16},B],I,d_{32})	21 (2/0/1)	23 (2/0/1)	22 (2/0/1)	24 (2/0/1)	28 (2/0/1)	26 (3/0/1)	28 (3/0/1)	29 (3/0/1)
([d_{32},B],I)	23 (2/0/1)	25 (2/0/1)	24 (2/0/1)	26 (2/0/1)	30 (2/0/1)	28 (3/0/1)	30 (3/0/1)	31 (3/0/1)
([d_{32},B],I,d_{16})	25 (2/0/1)	27 (2/0/1)	26 (2/0/1)	28 (2/0/1)	32 (2/0/1)	30 (3/0/1)	32 (3/0/1)	33 (3/0/1)
([d_{32},B],I,d_{32})	25 (2/0/1)	27 (2/0/1)	26 (2/0/1)	28 (2/0/1)	32 (2/0/1)	30 (3/0/1)	32 (3/0/1)	33 (3/0/0)

	Destination					
Source Address Mode	([d_{16},B],I)	([d_{16},B],I,d_{16})	([d_{16},B],I,d_{32})	([d_{32},B],I)	([d_{32},B],I,d_{16})	([d_{32},B],I,d_{32})
Rn	14 (1/0/1)	16 (1/0/1)	17 (1/0/1)	18 (1/0/1)	20 (1/0/1)	21 (1/0/1)
#<data>.B,W	14 (1/0/1)	16 (1/0/1)	17 (1/0/1)	18 (1/0/1)	20 (1/0/1)	21 (1/0/1)
#<data>.L	16 (1/0/1)	18 (1/0/1)	19 (1/0/1)	20 (1/0/1)	22 (1/0/1)	23 (1/0/1)
(An)	16 (2/0/1)	18 (2/0/1)	19 (2/0/1)	20 (2/0/1)	22 (2/0/1)	23 (2/0/1)
(An) +	16 (2/0/1)	18 (2/0/1)	19 (2/0/1)	20 (2/0/1)	22 (2/0/1)	23 (2/0/1)
– (An)	17 (2/0/1)	19 (2/0/1)	20 (2/0/1)	21 (2/0/1)	23 (2/0/1)	24 (2/0/1)
(d_{16},An) or (d_{16},PC)	17 (2/0/1)	19 (2/0/1)	20 (2/0/1)	21 (2/0/1)	23 (2/0/1)	24 (2/0/1)
(xxx).W	16 (2/0/1)	18 (2/0/1)	19 (2/0/1)	20 (2/0/1)	22 (2/0/1)	23 (2/0/1)
(xxx).L	16 (2/0/1)	18 (2/0/1)	19 (2/0/1)	20 (2/0/1)	22 (2/0/1)	23 (2/0/1)
(d_8,An,Xn) or (d_8,PC,Xn)	19 (2/0/1)	21 (2/0/1)	22 (2/0/1)	23 (2/0/1)	25 (2/0/1)	26 (2/0/1)
(d_{16},An,Xn) or (d_{16},PC,Xn)	19 (2/0/1)	21 (2/0/1)	22 (2/0/1)	23 (2/0/1)	25 (2/0/1)	26 (2/0/1)
(B)	19 (2/0/1)	21 (2/0/1)	22 (2/0/1)	23 (2/0/1)	25 (2/0/1)	26 (2/0/1)
(d_{16},B)	21 (2/0/1)	23 (2/0/1)	24 (2/0/1)	25 (2/0/1)	27 (2/0/1)	28 (2/0/1)
(d_{32},B)	25 (2/0/1)	27 (2/0/1)	28 (2/0/1)	29 (2/0/1)	31 (2/0/1)	32 (2/0/1)
([B],I)	24 (3/0/1)	26 (3/0/1)	27 (3/0/1)	28 (3/0/1)	30 (3/0/1)	31 (3/0/1)
([B],I,d_{16})	26 (3/0/1)	28 (3/0/1)	29 (3/0/1)	30 (3/0/1)	32 (3/0/1)	33 (3/0/1)
([B],I,d_{32})	26 (3/0/1)	28 (3/0/1)	29 (3/0/1)	30 (3/0/1)	32 (3/0/1)	33 (3/0/1)
([d_{16},B],I)	26 (3/0/1)	28 (3/0/1)	29 (3/0/1)	30 (3/0/1)	32 (3/0/1)	33 (3/0/1)
([d_{16},B],I,d_{16})	28 (3/0/1)	30 (3/0/1)	31 (3/0/1)	32 (3/0/1)	34 (3/0/1)	35 (3/0/1)
([d_{16},B],I,d_{32})	28 (3/0/1)	30 (3/0/1)	31 (3/0/1)	32 (3/0/1)	34 (3/0/1)	35 (3/0/1)
([d_{32},B],I)	30 (3/0/1)	32 (3/0/1)	33 (3/0/1)	34 (3/0/1)	36 (3/0/1)	37 (3/0/1)
([d_{32},B],I,d_{16})	32 (3/0/1)	34 (3/0/1)	35 (3/0/1)	36 (3/0/1)	38 (3/0/1)	39 (3/0/1)
([d_{32},B],I,d_{32})	32 (3/0/1)	34 (3/0/1)	35 (3/0/1)	36 (3/0/1)	38 (3/0/1)	39 (3/0/1)

TABLE D.8.

Worst Case

Source Address Mode	Destination							
	An	Dn	(An)	(An) +	− (An)	(d_{16},An)	(xxx).W	(xxx).L
Rn	3 (0/1/0)	3 (0/1/0)	5 (0/1/0)	5 (0/1/1)	6 (0/1/1)	7 (0/1/1)	7 (0/1/1)	9 (0/2/1)
#<data>.B,W	3 (0/1/0)	3 (0/1/0)	5 (0/1/0)	8 (0/1/1)	6 (0/1/1)	7 (0/1/1)	7 (0/1/1)	9 (0/2/1)
#<data>.L	5 (0/1/0)	5 (0/1/0)	7 (0/0/1)	7 (0/1/1)	8 (0/1/1)	9 (0/1/1)	9 (0/1/1)	11 (0/2/1)
(An)	7 (1/1/0)	7 (1/1/0)	9 (1/1/1)	9 (1/1/1)	9 (1/1/1)	11 (1/1/1)	11 (1/1/1)	13 (1/2/1)
(An) +	7 (1/1/0)	7 (1/1/0)	9 (1/1/1)	9 (1/1/1)	9 (1/1/1)	11 (1/1/1)	11 (1/1/1)	13 (1/2/1)
− (An)	8 (1/1/0)	8 (1/1/0)	10 (1/1/1)	10 (1/1/1)	10 (1/1/1)	12 (1/1/1)	12 (1/1/1)	14 (1/2/1)
(d_{16},An) or (d_{16},PC)	9 (1/2/0)	9 (1/2/0)	11 (1/2/1)	11 (1/2/1)	11 (1/2/1)	13 (1/2/1)	13 (1/2/1)	15 (1/3/1)
(xxx).W	8 (1/2/0)	8 (1/2/0)	10 (1/2/1)	10 (1/2/1)	10 (1/2/1)	12 (1/2/1)	12 (1/2/1)	14 (1/3/1)
(xxx).L	10 (1/2/0)	10 (1/2/0)	12 (1/2/1)	12 (1/2/1)	12 (1/2/1)	14 (1/2/1)	14 (1/2/1)	16 (1/3/1)
(d_8,An,Xn) or (d_8,PC,Xn)	11 (1/2/0)	11 (1/2/0)	13 (1/2/1)	13 (1/2/1)	13 (1/2/1)	15 (1/2/1)	15 (1/2/1)	17 (1/3/1)
(d_{16},An,Xn) or (d_{16},PC,Xn)	12 (1/2/0)	12 (1/2/0)	14 (1/2/1)	14 (1/2/1)	14 (1/2/1)	16 (1/2/1)	16 (1/2/1)	18 (1/3/1)
(B)	12 (1/2/0)	12 (1/2/0)	14 (1/2/1)	14 (1/2/1)	14 (1/2/1)	16 (1/2/1)	16 (1/2/1)	18 (1/3/1)
(d_{16},B)	15 (1/2/0)	15 (1/2/0)	17 (1/2/1)	17 (1/2/1)	17 (1/3/1)	19 (1/2/1)	19 (1/2/1)	21 (1/3/1)
(d_{32},B)	19 (1/3/0)	19 (1/3/0)	21 (1/3/1)	21 (1/3/1)	21 (1/3/1)	23 (1/3/1)	23 (1/3/1)	25d(1/4/1)
([B],I)	16 (2/2/0)	16 (2/2/0)	18 (2/2/1)	18 (2/2/1)	18 (2/2/1)	20 (2/2/1)	20 (2/2/1)	22d(2/3/1)
([B],I,d_{16})	19 (2/2/0)	19 (2/2/0)	21 (2/2/1)	21 (2/2/1)	21 (2/2/1)	23 (2/2/1)	23 (2/2/1)	25 (2/3/1)
([B],I,d_{32})	20 (2/3/0)	20 (2/3/0)	22 (2/3/1)	22 (2/3/1)	22 (2/3/1)	24 (2/3/1)	24 (2/3/1)	26 (2/4/1)
([d_{16},B],I)	19 (2/2/0)	19 (2/2/0)	21 (2/2/1)	21 (2/2/1)	21 (2/2/1)	23 (2/2/1)	23 (2/2/1)	25 (2/3/1)
([d_{16},B],I,d_{16})	22 (2/3/0)	22 (2/3/0)	24 (2/3/1)	24 (2/3/1)	24 (2/3/1)	26 (2/3/1)	26 (2/3/1)	28 (2/4/1)
([d_{16},B],I,d_{32})	23 (2/3/0)	23 (2/3/0)	25 (2/3/1)	25 (2/3/1)	25 (2/3/1)	27 (2/3/1)	27 (2/3/1)	29 (2/4/1)
([d_{32},B],I)	23 (2/3/0)	23 (2/3/0)	25 (2/3/1)	25 (2/3/1)	25 (2/3/1)	27 (2/3/1)	27 (2/3/1)	29 (2/4/1)
([d_{32},B],I,d_{16})	25 (2/3/0)	25 (2/3/0)	27 (2/3/1)	27 (2/3/1)	27 (2/3/1)	29 (2/3/1)	29 (2/3/1)	31 (2/4/1)
([d_{32},B],I,d_{32})	27 (2/4/0)	27 (2/4/0)	29 (2/4/1)	29 (2/4/1)	29 (2/4/1)	31 (2/4/1)	31 (2/4/1)	33 (2/5/1)

Source Address Mode	Destination							
	(d_8,An,Xn)	(d_{16},An,Xn)	(B)	(d_{16},B)	(d_{32},B)	([B],I)	([B],I,d_{16})	([B],I,d_{32})
Rn	9 (0/1/1)	12 (0/2/1)	10 (0/1/1)	14 (0/2/1)	19 (0/2/1)	14 (1/1/1)	17 (1/2/1)	20 (1/2/1)
#<data>.B,W	9 (0/1/1)	12 (0/2/1)	10 (0/1/1)	14 (0/2/1)	19 (0/2/1)	14 (1/1/1)	17 (1/2/1)	20 (1/2/1)
#<data>.L	11 (0/1/1)	14 (0/2/1)	12 (0/1/1)	16 (0/2/1)	21 (0/2/1)	16 (1/1/1)	19 (1/2/1)	22 (1/2/1)
(An)	11 (1/1/1)	14 (1/2/1)	12 (1/1/1)	16 (1/2/1)	21 (1/2/1)	12 (2/1/1)	19 (2/2/1)	22 (2/2/1)
(An) +	11 (1/1/1)	14 (1/2/1)	12 (1/1/1)	16 (1/2/1)	21 (1/2/1)	12 (2/1/1)	19 (2/2/1)	22 (2/2/1)
− (An)	12 (1/1/1)	15 (1/2/1)	13 (1/1/1)	17 (1/2/1)	22 (1/2/1)	13 (2/1/1)	20 (2/2/1)	23 (2/2/1)
(d_{16},An) or (d_{16},PC)	13 (1/2/1)	16 (2/3/1)	14 (1/2/1)	18 (1/3/1)	23 (1/3/1)	14 (2/2/1)	21 (2/3/1)	24 (2/3/1)
(xxx).W	12 (1/2/1)	15 (1/3/1)	13 (1/2/1)	17 (1/3/1)	22 (1/3/1)	13 (2/2/1)	20 (2/3/1)	23 (2/3/1)
(xxx).L	14 (1/2/1)	17 (1/3/1)	15 (1/2/1)	19 (1/3/1)	24 (1/3/1)	15 (2/2/1)	22 (2/3/1)	25 (2/3/1)
(d_8,An,Xn) or (d_8,PC,Xn)	15 (1/2/1)	18 (1/3/1)	16 (1/2/1)	20 (1/3/1)	25 (1/3/1)	16 (2/2/1)	23 (2/3/1)	26 (2/3/1)
(d_{16},An,Xn) or (d_{16},PC,Xn)	16 (1/2/1)	19 (1/3/1)	17 (1/2/1)	21 (1/3/1)	26 (1/3/1)	17 (2/2/1)	24 (2/3/1)	27 (2/3/1)
(B)	16 (1/2/1)	19 (1/3/1)	17 (1/2/1)	21 (1/3/1)	26 (1/3/1)	17 (2/2/1)	24 (2/3/1)	27 (2/3/1)
(d_{16},B)	19 (1/2/1)	22 (1/3/1)	20 (1/2/1)	24 (1/3/1)	29 (1/3/1)	20 (2/2/1)	27 (2/3/1)	30 (2/3/1)
(d_{32},B)	23 (1/3/1)	26 (1/4/1)	24 (1/3/1)	28 (1/4/1)	33 (1/4/1)	24 (2/3/1)	31 (2/4/1)	34 (2/4/1)
([B],I)	20 (2/2/1)	23 (2/3/1)	21 (2/2/1)	25 (2/3/1)	30 (2/3/1)	21 (3/2/1)	28 (3/3/1)	31 (3/3/1)
([B],I,d_{16})	23 (2/2/1)	26 (2/3/1)	24 (2/2/1)	28 (2/3/1)	33 (2/3/1)	24 (3/2/1)	31 (3/3/1)	34 (3/3/1)
([B],I,d_{32})	24 (2/3/1)	27 (2/4/1)	25 (2/3/1)	29 (2/4/1)	34 (2/4/1)	25 (3/3/1)	32 (3/4/1)	35 (3/4/1)
([d_{16},B],I)	23 (2/2/1)	26 (2/3/1)	24 (2/2/1)	28 (2/3/1)	33 (2/3/1)	24 (3/2/1)	31 (3/3/1)	34 (3/3/1)
([d_{16},B],I,d_{16})	26 (2/3/1)	29 (2/4/1)	27 (2/3/1)	31 (2/4/1)	36 (2/4/1)	27 (3/3/1)	34 (3/4/1)	37 (3/4/1)
([d_{16},B],I,d_{32})	27 (2/3/1)	30 (2/4/1)	28 (2/3/1)	32 (2/4/1)	37 (2/4/1)	28 (3/3/1)	35 (3/4/1)	38 (3/4/1)
([d_{32},B],I)	27 (2/3/1)	30 (2/4/1)	28 (2/3/1)	32 (2/4/1)	37 (2/4/1)	28 (3/3/1)	35 (3/4/1)	38 (3/4/1)
([d_{32},B],I,d_{16})	29 (2/3/1)	32 (2/4/1)	30 (2/3/1)	34 (2/4/1)	39 (2/4/1)	30 (3/3/1)	37 (3/4/1)	40 (3/4/1)
([d_{32},B],I,d_{32})	31 (2/4/1)	34 (2/5/1)	32 (2/4/1)	36 (2/5/1)	41 (2/5/1)	32 (3/4/1)	39 (3/5/1)	42 (3/5/1)

(continued)

TABLE D.8. (continued)

Worst Case

	Destination					
Source Address Mode	([d_{16},B],I)	([d_{16},B],I,d_{16})	([d_{16},B],I,d_{32})	([d_{32},B],I)	([d_{32},B],I,d_{16})	([d_{32},B],I,d_{32})
Rn	17 (1/2/1)	20 (1/2/1)	23 (1/3/1)	22 (1/2/1)	25 (1/3/1)	27 (1/3/1)
#<data>.B,W	17 (1/2/1)	20 (1/2/1)	23 (1/3/1)	22 (1/2/1)	25 (1/3/1)	27 (1/3/1)
#<data>.L	19 (1/2/1)	22 (1/2/1)	25 (1/3/1)	24 (1/2/1)	27 (1/3/1)	29 (1/3/1)
(An)	19 (2/2/1)	22 (2/2/1)	25 (2/3/1)	24 (2/2/1)	27 (2/3/1)	29 (2/3/1)
(An) +	19 (2/2/1)	22 (2/2/1)	25 (2/3/1)	24 (2/2/1)	27 (2/3/1)	29 (2/3/1)
– (An)	20 (2/2/1)	23 (2/2/1)	26 (2/3/1)	25 (2/2/1)	28 (2/3/1)	30 (2/3/1)
(d_{16},An) or (d_{16},PC)	21 (2/3/1)	24 (2/3/1)	27 (2/4/1)	26 (2/3/1)	29 (2/4/1)	31 (2/4/1)
(xxx).W	20 (2/3/1)	23 (2/3/1)	26 (2/4/1)	27 (2/3/1)	28 (2/4/1)	30 (2/4/1)
(xxx).L	22 (2/3/1)	25 (2/3/1)	28 (2/4/1)	29 (2/3/1)	30 (2/4/1)	32 (2/4/1)
(d_8,An,Xn) or (d_8,PC,Xn)	23 (2/3/1)	26 (2/3/1)	29 (2/4/1)	30 (2/3/1)	31 (2/4/1)	33 (2/4/1)
(d_{16},An,Xn) or (d_{16},PC,Xn)	24 (2/3/1)	27 (2/3/1)	30 (2/4/1)	31 (2/3/1)	32 (2/4/1)	34 (2/4/1)
(B)	24 (2/3/1)	27 (2/3/1)	30 (2/4/1)	31 (2/3/1)	32 (2/4/1)	34 (2/4/1)
(d_{16},B)	27 (2/3/1)	30 (2/3/1)	33 (2/4/1)	34 (2/3/1)	35 (2/4/1)	37 (2/4/1)
(d_{32},B)	31 (2/4/1)	34 (2/4/1)	37 (2/5/1)	38 (2/4/1)	39 (2/5/1)	41 (2/5/1)
([B],I)	28 (3/3/1)	31 (3/3/1)	34 (3/4/1)	35 (3/3/1)	36 (3/4/1)	38 (3/4/1)
([B],I,d_{16})	31 (3/3/1)	34 (3/3/1)	37 (3/4/1)	38 (3/3/1)	39 (3/4/1)	41 (3/4/1)
([B],I,d_{32})	32 (3/4/1)	35 (3/4/1)	38 (3/5/1)	39 (3/4/1)	40 (3/5/1)	42 (3/5/1)
([d_{16},B],I)	31 (3/3/1)	34 (3/3/1)	37 (3/4/1)	38 (3/3/1)	39 (3/4/1)	41 (3/4/1)
([d_{16},B],I,d_{16})	34 (3/4/1)	37 (3/4/1)	40 (3/5/1)	41 (3/4/1)	42 (3/5/1)	44 (3/5/1)
([d_{16},B],I,d_{32})	35 (3/4/1)	38 (3/4/1)	41 (3/5/1)	42 (3/4/1)	43 (3/5/1)	45 (3/5/1)
([d_{32},B],I)	35 (3/4/1)	38 (3/4/1)	41 (3/5/1)	42 (3/4/1)	43 (3/5/1)	45 (3/5/1)
([d_{32},B],I,d_{16})	37 (3/4/1)	40 (3/4/1)	43 (3/5/1)	44 (3/4/1)	45 (3/5/1)	47 (3/5/1)
([d_{32},B],I,d_{32})	39 (3/5/1)	42 (3/5/1)	45 (3/6/1)	46 (3/5/1)	47 (3/6/1)	49 (3/6/1)

Special Purpose MOVE Instruction

The special purpose MOVE timing table indicates the number of clock periods needed for the processor to fetch, calculate, and perform the special purpose MOVE operation on the control registers or specified effective address. The total number of clock cycles is outside the parentheses, the number of read, prefetch, and write cycles are given inside the parentheses as (r/p/w). They are included in the total clock cycle number.

TABLE D.9.

	Instruction		Best Case	Cache Case	Worst Case
	EXG	Ry,Rx	0 (0/0/0)	2 (0/0/0)	3 (0/1/0)
	MOVEC	Cr,Rn	3 (0/0/0)	6 (0/0/0)	7 (0/1/0)
	MOVEC	Rn,Cr	9 (0/0/0)	12 (0/0/0)	13 (0/1/0)
	MOVE	PSW,Rn	1 (0/0/0)	4 (0/0/0)	5 (0/1/0)
#	MOVE	PSW,Mem	5 (0/0/1)	5 (0/0/1)	7 (0/1/1)
*	MOVE	EA,CCR	4 (0/0/0)	4 (0/0/0)	5 (0/1/0)
*	MOVE	EA,SR	8 (0/0/0)	8 (0/0/0)	11 (0/2/0)
#*	MOVEM	EA,RL	8 + 4n (n/0/0)	8 + 4n (n/0/0)	9 + 4n (n/1/0)
#*	MOVEM	RL,EA	4 + 3n (0/0/n)	4 + 3n (0/0/n)	5 + 3n (0/1/n)
	MOVEP.W	Dn,(d_{16},An)	8 (0/0/2)	11 (0/0/2)	11 (0/1/2)
	MOVEP.L	Dn,(d_{16},An)	14 (0/0/4)	17 (0/0/4)	17 (0/1/4)

TABLE D.9. (continued)

	Instruction		Best Case	Cache Case	Worst Case
	MOVEP.W	(d_{16},An),Dn	10 (2/0/0)	12 (2/0/0)	12 (2/1/0)
	MOVEP.L	(d_{16},An),Dn	16 (4/0/0)	18 (4/0/0)	18 (4/1/0)
#*	MOVES	EA,Rn	7 (1/0/0)	7 (1/0/0)	8 (1/1/0)
#*	MOVES	Rn,EA	5 (0/0/1)	5 (0/0/1)	7 (0/1/1)
	MOVE	USP	0 (0/0/0)	2 (0/0/0)	3 (0/1/0)
	SWAP	Rx,Ry	1 (0/0/0)	4 (0/0/0)	4 (0/1/0)

n = number of registers to transfer
RL = Register List
* = Add Fetch Effective Address time
# = Add Calculate Effective Address time
#* = Add Calculate Immediate Address time

Arithmetic/Logical Operations

The arithmetic/logical operations timing table indicates the number of clock periods needed for the processor to perform the specified arithmetic/logical operation using the specified addressing mode. It also includes, in worst case, the amount of time needed to prefetch the instruction. Footnotes specify when to add either fetch address or fetch immediate effective address time. This sum gives the total effective execution time for the operation using the specified addressing mode. The total number of clock cycles is outside the parentheses, the number of read, prefetch, and write cycles are given inside the parentheses as (r/p/w). They are included in the total clock cycle number.

TABLE D.10.

	Instruction		Best Case	Cache Case	Worst Case
*	ADD	EA,Dn	0 (0/0/0)	2 (0/0/0)	3 (0/1/0)
*	ADD	EA,An	0 (0/0/0)	2 (0/0/0)	3 (0/1/0)
*	ADD	Dn,EA	3 (0/0/1)	4 (0/0/1)	6 (0/1/1)
*	AND	EA,Dn	0 (0/0/0)	2 (0/0/0)	3 (0/1/0)
*	AND	Dn,EA	3 (0/0/1)	4 (0/0/1)	6 (0/1/1)
*	EOR	Dn,Dn	0 (0/0/0)	2 (0/0/0)	3 (0/1/0)
*	EOR	Dn,Mem	3 (0/0/1)	4 (0/0/1)	6 (0/1/1)
*	OR	EA,Dn	0 (0/0/0)	2 (0/0/0)	3 (0/1/0)
*	OR	Dn,EA	3 (0/0/1)	4 (0/0/1)	6 (0/1/1)
*	SUB	EA,Dn	0 (0/0/0)	2 (0/0/0)	3 (0/1/0)
*	SUB	EA,An	0 (0/0/0)	2 (0/0/0)	3 (0/1/0)
*	SUB	Dn,EA	3 (0/0/1)	4 (0/0/1)	6 (0/1/1)
*	CMP	EA,Dn	0 (0/0/0)	2 (0/0/0)	3 (0/1/0)
*	CMP	EA,An	1 (0/0/0)	4 (0/0/0)	4 (0/1/0)
**	CMP2	EA,Rn	16 (1/0/0)	18 (1/0/0)	18 (1/1/0)
*	MUL.W	EA,Dn	25 (0/0/0)	27 (0/0/0)	28 (0/1/0)
**	MUL.L	EA,Dn	41 (0/0/0)	43 (0/0/0)	44 (0/1/0)
*	DIVU.W	EA,Dn	42 (0/0/0)	44 (0/0/0)	44 (0/1/0)
**	DIVU.L	EA,Dn	76 (0/0/0)	78 (0/0/0)	78 (0/1/0)
*	DIVS.W	EA,Dn	54 (0/0/0)	56 (0/0/0)	56 (0/1/0)
**	DIVS.L	EA,Dn	88 (0/0/0)	90 (0/0/0)	90 (0/1/0)

* Add Fetch Effective Address time
** Add Fetch Immediate Address time

Immediate Arithmetic/Logical Operations

The immediate arithmetic/logical operations timing table indicates the number of clock periods needed for the processor to fetch the source immediate data value, and perform the specified arithmetic/logical operation using the specified destination addressing mode. Footnotes indicate when to add appropriate fetch effective or fetch immediate effective address times. This computation will give the total execution time needed to perform the appropriate immediate arithmetic/logical operation. The total number of clock cycles is outside the parentheses, the number of read, prefetch, and write cycles are given inside the parentheses as (r/p/w). They are included in the total clock cycle number.

TABLE D.11.

	Instruction		Best Case	Cache Case	Worst Case
	MOVEQ	#<data>,Dn	0 (0/0/0)	2 (0/0/0)	3 (0/1/0)
	ADDQ	#<data>,Rn	0 (0/0/0)	2 (0/0/0)	3 (0/1/0)
*	ADDQ	#<data>,Mem	3 (0/0/1)	4 (0/0/1)	6 (0/1/1)
	SUBQ	#<data>,Rn	0 (0/0/0)	2 (0/0/0)	3 (0/1/0)
*	SUBQ	#<data>,Mem	3 (0/0/1)	4 (0/0/1)	6 (0/1/1)
**	ADDI	#<data>,Dn	0 (0/0/0)	2 (0/0/0)	3 (0/1/0)
**	ADDI	#<data>,Mem	3 (0/0/1)	4 (0/0/1)	6 (0/1/1)
**	ANDI	#<data>,Dn	0 (0/0/0)	2 (0/0/0)	3 (0/1/0)
**	ANDI	#<data>,Mem	3 (0/0/1)	4 (0/0/1)	6 (0/1/1)
**	EORI	#<data>,Dn	0 (0/0/0)	2 (0/0/0)	3 (0/1/0)
**	EORI	#<data>,Mem	3 (0/0/1)	4 (0/0/1)	6 (0/1/1)
**	ORI	#<data>,Dn	0 (0/0/0)	2 (0/0/0)	3 (0/1/0)
**	ORI	#<data>,Mem	3 (0/0/1)	4 (0/0/1)	6 (0/1/1)
**	SUBI	#<data>,Dn	0 (0/0/0)	2 (0/0/0)	3 (0/1/0)
**	SUBI	#<data>,Mem	3 (0/0/1)	4 (0/0/1)	6 (0/1/1)
**	CMPI	#<data>,EA	0 (0/0/0)	2 (0/0/0)	3 (0/1/0)

* Add Fetch Effective Address time
** Add Fetch Immediate Address time

Binary Coded Decimal Operations

The binary coded decimal operations table indicates the number of clock periods needed for the processor to perform the specified operation using the given addresing modes, with complete execution times given. No additional tables are needed to calculate total effective execution time for these instructions. The total number of clock cycles is outside the parentheses, the number of read, prefetch, and write cycles are given inside the parentheses as (r/p/w). They are included in the total clock cycle number.

TABLE D.12.

Instruction		Best Case	Cache Case	Worst Case
ABCD	Dn,Dn	4 (0/0/0	4 (0/0/0)	5 (0/1/0)
ABCD	–(An), –(An)	14 (2/0/1)	16 (2/0/1)	17 (2/1/1)
SBCD	Dn,Dn	4 (0/0/0)	4 (0/0/0)	5 (0/1/0)
SBCD	–(An), –(An)	14 (2/0/1)	16 (2/0/1)	17 (2/1/1)
ADDX	Dn,Dn	2 (0/0/0)	2 (0/0/0)	3 (0/1/0)
ADDX	–(An), –(An)	10 (2/0/1)	12 (2/0/1)	13 (2/1/1)
SUBX	Dn,Dn	2 (0/0/0)	2 (0/0/0)	3 (0/1/0)
SUBX	–(An), –(An)	10 (2/0/1)	12 (2/0/1)	13 (2/1/1)
CMPM	(An) +, (An)+	8 (2/0/0)	9 (2/0/0)	10 (2/1/0)
PACK	Dn,Dn#<data>	3 (0/0/0)	6 (0/0/0)	7 (0/1/0)
PACK	–(An), –(An),#<data>	11 (1/0/1)	13 (1/0/1)	13 (1/1/1)
UNPK	Dn,Dn,#<data>	5 (0/0/0)	8 (0/0/0)	9 (0/1/0)
UNPK	–(An), –(An),#<data>	11 (1/0/1)	13 (1/0/1)	13 (1/1/1)

Single Operand Instructions

The single operand instructions table indicates the number of clock periods needed for the processor to perform the specified operation on the given addressing mode. Footnotes indicate when it is necessary to add another table entry to calculate the total effective execution time for the instruction. The total number of clock cycles is outside the parentheses, the number of read, prefetch and write cycles are given inside the parentheses, as (r/p/w). They are included in the total clock cycle number.

Table D.13.

	Instruction		Best Case	Caches Case	Worst Case
	CLR	Dn	0 (0/0/0)	2 (0/0/0)	3 (0/1/0)
#	CLR	Mem	3 (0/0/1)	4 (0/0/1)	6 (0/1/1)
	NEG	Dn	0 (0/0/0)	2 (0/0/0)	3 (0/1/0)
*	NEG	Mem	3 (0/0/1)	4 (0/0/1)	6 (0/1/1)
	NEGX	Dn	0 (0/0/0)	2 (0/0/0)	3 (0/1/0)
*	NEGX	Mem	3 (0/0/1)	4 (0/0/1)	6 (0/1/1)
	NOT	Dn	0 (0/0/0)	2 (0/0/0)	3 (0/1/0)
*	NOT	Mem	3 (0/0/1)	4 (0/0/1)	6 (0/1/1)
	EXT	Dn	1 (0/0/0)	4 (0/0/0)	4 (0/1/0)
	NBCD	Dn	6 (0/0/0)	6 (0/0/0)	6 (0/1/0)
	Scc	Dn	1 (0/0/0)	4 (0/0/0)	4 (0/1/0)
#	Scc	Mem	6 (0/0/1)	6 (0/0/1)	6 (0/1/1)
	TAS	Dn	1 (0/0/0)	4 (0/0/0)	4 (0/1/0)
#	TAS	Mem	12 (1/0/1)	12 (1/0/1)	13 (1/1/1)
*	TST	EA	0 (0/0/0)	2 (0/0/0)	3 (0/1/0)

* Add Fetch Effective Address time

\# Add Calculate Effective Address time

Shift/Rotate Instructions

The shift/rotate instructions table indicates the number of clock periods needed for the processor to perform the specified operation on the given addressing mode. Footnotes indicate when it is necessary to add another table entry to calculate the total effective execution time for the instruction. The number of bits shifted does not affect execution time. The total number of clock cycles is outside the parentheses, the number of read, prefetch, and write cycles are given inside the parentheses as (r/p/w). They are included in the total clock cycle number.

TABLE D.14.

	Instruction		Best Case	Cache Case	Worst Case
	LSL	Dn (Static)	1 (0/0/0)	4 (0/0/0)	4 (0/1/0)
	LSR	Dn (Static)	1 (0/0/0)	4 (0/0/0)	4 (0/1/0)
	LSL	Dn (Dynamic)	3 (0/0/0)	6 (0/0/0)	6 (0/1/0)
	LSR	Dn (Dynamic)	3 (0/0/0)	6 (0/0/0)	6 (0/1/0)
*	LSL	Mem by 1	5 (0/0/1)	5 (0/0/1)	6 (0/1/1)
*	LSR	Mem by 1	5 (0/0/1)	5 (0/0/1)	6 (0/1/1)
	ASL	Dn	5 (0/0/0)	8 (0/0/0)	8 (0/1/0)
	ASR	Dn	3 (0/0/0)	6 (0/0/0)	6 (0/1/0)
*	ASL	Mem by 1	6 (0/0/1)	6 (0/0/1)	7 (0/1/1)
*	ASR	Mem by 1	5 (0/0/1)	5 (0/0/1)	6 (0/1/1)
	ROL	Dn	5 (0/0/0)	8 (0/0/0)	8 (0/1/0)
	ROR	Dn	5 (0/0/0)	8 (0/0/0)	8 (0/1/0)
*	ROL	Mem by 1	7 (0/0/1)	7 (0/0/1)	7 (0/1/1)
*	ROR	Mem by 1	7 (0/0/1)	7 (0/0/1)	7 (0/1/1)
	ROXL	Dn	9 (0/0/0)	12 (0/0/0)	12 (0/1/0)
	ROXR	Dn	9 (0/0/0)	12 (0/0/0)	12 (0/1/0)
*	ROXd	Mem by 1	5 (0/0/1)	5 (0/0/1)	6 (0/1/1)

* Add Fetch Effective Address time
d Is direction of shift/rotate; L or R

Bit Manipulation Instructions

The bit manipulation instructions table indicates the number of clock periods needed for the processor to perform the specified bit operation on the given addressing mode. Footnotes indicate when it is necessary to add another table entry to calculate the total effective execution time for the instruction. The total number of clock cycles is outside the parentheses, the number of read, prefetch, and write cycles are given inside the parentheses as (r/p/w). The are included in the total clock cycle number.

TABLE D.15.

	Instruction		Best Case	Cache Case	Worst Case
	BTST	#<data>,Dn	1 (0/0/0)	4 (0/0/0)	5 (0/1/0)
	BTST	Dn,Dn	1 (0/0/0)	4 (0/0/0)	5 (0/1/0)
**	BTST	#<data>,Mem	4 (0/0/0)	4 (0/0/0)	5 (0/1/0)
*	BTST	Dn,Mem	4 (0/0/0)	4 (0/0/0)	5 (0/1/0)
	BCHG	#<data>,Dn	1 (0/0/0)	4 (0/0/0)	5 (0/1/0)

TABLE D.15. (continued)

	Instruction		Best Case	Cache Case	Worst Case
	BCHG	Dn,Dn	1 (0/0/0)	4 (0/0/0)	5 (0/1/0)
**	BCHG	#<data>,Mem	4 (0/0/1)	4 (0/0/1)	5 (0/1/1)
*	BCHG	Dn,Mem	4 (0/0/1)	4 (0/0/1)	5 (0/1/1)
	BCLR	#<data>,Dn	1 (0/0/0)	4 (0/0/0)	5 (0/1/0)
	BCLR	Dn,Dn	1 (0/0/0)	4 (0/0/0)	5 (0/1/0)
**	BCLR	#<data>,Mem	4 (0/0/1)	4 (0/0/1)	5 (0/1/1)
*	BCLR	#<data>,Mem	4 (0/0/1)	4 (0/0/1)	5 (0/1/1)
	BSET	#<data>,Dn	1 (0/0/0)	4 (0/0/0)	5 (0/1/0)
	BSET	Dn,Dn	1 (0/0/0)	4 (0/0/0)	5 (0/1/0)
**	BSET	#<data>,Mem	4 (0/0/1)	4 (0/0/1)	5 (0/1/1)
*	BSET	Dn,Mem	4 (0/0/1)	4 (0/0/1)	5 (0/1/1)

* Add Fetch Effective Address time
** Add Fetch Immediate Address time

Bit Field Manipulation Instructions

The bit field manipulation instructions table indicates the number of clock periods needed for the processor to perform the specified bit field operation using the given addressing mode. Footnotes indicate when it is necessary to add another table entry to calculate the total effective execution time for the instruction. The total number of clock cycles is outside the parentheses, the number of read, prefetch, and write cycles are given inside the parentheses as (r/p/w). They are included in the total clock cycle number.

TABLE D.16.

	Instruction		Best Case	Cache Case	Worst Case
	BFTST	Dn	3 (0/0/0)	6 (0/0/0)	7 (0/1/0)
#*	BFTST	Mem (<5 bytes)	11 (1/0/0)	11 (1/0/0)	12 (1/1/0)
#*	BFTST	Mem (5 bytes)	15 (2/0/0)	15 (2/0/0)	16 (2/1/0)
	BFCHG	Dn	9 (0/0/0)	12 (0/0/0)	12 (0/1/0)
#*	BFCHG	Mem (<5 bytes)	16 (1/0/1)	16 (1/0/1)	16 (1/1/1)
#*	BFCHG	Mem (5 bytes)	24 (2/0/2)	24 (2/0/2)	24 (2/0/2)
	BFCLR	Dn	9 (0/0/0)	12 (0/0/0)	12 (0/1/0)
#*	BFCLR	Mem (<5 bytes)	16 (1/0/1)	16 (1/0/1)	16 (1/1/1)
#*	BFCLR	Mem (5 bytes)	24 (2/0/2)	24 (2/0/2)	24 (2/0/2)
	BFSET	Dn	9 (0/0/0)	12 (0/0/0)	12 (0/1/0)
#*	BFSET	Mem (<5 bytes)	16 (1/0/1)	16 (1/0/1)	16 (1/1/1)
#*	BFSET	Mem (5 bytes)	24 (2/0/2)	24 (2/0/2)	24 (2/0/2)
	BFEXTS	Dn	5 (0/0/0)	8 (0/0/0)	8 (0/1/0)
	BFEXTS	Mem (<5 Bytes)	13 (1/0/0)	13 (1/0/0)	13 (1/1/0)
	BFEXTS	MEM (5 Bytes)	18 (2/0/0)	18 (2/0/0)	18 (2/1/0)
	BFEXTU	Dn	5 (0/0/0)	8 (0/0/0)	8 (0/1/0)
	BFEXTU	Mem (<5 Bytes)	13 (1/0/0)	13 (1/0/0)	13 (1/1/0)
	BFEXTU	Mem (5 Bytes)	18 (2/0/0)	18 (2/0/0)	18 (2/1/0)
	BFINS	Dn	7 (0/0/0)	10 (0/0/0)	10 (0/1/0)
	BFINS	Mem (<5 Bytes)	14 (1/0/1)	14 (1/0/1)	15 (1/1/1)
	BFINS	Mem (5 Bytes)	20 (2/0/2)	20 (2/0/2)	21 (2/1/2)

(continued)

TABLE D.16. (continued)

Instruction		Best Case	Cache Case	Worst Case
BFFFO	Dn	15 (0/0/0)	18 (0/0/0)	18 (0/1/0)
BFFFO	Mem (<5 Bytes)	24 (1/0/0)	24 (1/0/0)	24 (1/1/0)
BFFFO	Mem (5 Bytes)	32 (2/0/0)	32 (2/0/0)	32 (2/1/0)

#* Add Calculate Immediate Address time
NOTE: A bit field of 32 bits may span 5 bytes that requires two operand cycles to access, or may span 4 bytes that requires only one operand cycle to access.

Conditional Branch Instructions

The conditional branch instructions table indicates the number of clock periods needed for the processor to perform the specified branch on the given branch size, with complete execution times given. No additional tables are needed to calculate total effective execution time for these instructions. The total number of clock cycles is outside the parentheses, the number of read, prefetch, and write cycles are given inside the parentheses as (r/p/w). They are included in the total clock cycle number.

TABLE D.17.

Instruction	Best Case	Cache Case	Worst Case
Bcc (taken)	3 (0/0/0)	6 (0/0/0)	9 (0/2/0)
Bcc.B (not taken)	1 (0/0/0)	4 (0/0/0)	5 (0/1/0)
Bcc.W (not taken)	3 (0/0/0)	6 (0/0/0)	7 (0/1/0)
Bcc.L (not taken)	3 (0/0/0)	6 (0/0/0)	9 (0/2/0)
DBcc (cc=false, count not expired)	3 (0/0/0)	6 (0/0/0)	9 (0/2/0)
DBcc (cc=false, count expired)	7 (0/0/0)	10 (0/0/0)	10 (0/3/0)
DBcc (cc=true)	3 (0/0/0)	6 (0/0/0)	7 (0/1/0)

Control Instructions

The control instructions table indicates the number of clock periods needed for the processor to perform the specified operation. Footnotes specify when it is necessary to add an entry from another table to calculate the total effective execution time for the given instruction. The total number of clock cycles is outside the parentheses, the number of read, prefetch, and write cycles are given inside the parentheses as (r/p/w). They are included in the total clock cycle number.

TABLE D.18.

Instruction	Best Case	Cache Case	Worst Case
ANDI to SR	9 (0/0/0)	12 (0/0/0)	15 (0/2/0)
EORI to SR	9 (0/0/0)	12 (0/0/0)	15 (0/2/0)
ORI to SR	9 (0/0/0)	12 (0/0/0)	15 (0/2/0)
ANDI to CCR	9 (0/0/0)	12 (0/0/0)	15 (0/2/0)
EORI to CCR	9 (0/0/0)	12 (0/0/0)	15 (0/2/0)

TABLE D.18. (continued)

	Instruction	Best Case	Cache Case	Worst Case
	ORI to CCR	9 (0/0/0)	12 (0/0/0)	15 (0/2/0)
	BSR	5 (0/0/1)	7 (0/0/1)	13 (0/2/1)
**	CALLM (type 0)	28 (2/0/6)	30 (2/0/6)	36 (2/2/6)
**	CALLM (type 1) –no stack copy–	48 (5/0/8)	50 (5/0/8)	56 (5/2/8)
**	CALLM (type 1) –no stack copy–	55 (6/0/8)	57 (6/0/8)	64 (6/2/8)
**	CALLM (type 1) –stack copy	63 + 6n (7 + n/0/8 +n)	65 + 6n (7 + n/0/8 + n)	71 + 6n (7 + n/2/8 + n)
#*	CAS (successful compare)	15 (1/0/1)	15 (1/0/1)	16 (1/1/1)
	CAS (unsuccessful compare)	12 (1/0/0)	12 (1/0/0)	13 (1/1/0)
	CAS2 (successful compare)	23 (2/0/2)	25 (2/0/2)	28 (2/2/2)
	CAS2 (unsuccess-ful compare)	19 (2/0/0)	22 (2/0/0)	25 (2/2/0)
*	CHK	8 (0/0/0)	8 (0/0/0)	8 (0/1/0)
**	CHK2 EA,Rn	16 (2/0/0)	18 (2/0/0)	18 (2/1/0)
%	JMP	1 (0/0/0)	4 (0/0/0)	7 (0/2/0)
%	JSR	3 (0/0/1)	5 (0/0/1)	11 (0/2/1)
#	LEA	2 (0/0/0)	2 (0/0/0)	3 (0/2/0)
	LINK.W	3 (0/0/1)	5 (0/0/1)	7 (0/1/1)
	LINK.L	4 (0/0/1)	6 (0/0/1)	10 (0/2/1)
	NOP	2 (0/0/0)	2 (0/0/0)	3(0/1/0)
#	PEA	3 (0/0/1)	5 (0/0/1)	6 (0/1/1)
	RTD	9 (1/0/0)	10 (1/0/0)	12 (1/2/0)
	RTM (type 0)	18 (4/0/0)	19 (4/0/0)	22 (4/2/0)
	RTM (type 1)	31 (6/0/1)	32 (6/0/1)	35 (6/2/1)
	RTR	13 (2/0/0)	14 (2/0/0)	15 (2/2/0)
	RTS	9 (1/0/0)	10 (1/0/0)	12 (1/2/0)
	UNLK	5 (1/0/0)	6 (1/0/0)	7 (1/1/0)

n Number of operand transfers required
* Add Fetch Effective Address time
\# Add Calculate Effective Address time
% Add Jump Effective Address time
** Add Fetch Immediate Address time
#* Add Calculate Immediate Address time

Exception Related Instructions

The exception related instructions table indicates the number of clock periods needed for the processor to perform the specified exception related action. Footnotes specify when it is necessary to add the entry from another table to calculate the total effective execution time for the given instruction. The total number of clock cycles is outside the parentheses, the number of read, prefetch, and write cycles are given inside the parentheses as (r/p/w). They are included in the total clock cycle number.

TABLE D.19.

Instruction	Best Case	Cache Case	Worst Case
BKPT	9 (1/0/0)	10 (1/0/0)	10 (1/0/0)
Interrupt (I-stack)	26 (2/0/4)	26 (2/0/4)	33 (2/2/4)
Interrupt (M-stack)	41 (2/0/8)	41 (2/0/8)	48 (2/2/8)
RESET Instruction	518 (0/0/0)	518 (0/0/0)	519 (0/1/0)
STOP	8 (0/0/0)	8 (0/0/0)	8 (0/0/0)
Trace	25 (1/0/5)	25 (1/0/5)	32 (1/2/5)
TRAP #n	20 (1/0/4)	20 (1/0/4)	27 (1/2/4)
Illegal Instruction	20 (1/0/4)	20 (1/0/4)	27 (1/2/4)
A-Line Trap	20 (1/0/4)	20 (1/0/4)	27 (1/2/4)
F-Line Trap	20 (1/0/4)	20 (1/0/4)	27 (1/2/4)
Privilege Violation	20 (1/0/4)	20 (1/0/4)	27 (1/2/4)
TRAPcc (trap)	23 (1/0/5)	25 (1/0/5)	32 (1/2/5)
TRAPcc (no trap)	1 (0/0/0)	4 (0/0/0)	5 (0/1/0)
TRAPcc.W (trap)	23 (1/0/5)	25 (1/0/5)	33 (1/3/5)
TRAPcc.W (no trap)	3 (0/0/0)	6 (0/0/0)	7 (0/1/0)
TRAPcc.L (trap)	23 (1/0/5)	25 (1/0/5)	33 (1/3/5)
TRAPcc.L (no trap)	5 (0/0/0)	8 (0/0/0)	10 (0/2/0)
TRAPV (trap)	23 (1/0/5)	25 (1/0/5)	32 (1/2/5)
TRAPV (no trap)	1 (0/0/0)	4 (0/0/0)	5 (0/1/0)

Save and Restore Operations

The save and restore operations table indicates the number of clock periods needed for the processor to perform the specified state save, or return from exception, with complete execution times and stack length given. No additional tables are needed to calculate total effective execution time for these operations. The total number of clock cycles is outside parentheses, the number of read, prefetch, and write cycles are given inside the parentheses as (r/p/w). They are included in the total clock cycle number.

TABLE D.20.

Operation	Best Case	Cache Case	Worst Case
Bus Cycle Fault (Short)	42 (1/0/10)	43 (1/0/10)	50 (1/2/10)
Bus Cycle Fault (Long)	79 (1/0/24)	79 (1/0/24)	86 (1/2/24)
RTE (Normal)	20 (4/0/0)	21 (4/4/0)	24 (4/2/0)
RTE (Six-Word)	20 (4/0/0)	21 (4/0/0)	24 (4/2/0)
RTE (Throwaway)*	15 (4/0/0)	16 (4/0/0)	39 (4/0/0)
RTE (Coprocessor)	31 (7/0/0)	32 (7/0/0)	33 (7/1/0)
RTE (Short Fault)	42 (10/0/0)	43 (10/0/0)	45 (10/2/0)
RTE (Long Fault)	91 (24/0/0)	92 (24/0/0)	94 (24/2/0)

* Add the time for RTE on second stack frame.

TIMING TABLES FOR MC68030

The MC68030's timing information is based on the following premises:

- All memory accesses take two clock bus cycles without wait states.
- All operands in memory, including the system stack are long word aligned.
- A 32-bit bus is used for communications between the MC68030 and system memory.
- The data cache is not enabled.
- No exceptions occur (except as specified).
- Required address translations for all external bus cycles are resident in the address translation cache.

Four values are listed for each instruction and effective address and they are:

- Head
- Tail
- Instruction cache case (CC) when the instruction is in the cache but has no overlap.
- Average no-cache case (NCC), when the instruction is not in the cache or the cache is disabled and there is no instruction overlap.

The only instances for which the size of the operand has any effect are the instructions with immediate operands and the ADDA and SUBA instructions. Unless specified otherwise, immediate byte and word operands have identical execution times.

Like MC68020's timing table, four sets of numbers (three numbers are enclosed in parentheses) describe the execution time for cache case and no-cache case. An example of an instruction timing table is:

19 (1/ 3../ 1) = MOVE EA,([d_{16}B], d_{32}

```
Total no. of clocks.....  |       |     |     |
                                   |     |     |
No. of read cycles..............  |     |     |
                                         |     |
Max no. of instruction                  |     |
        access cycles...................|     |
                                               |
No. of write cycles............................|
```

The first five tables deal exclusively with fetching and calculating effective addresses and immediate operands. The rests deal with instruction and operation timings. The timing tables are based on the following premises; all read and write accesses are assumed to take two clock periods.

Fetch Effective Address (FEA)

The fetch effective address table indicates the number of clock periods needed for the processor to calculate and fetch the specified effective address. The effective addresses are divided by their formats (refer to **Effective Address Encoding Summary**). For instruction-cache case and for no-cache case, the total number of clock cycles is outside the parentheses. The number of read, prefetch, and write cycles are given inside the parentheses as (r/p/w). The read, prefetch, and write cycles are included in the total clock cycle number.

All timing data assumes two clock reads and writes.

TABLE D.21.

	Address Mode	Head	Tail	I-Cache Case	No-Cache Case
Single Effective Address Instruction Format					
%	Dn	—	—	0 (0/0/0)	0 (0/0/0)
%	An	—	—	0 (0/0/0)	0 (0/0/0)
	(An)	1	1	3 (1/0/0)	3 (1/0/0)
	(An) +	0	1	3 (1/0/0)	3 (1/0/0)
	–(An)	2	2	4 (1/0/0)	4 (1/0/0)
	(d_{16},An) or (d_{16},PC)	2	2	4 (1/0/0)	4 (1/1/0)
	(xxx).W	2	2	4 (1/0/0)	4 (1/1/0)
	(xxx).L	1	0	4 (1/0/0)	5 (1/1/0)
	#<data>.B	2	0	2 (0/0/0)	2 (0/1/0)
	#<data>.W	2	0	2 (0/0/0)	2 (0/1/0)
	#<data>.L	4	0	4 (0/0/0)	4 (0/1/0)
Brief Format Extension Word					
	(dg,An,Xn) or (dg,PC,Xn)	4	2	6 (1/0/0)	6 (1/1/0)
Full Format Extension Word(s)					
	(d_{16},An) or (d_{16},PC)	2	0	6 (1/0/0)	7 (1/1/0)
	(d_{16},An,Xn) or (d_{16},PC,Xn)	4	0	6 (1/0/0)	7 (1/1/0)
	([d_{16},An]) or ([d_{16},PC])	2	0	10 (2/0/0)	10 (2/1/0)
	([d_{16},An],Xn) or ([d_{16},PC],Xn)	2	0	10 (2/0/0)	10 (2/1/0)
	([d_{16},An],d_{16}) or ([d_{16},PC],d_{16})	2	0	12 (2/0/0)	13 (2/2/0)
	([d_{16},An],Xn,d_{16}) or ([d_{16},PC],Xn,d_{16})	2	0	12 (2/0/0)	13 (2/2/0)
	([d_{16},An]d_{32}) or ([d_{16},PC],d_{32})	2	0	12 (2/0/0)	14 (2/2/0)
	([d_{16},An],Xn,d_{32}) or ([d_{16},PC],Xn,d_{32})	2	0	12 (2/0/0)	14 (2/2/0)
	(B)	4	0	6 (1/0/0)	7 (1/1/0)
	(d_{16},B)	4	0	8 (1/0/0)	10 (1/1/0)
	(d_{32},B)	4	0	12 (1/0/0)	13 (1/2/0)
	([B])	4	0	10 (2/0/0)	10 (2/1/0)
	([B],I)	4	0	10 (2/0/0)	10 (2/1/0)
	([B],d_{16})	4	0	12 (2/0/0)	13 (2/1/0)
	([B],I,d_{16})	4	0	12 (2/0/0)	13 (2/1/0)
	([B],d_{32})	4	0	12 (2/0/0)	14 (2/2/0)
	([B]),I,d_{32})	4	0	12 (2/0/0)	14 (2/2/0)
	([d_{16},B])	4	0	12 (2/0/0)	13 (2/1/0)
	([d_{16},B],I)	4	0	12 (2/0/0)	13 (2/1/0)
	([d_{16},B],d_{16})	4	0	14 (2/0/0)	16 (2/2/0)
	([d_{16},B],I,d_{16})	4	0	14 (2/0/0)	16 (2/2/0)
	([d_{16},B],d_{32})	4	0	14 (2/0/0)	17 (2/2/0)
	([d_{16},B],I,d_{32})	4	0	14 (2/0/0)	17 (2/2/0)
	([d_{32},B])	4	0	16 (2/0/0)	17 (2/2/0)
	([d_{32},B],I)	4	0	16 (2/0/0)	17 (2/2/0)
	([d_{32},B],d_{16})	4	0	18 (2/0/0)	20 (2/2/0)

TABLE D.21. (continued)

Address Mode	Head	Tail	I-Cache Case	No-Cache Case
$([d_{32},B],I,d_{16})$	4	0	18 (2/0/0)	20 (2/2/0)
$([d_{32},B],d_{32})$	4	0	18 (2/0/0)	21 (2/3/0)
$([d_{32},B],I,d_{32})$	4	0	18 (2/0/0)	21 (2/3/0)

B = Base Address; 0, An, PC, Xn, An + Xn, PC + Xn. Form does not affect timing.
I = Index; 0, Xn
% = No Clock Cycles incurred by Effective Address Fetch.
NOTE: Xn cannot be in B and I at the same time. Scaling and size of Xn does not affect timing.

Fetch Immediate Effective Address (FIEA)

The fetch immediate effective address table indicates the number of clock periods needed for the processor to fetch the immediate source operand and to calculate and fetch the specified destination operand. In the case of two word instruction, this table indicates the number of clock periods needed for the procesor to fetch the second word of the instruction and to calculate and fetch the specified source operand or single operand. The effective addresses are divided by their formats (refer to **Effective Address Encoding Summary**). For instruction-cache case and for no-cache case, the total number of clock cycles is outside the parentheses. The number of read, prefetch, and write cycles are given inside the parentheses as (r/p/w). The read, prefetch, and write cycles are included in the total clock cycle number.

All timing data assumes two clock reads and writes.

TABLE D.22.

	Address Mode	Head	Tail	I-Cache Case	No-Cache Case
	Single Effective Address Instruction Format				
%	#<data>.W,Dn	2 + op head	0	2 (0/0/0)	2 (0/1/0)
%	#<data>.L,Dn	4 + op head	0	4 (0/0/0)	4 (0/1/0)
	#<data>.W,(An)	1	1	3 (1/0/0)	4 (1/1/0)
	#<data>.L,(An)	1	0	4 (1/0/0)	5 (1/1/0)
	#<data>.W,(An)+	2	1	5 (1/0/0)	5 (1/1/0)
	#<data>.L,(An)+	4	1	7 (1/0/0)	7 (1/1/0)
	#<data>.W,–(An)	2	2	4 (1/0/0)	4 (1/1/0)
	#<data>.L,–(An)	2	0	4 (1/0/0)	5 (1/1/0)
	#<data>.W,$(D_{16}$,An)	2	0	4 (1/0/0)	5 (1/1/0)
	#<data>.L,(d16,An)	4	0	6 (1/0/0)	8 (1/2/0)
	#<data>.W,$XXX.W	4	2	6 (1/0/0)	6 (1/1/0)
	#<data>.L,$XXX.W	6	2	8 (1/0/0)	8 (1/2/0)
	#<data>.W,$XXX.L	3	0	6 (1/0/0)	7 (1/2/0)
	#<data>.L,$XXX.L	5	0	8 (1/0/0)	9 (1/2/0)
	#<data>.W,#<data>.L	6 + op head	0	6 (0/0/0)	6 (0/2/0)
	Brief Format Extension Word				
	#<data>.W,(dg.An,Xn) or (dg,PC,Xn)	6	2	8 (1/0/0)	8 (1/2/0)
	#<data>.L,(dg,An,Xn) or (dg,PC,Xn)	8	2	10 (1/0/0)	10 (1/2/0)

(continued)

TABLE D.22. (continued)

Address Mode	Head	Tail	I-Cache Case	No-Cache Case
Full Format Extension Word(s)				
#<data>.W,(d16,An) or (d16,PC)	4	0	8 (1/0/0)	9 (1/2/0)
#<data>.L,(d16,An) or (d16,PC)	6	0	10 (1/0/0)	11 (1/2/0)
#<data>.W,(d_{16},An,Xn) or d_{16},PC,Xn)	6	0	8 (1/0/0)	9 (1/2/0)
#<data>.L,(d_{16},An,Xn) or (d_{16},PC,Xn)	8	0	10 (1/0/0)	11 (1/2/0)
#<data>.W,([d_{16},An]) or ([d_{16},PC])	4	0	12 (2/0/0)	12 (2/2/0)
#<data>.L([d_{16},An]) or ([d_{16},PC])	6	0	14 (2/0/0)	14 (2/2/0)
#<data>.W,([d_{16},An],Xn) or ([d_{16},PC],Xn)	4	0	12 (2/0/0)	12 (2/2/0)
#<data>.L,([d_{16},An],Xn) or ([d_{16},PC],Xn)	6	0	14 (2/0/0)	14 (2/2/0)
#<data>.W,([d_{16},An],d_{16}) or ([d_{16},PC]d_{16})	4	0	14 (2/0/0)	15 (2/2/0)
#<data>.L,([d_{16},An],d_{16}) or ([d_{16},PC],d_{16})	6	0	16 (2/0/0)	17 (2/3/0)
#<data>.W,([d_{16},An],Xn,d_{16}) or ([d_{16},PC],Xn,d_{16})	4	0	14 (2/0/0)	15 (2/2/0)
#<data>.L,([d_{16},An],Xn,d_{16}) or ([d_{16},PC],Xn,d_{16})	6	0	16 (2/0/0)	17 (2/3/0)
#<data>.W,([d_{16},An],d_{32}) or ([d_{16},PC],d_{32})	4	0	14 (2/0/0)	16 (2/3/0)
#<data>.L,([d_{16},An],d_{32}) or ([d_{16},PC],d_{32})	6	0	16 (2/0/0)	18 (2/3/0)
#<data>.W,([d_{16},An],Xn,d_{32}) or ([d_{16},PC],Xn,d_{32})	4	0	14 (2/0/0)	16 (2/3/0)
#<data>.L,([d_{16},An],Xn,d_{32}) or ([d_{16},PC],Xn,d_{32})	6	0	16 (2/0/0)	18 (2/3/0)
#<data>.W,(B)	6	0	8 (1/0/0)	9 (1/1/0)
#<data>.L,(B)	8	0	10 (1/0/0)	11 (1/2/0)
#<data>.W,(d_{16},B)	6	0	10 (1/0/0)	12 (1/2/0)
#<data>.L,(d_{16},B)	8	0	12 (1/0/0)	14 (1/2/0)
#<data>.W,(d_{32},B)	10	0	14 (1/0/0)	16 (1/2/0)
#<data>.L,(d_{32},B)	12	0	16 (1/0/0)	18 (1/3/0)
#<data>.W,([B])	6	0	12 (2/0/0)	12 (2/1/0)
#<data>.L,([B])	8	0	14 (2/0/0)	14 (2/2/0)
#<data>.W,([B],I)	6	0	12 (2/0/0)	12 (2/1/0)
#<data>.L,([B],I)	8	0	14 (2/0/0)	14 (2/2/0)
#<data>.W,([B],d_{16})	6	0	14 (2/0/0)	15 (2/2/0)
#<data>.L,([B],d_{16})	8	0	16 (2/0/0)	17 (2/2/0)
#<data>.W,([B],I,d_{16})	6	0	14 (2/0/0)	15 (2/2/0)
#<data>.L,([B],I,d_{16})	8	0	16 (2/0/0)	17 (2/2/0)
#<data>.W,([B],d_{32})	6	0	14 (2/0/0)	16 (2/2/0)
#<data>.L,([B],d_{32})	8	0	16 (2/0/0)	18 (2/3/0)
#<data>.W,([B],I,d_{32})	6	0	14 (2/0/0)	16 (2/2/0)
#<data>.L,([B],I,d_{32})	8	0	16 (2/0/0)	18 (2/3/0)
#<data>.W,([d_{16},B])	6	0	14 (2/0/0)	15 (2/2/0)
#<data>.L,([d_{16},B])	8	0	16 (2/0/0)	17 (2/2/0)
#<data>.W,([d_{16},B],I)	6	0	14 (2/0/0)	15 (2/2/0)
#<data>.L,([d_{16},B],I)	8	0	16 (2/0/0)	17 (2/2/0)
#<data>.W,([d_{16},B]d_{16})	6	0	16 (2/0/0)	18 (2/2/0)
#<data>.L,([d_{16},B],d_{16})	8	0	18 (2/0/0)	20 (2/3/0)
#<data>.W,([d_{16},B],I,d_{16})	6	0	16 (2/0/0)	18 (2/2/0)
#<data>.L,([d_{16},B],I,d_{16})	8	0	18 (2/0/0)	20 (2/3/0)
#<data>.W,([d_{16},B],d_{32})	6	0	16 (2/0/0)	19 (2/3/0)
#<data>.L,([d_{16},B],d_{32})	8	0	18 (2/0/0)	21 (2/3/0)
#<data>.W,([d_{16},B],I,d_{32})	6	0	16 (2/0/0)	19 (2/3/0)
#<data>.L,([d_{16},B],I,d_{32})	8	0	18 (2/0/0)	21 (2/3/0)

TABLE D.22. (continued)

Address Mode	Head	Tail	I-Cache Case	No-Cache Case
Full Format Extension Word(s)				
#<data>.W,([d_{32},B])	6	0	18 (2/0/0)	19 (2/2/0)
#<data>.L,([d_{32},B])	8	0	20 (2/0/0)	21 (2/3/0)
#<data>.W,([d_{32},B],I)	6	0	18 (2/0/0)	19 (2/2/0)
#<data>.L,([d_{32},B],I)	8	0	20 (2/0/0)	21 (2/3/0)
#<data>.W,([d_{32},B],d_{16})	6	0	20 (2/0/0)	22 (2/3/0)
#<data>.L,([d_{32},B],d_{16})	8	0	22 (2/0/0)	24 (2/3/0)
#<data>.W,([d_{32},B],I,d_{16})	6	0	20 (2/0/0)	22 (2/3/0)
#<data>.L,([d_{32},B],I,d_{16})	8	0	22 (2/0/0)	24 (2/3/0)
#<data>.W,([d_{32},B],d_{32})	6	0	20 (2/0/0)	23 (2/3/0)
#<data>.L,([d_{32},B],d_{32})	8	0	22 (2/0/0)	25 (2/4/0)
#<data>.W,([d_{32},B],I,d_{32})	6	0	20 (2/0/0)	23 (2/3/0)
#<data>.L,([d_{32},B],I,d_{32})	8	0	22 (2/0/0)	25 (2/4/0)

B = Base Address: 0, An, PC, Xn, An + Xn, PC + Xn. Form does not affect timing.
I = Index: 0, Xn
% = Total Head for Fetch Immediate Effective Address timing includes the Head Time for the Operation.
NOTE: Xn cannot be in B and I at the same time. Scaling and size of Xn does not affect timing.

Calculate Effective Address (CEA)

The calculate effective address table indicates the number of clock periods needed for the processor to calculate the specified effective address. Fetch time is only included for the first level of indirection on memory indirect addressing modes. The effective addresses are divided by their formats (refer to **Effective Address Encoding Summary**). For instruction-cache case and for no-cache case, the total number of clock cycles is outside the parentheses. The number of read, prefetch, and write cycles are given inside the parentheses as (r/p/w). The read, prefetch, and write cycles are included in the total clock cycle number.

All timing data assumes two clock reads and writes.

TABLE D.23.

	Address Mode	Head	Tail	I-Cache Case	No-Cache Case
	Single Effective Address Instruction Format				
%	Dn	—	—	0 (0/0/0)	0 (0/0/0)
%	An	—	—	0 (0/0/0)	0 (0/0/0)
	(An)	2 + op head	0	2 (0/0/0)	2 (0/0/0)
	(An)+	0	0	2 (0/0/0)	2 (0/0/0)
	–(An)	2 + op head	0	2 (0/0/0)	2 (0/0/0)
	(d_{16},An) or (d_{16},PC)	2 + op head	0	2 (0/0/0)	2 (0/1/0)
	(xxx).W	2 + op head	0	2 (0/0/0)	2 (0/1/0)
	(xxx).L	4 + op head	0	4 (0/0/0)	4 (0/1/0)

(*continued*)

TABLE D.23. (continued)

Address Mode	Head	Tail	I-Cache Case	No-Cache Case
Brief Format Extension Word				
(dg,An,Xn) or (dg,PC,Xn)	4 + op head	0	4 (0/0/0)	4 (0/1/0)
Full Format Extension Word(s)				
(d_{16},An) or (d_{16},PC)	2	0	6 (0/0/0)	6 (0/1/0)
(d_{16},An,Xn) or (d_{16},PC,Xn)	6 + op head	0	6 (0/0/0)	6 (0/1/0)
([d_{16},An]) or ([d_{16},PC])	2	0	10 (1/0/0)	10 (1/1/0)
([d_{16},An],Xn) or ([d_{16},PC],Xn)	2	0	10 (1/0/0)	10 (1/1/0)
([d_{16},An],d_{16}) or ([d_{16},PC],d_{16})	2	0	12 (1/0/0)	13 (1/2/0)
([d_{16},An],Xn,d_{16}) or ([d_{16},PC],Xn,d_{16})	2	0	12 (1/0/0)	13 (1/2/0)
([d_{16},An],d_{32}) or ([d_{16},PC],d_{32})	2	0	12 (1/0/0)	13 (1/2/0)
([d_{16},An],Xn,d_{32}) or ([d_{16},PC],Xn,d_{32})	2	0	12 (1/0/0)	13 (1/2/0)
(B)	6 + op head	0	6 (0/0/0)	6 (0/1/0)
(d_{16},B)	4	0	8 (0/0/0)	9 (0/1/0)
(d_{32},B)	4	0	12 (0/0/0)	12 (0/2/0)
([B])	4	0	10 (1/0/0)	10 (1/1/0)
([B],I)	4	0	10 (1/0/0)	10 (1/1/0)
([B],d_{16})	4	0	12 (1/0/0)	13 (1/1/0)
([B]),I,d_{16})	4	0	12 (1/0/0)	13 (1/1/0)
([B],d_{32})	4	0	12 (1/0/0)	13 (1/2/0)
([B],Id_{32})	4	0	12 (2/0/0)	13 (1/2/0)
([d_{16},B])	4	0	12 (1/0/0)	13 (1/1/0)
([d_{16},B],I)	4	0	12 (1/0/0)	13 (1/1/0)
([d_{16},B],d_{16})	4	0	14 (1/0/0)	16 (1/2/0)
([d_{16},B],I,d_{16})	4	0	14 (1/0/0)	16 (1/2/0)
([d_{16},B],d_{32})	4	0	14 (1/0/0)	16 (1/2/0)
([d_{16},B],I,d_{32})	4	0	14 (1/0/0)	16 (1/2/0)
([d_{32},B])	4	0	16 (1/0/0)	17 (1/2/0)
([d_{32},B],I)	4	0	16 (1/0/0)	17 (1/2/0)
(d_{32},B],d_{16})	4	0	18 (1/0/0)	20 (1/2/0)
([d_{32},B],I,d_{16})	4	0	18 (1/0/0)	20 (1/2/0)
([d_{32},B],d_{32})	4	0	18 (1/0/0)	20 (1/2/0)
([d_{32},B],I,d_{32})	4	0	18 (1/0/0)	20 (1/3/0)

B = Base address; 0, An, PC, Xn, An + Xn, PC + Xn. Form does not affect timing.
I = Index; 0, Xn
% = No clock cycles incurred by Effective Address Calculation.
NOTE: Xn cannot be in B and I at the same time. Scaling and size of Xn does not affect timing.

Calculate Immediate Effective Address Mode (CIEA)

The calculate immediate effective address table indicates the number of clock periods needed for the processor to fetch the immediate source operand and calculate the specified destination effective address. In the case of two word instructions, this table indicates the number of clock periods needed for the processor to fetch the second word of the instruction and calculate the specified source operand or single operand. Fetch time is only included for the first level of indirection on memory indirect addressing modes.

The effective addresses are divided by their formats (refer to **Effective Address Encoding Summary**). For instruction-cache case and for no-cache case, the total number of clock cycles is outside the parentheses. the number of read, prefetch, and write cycles are given inside the parentheses as (r/p/w). The read, prefetch, and write cycles are included in the total clock cycle number.

All timing data assumes two clock reads and writes.

Table D.24.

Address Mode	Head	Tail	I-Cache Case	No-Cache Case
Single Effective Address Instruction Format				
% #<data>.W,Dn	2 + op head	0	2 (0/0/0)	2 (0/1/0)
% #<data>.L,Dn	4 + op head	0	4 (0/0/0	4 (0/1/0)
% #<data>.W,(An)	2 + op head	0	2 (0/0/0	2 (0/1/0)
% #<data>.L,(An)	4 + op head	0	4 (0/0/0	4 (0/1/0)
#<data>.W,(An)+	2	0	4 (0/0/0)	4 (0/1/0)
#<data>.L,(An)+	4	0	6 (0/0/0)	6 (0/1/0)
% #<data>.W,–(An)	2 + op head	0	2 (0/0/0)	2 (0/1/0)
% #<data>.L,–(An)	4 + op head	0	4 (0/0/0)	4 (0/1/0)
% #<data>.W,(d_{16},An)	4 + op head	0	4 (0/0/0)	4 (0/1/0)
% #<data>.L,(d_{16},An)	6 + op head	0	6 (0/0/0)	7 (0/2/0)
% #<data>.W,$XXX.W	4 + op head	0	4 (0/0/0)	4(0/1/0)
% #<data>.L,$XXX.W	6 + op head	0	6 (0/0/0)	6 (0/2/0)
% #<data>.W,$XXX.L	6 + op head	0	6 (0/0/0)	6 (0/2/0)
% #<data>.L,$XXX.L	8 + op head	0	8 (0/0/0)	8 (0/2/0)
Brief Format Extension Word				
% #<data>.W,(dg,An,Xn) or (dg,PC,Xn)	6 + op head	0	6 (0/0/0)	6 (0/2/0)
% #<data>.L,(dg,An,Xn) or (dg,PC,Xn)	8 + op head	0	8 (0/0/0)	8 (0/2/0)
Full Format Extension Word(s)				
#<data>.W,(d_{16},An) or (d_{16},PC)	4	0	8 (0/0/0)	8 (0/2/0)
#<data>.L,(d_{16},An) or (d_{16},PC)	6	0	10 (0/0/0)	10 (0/2/0)
% #<data>.W,(d_{16},An,Xn) or (d_{16},PC,Xn)	8 + op head	0	8 (0/0/0)	8 (0/2/0)
% #<data>.L,(d_{16},An,Xn) or (d_{16},PC,Xn)	10 + op head	0	10 (0/0/0)	10 (0/2/0)
#<data>.W,([d_{16},An]) or ([d_{16},PC])	4	0	12 (1/0/0)	12 (1/2/0)
#<data>.L([d_{16},An]) or ([d_{16},PC])	6	0	14 (1/0/0)	14 (1/1/0)
#<data>.W,([d_{16},An],Xn) or ([d_{16},PC],Xn)	4	0	12 (1/0/0)	12 (1/2/0)
#<data>.L([d_{16},An],Xn) or ([d_{16},PC],Xn	6	0	14 (1/0/0)	14 (1/1/0)
#<data>.W,([d_{16},An],d_{16}) or ([d_{16},PC],d_{16})	4	0	14 (1/0/0)	15 (1/2/0)
#<data>.L,([d_{16},An],d_{16}) or ([d_{16},PC],d_{16})	6	0	16 (1/0/0)	17 (1/3/0)
#<data>.W,([d_{16},An],d_{16}) or ([d_{16},PC], Xn,d_{16})	4	0	14 (1/0/0)	15 (1/2/0)
#<data>.L,([d_{16},An],Xn,d_{16}) or ([d_{16},PC],Xn,d_{16})	6	0	16 (1/0/0)	17 (1/3/0)
#<data>.W,([d_{16},An],d_{32}) or ([d_{16},PC],d_{32})	4	0	14 (1/0/0)	16 (1/3/0)
#<data>.L,([d_{16},An],d_{32}) or ([d_{16},PC],d_{32})	6	0	16 (1/0/0)	17 (1/3/0)
#<data>.W,([d_{16},An],Xn,d_{32}) or ([d_{16},PC],Xn,d_{32})	4	0	14 (1/0/0)	15 (1/3/0)
#<data>.L,([d_{16},An],Xn,d_{32}) or ([d_{16},PC],Xn,d_{32})	6	0	16 (1/0/0)	17 (1/3/0)
% #<data>.W,(B)	8 + op head	0	8 (0/0/0)	8 (0/1/0)

(continued)

TABLE D.24.

Address Mode	Head	Tail	I-Cache Case	No-Cache Case
Full Format Extension Word(s)				
% #<data>.L,(B)	10 + op head	0	10 (0/0/0)	10 (0/2/0)
#<data>.W,(d_{16},B)	6	0	10 (0/0/0)	11 (0/2/0)
#<data>.L,(d_{16},B)	8	0	12 (0/0/0)	13 (0/2/0)
#<data>.W,(d_{32},B)	6	0	14 (0/0/0)	15 (0/2/0)
#<data>.L,(d_{32},B)	8	0	16 (0/0/0)	17 (0/3/0)
#<data>.W,([B])	6	0	12 (1/0/0)	12 (1/1/0)
#<data>.L,([B])	8	0	14 (1/0/0)	14 (1/2/0)
#<data>.W,([B]),I)	6	0	12 (1/0/0)	12 (1/1/0)
#<data>.L,([B],I)	8	0	14 (1/0/0)	14 (1/2/0)
#<data>.W,([B],d_{16})	6	0	14 (1/0/0)	15 (1/2/0)
#<data>.L,([B],d_{16})	8	0	16 (1/0/0)	17 (1/2/0)
#<data>.W,([B],I,d_{16})	6	0	14 (1/0/0)	15 (1/2/0)
#<data>.L,([B],I,d_{16})	8	0	16 (2/0/0)	17 (1/2/0)
#<data>.W,([B],d_{32})	6	0	14 (1/0/0)	15 (1/2/0)
#<data>.L,([B],d_{32})	8	0	16 (1/0/0)	17 (1/3/0)
#<data>.W,([B],I,d_{32})	6	0	14 (1/0/0)	15 (1/2/0)
#<data>.L,([B],I,d_{32})	8	0	16 (1/0/0)	17 (1/3/0)
#<data>.W,([d_{16},B])	6	0	14 (1/0/0)	15 (1/2/0)
#<data>.L,([d_{16},B])	8	0	16 (1/0/0)	17 (1/2/0)
#<data>.W,([d_{16},B],I)	6	0	14 (1/0/0)	15 (1/2/0)
#<data>.L,([d_{16},B],I)	8	0	16 (1/0/0)	17 (1/2/0)
#<data>.W,([d_{16},B],d_{16})	6	0	16 (1/0/0)	18 (1/2/0)
#<data>.L,([d_{16},B],d_{16})	8	0	18 (1/0/0)	20 (1/3/0)
#<data>.W,([d_{16},B],I,d_{16})	6	0	16 (1/0/0)	18 (1/2/0)
#<data>.L,([d_{16},B],I,d_{16})	8	0	18 (1/0/0)	20 (1/3/0)
#<data>.W,([d_{16},B],d_{32})	6	0	16 (1/0/0)	18 (1/3/0)
#<data>.L,([d_{16},B],d_{32})	8	0	18 (1/0/0)	20 (1/3/0)
#<data>.W, ([d_{16},B],I,d_{32})	6	0	16 (1/0/0)	18 (1/3/0)
#<data>.L,([d_{16},B],I,d_{32})	8	0	18 (1/0/0)	20 (1/3/0)
#<data>.W,([d_{32},B])	6	0	18 (1/0/0)	19 (1/2/0)
#<data>.L,([d_{32},B])	8	0	20 (1/0/0)	21 (1/3/0)
#<data>.W,([d_{32},B],I)	6	0	18 (1/0/0)	19 (1/2/0)
#<data>.L,([d_{32},B],I)	8	0	20 (1/0/0)	21 (1/3/0)
#<data>.W,([d_{32},B],d_{16})	6	0	20 (1/0/0)	22 (1/3/0)
#<data>.L,([d_{32},B],d_{16})	8	0	22 (1/0/0)	24 (1/3/0)
#<data>.W,([d_{32},B],I,d_{16})	6	0	20 (1/0/0)	22 (1/3/0)
#<data>.L,([d_{32},B],I,d_{16})	8	0	22 (1/0/0)	24 (1/3/0)
#<data>.W,([d_{32},B],d_{32})	6	0	20 (1/0/0)	22 (1/3/0)
#<data>.L,([d_{32},B],d_{32})	8	0	22 (1/0/0)	24 (1/4/0)
#<data>.W,([d_{32},B],I,d_{32})	6	0	20 (1/0/0)	22 (1/3/0)
#<data>.L,([d_{32},B],I,d_{32})	8	0	22 (1/0/0)	24 (1/4/0)

B = Base address; 0, An, PC, An, An + Xn, PC + Xn. Form does not affect timing.
I = Index; 0, Xn
% = Total Head for Address Timing includes the Head Time for the Operation.
NOTE: Xn cannot be in B and I at the same time. Scaling and size of Xn does not affect timing

Jump Effective Address Mode

The jump effective address table indicates the number of clock periods needed for the processor to calculate the specified effective address for the JMP or JSR instructions. Fetch time is only included for the first level of indirection on memory indirect addressing modes. The effective addresses are divided by their formats (refer to **Effective Address Encoding Summary**). For instruction-cache case and for no-cache case, the total number of clock cycles is outside the parentheses. The number of read, prefetch, and write cycles are included in the total clock cycle number.

All timing data assumes two clock reads and writes.

TABLE D.25.

	Address Mode	Head	Tail	I-Cache Case	No-Cache Case
Single Effective Address Instruction Format					
%	(An)	2 + op head	0	2 (0/0/0)	2 (0/0/0)
%	(d_{16},An)	4 + op head	0	4 (0/0/0)	4 (0/0/0)
%	(xxx).W	2 + op head	0	2 (0/0/0)	2 (0/0/0)
%	(xxx).L	2 + op head	0	2 (0/0/0)	2 (0/0/0)
Brief Format Extension Word					
%	(d_8,.An,Xn) or (d_8,PC,Xn)	6 + op head	0	6 (0/0/0)	6 (0/0/0)
Full Format Extension Word(s)					
	(d_{16},An) or (d_{16},PC)	2	0	6 (0/0/0)	6 (0/0/0)
%	(d_{16},An,Xn) or (d_{16},PC,Xn)	6 + op head	0	6 (0/0/0)	6 (0/0/0)
	([d_{16},An]) or ([d_{16},PC])	2	0	10 (1/0/0)	10 (1/1/0)
	([d_{16},An],Xn) or ([d_{16},PC],Xn)	2	0	10 (1/0/0)	10 (1/1/0)
	([d_{16},An],d_{16}) or ([d_{16},PC],d_{16})	2	0	12 (1/0/0)	12 (1/1/0)
	([d_{16},An],Xn,d_{16}) or ([d_{16},PC],Xn,d_{16})	2	0	12 (1/0/0)	12 (1/1/0)
	([d_{16},An],d_{32}) or ([d_{16},PC],d_{32})	2	0	12 (1/0/0)	12 (1/1/0)
	([d_{16},An],Xn,d_{32}) or ([d_{16},PC],Xn,d_{32})	2	0	12 (1/0/0)	12 (1/1/0)
%	(B)	6 + op head	0	6 (0/0/0)	6 (0/0/0)
	(d_{16},B)	4	0	8 (0/0/0)	9 (0/1/0)
	(d_{32},B)	4	0	12 (0/0/0)	13 (0/1/0)
	([B])	4	0	10 (1/0/0)	10 (1/1/0)
	([B],I)	4	0	10 (1/0/0)	10 (1/1/0)
	([B],d_{16})	4	0	12 (1/0/0)	12 (1/1/0)
	([B],I,d_{16})	4	0	12 (1/0/0)	12 (1/1/0)
	([B],d_{32})	4	0	12 (1/0/0)	12 (1/1/0)
	([B],d_{32})	4	0	12 (1/0/0)	12 (1/1/0)
	([B],I,d_{32})	4	0	12 (1/0/0)	12 (1/1/0)
	([d_{16},B])	4	0	12 (1/0/0)	13 (1/1/0)
	([d_{16},B],I)	4	0	12 (1/0/0)	13 (1/1/0)
	([d_{16},B],d_{16})	4	0	14 (1/0/0)	15 (1/1/0)
	([d_{16},B],I,d_{16})	4	0	14 (1/0/0)	15 (1/1/0)
	([d_{16},B],d_{32})	4	0	14 (1/0/0)	15 (1/1/0)

(continued)

TABLE D.25. (continued)

Address Mode	Head	Tail	I-Cache Case	No-Cache Case
Full Format Extension Word(s)				
([d_{16},B]),I,d_{32})	4	0	14 (1/0/0)	15 (1/1/0)
([d_{32},B])	4	0	16 (1/0/0)	17 (1/2/0)
([d_{32},B],I)	4	0	16 (1/0/0)	17 (1/2/0)
([d_{32},B],d_{16})	4	0	18 (1/0/0)	19 (1/2/0)
([d_{32},B],I,d_{16})	4	0	18 (1/0/0)	19 (1/2/0)
([d_{32},B],d_{32})	4	0	18 (1/0/0)	19 (1/2/0)
([d_{32},B]I,d_{32})	4	0	18 (1/0/0)	19 (1/2/0)

B = Base address; 0, An, PC, Xn, An + Xn, PC + Xn. Form does not affect timing.
I = Index; 0, Xn
% = Total Head for Address Timing includes the Head Time for the Operation.
NOTE: Xn cannot be in B and I at the same time. Scaling and size of Xn does not affect timing

MOVE Instruction

The MOVE instruction timing table indicates the number of clock periods needed for the processor to calculate the destination effective address and perform the MOVE or MOVEA instruction, including the first level of indirection on memory indirect addressing modes. The fetch effective address table is needed on most MOVE operations (source, destination dependent). The destination effective addresses are divided by their formats (refer to **Effective Address Encoding Summary**). For instruction cache case and no-cache case, the total number of clock cycles is outside the parentheses. The number of read, prefetch, and write cycles are given inside the parentheses as (r/p/w). The read, prefetch, and write cycles are included in the total clock cycle number.

All timing data assumes two clock reads and writes.

TABLE D.26.

	MOVE Source,Destination	Head	Tail	I-Cache Case	No-Cache Case
	Single Effective Address Instruction Format				
	MOVE Rn, Dn	2	0	2 (0/0/0)	2 (0/1/0)
	MOVE Rn, An	2	0	2 (0/0/0)	2 (0/1/0)
*	MOVE EA,An	0	0	2 (0/0/0)	2 (0/1/0)
*	MOVE EA,Dn	0	0	2 (0/0/0)	2 (0/1/0)
	MOVE Rn,(An)	0	1	3 (0/0/1)	4 (0/1/1)
*	MOVE SOURCE, (An)	2	0	4 (0/0/1)	5 (0/1/1)
	MOVE Rn,(An)+	0	1	3 (0/0/1)	4 (0/1/1)
*	MOVE SOURCE, (An)+	2	0	4 (0/0/1)	5 (0/1/1)
	MOVE Rn,–(An)	0	2	4 (0/0/1)	4 (0/1/1)
*	MOVE SOURCE, –(An)	2	0	4 (0/0/1)	5 (0/1/1)
*	MOVE EA, (d_{16},An)	2	0	4 (0/0/1)	5 (0/1/1)
*	MOVE EA,XXX.W	2	0	4 (0/0/1)	5 (0/1/1)
*	MOVE EA,XXX.L	0	0	6 (0/0/1)	7 (0/2/1)

TABLE D.26. (continued)

Brief Format Extension Word					
*	MOVE EA, (d_8,An,Xn)	4	0	6 (0/0/1)	7 (0/1/1)
Full Format Extension Word(s)					
*	MOVE EA, (d_{16},An) or (d_{16},PC)	2	0	8 (0/0/1)	9 (0/2/1)
*	MOVE EA, (d_{16},An,Xn) or (d_{16},PC,Xn)	2	0	8 (0/0/1)	9 (0/2/1)
*	MOVE EA, ([d_{16},An],Xn) or ([d_{16},PC],Xn)	2	0	10 (1/0/1)	11 (1/2/1)
*	MOVE EA,([d_{16},An],d_{16}) or ([d_{16},PC],d_{16})	2	0	12 (1/0/1)	14 (1/2/1)
*	MOVE EA, ([d_{16},An],Xn,d_{16}) or ([d_{16},PC],Xn,d_{16})	2	0	12 (1/0/1)	14 (1/2/1)
*	MOVE EA,([d_{16},An],Xn,d_{16}) or ([d_{16},PC],Xn,d_{16})	2	0	12 (1/0/1)	14 (1/2/1)
*	MOVE EA,([d_{16},An],d_{32}) or ([d_{16},PC],d_{32})	2	0	14 (1/0/1)	16 (1/3/1)
*	MOVE EA,([d_{16},An],Xn,d_{32}) or ([d_{16},PC],Xn,d_{32})	2	0	14 (1/0/1)	16 (1/3/1)
*	MOVE EA,(B)	4	0	8 (0/0/1)	9 (0/1/1)
*	MOVE EA,(d_{16},B)	4	0	10 (0/0/1)	12 (0/2/1)
*	MOVE EA,(d_{32},B)	4	0	14 (0/0/1)	16 (0/2/1)
*	MOVE EA,([B])	4	0	10 (1/0/1)	11 (1/1/1)
*	MOVE EA,([B],I)	4	0	10 (1/0/1)	11 (1/1/1)
*	MOVE EA,([B],d_{16})	4	0	12 (1/0/1)	14 (1/2/1)
*	MOVE EA,([B],I,d_{16})	4	0	12 (1/0/1)	14 (1/2/1)
*	MOVE EA,([B],d_{32})	4	0	14 (1/0/1)	16 (1/2/1)
*	MOVE EA,([B],I,d_{32})	4	0	14 (1/0/1)	16 (1/2/1)
*	MOVE EA,([d_{16},B])	4	0	12 (1/0/1)	14 (1/2/1)
*	MOVE EA,([d_{16},B]),I	4	0	12 (1/0/1)	14 (1/2/1)
*	MOVE EA,([d_{16},B],d_{16})	4	0	14 (1/0/1)	17 (1/2/1)
*	MOVE EA,([d_{16},B],I,d_{16})	4	0	14 (1/0/1)	17 (1/2/1)
*	MOVE EA,([d_{16},B],d_{32})	4	0	16 (1/0/1)	19 (1/3/1)
*	MOVE EA,([d_{16},B],I,d_{32})	4	0	16 (1/0/1)	19 (1/3/1)
*	MOVE EA,([d_{32},B])	4	0	16 (1/0/1)	18 (1/2/1)
*	MOVE EA,([d_{32},B],I)	4	0	16 (1/0/1)	18 (1/2/1)
*	MOVE EA,([d_{32},B],d_{16})	4	0	18 (1/0/1)	21 (1/3/1)
*	MOVE EA,([d_{32},B],I,d_{16})	4	0	18 (1/0/1)	21 (1/3/1)
*	MOVE EA,([d_{32},B],d_{32})	4	0	20 (1/0/1)	23 (1/3/1)
*	MOVE EA,([d_{32},B],I,d_{32})	4	0	20 (1/0/1)	23 (1/3/1)

*Add Fetch Effective Address Time
Rn Is a Data or Address Register
SOURCE is Memory or Immediate Data Address Mode
EA is any Effective Address

Special Purpose MOVE Instruction

The special purpose MOVE timing table indicates the number of clock periods needed for the processor to fetch, calculate, and perform the special purpose MOVE operation on the control registers or specified effective address. Footnotes indicate when to account for the appropriate effective address times. The total number of clock cycles is outside the parentheses. The number of read, prefetch, and write cycles are given inside the parentheses as (r/p/w). The read, prefetch, and write cycles are included in the total clock cycle number.

All timing data assumes two clock reads and writes.

TABLE D.27.

	Instruction		Head	Tail	I-Cache Case	No-Cache Case
	EXG	Ry,Rx	4	0	4 (0/0/0)	4 (0/1/0)
	MOVEC	Cr,Rn	6	0	6 (0/0/0)	6 (0/1/0)
	MOVEC	Rn,Cr–A	6	0	6 (0/0/0)	6 (0/1/0)
	MOVEC	Rn,Cr–B	4	0	12 (0/0/0)	12 (0/1/0)
	MOVE	CCR,Dn	2	0	4 (0/0/0)	4 (0/1/0)
*	MOVE	CCR,Mem	2	0	4 (0/0/1)	5 (0/1/1)
	MOVE	Dn,CCR	4	0	4 (0/0/0)	4 (0/1/0)
*	MOVE	EA,CCR	0	0	4 (0/0/0)	4 (0/1/0)
	MOVE	SR,Dn	2	0	4 (0/0/0)	4 (0/1/0)
*	MOVE	SR,Mem	2	0	4 (0/0/1)	5 (0/1/1)
#	MOVE	EA,SR	0	0	8 (0/0/0)	10 (0/2/0)
% +	MOVE M	EA,RL	2	0	8 + 4n (n/0/0)	8 + 4n (n/1/0)
%+	MOVEM	RL,EA	2	0	4 + 2n (0/0/n)	4 + 2n (0/1/n)
	MOVEP.W	Dn,(d_{16},An)	4	0	10 (0/0/2)	10 (0/1/2)
	MOVEP.W	(d_{16},An),Dn	2	0	10 (2/0/0)	10 (2/1/0)
	MOVEP.L	Dn,(d_{16},An)	4	0	14 (0/0/4)	14 (0/1/4)
	MOVEP.L	(d_{16},An),Dn	2	0	14 (4/0/0)	14 (4/1/0)
%	MOVES	EA,Rn	3	0	7 (1/0/0)	7 (1/1/0)
%	MOVES	Rn,EA	2	1	5 (0/0/1)	6 (0/1/1)
	MOVE	USP,An	4	0	4 (0/0/0)	4 (0/1/0)
	MOVE	An,USP	4	0	4 (0/0/0)	4 (0/1/0)
	SWAP	Dn	4	0	4 (0/0/0)	4 (0/1/0)

CR–A Control Registers USP, VBR, CAAR, MSP, and ISP +
CR–B Control Registers SFC, DFC, and CACR
n Number of Register to Transfer (n>0)
RL Register List
* Add Calculate Effective Address Time
# Add Fetch Effective Address Time
% Add Calculate Immediate Address Time
MOVEM EA,RL—For n Registers (n >0) and w Wait States
I-Cache Case Timing = $w \le 2$: $(8 + 4n)$
$w > 2$: $(8 + 4n) + (w - 2)n$
Tail = 0 for all Wait States
MOVEM RL,EA—For n Registers (n >0) and w Wait States
I-Cache Timing = $w \le 2$: $(4 + 2n) + (n-1)w$
$w > 2$: $(4 + 2n) + (n-1)w + (w - 2)$
Tail = $w < 2$: $(n - 1)w$
$w > 2$: $(n)w + (n)(w - 2)$

Arithmetical/Logical Instructions

The arithmetical/logical operation timing table indicates the number of clock periods needed for the processor to perform the specified arithmetical/logical instruction using the specified addressing mode. Footnotes indicate when to account for the appropriate fetch effective address or fetch immediate effective address times. For instruction cache case and for no-cache case, the total number of clock cycles is outside the parentheses. The number of read, prefetch, and write cycles are given inside the parentheses as (r/p/w). The read, prefetch, and write cycles are included in the total clock cycle number.

All timing assumes two clock reads and writes.

TABLE D.28.

	Instruction		Head	Tail	I-Cache Case	No-Cache Case
	ADD	Rn,Dn	2	0	2 (0/0/0)	2 (0/1/0)
	ADDA.W	Rn,An	4	0	4 (0/0/0)	4 (0/1/0)
	ADDA.L	Rn,An	2	0	2 (0/0/0)	2 (0/1/0)
*	ADD	EA,Dn	0	0	2 (0/0/0)	2 (0/1/0)
*	ADD.W	EA,An	0	0	4 (0/0/0)	4 (0/1/0)
*	ADDA.L	EA,An	0	0	2 (0/0/0)	2 (0/1/0)
*	ADD	Dn,EA	0	1	3 (0/0/1)	4 (0/1/1)
	AND	Dn,Dn	2	0	2 (0/0/0)	2 (0/1/0)
*	AND	EA,Dn	0	0	2 (0/0/0)	2 (0/1/0)
*	AND	Dn,EA	0	1	3 (0/0/1)	4 (0/1/1)
	EOR	Dn,Dn	2	0	2 (0/0/0)	2 (0/1/0)
*	EOR	Dn,EA	0	1	3 (0/0/1)	4 (0/1/1)
	OR	Dn,Dn	2	0	2 (0/0/0)	2 (0/1/0)
*	OR	EA,Dn	0	0	2 (0/0/0)	2 (0/1/0)
*	OR	Dn,EA	0	1	3 (0/0/1)	4 (0/1/1)
	SUB	Rn,Dn	2	0	2 (0/0/0)	2 (0/1/0)
*	SUB	EA,Dn	0	0	2 (0/0/0)	2 (0/1/0)
*	SUB	Dn,EA	0	1	3 (0/0/1)	4 (0/1/1)
	SUBA.W	Rn,An	4	0	4 (0/0/0)	4 (0/1/0)
	SUBA.L	Rn,An	2	0	2 (0/0/0)	2 (0/1/0)
*	SUBA.W	EA,An	0	0	4 (0/0/0)	4 (0/1/0)
*	SUBA.L	EA,An	0	0	2 (0/0/0)	2 (0/1/0)
	CMP	Rn,Dn	2	0	2 (0/0/0)	2 (0/1/0)
*	CMP	EA,Dn	0	0	2 (0/0/0)	2 (0/1/0)
	CMPA	Rn,An	4	0	4 (0/0/0)	4 (0/1/0)
*	CMPA	EA,An	0	0	4 (0/0/0)	4 (0/1/0)
**+	CMP2	EA,Rn	2	0	20 (1/0/0)	20 (1/1/0)
*+	MULS.W	EA.Dn	2	0	28 (0/0/0)	28 (0/1/0)
**+	MULS.L	EA,Dn	2	0	44 (0/0/0)	44 (0/1/0)
*+	MULU.W	EA,Dn	2	0	28 (0/0/0)	28 (0/1/0)
**+	MULU.L	EA,Dn	2	0	44 (0/0/0)	44 (0/1/0)
+	DIVS.W	Dn,Dn	2	0	56 (0/0/0)	56 (0/1/0)
*+	DIVS.W	EA,Dn	0	0	56 (0/0/0)	56 (0/1/0)
**+	DIVS.L	Dn,Dn	6	0	90 (0/0/0)	90 (0/1/0)
**+	DIVS.L	EA,Dn	0	0	90 (0/0/0)	90 (0/1/0)
+	DIVU.W	Dn,Dn	2	0	44 (0/0/0)	44 (0/1/0)
*+	DIVU.W	EA,Dn	0	0	44 (0/0/0)	44 (0/1/0)
**+	DIVU.L	Dn,Dn	6	0	78 (0/0/0)	78 (0/1/0)
**+	DIVU.L	EA,Dn	0	0	78 (0/0/0)	78 (0/1/0)

*Add Fetch Effective Address Time
**Add Fetch Immediate Effective Address Time
+Indicates Maximum Time (Actual time is data dependent)

Immediate Arithmetical/Logical Instructions

The immediate arithmetical/logical operation timing table indicates the number of clock periods needed for the processor to fetch the source immediate data value, and perform the specified arithmetic/logical operation using the specified destination addressing mode. Footnotes indicate when to account for the appropriate fetch effective or fetch immediate effective address times. For instruction-cache case and for no-cache case, the total number of clock cycles is outside the parentheses. the number of read, prefetch, and write cycles are given inside the parentheses as (r/p/w). The read, prefetch, and write cycles are included in the total clock cycle number.

All timing data assumes two clock reads and writes.

TABLE D.29.

	Instructions		Head	Tail	I-Cache Case	No-Cache Case
	MOVEQ	#<data>,Dn	2	0	2 (0/0/0)	2 (0/1/0)
	ADDQ	#<data>,Rn	2	0	2 (0/0/0)	2 (0/1/0)
*	ADDQ	#<data>,Mem	0	1	3 (0/0/1)	4 (0/1/1)
	SUBQ	#<data>,Rn	2	0	2 (0/0/0)	2 (0/1/0)
*	SUBQ	#<data>,Mem	0	1	3 (0/0/1)	4 (0/1/1)
**	ADDI	#<data>,Dn	2	0	2 (0/0/0)	2 (0/1/0)
**	ADDI	#<data>,Mem	0	1	3 (0/0/1)	4 (0/1/1)
**	ANDI	#<data>,Dn	2	0	2 (0/0/0)	2 (0/1/0)
**	ANDI	#<data>,Mem	0	1	3 (0/0/1)	4 (0/1/1)
**	EORI	#<data>,Dn	2	0	2 (0/0/0)	2 (0/1/0)
**	EORI	#<data>,Mem	0	1	3 (0/0/1)	4 (0/1/1)
**	ORI	#<data>,Dn	2	0	2 (0/0/0)	2 (0/1/0)
**	ORI	#<data>,Mem	0	1	3 (0/0/1)	4 (0/1/1)
**	SUBI	#<data>,Dn	2	0	2 (0/0/0)	2 (0/1/0)
**	SUBI	#<data>,Mem	0	1	3 (0/0/1)	4 (0/1/1)
**	CMPI	#<data>,Dn	2	0	2 (0/0/0)	2 (0/1/0)
**	CMPI	#<data>,Mem	0	0	2 (0/0/0)	2 (0/1/0)

*Add Fetch Effective Address Time
**Add Fetch Immediate Effective Address Time

Binary Coded Decimal and Extended Instructions

The binary coded decimal and extended instruction table indicates the number of clock periods needed for the processor to perform the specified operation using the given addressing modes. No additional tables are needed to calculate total effective execution time for these instructions. For instruction-cache case and for no-cache case, the total number of clock cycles is outside the parentheses. The number of read, prefetch, and write cycles are given inside the parentheses as (r/p/w). The read, prefetch, and write cycles are included in the total clock cycle number.

All timing data assumes two clock reads and writes.

TABLE D.30.

Instruction		Head	Tail	I-Cache Case	No-Cache Case
ABCD	Dn,Dn	0	0	4 (0/0/0)	4 (0/1/0)
ABCD	–(An),–(An)	2	1	13 (2/0/1)	14 (2/1/1)
SBCD	Dn,Dn	0	0	4 (0/0/0)	4 (0/1/0)
SBCD	–(An),–(An)	2	1	13 (2/0/1)	14 (2/1/1)
ADDX	Dn,Dn	2	0	2 (0/0/0)	2 (0/1/0)
ADDX	–(An),–(An)	2	1	9 (2/0/1)	10 (2/1/1)
SUBX	Dn,Dn	2	0	2 (0/0/0)	2 (0/1/0)
SUBX	–(An),–(An)	2	1	9 (2/0/1)	10 (2/1/1)
CMPM	(An)+,(An)+	0	0	8 (2/0/0)	8 (2/1/0)
PACK	Dn,Dn,#<data>	6	0	6 (0/0/0)	6 (0/1/0)
PACK	–(An),–(An),#<data>	2	1	11 (1/0/1)	11 (1/1/1)
UNPK	Dn,Dn,#<data>	8	0	8 (0/0/0)	8 (0/1/0)
UNPK	–(An), (An),#<data>	2	1	11 (1/0/1)	11 (1/1/1)

Single Operand Instructions

The single operand instructions table indicates the number of clock periods needed for the processor to perform the specified operation on the given addressing mode. Footnotes indicate when it is necessary to account for the appropriate effective address time. For instruction-cache case and for no-cache case, the total number of clock cycles is outside the parentheses. The number of read, prefetch, and write cycles are given inside the parentheses as (r/p/w). The read, prefetch, and write cycles are included in the total clock cycle number.

All timing data assumes two clock reads and writes.

TABLE D.31.

		Instruction	Head	Tail	I-Cache Case	No-Cache Case
	CLR	Dn	2	0	2 (0/0/0)	2 (0/1/0)
*	CLR	Mem	0	1	3 (0/0/1)	4 (0/1/1)
	NEG	Dn	2	0	2 (0/0/0)	2 (0/1/0)
*	NEG	Mem	0	1	3 (0/0/1)	4 (0/1/1)
	NEGX	Dn	2	0	2 (0/0/0)	2 (0/1/0)
*	NEGX	Mem	0	1	3 (0/0/1)	4 (0/1/1)
	NOT	Dn	2	0	2 (0/0/0)	2 (0/1/0)
*	NOT	Mem	0	1	3 (0/0/1)	4 (0/1/1)
	EXT	Dn	4	0	4 (0/0/0)	4 (0/1/0)
	NBCD	Dn	0	0	6 (0/0/0)	6 (0/1/0)
	Scc	Dn	4	0	4 (0/0/0)	4 (0/1/0)
**	Scc	Mem	0	1	5 (0/0/1)	5 (0/1/1)
	TAS	Dn	4	0	4 (0/0/0)	4 (0/1/0)
**	TAS	Mem	3	0	12 (1/0/1)	12 (1/1/1)
	TST	Dn	0	0	2 (0/0/0)	2 (0/1/0)
*	TST	Mem	0	0	2 (0/0/0)	2 (0/1/0)

* Add Fetch Effective Address Time
** Add Calculate Effective Address Time

Shift/Rotate Instructions

The shift/rotate instruction table indicates the number of clock periods needed for the processor to perform the specified operation on the given addressing mode. Footnotes indicate when it is necessary to account for the appropriate effective address time. The number of bits shifted does not affect the execution time, unless noted. For instruction-cache case and for no-cache case, the total number of clock cycles is outside the parentheses. The number of read, prefetch, and write cycles are given inside the parentheses as (r/p/w). The read, prefetch, and write cycles are included in the total clock cycle number.

All timing data assumes two clock read and writes.

TABLE D.32.

		Instruction	Head	Tail	I-Cache Case	No-Cache Case
	LSd	#<data>,Dy	4	0	4 (0/0/0)	4 (0/1/0)
%	LSd	Dx,Dy	6	0	6 (0/0/0)	6 (0/1/0)
+	LSd	Dx,Dy	8	0	8 (0/0/0)	8 (0/1/0)
*	LSd	Mem by 1	0	0	4 (0/0/1)	4 (0/1/1)
	ASL	#<data>,Dy	2	0	6 (0/0/0)	6 (0/1/0)
	ASL	Dx,Dy	4	0	8 (0/0/0)	8 (0/1/0)
*	ASL	Mem by 1	0	0	6 (0/0/1)	6 (0/1/1)
	ASR	#<data>,Dy	4	0	4 (0/0/0)	4 (0/1/0)
%	ASR	Dx,Dy	6	0	6 (0/0/0)	6 (0/1/0)
+	ASR	Dx,Dy	10	0	10 (0/0/0)	10 (0/1/0)
*	ASR	Mem by 1	0	0	4 (0/0/1)	4 (0/1/1)
	ROd	#<data>,Dy	4	0	6 (0/0/0)	6 (0/1/0)
	ROd	Dx,Dy	6	0	8 (0/0/0)	8 (0/1/0)
*	ROd	Mem by 1	0	0	6 (0/0/1)	6 (0/1/1)
	ROXd	Dn	10	0	12 (0/0/0)	12 (0/1/0)
*	ROXd	Mem by 1	0	0	4 (0/0/0)	4 (0/1/0)

d Is direction of shift/rotate; L or R
* Add Fetch Effective Address Time
% Indicates shift count is less than or equal to the size of data
+ Indicates shift count is greater than size of data

Bit Manipulation Instructions

The bit manipulation instructions table indicates the number of clock periods needed for the processor to perform the specified bit operation on the given addressing mode. Footnotes indicate when it is necessary to account for the appropriate effective address time. For instruction-cache case and for no-cache case, the total number of clock cycles is outside the parentheses. The number of read, prefetch, and write cycles are given inside the parentheses as (r/p/w). The read, prefetch, and write cycles are included in the total clock cycle number.

All timing data assumes two clock reads and writes.

TABLE D.33.

	Instruction		Head	Tail	I-Cache Case	No-Cache Case
	BTST	#<data>,Dn	4	0	4 (0/0/0)	4 (0/1/0)
	BTST	Dn,Dn	4	0	4 (0/0/0)	4 (0/1/0)
#	BTST	#<data>,Mem	0	0	4 (0/0/0)	4 (0/1/0)
*	BTST	Dn,Mem	0	0	4 (0/0/0)	4 (0/1/0)
	BCHG	#<data>,Dn	6	0	6 (0/0/0)	6 (0/1/0)
	BCHG	Dn,Dn	6	0	6 (0/0/0)	6 (0/1/0)
#	BCHG	#<data>,Mem	0	0	6 (0/0/1)	6 (0/1/1)
*	BCHG	Dn,Mem	0	0	6 (0/0/1)	6 (0/1/1)
	BCLR	#<data>,Dn	6	0	6 (0/0/0)	6 (0/1/0)
	BCLR	Dn,Dn	6	0	6 (0/0/0)	6 (0/1/0)
#	BCLR	#<data>,Mem	0	0	6 (0/0/1)	6 (0/1/1)
*	BCLR	Dn,Mem	0	0	6 (0/0/1)	6 (0/1/1)
	BSET	#<data>,Dn	6	0	6 (0/0/0)	6 (0/1/0)
	BSET	Dn,Dn	6	0	6 (0/0/0)	6 (0/1/0)
#	BSET	#<data>,Mem	0	0	6 (0/0/1)	6 (0/1/1)
*	BSET	Dn,Mem	0	0	6 (0/0/1)	6 (0/1/1)

* Add Fetch Effective Address time

\# Add Fetch Immediate Effective Address time

Bit Field Manipulation Instructions

The bit field manipulation instructions table indicates the number of clock periods needed for the processor to perform the specified bit field operation using the given addressing mode. Footnotes indicate when it is necessary to account for the appropriate effective address time. For instruction-cache case and for no-cache case, the total number of clock cycles is outside the parentheses. The number of read, prefetch, and write cycles are included in the total clock cycle number.

All timing data assumes two clock reads and writes.

TABLE D.34.

	Instruction		Head	Tail	I-Cache Case	No-Cache Case
	BFTST	Dn	8	0	8 (0/0/0)	8 (0/1/0)
*	BFTST	Mem (<5 Bytes)	6	0	10 (1/0/0)	10 (1/1/0)
*	BFTST	Mem (5 Bytes)	6	0	14 (2/0/0)	14 (2/1/0)
	BFCHG	Dn	14	0	14 (0/0/0)	14 (0/1/0)
*	BFCHG	Mem (<5 Bytes)	6	0	14 (1/0/1)	14 (1/1/1)
*	BFCHG	Mem (5 Bytes)	6	0	22 (2/0/2)	22 (2/1/2)
	BFCLR	Dn	14	0	14 (0/0/0)	14 (0/1/0)
*	BFCLR	Mem (<5 Bytes)	6	0	14 (1/0/1)	14 (1/1/1)
*	BFCLR	Mem (5 Bytes)	6	0	22 (2/0/2)	22 (2/1/2)
	BFSET	Dn	14	0	14 (0/0/0)	14 (0/1/0)
*	BFSET	Mem (<5 Bytes)	6	0	14 (1/0/1)	14 (1/1/1)
*	BFSET	Mem (5 Bytes)	6	0	22 (2/0/2)	22 (2/1/2)
	BFEXTS	Dn	10	0	10 (0/0/0)	10 (0/1/0)
*	BFEXTS	Mem (<5 Bytes)	6	0	12 (1/0/0)	12 (1/1/0)

(continued)

TABLE D.34. (continued)

	Instruction	Head	Tail	I-Cache Case	No-Cache Case
* BFEXTS	Mem (5 Bytes)	6	0	18 (2/0/0)	18 (2/1/0)
BFEXTU	Dn	10	0	10 (0/0/0)	10 (0/1/0)
* BFEXTU	Mem (<5 Bytes)	6	0	12 (1/0/0)	12 (1/1/0)
* BFEXTU	Mem (5 Bytes)	6	0	18 (2/0/0)	18 (2/1/0)
BFINS	Dn	12	0	12 (0/0/0)	12 (0/1/0)
* BFINS	Mem (<5 Bytes)	6	0	12 (1/0/1)	12 (1/1/1)
* BFINS	Mem (5 Bytes)	6	0	18 (2/0/2)	18 (2/1/2)
BFFFO	Dn	20	0	20 (0/0/0)	20 (0/1/0)
* BFFFO	Mem (<5 Bytes)	6	0	22 (1/0/0)	22 (1/1/0)
* BFFFO	Mem (5 Bytes)	6	0	28 (2/0/0)	28 (2/1/0)

* Add Calculate Immediate Effective Address time
NOTE: A bit field of 32 bits may span 5 bytes that require two operand cycles to access, or may span 4 bytes that require only one operand cycle to access.

Conditional Branch Instructions

The conditional branch instructions table indicates the number of clock periods needed for the processor to perform the specified branch on the given branch size, with complete execution times given. No additional tables are needed to calculate total effective execution time for these instructions. For instruction-cache case and for no-cache case, the total number of clock cycles is outside the parentheses. The number of read, prefetch, and write cycles are given inside the parentheses as (r/p/w). The read, prefetch, and write cycles are included in the total clock cycle number.

All timing data assumes two clock reads and writes.

TABLE D.35.

	Instruction	Head	Tail	I-Cache Case	No-Cache Case
Bcc	(Taken)	6	0	6 (0/0/0)	8 (0/2/0)
Bcc.B	(Not Taken)	4	0	4 (0/0/0)	4 (0/1/0)
Bcc.W	(Not Taken)	6	0	6 (0/0/0)	6 (0/1/0)
Bcc.L	(Not Taken)	6	0	6 (0/0/0)	8 (0/2/0)
DBcc	(cc = False, Count Not Expired)	6	0	6 (0/0/0)	8 (0/2/0)
DBcc	(cc = False, Count Expired)	10	0	10 (0/0/0)	13 (0/3/0)
DBcc	(cc = True)	6	0	6 (0/0/0)	8 (0/1/0)

Control Instructions

The control instructions table indicates the number of clock periods needed for the processor to perform the specified operation. Footnotes indicate when it is necessary to account for the appropriate effective address time. For instruction-cache case and for no-cache case, the total number of clock cycles is outside the parentheses. The number of read, prefetch, and write cycles are given inside the parentheses as (r/p/w). The read, prefetch, and write cycles are included in the total clock cycle number.

All timing data assumes two clock reads and writes.

TABLE D.36.

	Instruction			Head	Tail	I-Cache Case	No-Cache Case
	ANDI to SR			4	0	12 (0/0/0)	14 (0/2/0)
	EORI to SR			4	0	12 (0/0/0)	14 (0/2/0)
	ORI to SR			4	0	12 (0/0/0)	14 (0/2/0)
	ANDI to CCR			4	0	12 (0/0/0)	14 (0/2/0)
	EORI to CCR			4	0	12 (0/0/0)	14 (0/2/0)
	ORI to CCR			4	0	12 (0/0/0)	14 (0/2/0)
	BSR			2	0	6 (0/0/1)	9 (0/2/1)
##	CAS		(Successful Compare)	1	0	13 (1/0/1)	13 (1/1/1)
##	CAS		(Unsuccessful Compare)	1	0	11 (1/0/0)	11 (1/1/0)
+	CAS2		(Successful Compare)	2	0	24 (2/0/2)	26 (2/2/2)
+	CAS2		(Unsuccessful Compare)	2	0	24 (2/0/0)	24 (2/2/0)
	CHK	Dn,Dn	(No Exception)	8	0	8 (0/0/0)	8 (0/1/0)
+	CHK	Dn,Dn	(Exception Taken)	4	0	28 (1/0/4)	30 (1/3/4)
*	CHK	EA,Dn	(No Exception)	0	0	8 (0/0/0)	8 (0/1/0)
*+	CHK	EA,Dn	(Exception Taken)	0	0	28 (1/0/4)	30 (1/3/4)
#+	CHK2	Mem,Rn	(No Exception)	2	0	18 (1/0/0)	18 (1/1/0)
#+	CHK2	Mem,Rn	(Exception Taken)	2	0	40 (2/0/4)	42 (2/3/4)
%	JMP			4	0	4 (0/0/0)	6 (0/2/0)
%	JSR			0	0	4 (0/0/1)	7 (0/2/1)
**	LEA			2	0	2 (0/0/0)	2 (0/1/0)
	LINK.W			0	0	4 (0/0/1)	5 (0/1/1)
	LINK.L			2	0	6 (0/0/1)	7 (0/2/1)
	NOP			0	0	2 (0/0/0)	2 (0/1/0)
**	PEA			0	2	4 (0/0/1)	4 (0/1/1)
	RTD			2	0	10 (1/0/0)	12 (1/2/0)
	RTR			1	0	12 (2/0/0)	14 (2/2/0)
	RTS			1	0	9 (1/0/0)	11 (1/2/0)
	UNLK			0	0	5 (1/0/0)	5 (1/1/0)

+ Indicates Maximum Time
* Add Fetch Effective Address Time
** Add Calculate Effective Address Time
# Add Fetch Immediate Address Time
Add Calculate Immediate Address Time
% Add Jump Effective Address Time

Exception Related Instructions and Operations

The exception related instructions and operations table indicates the number of clock periods needed for the processor to perform the specified exception related action. No additional tables are needed to calculate total effective execution time for these operations. For instruction-cache case and for no-cache case, the total number of clock cycles is outside the parentheses. The number of read, prefetch, and write cycles are included in the total clock cycle number.

All timing data assumes two clock reads and writes.

TABLE D.37.

Instruction	Head	Tail	I-Cache Case	No-Cache Case
BKPT	1	0	9 (1/0/0)	9 (1/0/0)
Interrupt (I-Stack)	0	0	23 (2/0/4)	24 (2/2/4)
Interrupt (M-Stack)	0	0	33 (2/0/8)	34 (2/2/8)
RESET Instruction	0	0	518 (0/0/0)	518 (0/1/0)
STOP	0	0	8 (0/0/0)	8 (0/2/0)
TRACE	0	0	22 (1/0/5)	24 (1/2/5)
TRAP #n	0	0	18 (1/0/4)	20 (1/2/4)
Illegal Instruction	0	0	18 (1/0/4)	20 (1/2/4)
A-Line Trap	0	0	18 (1/0/4)	20 (1/2/4)
F-Line Trap	0	0	18 (1/0/4)	20 (1/2/4)
Privilege Violation	0	0	18 (1/0/4)	20 (1/2/4)
TRAPcc (Trap)	2	0	22 (1/0/5)	24 (1/2/5)
TRAPcc (No Trap)	4	0	4 (0/0/0)	4 (0/1/0)
TRAPcc.W (Trap)	5	0	24 (1/0/5)	26 (1/3/5)
TRAPcc.W (No Trap)	6	0	6 (0/0/0)	6 (0/1/0)
TRAPcc.L (Trap)	6	0	26 (1/0/5)	28 (1/3/5)
TRAPcc.L (No Trap)	8	0	8 (0/0/0)	8 (0/2/0)
TRAPV (Trap)	2	0	22 (1/0/5)	24 (1/2/5)
TRAPV (No Trap)	4	0	4 (0/0/0)	4 (0/1/0)

Save and Restore Operations

The save and restore operations table indicates the number of clock periods needed for the processor to perform the specified state save, or return from exception, with complete execution times and stack length given. No additional tables are needed to calculate total effective execution time for these operations. For instruction-cache case and for no-cache case, the total number of clock cycles is outside the parentheses. The number of read, prefetch, and write cycles are given inside the parentheses as (r/p/w). The read, prefetch, and write cycles are included in the total clock cycle number.

All timing data assumes two clock reads and writes.

TABLE D.38.

Operation	Head	Tail	I-Cache Case	No-Cache Case
Bus Cycle Fault (Short)	0	0	36 (1/0/10)	38 (1/2/10)
Bus Cycle Fault (Long)	0	0	62 (1/0/24)	64 (1/2/24)
RTE (Normal-4 Word)	1	0	18 (4/0/0)	20 (4/2/0)
RTE (Six-Word)	1	0	18 (4/0/0)	20 (4/2/0)
RTE (Throwaway)	1	0	12 (4/0/0)	12 (4/0/0)
RTE (Coprocessor)	1	0	26 (7/0/0)	26 (7/2/0)
RTE (Short Fault)	1	0	36 (10/0/0)	26 (10/2/0)
RTE (Long Fault)	1	0	76 (25/0/0)	76 (25/2/0)

Appendix E

Foot Print and Electrical Data

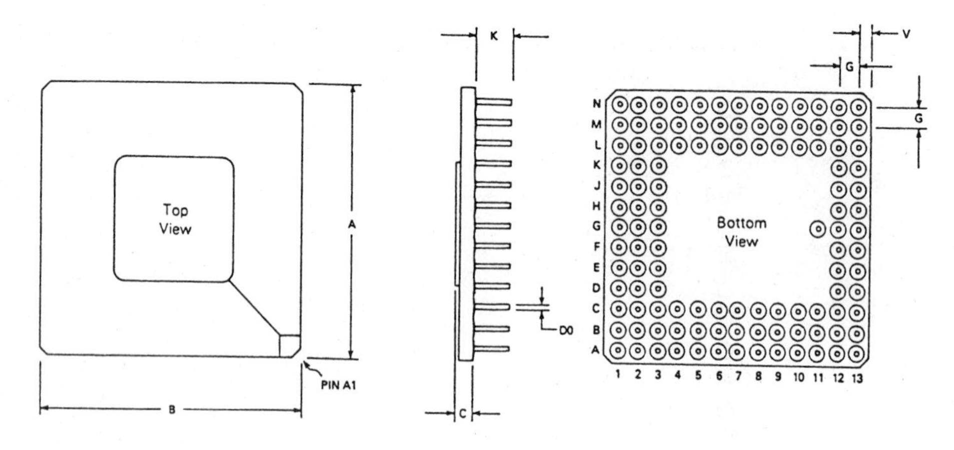

Pin Number	Function
A1	$\overline{BGACK}$
A2	A1
A3	A31
A4	A28
A5	A26
A6	A23
A7	A22
A8	A19
A9	V_{CC}
A10	GND
A11	A14
A12	A11
A13	A8
B1	GND
B2	$\overline{BG}$
B3	$\overline{BR}$
B4	A30
B5	A27
B6	A24
B7	A20
B8	A18
B9	GND
B10	A15
B11	A13
B12	A10
B13	A6
C1	$\overline{RESET}$
C2	CLOCK
C3	GND
C4	A0
C5	A29
C6	A25
C7	A21
C8	A17
C9	A16
C10	A12
C11	A9
C12	A7
C13	A5

Pin Number	Function
D1	V_{CC}
D2	V_{CC}
D3	V_{CC}
D4-D11	–
D12	A4
D13	A3
E1	FC0
E2	$\overline{RMC}$
E3	V_{CC}
E12	A2
E13	$\overline{OCS}$
F1	SIZ0
F2	FC2
F3	FC1
F12	GND
F13	$\overline{IPEND}$
G1	$\overline{ECS}$
G2	SIZ1
G3	$\overline{DBEN}$
G11	V_{CC}
G12	GND
G13	V_{CC}
H1	$\overline{CDIS}$
H2	$\overline{AVEC}$
H3	$\overline{DSACK0}$
H12	$\overline{IPL2}$
H13	GND
J1	$\overline{DSACK1}$
J2	$\overline{BERR}$
J3	GND
J12	$\overline{IPL0}$
J13	$\overline{IPL1}$

Pin Number	Function
K1	GND
K2	$\overline{HALT}$
K3	GND
K12	D1
K13	D0
L1	$\overline{AS}$
L2	R/$\overline{W}$
L3	D30
L4	D27
L5	D23
L6	D19
L7	GND
L8	D15
L9	D11
L10	D7
L11	GND
L12	D3
L13	D2
M1	$\overline{DS}$
M2	D29
M3	D26
M4	D24
M5	D21
M6	D18
M7	D16
M8	V_{CC}
M9	D13
M10	D10
M11	D6
M12	D5
M13	D4
N1	D31
N2	D28
N3	D25
N4	D22
N5	D20
N6	D17
N7	GND
N8	V_{CC}
N9	D14
N10	D12
N11	D9
N12	D8
N13	V_{CC}

FIGURE E.1. Pin Assignment Diagram for MC68020

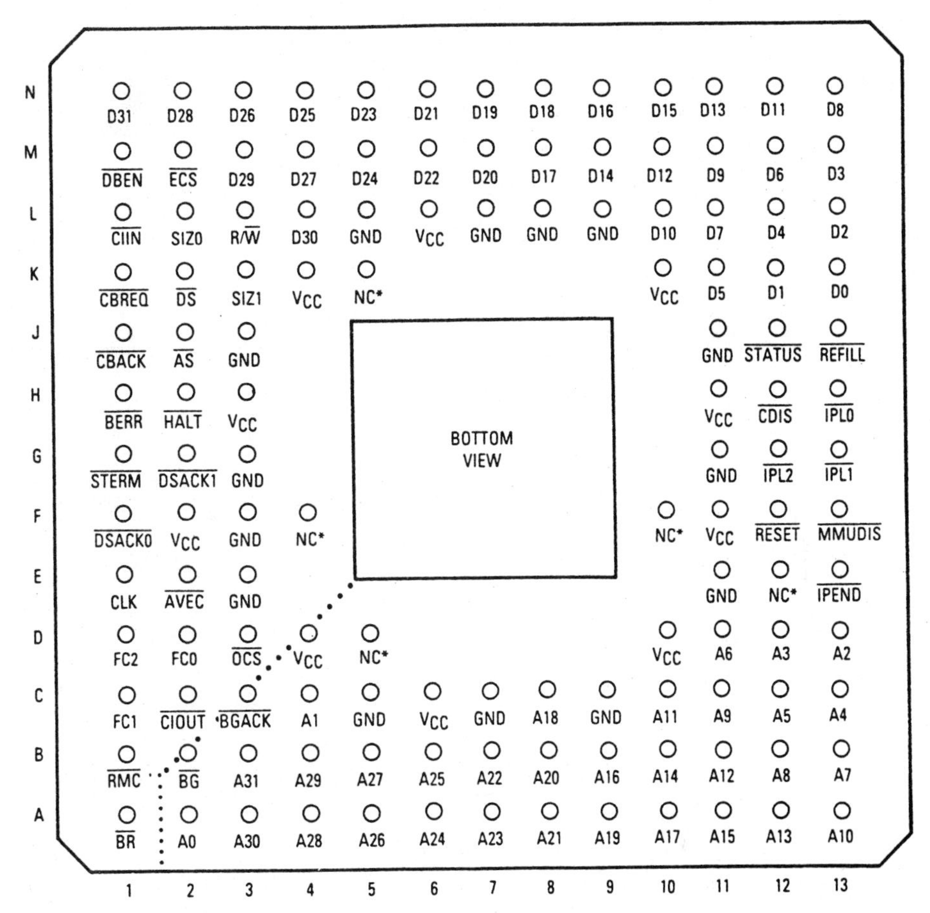

*NC — Do not connect to this pin.

Pin Group	V_{CC}	GND
Address Bus	C6, D10	C5, C7, C9, E11
Data Bus	L6, K10	J11, L9, L7, L5
ECS, SIZx, DS, AS, DBEN, CBREQ, R/W	K4	J3
FC0-FC2, RMC, OCS, CIOUT, BG	D4	E3
Internal Logic, RESET, STATUS, REFILL, Misc.	H3, F2, F11, H11	L8, G3, F3, G11

FIGURE E.2. Pin Assignment Diagram for MC68030

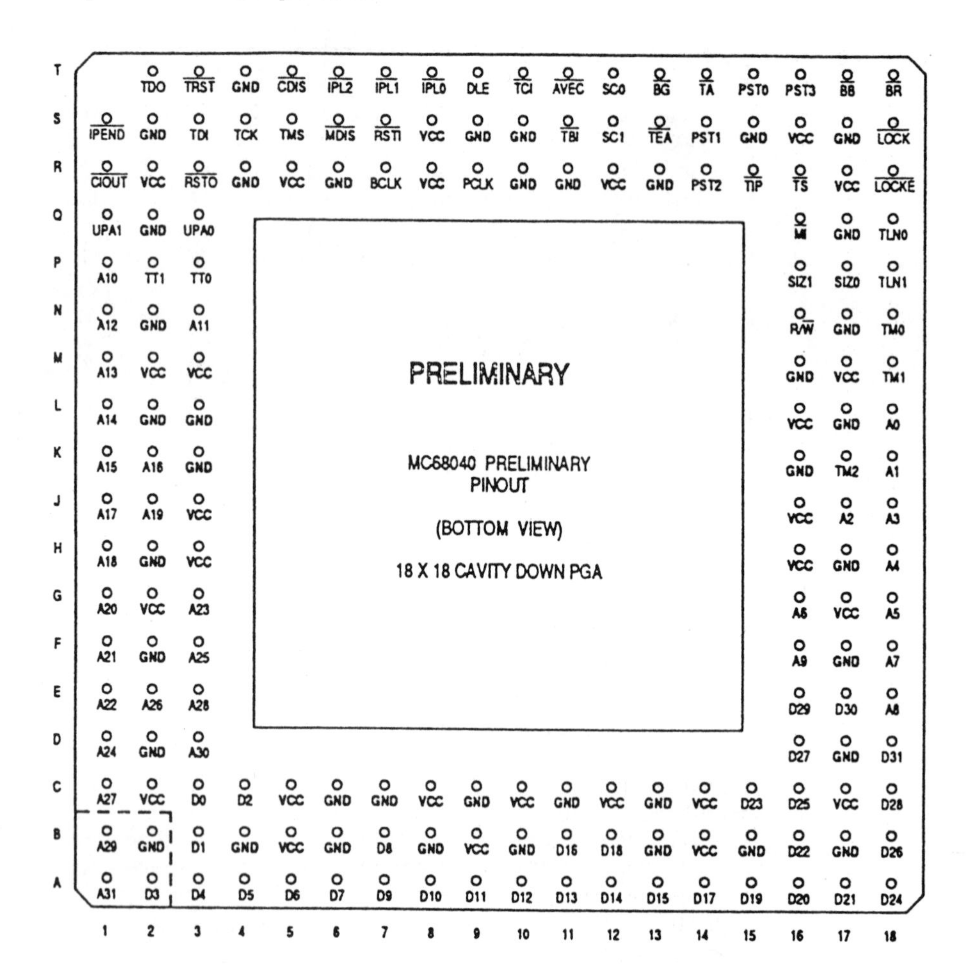

	GND	VCC
PLL	S9, R6, R10	R8, S8
Internal Logic	C6, C7, C9, C11, C13, K3, K16, L3, M16, R4, R11, R13, S10, T4	C5, C8, C10, C12, C14, H3, H16, J3, J16, L16, M3, R5, R12
Output Drivers	B2, B4, B6, B8, B10, B13, B15, B17, D2, D17, F2, F17, H2, H17, L2, L17, N2, N17, Q2, Q17, S2, S15, S17	B5, B9, B14, C2, C17, G2, G17, M2, M17, R2, R17, S16

FIGURE E.3. Pin Assignment Diagram for MC68040

References

[Abramovici 90] Abramovici, Miron, Breuer, Melvin A., and Friedman, Arthur D., 1990. *Digital Systems Testing and Testable Design.* New York: Computer Science Press.

[Balakrishnan 84] Balakrishnan, R. V., 1984. *The Proposed IEEE896 Futurebus—A Solution to the Bus Driving Problem.* IEEE Micro, pp. 23–27, August 1984.

[Barthmaler 79] Barthmaler, Joe, 1979. *Intel Multibus Interfacing.* San Jose, CA: Intel Corporation AP-28A.

[Borrill 84] Borrill, Paul and Theus, John, 1984. *An Advanced Communication Protocol For The Proposed IEEE-896 Futurebus.* IEEE Micro, pp. 42–56.

[Deitel 83] Deitel, H. M., 1983. *An Introduction to Operating Systems.* Reading, MA: Addison-Wesley Publishing Co.

[Dexter 86] Dexter, L. Arthur, 1986. *Microcomputer Bus Structures and Bus Interface Design.* New York: Marcel Dekker, Inc.

[Ford 87] Ford, William and Topp, William, 1987. *MC68000 Assembly Language and System Programming.* Lexington, MA: D.C. Heath and Company.

[Harman 85] Harman, L. Thomas and Lawson, Barbara, 1985. *The Motorola MC68000 Microprocessor Family: Assembly Language, Interface Design and System Design.* Englewood Cliffs, NJ: Prentice-Hall Inc.

[Holden 87] Holden, Kirk, Mothersole, David, and Vegesna, Raju, 1987. *Memory Management In The MC68030 Microprocessor.* VLSI System Design.

[Howard 91a] Howard, David, 1991. *Using An MMU.* Embedded Systems Programming, February 1991.

[Howard 91b] Howard, David, 1991. *Programming An MMU.* Embedded Systems Programming, March 1991.

[Hwang 84] Hwang, Kai, and Briggs, A. Faye, 1984. *Computer Architecture and Parallel Processing,* New York: McGraw-Hill Book Company.

[IEEE P896] IEEE, 1989. *IEEE P896.2 Physical Layer and Profile Specifications.* New York: IEEE Publication.

[IEEE p1149.1–1990], IEEE Standard 1149.1–1990, 1990. *IEEE Standard Test Access Port and Boundary Scan Architecture.* New York: IEEE Standards Boards.

[Johnson 84] Johnson, James B. and Kassel, Steve, 1984. *The Multibus Design Guidebook: Structures, Architectures and Applications.* New York: McGraw-Hill Book Company.

[Kane 81] Kane, Gerry, Hawkins, Doug, and Leventhal, Lance, 1981. *68000 Assembly Language Programming.* Berkeley, CA: Osborne/McGraw-Hill.

[Kelly-Bottle 85] Kelly-Bottle, Stan, and Fowler, Bob, 1985. *68000, 68010, 68020 Primer.* Indianapolis, IN: Howard W. Sams & Co. Inc.

[LEA 80] Lea, A. Wayne, 1980. *Trends in Speech Recognition.* Englewood Cliffs, NJ: Prentice-Hall Inc.

[Lindhorst 91] Lindhorst, Greg, Anderson, Andrew, and Dahms, David, 1991. *Programming The 68040.* Byte, pp. 121–128, August, 1991.

[Lioupis 83] Lioupis, D., 1983. *The RISC-II.* Berkeley, CA: Internal U.C. Berkeley Working Paper, June 1983.

[MC68030 89] Motorola Inc., 1989. *MC68030 Enhanced 32-Bit Microprocessor User's Manual.* Englewood Cliffs, NJ: Prentice-Hall Inc.

[MC68040 89] Motorola Inc. 1989. *MC68040 32-Bit Microprocessor.* Arizona: Motorola Literature Distribution.

[Milutinovic 86] Milutinovic, Veljko (ed.), 1986. *Tutorial on: Advanced Microprocessors and High-Level Language Computer Architecture.* Los Alamitos, CA: IEEE Computer Society Press.

[Motorola 86] Motorola Inc. 1986. *MC68851 Paged Memory Management Unit User's Manual.* Englewood Cliffs, NJ: Prentice-Hall Inc.

[Motorola 85] Motorola Inc. 1985. *MC68881 Floating-Point Coprocessor User's Manual.* Englewood Cliffs, NJ: Prentice-Hall Inc.

[Motorola 85a] Motorola Inc., 1985. *MC68020 32-Bit Microprocessor User's Manual.* Englewood Cliffs, NJ: Prentice-Hall, Inc.

[Motorola 68000] Motorola Inc., 1988. *MC68000 Family Reference.* Phoenix, AZ: Semiconductor Products Sectors, USA Literature Distribution Center.

[Motorola AN-853] Scherer, Victor A. and Peterson, William, 1982. *The MC68230 Parallel Interface/Timer Provides An Effective Printer Interface.* Austin, TX: Motorola Application Note AN-854.

[MTTA2 87] Motorola Inc. 1987. *MC68020 Audio Cassette Course.* Phoenix, AZ: Motorola Semiconductor Products Sector Technical Operations.

[MTTA3 87] Motorola Inc. 1987. *MC68030 Audio Cassette Course.* Phoenix, AZ: Motorola Semiconductor Products Sector Technical Operations.

[NEC 91] NEC Electronics Inc, 1991. *1991 Memory Products Data Book.* Document No. 60105, Mountain View, CA.

[Nikitas 84] Alexandridas, A. Nikitas, 1984. *Microprocessor System Design Concepts.* Rockville, MD: Computer Science Press.

[Patterson 82] Patterson, D. A. and Seguin, Charles H. 1982. *A VLSI RISC.* IEEE Computer Magazine, Vol. 15, No. 9, pp. 8–21, Sept. 1982.

[Papamichalis 87] Papamichalis, E. Panos, 1987. *Practical Approaches To Speech Coding.* Englewood Cliffs, NJ: Prentice-Hall Inc.

[Radin 82] Radin, G., 1982. *The 901 Minicomputer.* Proceedings of Symposium for Programming Languages and Operating System Support, pp. 39–47.

[Scanlon 81] Scanlon, J. Leo, 1981. *The 68000: Principles and Programming.* Indianapolis, IN: Howard W. Sams & Co. Inc.

[Stone 87] Stone, S. Harold,1987. *High-Performance Computer Architecture.* Reading, MA: Addison-Wesley Publishing Company.

[Taub 84] Taub, D. M., 1984 *Arbitration and Control Acquisition in the Proposed IEEE896 Futurebus.* IEEE Micro, pp. 28–41, August 1984.

[Triebel 86] Triebel, A. Walter and Singh, Avtar, 1986. *The 68000 Microprocessor: Architecture, Software and Interfacing Techniques.* Englewood Cliffs, NJ: Prentice-Hall Inc.

[Tanenbaum 84] Tanenbaum, S. Andrew, 1984. *Structured Computer Organization.* Englewood Cliffs, NJ: Prentice-Hall, Inc.

[VMEbus 85] Motorola Inc. 1985. *The VMEbus Specification.* Phoenix, AZ: Micrology PBT.

[Vegesna 86] Vegesna, Raju, Hoffman, Bruce, Stanphill, Russell, Scales, Hunter, and Tietjan, Don, 1986. *The High Performance Integrated Processor MC68030.* International Conference On Computers.

[Zoch 86] Mitchell, H. J. (ed.), 1986. *32-Bit Microprocessors.* New York: McGraw-Hill Company.

Index

F

H

I